Basic Methods in Microscopy

Protocols and Concepts from *Cells: A Laboratory Manual*

ALSO FROM COLD SPRING HARBOR LABORATORY PRESS

RELATED LABORATORY MANUALS

Live Cell Imaging: A Laboratory Manual
Cells: A Laboratory Manual CD
Molecular Cloning: A Laboratory Manual
Imaging in Neuroscience and Development: A Laboratory Manual

OTHER RELATED TITLES

Lab Math: A Handbook of Measurements, Calculations, and Other Quantitative Skills for Use at the Bench
Lab Ref: A Handbook of Recipes, Reagents, and Other Reference Tools for Use at the Bench
Lab Dynamics: Management Skills for Scientists
At the Bench: A Laboratory Navigator
At the Helm: A Laboratory Navigator

Basic Methods in Microscopy

Protocols and Concepts from *Cells: A Laboratory Manual*

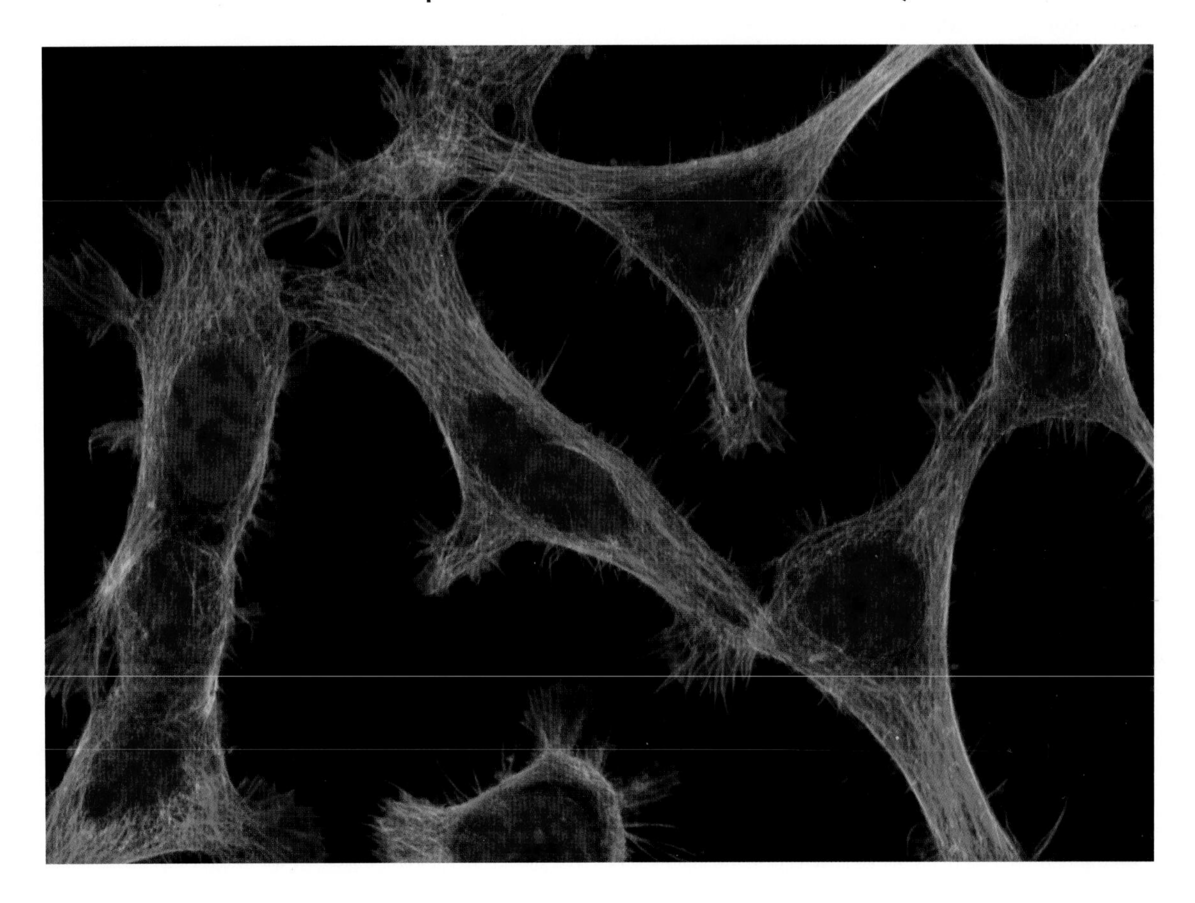

EDITED BY

David L. Spector

Cold Spring Harbor Laboratory

Robert D. Goldman

Northwestern University Medical School

COLD SPRING HARBOR LABORATORY PRESS
Cold Spring Harbor, New York

Basic Methods in Microscopy
Protocols and Concepts from *Cells: A Laboratory Manual*

Printed in China

Publisher	John Inglis
Acquisition Editor	David Crotty
Development Manager	Jan Argentine
Project Coordinator	Mary Cozza
Permissions Coordinator	Maria Falasca
Production Manager	Denise Weiss
Production Editor	Kathleen Bubbeo
Desktop Editor	Susan Schaefer
Cover Designer	Ed Atkeson

Front Cover and Title Page Image: Two-photon fluorescence image of cultured HeLa cells labeled for microtubules (*green*), actin (*red*), and cell nuclei (*blue*). Image courtesy of Thomas Deerinck and Mark Ellisman, the National Center for Microscopy and Imaging Research, UCSD.
Back Cover: Differential interference contrast image of a mitotic Rat2 cell overlayed with a DAPI-stained image of the same cell to show the chromosomes. Photo provided by Satya Khoun and Robert D. Goldman, Northwestern University Medical School, Chicago, IL.

Library of Congress Cataloging-in-Publication Data

Basic methods in microscopy : protocols and concepts from cells : a laboratory manual/ edited by David L. Spector, Robert D. Goldman.
p. cm.
Includes bibliographical references and index.
ISBN 0-87969-751-2 (pbk. : alk. paper) — ISBN 0-87969-747-4 (cloth : alk. paper)
1. Microscopy. 2. Cytology. 3. Microscope slides. 4. Mounting of microscope specimens.
I. Spector, David L. II. Goldman, Robert D., 1939-
QH207.B27 2005
570'.28'2—dc22

2004030135

10 9 8 7 6 5 4 3 2

All Cold Spring Harbor Laboratory Press publications may be ordered directly from Cold Spring Harbor Laboratory Press, 500 Sunnyside Blvd., Woodbury, NY 11797-2924. Phone: 1-800-843-4388 in Continental U.S. and Canada. All other locations: (516) 422-4100. FAX: (516) 422-4097. E-mail: cshpress@cshl.edu. For a complete catalog of all Cold Spring Harbor Laboratory Press publications, visit our World Wide Web site http://www.cshlpress.com/

We wish to dedicate this book to our wives
Mona Spector and Anne Goldman
for their constant support, devotion, and
understanding throughout the preparation
of this book.

Contents

Preface

The use of microscopic approaches to assess cell structure and function has continued to gain widespread popularity. The chapters in *Basic Methods in Microscopy* cover the techniques of microscopy that are essential for both the operation of microscopes and the interpretation of microscopic images. Examples of methods used for the preparation of cells and tissues for microscopy are also presented. The information contained in this manual has withstood the test of time and should be useful to researchers, students, and hospital laboratory technologists. For protocols used to study living cells and organisms, the reader is referred to the companion book *Live Cell Imaging: A Laboratory Manual.*

The chapters are derived, for the most part, from the popular *Cells: A Laboratory Manual*, which is now out of print. The procedures cover the classic applications of light and electron microscopy as well as the more recent imaging techniques such as confocal, deconvolution, and multiphoton microscopy. Optimized protocols are presented to localize proteins or nucleic acids using fluorescence-based technology in various cells and organisms. All of the protocols in this manual have been tried and tested in laboratories around the world.

We wish to express our thanks and gratitude to all of the authors and their research groups throughout the world who made this volume a reality. We thank Dr. Leong Chew for help with Appendix 1. We are grateful to the staff members at Cold Spring Harbor Laboratory Press who were instrumental in the production of this volume. In particular, we thank our Publisher John Inglis and our Acquisition Editor David Crotty for coordinating efforts and keeping this book in the forefront of our minds. We thank Project Coordinator, Mary Cozza; Production Manager, Denise Weiss; Production Editor, Kathleen Bubbeo; and Desktop Editor, Susan Schaefer.

David L. Spector
Robert D. Goldman

CHAPTER 1

Light Microscopy

Ernst Keller[1] and Robert D. Goldman[2]

[1]*Carl Zeiss, Inc., Thornwood, New York*
[2]*Northwestern University Medical School, Chicago, Illinois*

INTRODUCTION

The light microscope, often the symbol of research and scientific discovery, has evolved over the last 350 years from Antonie van Leeuwenhoek's simple magnifier to the more sophisticated instruments of today. Studies of biological structures and processes on both fixed and live specimens have advanced light microscopy into an indispensable tool for cell and molecular biologists.

This chapter provides an overview of light microscopy, including the principles and equipment as well as practical guidelines for achieving the best results. It will not replace the specific instructions provided for a given microscope. For more in-depth information, see the Reference list at the end of this chapter. Other aspects of and systems for microscopy are discussed elsewhere in this manual, for example, confocal microscopy (Chapter 2), preparation of cells and tissues for microscopy (Chapter 4), and scanning and transmission electron microscopy (Chapters 19–21).

The light microscope creates a magnified, detailed image of seemingly invisible objects or specimens, based on the principles of transmission, absorption, diffraction, and refraction of light waves. The various types of microscopes produce images of objects employing different strategies. In all instances (e.g., bright field, phase contrast, and fluorescence), production of a clear and informative image is dependent on the magnification of the object, its contrast with respect to its internal or external surroundings, and the ability to resolve structural details.

In the microscope, objects are enlarged or magnified with a convex lens that bends light rays by refraction. Diverging rays from points within the object (object points) are made to converge behind the convex lens and cross over each other to form image points (i.e., a focused image). The distance of the object from the lens divided into the distance of the focused image from the lens determines the magnification. In the compound microscope there are usually two magnifying systems in tandem, one defined by the objective and the other defined by the eyepiece. Another important property of a lens is its focal length, which is defined by the distance from the lens at which parallel rays of light are focused.

The visibility of the magnified object depends on contrast and resolution. In general, the contrast or differences in light intensity between an object and its background or surroundings render the object distinct. For colorless specimens, as is the case for most biological material, contrast is achieved in various ways. The object itself or selected portions of it may be stained, thus reducing the amplitude of certain light waves passing through the stained areas. However, this usually requires the killing or fixation and staining of cells. Such stained specimens are typically observed using bright-field microscopy (see p. 16). Alternatively, several kinds of specially developed microscope systems may be used that can enhance the contrast of live specimens. These systems, described in this section, include the following:

- Oblique illumination
- Dark field

- Phase contrast
- Polarized light
- Nomarski or differential interference contrast
- Reflection interference
- Fluorescence
- Video microscopy

Table 1.1 summarizes these various systems and their respective applications and Figure 1.1 illustrates the visualization of tissue (stained or unstained) using either bright-field or phase-contrast optics. The degree of structural detail revealed within a cell studied in the light microscope is determined by the "resolving power" of the entire microscope lens system. Resolution is defined as the limiting distance between two points at which they are perceived as distinct from one another. Superior quality objective lenses with high resolving power are critical for producing clear and precise images. The resolving power of a microscope also depends to a great extent on the condenser that delivers light to the specimen. These considerations are discussed in greater detail below.

KÖHLER ILLUMINATION: PRINCIPLES OF LIGHT MICROSCOPY AND FACTORS RELATED TO RESOLUTION

The light microscope is a critical tool in studies ranging from subcellular structure and function to pathology, embryology, gene expression, and gene mapping. For many of these purposes, the limits of resolution of the light microscope must be exploited to the fullest potential. For optimal results in a given application, the microscope should be equipped with high-quality optics (objectives, eyepieces, and condensers), be precisely aligned, and make use of the appropriate light sources, filters, and contrast enhancement devices (e.g., phase contrast).

The first and most critical step in setting up a microscope for optimal resolution involves the mechanics of Köhler illumination. Köhler illumination was first described in 1893 by August

TABLE 1.1 A variety of microscopic techniques exploit light properties to enhance contrast

Contrast mode	Mechanism	Comments
Bright field	contrast depends on light absorption	usually used in conjunction with histological stains to boost contrast
Phase contrast	converts optical path differences to intensity differences	contrast proportional to local "phase dense" objects including mitochondria, lysosomes, chromosomes, nucleoli, and stress fibers
Differential interference contrast (DIC)	converts rate of change of optical path across specimen	cell and organelle edges where optical path abruptly changes stand out in relief
Dark field	scattered light observed	produces images of cell and organelle edges
Interference reflection (IRM)	contrast depends on interference between closely spaced surfaces	used to visualize zones of cell-substratum contact in cultured cells
Polarization	detects birefringence caused by supramolecular organization below optical resolution	used to study oriented arrays such as cytoskeletal structures (e.g., microtubules in the mitotic apparatus and stress fibers); also used to study membranes
Fluorescence	contrast depends on absorption of light by fluorophore and its quantum yield	limited only by appropriate fluorescent probes

A

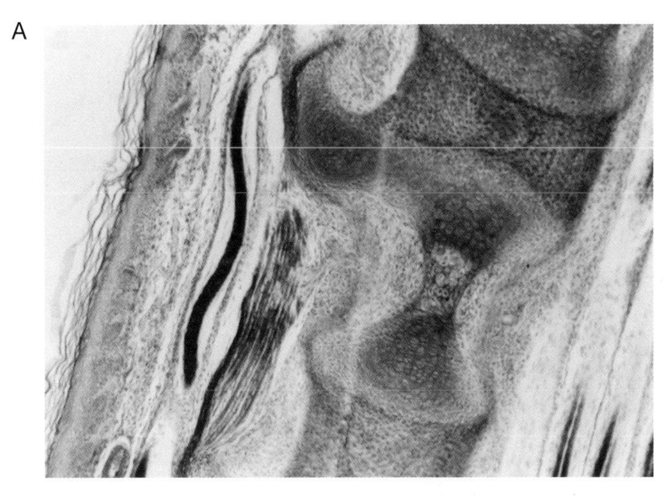

B

C

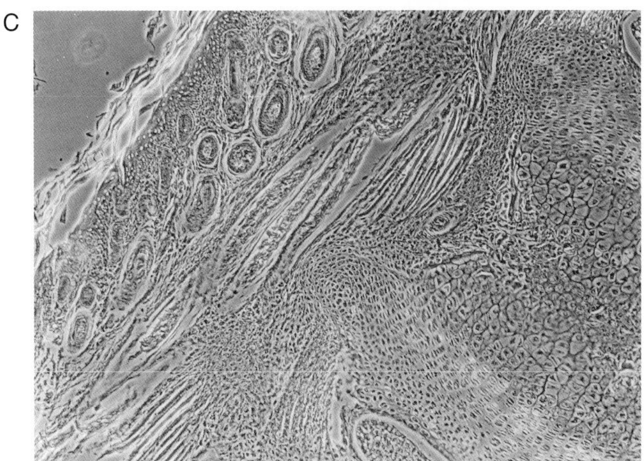

FIGURE 1.1

(*A*) Bright-field microscope photomicrograph of a section of a paraffin-embedded late-stage mouse embryo. The section is through the proximal region of the tail. It has been deparaffinized and stained with hematoxylin and eosin. The skin is located on the left side where the stratum corneum is evident at the surface. Many cell types are evident and are readily observed because of the color-generated contrast. (*B*) A section that has been prepared exactly as in *A* through the same region of a mouse embryo. The only difference is that the section has not been stained. The skin is located in the same position at the left. The section and the various tissue cells are essentially invisible (without the color contrast generated by staining) when the microscope is arranged for optimal bright field with Köhler illumination (see below). (*C*) The same section as in *B*, but observed with phase-contrast optics (see below). Even in the absence of color-generated contrast, the various regions of the tissue such as the stratum corneum of the skin (on the left side) are obvious. (Photos provided by R.D. Goldman, Northwestern University, and H.E. Keller, Carl Zeiss, Inc.)

Köhler, a young zoologist in Giessen, Germany, who later joined Carl Zeiss. It provides efficient, bright, and even illumination in the specimen field, minimizes internal stray light, and allows for control of contrast and depth.

A look at the components of the microscope and at the path of light rays helps in understanding the underlying principle and assists in the alignment of the instrument for best performance. The basic components and image locations of the typical modern microscope, from light source to final image formation in either the eye, camera, or other detector, are displayed in Figure 1.2. The two geometric optical ray paths, the imaging and illuminating paths, shown in Figure 1.3, are depicted for Köhler illumination in both transmitted and reflected or incident light systems.

For the illumination ray path, the angle of radiation is depicted from a single point on the light source (Fig. 1.3A, L1) that is received by the lamp collector, which then images this point from the source onto the front focal plane of the condenser (location of condenser aperture diaphragm; see L2). From here, the source point is projected by the condenser to infinity and evenly illuminates the specimen. The objective receives the parallel, infinity-projected source rays and forms an image of the source in its back focal plane (exit pupil; L3). This image of the light source is then transferred to the exit pupil of the eyepiece, also called the eyepoint (L4). Therefore, from original light source to eyepoint, there are four images of the light source ("source-conjugated" images). The final source image in the exit pupil of the microscope eyepiece is located in the same plane as the entrance pupil of the observer's eye.

A

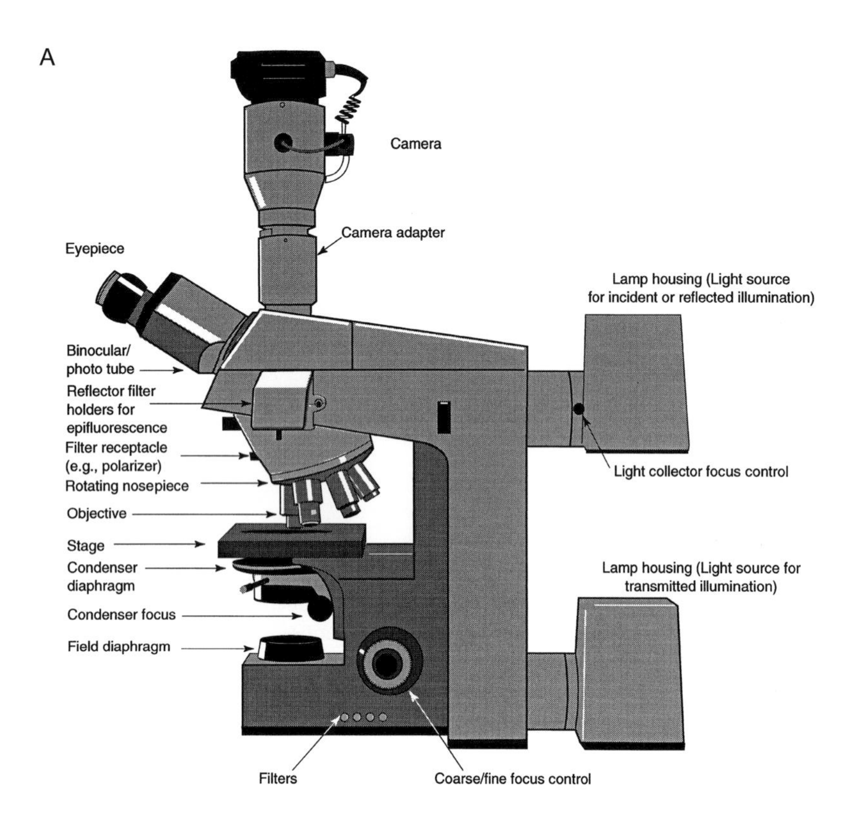

B

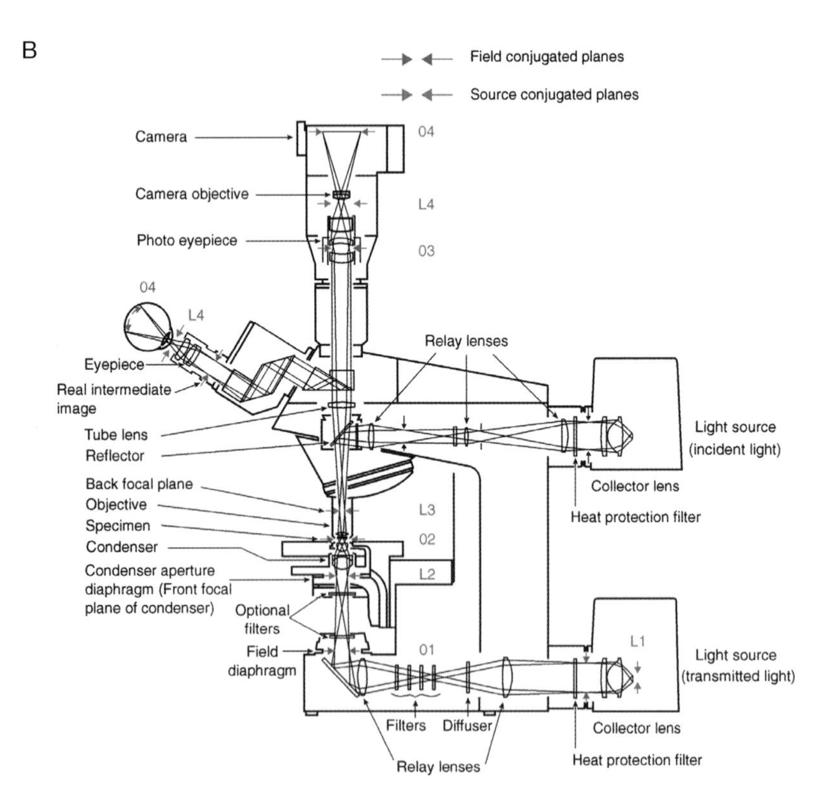

FIGURE 1.2

The light microscope. (*A*) Basic components of the light microscope arranged for transmitted and incident illumination. (*B*) Diagrammatic representation of the transmitted and incident light paths. Light from the source to final image either in the camera or on the human retina is shown. Four field-conjugated planes (represented by *red arrows*) and four source-conjugated planes (represented by *green arrows*) are within the optical system of the microscope. The last field-conjugated plane is the final image in the camera or on the retina. (For definitions of 01, 02, 03, 04 and L1, L2, L3, L4, see Fig. 1.3A.)

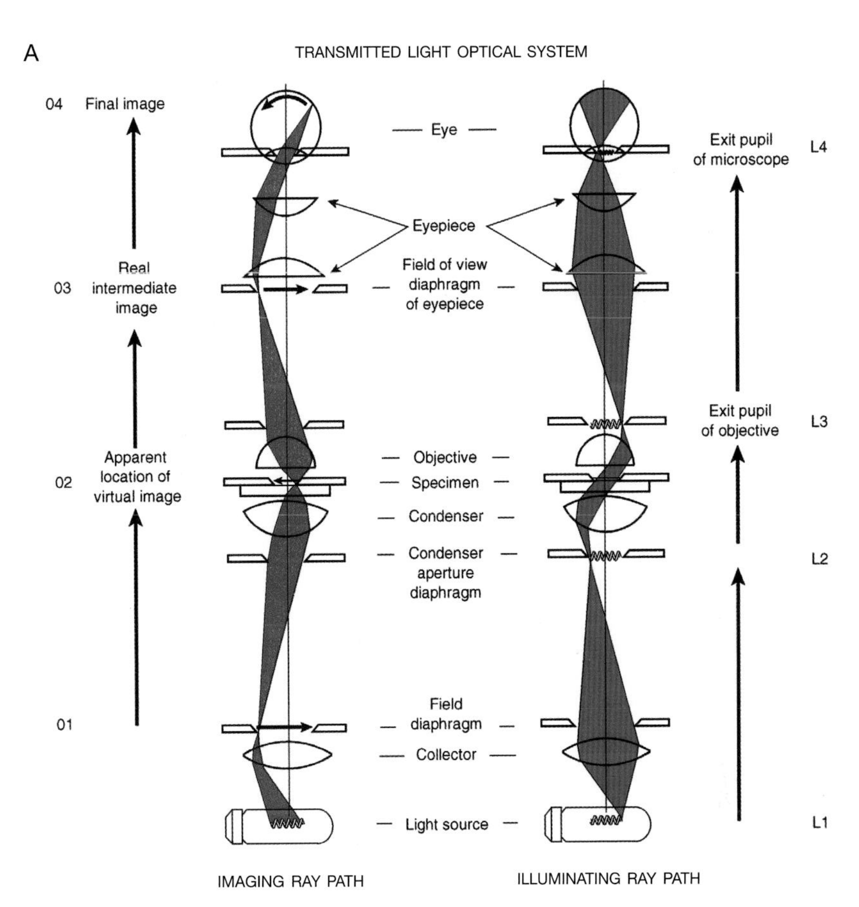

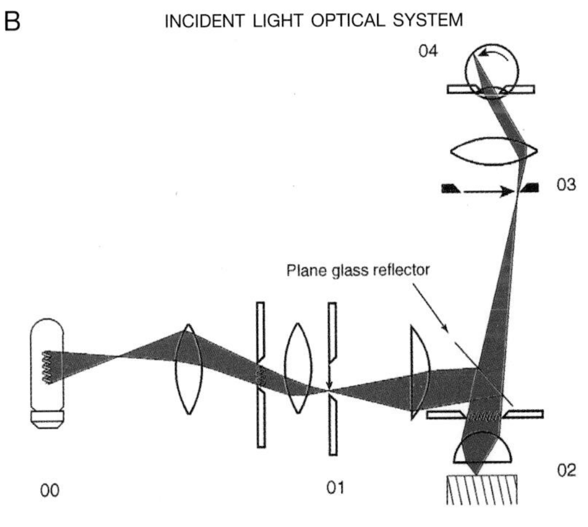

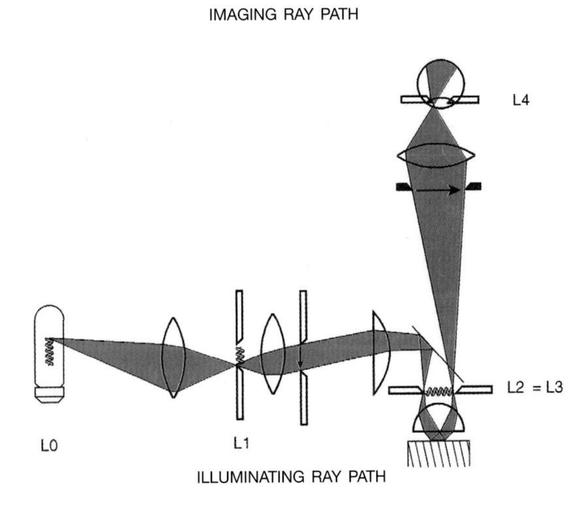

FIGURE 1.3

Köhler illumination. (*A*) Ray paths in Köhler transmitted light for a finitely corrected microscope. In the imaging ray path, 01, 02, 03, and 04 represent the image-conjugated planes (see also Fig. 1.2B). The arrows in the imaging ray path indicate image orientation. In the illuminating ray path, L1, L2, L3, and L4 represent the source-conjugated planes (see Fig. 1.2B). (*B*) Ray path in incident light Köhler illumination (epifluorescence). In the imaging ray path, 00, 01, 02, 03, and 04 represent the image-conjugated planes. In the illuminating ray path, L0, L1, L2, L3, and L4 represent the source-conjugated planes. Because the objective serves also as condenser, L2 and L3 are coincident.

The imaging path shows four "image-conjugated" planes, starting with the luminous field diaphragm (field stop; Fig. 1.3A, 01), next the specimen (02), and then the real intermediate image formed by the objective (03). This real intermediate image is located in the focal plane of the eyepiece, from which it is projected into infinity and received by the relaxed and infinity-adjusted eye of the observer (04). Thus, from field stop to final image, there are again four "image- or specimen-conjugated" planes.

In visual observation through the microscope, an enlarged virtual image apparently suspended in space can be seen. This virtual image is formed by extension of the image-forming rays from the eyepiece and appears ~10 inches below the eyepiece, just below the microscope stage. For further details of this phenomenon, see p. 11, The Finite versus the Infinity-corrected Objective.

In the incident light ray path (e.g., the path used in epifluorescence; see Figs. 1.2 and 1.3B), the only difference is the addition of a "source-conjugated" plane in which an aperture diaphragm permits control of the illumination aperture (L1) outside of the imaging ray path. This is necessary because in incident light, the objective becomes its own condenser.

The advantages of Köhler illumination are listed below.

- Only the specimen area viewed by a given objective/eyepiece combination is illuminated; no stray light or "noise" is generated inside the microscope.
- Even, uniform illumination of the specimen area is achieved by distributing the energy of each source point over the full field.
- Full control of the illumination aperture (condenser iris diaphragm; see below) provides for best resolution, best contrast, and optimal depth of field.

Adjusting the microscope for Köhler illumination is explained in the section on bright-field microscopy (see p. 16).

The Objective

Although illumination of the specimen is important, the *microscope objective* is the single most critical component of the microscope. Its properties largely determine depth of focus, resolution, and contrast of the specimen. The eyepiece and/or other so-called transfer optical devices simply magnify the resolved detail in the real intermediate image formed by the objective, which permits a detector (eye or camera) to record what has been resolved.

To understand the resolution limits of an objective, it is helpful to review the "wave concept" of light. Electric and magnetic vectors oscillate as a sine wave around the direction of propagation, defined by the light "ray." The wavelength in a vacuum or the frequency of light determines the color of light, whereas the amplitude (height) of the wave determines its intensity. Thus, the wavelengths of light in the visible spectrum range from about 400 nm for blue light to about 700 nm for red light; white light represents a mixture of all wavelengths. With light that is "coherent" or coming from the same source point, these waves can constructively or destructively interfere with each other. Wavefronts traveling at the speed of light are diffracted when they strike an object in their path or pass through a very small opening. When this occurs, Huygens' "wavelets," contained in the wavefront, form new spherical wavefronts at the obstruction. Rays of light are also bent by refraction, which occurs when light encounters an object of different density at an angle (as when light passes from air into water). The ratio of the speed of light traveling through a vacuum to its velocity through a particular object or medium is known as the refractive index.

Diffraction and Resolution: The Rayleigh Limit and the Airy Disk

Diffraction on the objective's aperture converts infinitely small self-luminous points within an object into so-called Airy disks within the image. The Airy disk is a bright disk, surrounded by concentric rings, that has a negative impact on resolution (see Fig. 1.4). The diameter (D) of this disk translated into the object itself is described as

$$D = \frac{1.22 \times \lambda}{n \sin \alpha} \quad \text{or} \quad D = \frac{1.22 \times \lambda}{(\text{NA})}$$

where λ = wavelength of light
α = half of collection angle of objective
n = refractive index of medium between object and objective
NA = numerical aperture = $n \sin \alpha$

The numerical aperture (NA) of the lens is a function of the light-collecting ability of the lens, or a measure of the "cone of light" entering the objective from a fixed object distance. The refractive index (n) of a material represents the optical density (e.g., the speed of propagation of light rays) between materials such as glass and air. Typically, the space between the objective lens and the specimen is air, which has a refractive index of about 1. Special lenses called oil immersion lenses function with oil rather than air in this space. Immersion oil has a refractive index of about 1.5. Therefore, with oil immersion lenses, resolution is increased (see below). Furthermore, because glass and immersion oil have the same refractive index, no light is lost through reflection from the surface of the lens and the cover glass. Therefore, the higher the refractive index, the smaller the resolvable distance between two points and the better the resolution. This can be explained as follows. In Figure 1.4, the dark ring around the central bright disk represents a zone of destructively interfering diffracted wavefronts and defines the diameter of the disk. Using this diameter, Rayleigh set the limit for the smallest resolvable distance between two points at

$$d = \frac{1.22 \times \lambda}{2 \times \text{NA}}$$

This resolution limit for self-luminous objects (fluorescence) can be exceeded somewhat by confocal microscopes (see Chapter 2) and by electronic image processing, but it provides a good rule of thumb for visual observation and photomicrography.

The Abbé Limit of Resolution for Illuminated Objects

Diffraction also takes place within the object when it is illuminated in the microscope. Constructive interference between two diffracted wavefronts from adjacent points within the object, separated by a distance d (Fig. 1.5A), generates new plane waves (diffraction orders) at the diffraction angle (Fig. 1.5B). The following relationship develops from this phenomenon:

$$\text{NA} = \frac{\lambda}{d} \quad \text{or} \quad d = \frac{\lambda}{\text{NA}}$$

This relationship assumes illumination parallel to the optical axis.

Abbé postulated that for two points to be resolved, at least two adjacent orders of diffracted light (see Fig. 1.5) produced by their spacing d must be collected by the objective. The NA of an objective directly determines its ability to collect diffracted wavefronts that emanate from the object.

Either diffracted and nondiffracted wavefronts or two orders of diffracted light are required to resolve structural detail. It is the constructive or destructive interference between the two in the intermediate image plane that permits the specimen detail to be resolved (see Fig. 1.6).

If a condenser is placed in the optical path and is to illuminate fully the objective's aperture, it effectively doubles the diffraction angle an objective can receive, especially when compared to the situation without a condenser (see Fig. 1.7).

Then the point-to-point resolution becomes

$$\frac{\lambda}{\text{NA}_{\text{Obj.}} + \text{NA}_{\text{Cond.}}}$$

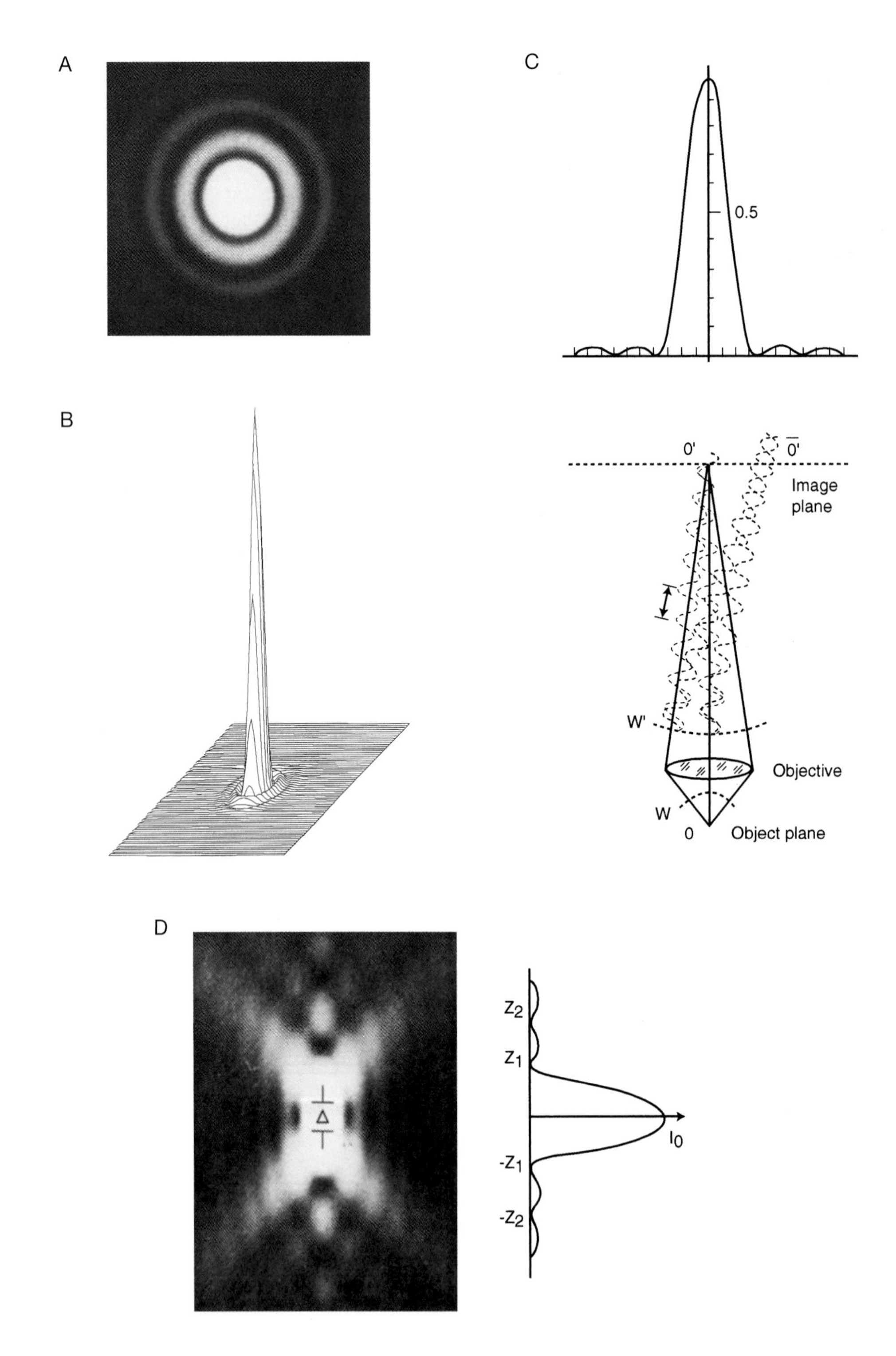

FIGURE 1.4

Generation of the Airy disk. It is the Airy disk phenomenon that limits resolution in light microscopy. With the highest numerical aperture objective (e.g., 1.4), the Airy disk phenomenon limits resolution to ~0.2 µm. (*A*) Micrograph of the Airy disk generated by a 0.2-µm pinhole. This represents a cross-sectional view through the Airy body in the image plane. (*B*) Light intensity distribution across an Airy disk. (*C*) Diagrammatic representation of wavefronts diffracted by the objective aperture and their constructive (0′) and destructive ($\overline{0}'$) interference. This results in the dark and light concentric rings seen in *A*. The light intensity plot across the image plane is seen above the dotted line. (*D*) This diagram represents a section through the Airy body in the optical (*Z*) axis (perpendicular to the image plane), and its intensity distribution.

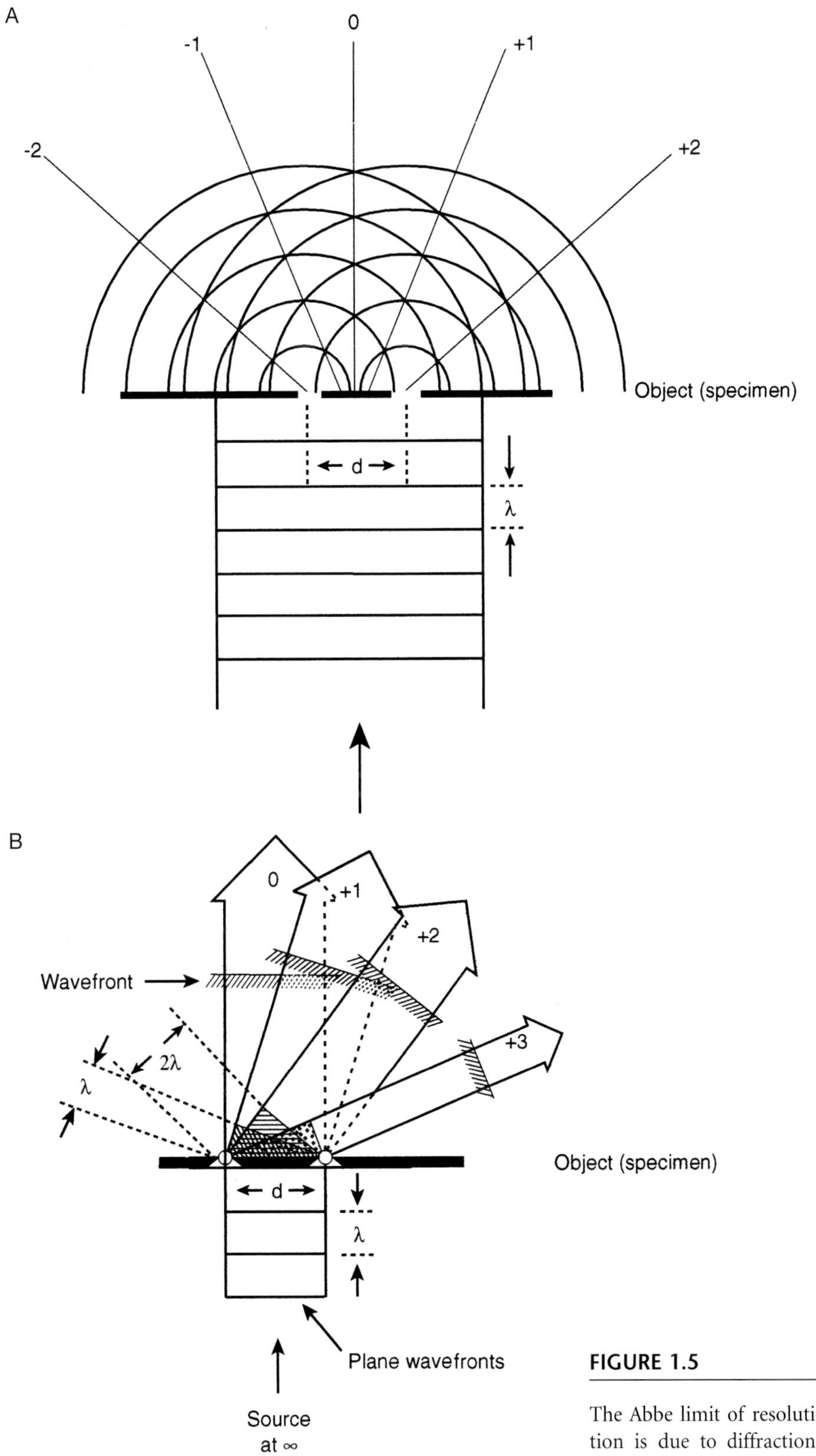

FIGURE 1.5

The Abbe limit of resolution. The limit of resolution is due to diffraction in the object and the numerical aperture of the lens. (*A*) Diffraction orders (+1, +2, and −1,−2) generated by two points (slits) separated by d. (*B*) Three diffraction orders (+1, +2, and +3) shown in the plus direction only.

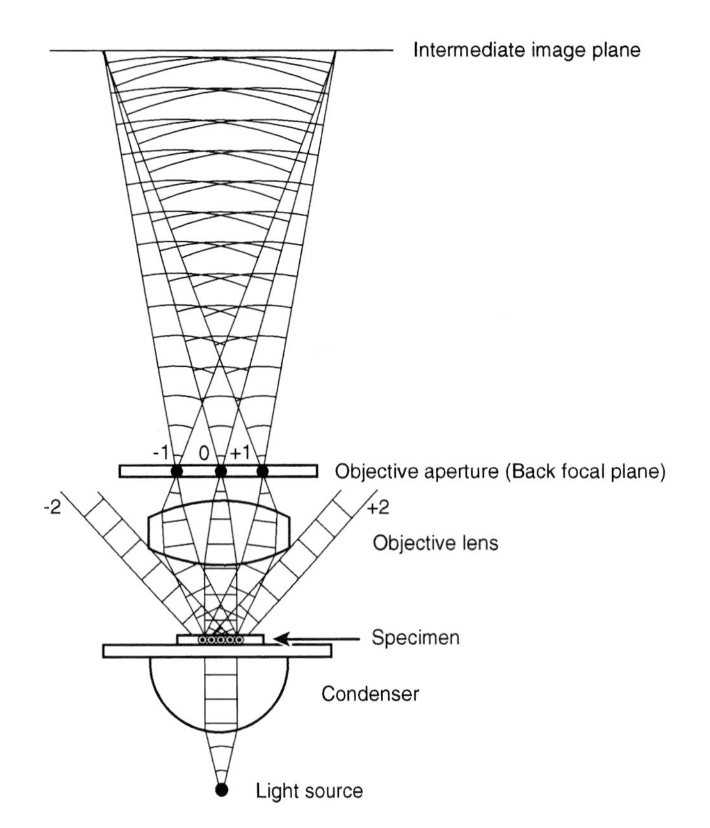

FIGURE 1.6

Nondiffracted (0 order) and diffracted (+1 and −1, etc., orders) wavefronts interfere in the image plane to resolve structural spacing in the specimen. Either 0 and +1 or 0 and −1 are the two orders of diffraction required to resolve the detail within the object (specimen) in the intermediate image plane. Note that the diffracted orders +2 and −2 do not contribute to the resolution of structures within the object (d in Fig. 1.5).

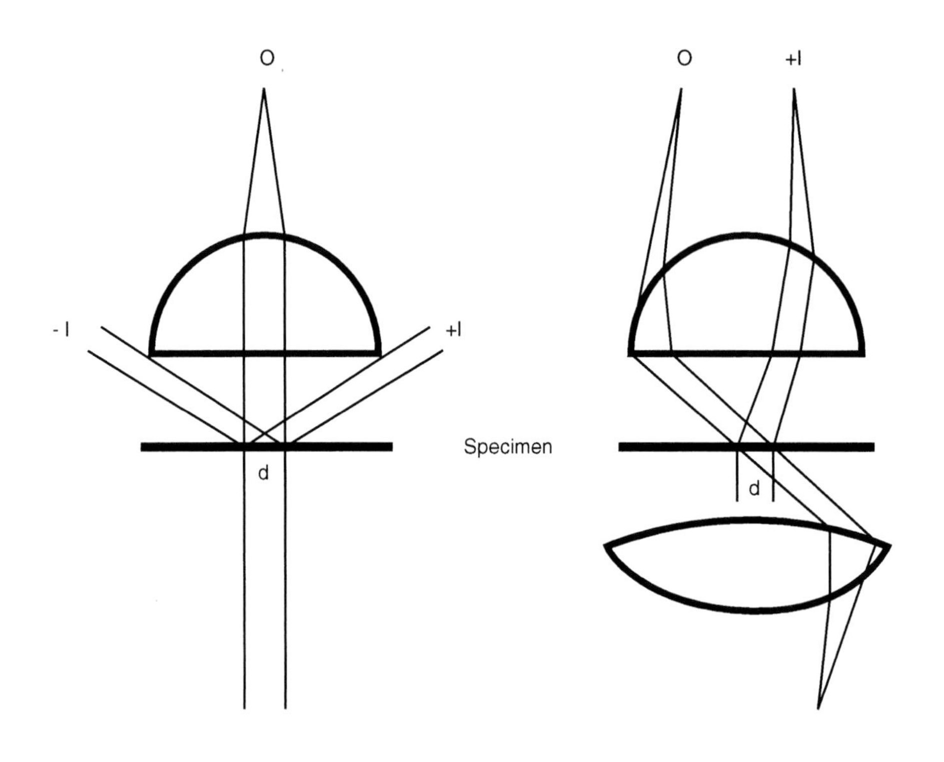

FIGURE 1.7

How the condenser enhances resolution. (*Left*) Without condenser, the first order of diffraction (−1 and +1) for spacing d is not collected by the objective. (*Right*) With condenser, the zeroth order enters obliquely and the first order (+1) of diffraction is collected.

Other Parameters of the Microscope Objective

The use of high-quality objective lenses is necessary for obtaining the maximum amount of information while studying a specimen. The ideal objective lens system must have both a high resolving power and effective correction for spherical and chromatic lens aberrations. Spherical aberration is caused by the spherically curved surfaces of the lens. This latter aberration can also be due to the use of cover glasses of the incorrect thickness or refractive index mismatch (e.g., the use of improper immersion oil or aqueous medium in which cells are immersed).

Microscopes used in the early 1800s, with either single or compound lens systems, were unable to resolve fine detail because of chromatic aberration: White light, broken up into its constituent colors, resulted in halos of colors around small objects. After about 1820, simple achromatic objective lenses were developed that corrected for spherical aberrations in the middle of the light spectrum while imaging blue and red into the same plane. The more recently developed and more complex Plan-achromatic objectives produce much less curvature of field aberration than ordinary achromatic objectives (see Table 1.2). The sophisticated Plan-apochromatic objectives are complex flat-field objectives that provide the best correction for all aberrations. From Achromats to Fluorites and Plan-apochromats, a wide range of basic performance criteria is always clearly marked on the body tube of objectives (Fig. 1.8). These include the magnification, the NA, finite or infinity designed, immersion properties (e.g., oil), correction collar (to compensate for differences in cover-glass thickness), Ph (phase contrast) or Pol (strain-free for polarized light), and occasionally an iris diaphragm to reduce the NA as required for dark-field microscopy and in instances where there is too much light for fluorescence microscopy (see Fig. 1.8). All of these properties should be understood in order to achieve the maximum resolution of a given lens.

The light microscope is typically equipped with 10× eyepieces. A low-power 10× or 20× dry objective is used to scan and locate the object or specimen (e.g., cells or chromosomes), an intermediate-power 40× to 63× objective is used to gather more detailed information, and a high-power 100× oil immersion objective provides maximum analysis of detail within a given specimen.

The objective magnification times the eyepiece magnification times possible magnification changers (e.g., optovars and projection lenses) results in the total magnification. For visual observation, this total magnification should not exceed 500–1000× the NA (useful magnification range). Below 500× NA, the eye will not be able to see resolved detail in the image; above 1000× NA, "empty magnification" develops, resulting in blurred images.

The Finite versus the Infinity-corrected Objective

The finitely corrected objective directly projects a real intermediate image within the microscope (see Fig. 1.3A), whereas the infinity-corrected objective sends this image to infinity, requiring a tube lens to form the intermediate image. The direct availability of "infinity space" offers advantages whenever elements, such as reflectors, DIC prisms, color filters, and filters composed of polaroid that are used as analyzers or compensators for polarized light and DIC optical systems,

TABLE 1.2 Common objective descriptions

Achromats	corrects chromatic aberration for blue and red wavelengths; also corrects spherical aberration for green
Fluorites	corrects chromatic aberration for blue and red wavelengths; also corrects spherical aberration for two colors
Apochromats	corrects chromatic aberration for blue, green, and red wavelengths; also corrects spherical aberration for two colors
Plan	corrected to provide flat field

FIGURE 1.8

Lenses with performance criteria. Typical lenses with their properties etched into the metal body tubes. (Photo provided by H.E. Keller, Carl Zeiss, Inc.)

need to be inserted into the observation beam path. The use of these infinity-corrected lenses reduces the number of lenses required in the optical path (e.g., making telan lenses superfluous; see Fig. 1.9), thereby yielding more light and fewer internal reflections, limiting lens aberrations, and enhancing contrast. In these cases, no axial or lateral image shift (resulting in image deterioration) occurs, as long as the components of these elements are plane parallel (Fig. 1.9).

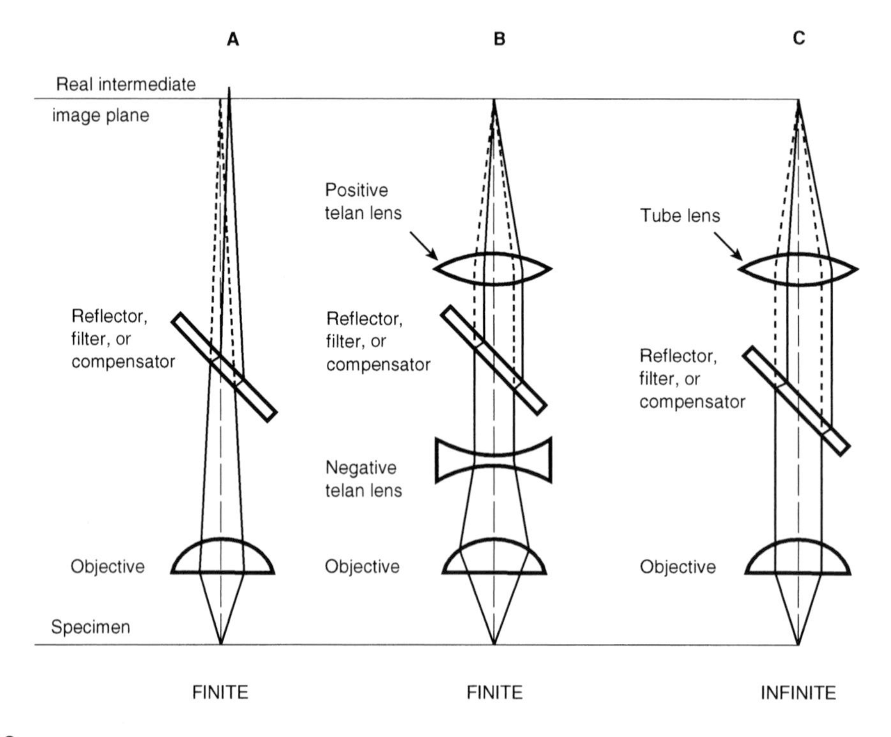

FIGURE 1.9

Benefits of infinity space. (*A*) Insertion of reflector or filter causes lateral and axial shift. (*B*) Two telan lenses generate infinity space to eliminate shift. (*C*) Objective directly provides infinity space.

More about Aberrations

"Refractive" optics can have wavelength-independent monochromatic aberrations as well as chromatic deficiencies (Figs. 1.10–1.14). Among the monochromatic aberrations are flatness of field, astigmatism, coma (radial distortion of the image point), and distortion. Chromatic aberrations are composed of longitudinal as well as lateral chromatic distortions. Of all of the possible aberrations, spherical aberration, which has a monochromatic as well as a chromatic component, is the most critical (see Fig. 1.10). Even the most highly corrected Plan-apochromat can render fuzzy images because of spherical aberration if coverslip thickness, immersion medium, mounting medium, and tube length, or the axial location of the intermediate image, do not meet the objective's design specifications. Therefore, it is essential to provide conditions that meet the specifications of an objective.

Dry objectives of high NA are designed for use with cover glasses of 0.17-mm thickness (#1.5) to render optimal images. Bear in mind that cover-glass tolerances vary greatly for different optical systems. For critical work, cover-glass thickness should be measured, for example, by a micrometer or by using the calibrated fine-focus control of the microscope, i.e., going from the top of the coverslip to the substrate (slide) on the edge of the cover glass through air. Using coverslips of the proper thickness, cultured cells attached to their surface can be optimally resolved. However, live specimens in aqueous media several micrometers below the cover glass may render poor images even with a high NA oil objective because of refractive index mismatches. The refractive indices and dispersions of all media between object and objective influence the size and intensity distribution in the Airy disk and consequently the contrast, resolution, and sharpness of the image.

For the best performance, all optical components from collector to eyepiece must be clean. Lens paper, cotton, lens cleaner, distilled water, and so forth are all acceptable tools and solvents for cleaning these components. The less rubbing needed, the better. As the whole microscope is a precision tool, so is each component. The objective, in particular, where top performance can depend on precise axial and lateral alignment of all lenses to submicron tolerances, must be treated carefully, gently, and with great respect.

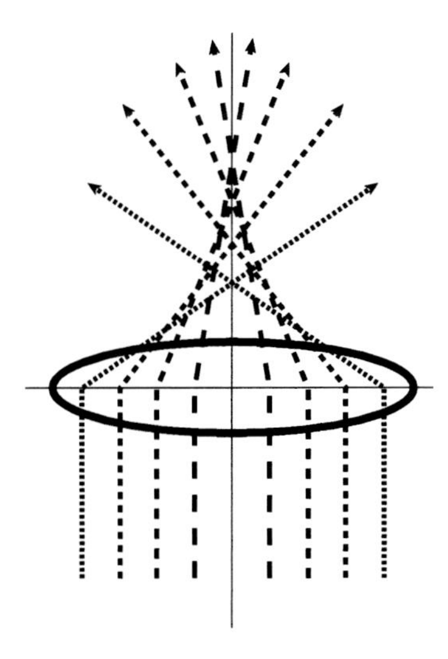

FIGURE 1.10

Spherical aberration. Peripheral rays (dotted lines) passing through a spherically shaped lens are focused closer to the lens than are paraxial rays (those rays entering the objective closest to the optical axis). Nowhere can a sharply focused image be found.

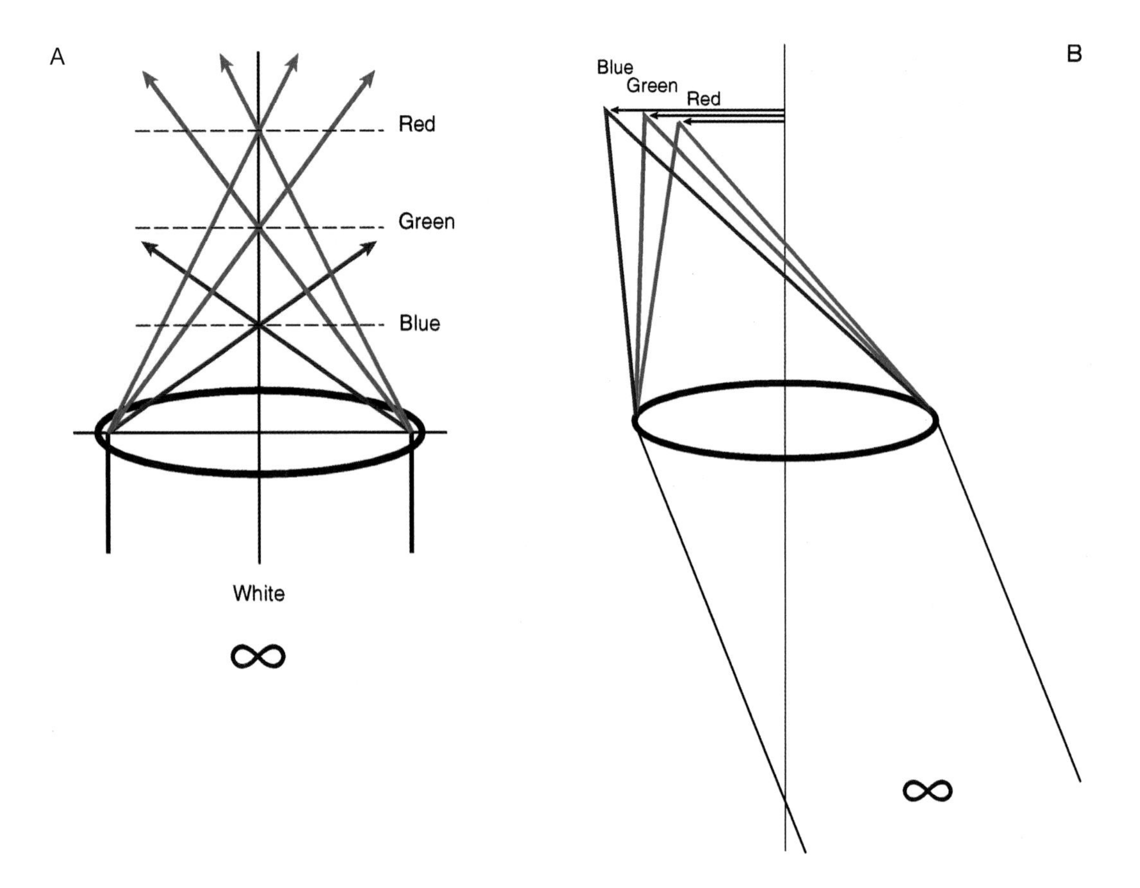

FIGURE 1.11

Chromatic aberration. (*A*) Longitudinal chromatic aberration: Blue light passing through a lens is focused closer to the lens than are green and red light. (*B*) Lateral chromatic aberration: The blue image is larger than the red image. Peripheral color fringing occurs.

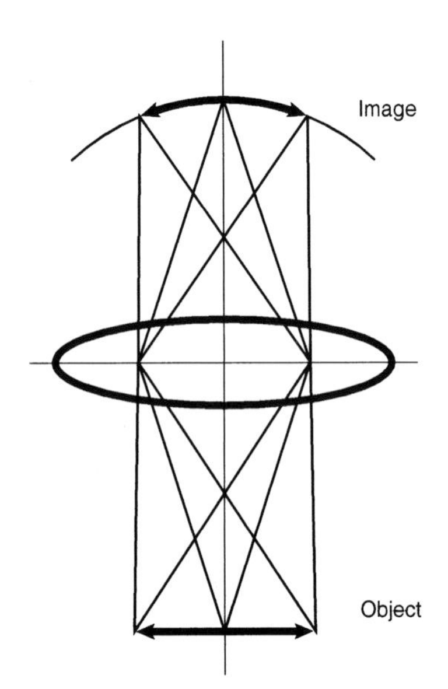

FIGURE 1.12

Curvature of field. Peripheral object points are focused closer to the lens than are paraxial ones. An image "dish" results.

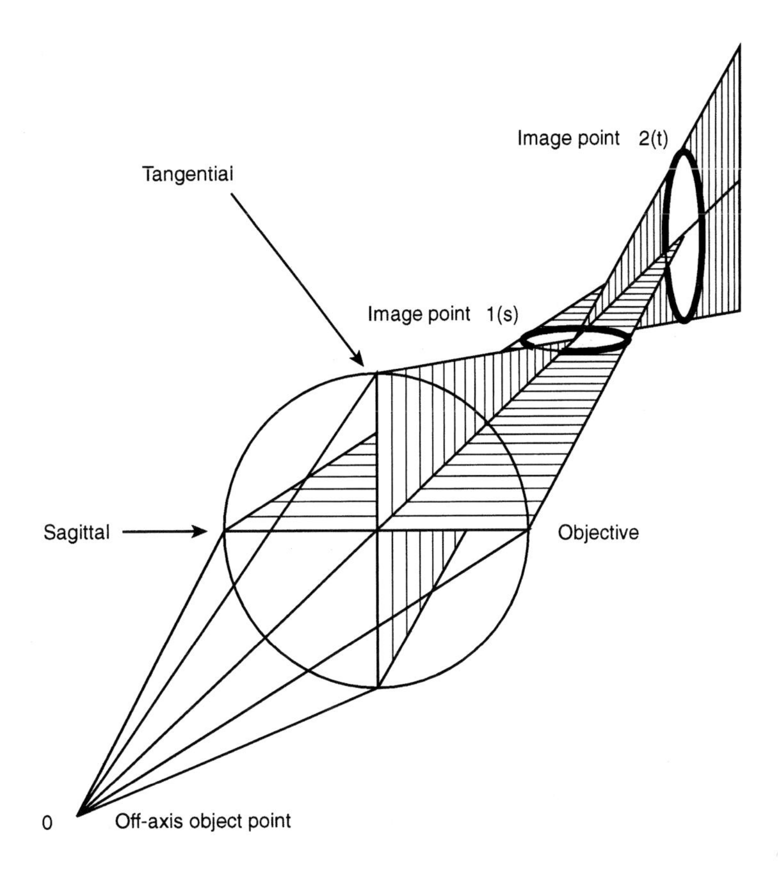

FIGURE 1.13

Astigmatism. Tangential (t) and sagittal (s) cross sections through the objective form different image locations from the optical axis for off-axis points. Depending on the focus, the point becomes tangentially or radially distorted. Note that the further off axis an object point is, the more distorted its image becomes.

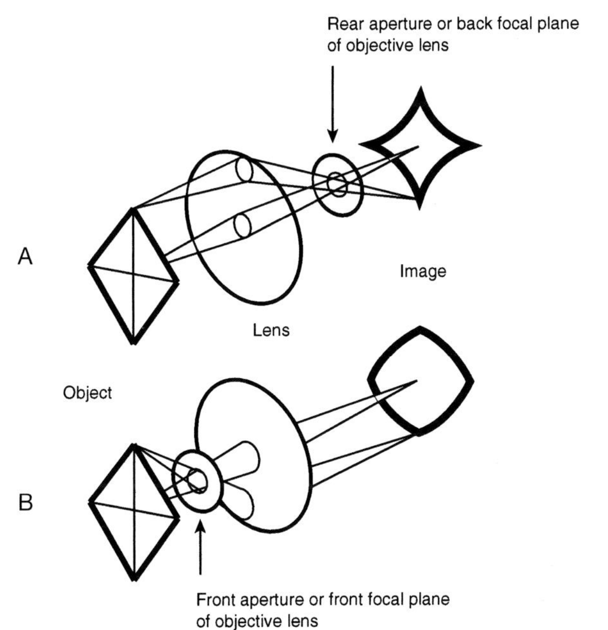

FIGURE 1.14

Distortion. Nonlinear magnification from center to edge of field can result in (*A*) pincushion or (*B*) barrel distortion.

Below is a summary of useful formulas†:

Numerical aperture: $\mathrm{NA} = n \sin \alpha$

Resolution: $$d = \frac{\lambda}{\mathrm{NA}_{\mathrm{Obj.}} + \mathrm{NA}_{\mathrm{Cond.}}}$$

Useful magnification $= 500\text{–}1000\text{x NA}$

Field of view (mm in specimen) $$= \frac{\text{field of view number * of eyepiece}}{\text{Magn. obj.} \times \text{mag. Ch' ger (e.g., optovar)}}$$

Depth of field: $$T\,(\mathrm{mm}) = \frac{1000}{7 \times \mathrm{NA}_{\mathrm{Obj.}} \times \text{mag. total}} + \frac{\lambda}{2 \times \mathrm{NA}^{2}_{\mathrm{Obj.}}}$$

†See p. 7 for definitions of n, α, and λ.

*The diameter of the field-limiting fixed stop in the eyepiece in millimeters is usually marked on the eyepiece after the magnification.

BRIGHT-FIELD MICROSCOPY

Bright-field microscopy (BF), the most commonly used light microscope technique, is ideal for fixed, stained specimens or for objects with high natural absorption or so-called amplitude specimens (e.g., plant cells with chloroplasts and pigmented epithelial cells). The amplitude or intensity of the illuminating wavefronts is reduced; the phase relationship between source and diffracted waves is $\lambda/2$, resulting in destructive interference-producing contrast for dark (or colored) specimen features.

In BF, the light source and condenser are arranged to fill the objective's aperture fully or partially with bundles of rays symmetrical to the optical axis. Both side bands (+ and − orders of diffraction) of diffracted light participate with the nondiffracted "source" waves in image formation (see Fig. 1.5). The background is bright, and absorbing structures within the specimen appear darker.

Alignment of the microscope for Köhler illumination is the first step in setting up BF. This basic adjustment is also used for all other light-microscopic techniques.

1 Center and focus the light source: Remove light diffuser (if applicable) and, without condenser, project the image of the source onto a piece of paper placed in the condenser carrier. Fill the circular area with the image of the light source filament and (if applicable) the filament's mirror image (see manufacturer's instructions as lamps and their housings differ). Proper use of these mirrors mounted behind the light source increases the brightness of the image by reflecting light "lost" to the back of the lamp housing back into the illuminating path. These mirrors also help to provide an evenly illuminated field of view.

2 Insert the condenser and light diffuser (if applicable). Use a medium-power objective. Open condenser aperture diaphragm (Fig. 1.2B). Close down the field diaphragm (Fig. 1.2).

3 Focus on any specimen feature.

4 Focus (raise) condenser until the image of the field diaphragm is sharp. Make certain that turret-type condensers are in the BF position.

5 If necessary, move condenser (with condenser centering adjustments) to center the image of the field diaphragm in the field of view. Now open the field diaphragm to just fill the field of view.

6 Slowly close condenser aperture diaphragm for best contrast and resolution. This controls the angle of the cone of light illuminating the object (see section on diffraction and resolution below). The optimal setting for this condenser diaphragm is largely dependent on contrast features inherent in the specimen.

Figure 1.15 illustrates the consequences of the proper and improper alignment for bright-field microscopy.

Notes

- Never use the condenser diaphragm to control light intensity. Resolution and contrast are directly affected.
- When the condenser aperture is fully opened, the largest diffraction angle can be received but the amplitude of the source waves predominates, and the relatively weak object waves cannot interfere sufficiently. Full resolution but poor contrast are the results.
- Closing the aperture in effect attenuates the light source wave, thereby equalizing the relative amplitudes between source and object waves for improved interference conditions. At the same time, however, the coherence of the illumination increases and more and more of the diffracted wavefronts generated in the specimen can and will interfere, resulting in potential artifacts, reduced fidelity, and poor resolution.
- Specific color filters can also help to enhance contrast. A weak red stain, for example, will stand out more clearly when transilluminated with blue light.

Oblique or Anaxial Illumination

Most cells have little or no contrast in bright field. Only a few cell types, such as pigmented retinal epithelial cells, epithelial melanocytes, plant cells with green chloroplasts, and blue-green algae, generate contrast sufficient to be seen in the BF microscope. Thus, most cells do not change the amplitude of light sufficiently to generate contrast. One very simple technique to generate contrast and reveal structures in these kinds of cells involves the deliberate although slight misalignment of the microscope. This provides an economical method for obtaining oblique or anaxial illumination (OI). This type of illumination enhances contrast in unstained or lightly stained specimens. When combined with electronic image processing, oblique illumination can be a very useful technique for revealing morphological details within a specimen.

Following alignment for Köhler illumination (see above), move the condenser diaphragm off the optical axis by slightly rotating the condenser turret. Nondiffracted source waves (zero order) are attenuated and only one side band (either + or − orders of diffraction; see Fig. 1.5) of the diffracted orders can participate in image formation (see Fig. 1.16). The result can be a striking relief-like effect enhancing edges and density or thickness gradients in the specimen (see Fig. 1.17).

Hoffman Modulation Contrast

This method is widely used in techniques employing micromanipulation or microinjection (see Chapter 83 of Spector et al. 1998) of eggs, such as in vitro fertilization (IVF), and of cultured cells. The advantage provided by Hoffman modulation contrast (MC) is the production of an image with a quasi three-dimensional appearance. The relative insensitivity of this method to depolarization of light caused by birefringent (stressed) plastic dishes, for example, makes this technique highly useful for visualizing cells while microinjecting in sterile petri dishes.

A slit aperture in the condenser is positioned such that nondiffracted source waves (zero order) enter the objective at close to its maximum aperture angle (Fig. 1.18). The off-axis slit is then imaged onto an off-axis modulator in the conjugate back focal plane of the objective (see Fig. 1.18).

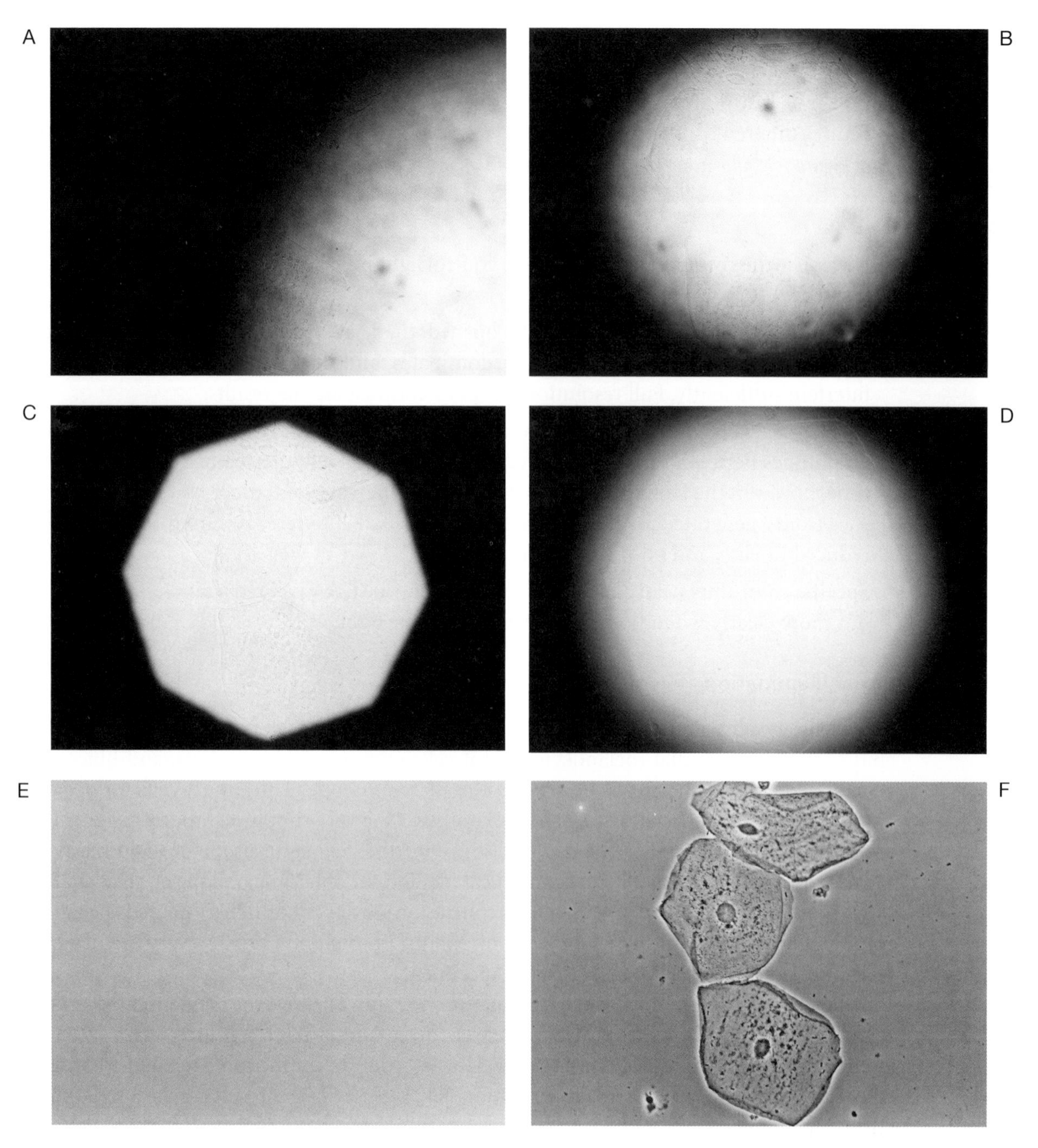

FIGURE 1.15

A series of micrographs of the same field of unfixed and unstained human cheek epithelial cells is used to demonstrate the proper setup for Köhler illumination both for bright-field and phase contrast. (*A*) The field diaphragm is out of focus and off center as indicated by the illuminated area at right. A barely detectable cell is located at the lower left of the illuminated area. (*B*) The same field after the field diagram has been centered using the condenser centering adjustment. However, the field diaphragm remains out of focus. (*C*) The condenser focusing control has been adjusted to bring the field diaphragm into sharp focus. This should be done with the condenser aperture closed down. The cells are just detectable under these conditions. (*D*) Same as *C* with the condenser aperture opened until the light just "spills over the edge" of the field diaphragm. The system is now set for optimal Köhler illumination and resolution. (*E*) Same as *D* with the field diaphragm opened to fill the microscope field with light. The live cheek cells are not apparent under these optimal bright-field conditions. (*F*) Same as *E* with phase-contrast optics in place. Note details of cell structure not seen in the bright-field image. (Photos provided by R.D. Goldman, Northwestern University.)

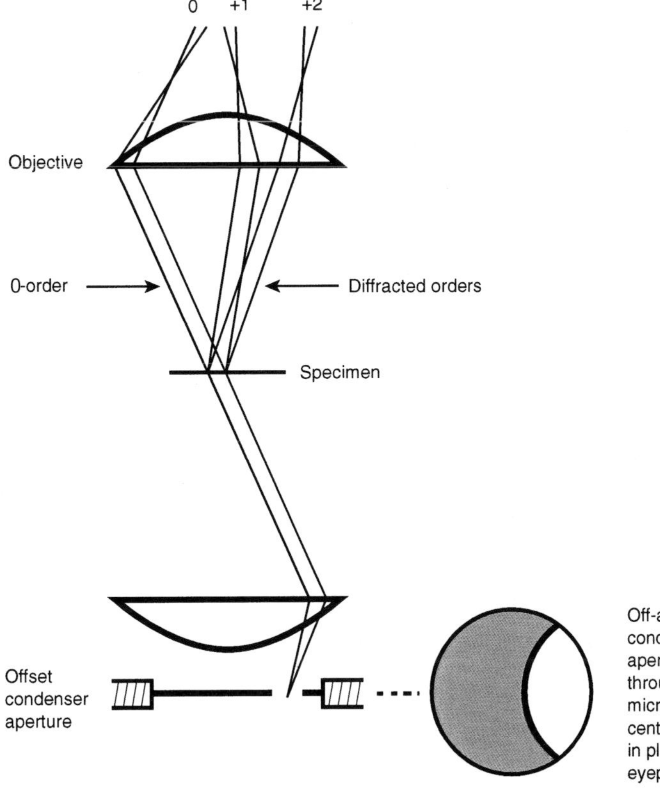

FIGURE 1.16

Oblique illumination. Oblique illumination permits the generation of contrast in the living cell. Attenuating the 0-order rays and allowing only one side band (either +1 or −1) to participate with full intensity in image formation improves the interference conditions and thus greatly enhances contrast.

This modulator attenuates the zero-order rays and preferentially permits one side band of diffracted light (+1, +2; see Fig. 1.18) to participate in the image formation. The resulting image is very similar to that obtained by oblique illumination, but it is enhanced, reproducible, and more uniform over the full field of view. Refractive or thickness gradients in the off-axis direction of the illuminating slit are optimally contrasted in opposing gray levels.

Next to the major slit aperture in the condenser is another slit, covered by a fixed polarizing filter. Combined with another rotatable polarizer mounted under the condenser, the amount of light

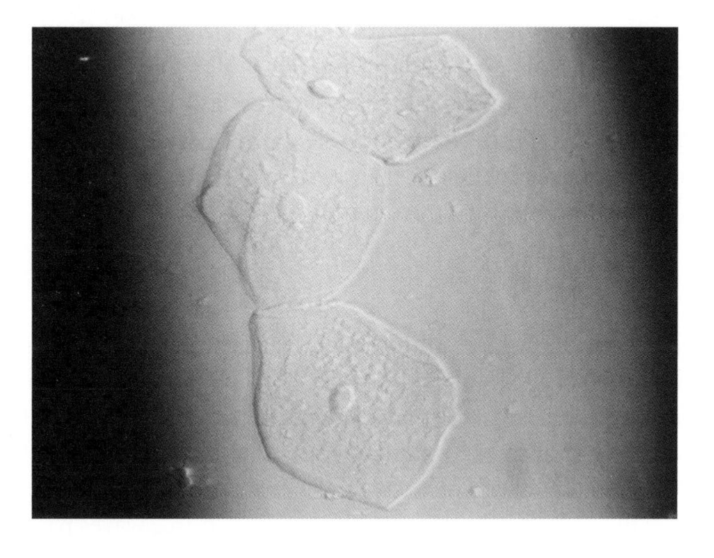

FIGURE 1.17

Photomicrograph of the same human cheek epithelial cells seen in Fig. 1.15. The microscope is set optimally for bright field and then the condenser is deliberately moved slightly off center (this is apparent from the shadows seen on both sides of the field of view). This results in the oblique or anaxial illumination of the cells. The cells are clearly seen with apparent bas-relief structure. This is an inexpensive way to generate contrast. However, this is a low-resolution image and should not be confused with the high-resolution images produced with DIC (see Fig. 1.24). (Photo provided by R.D. Goldman, Northwestern University.)

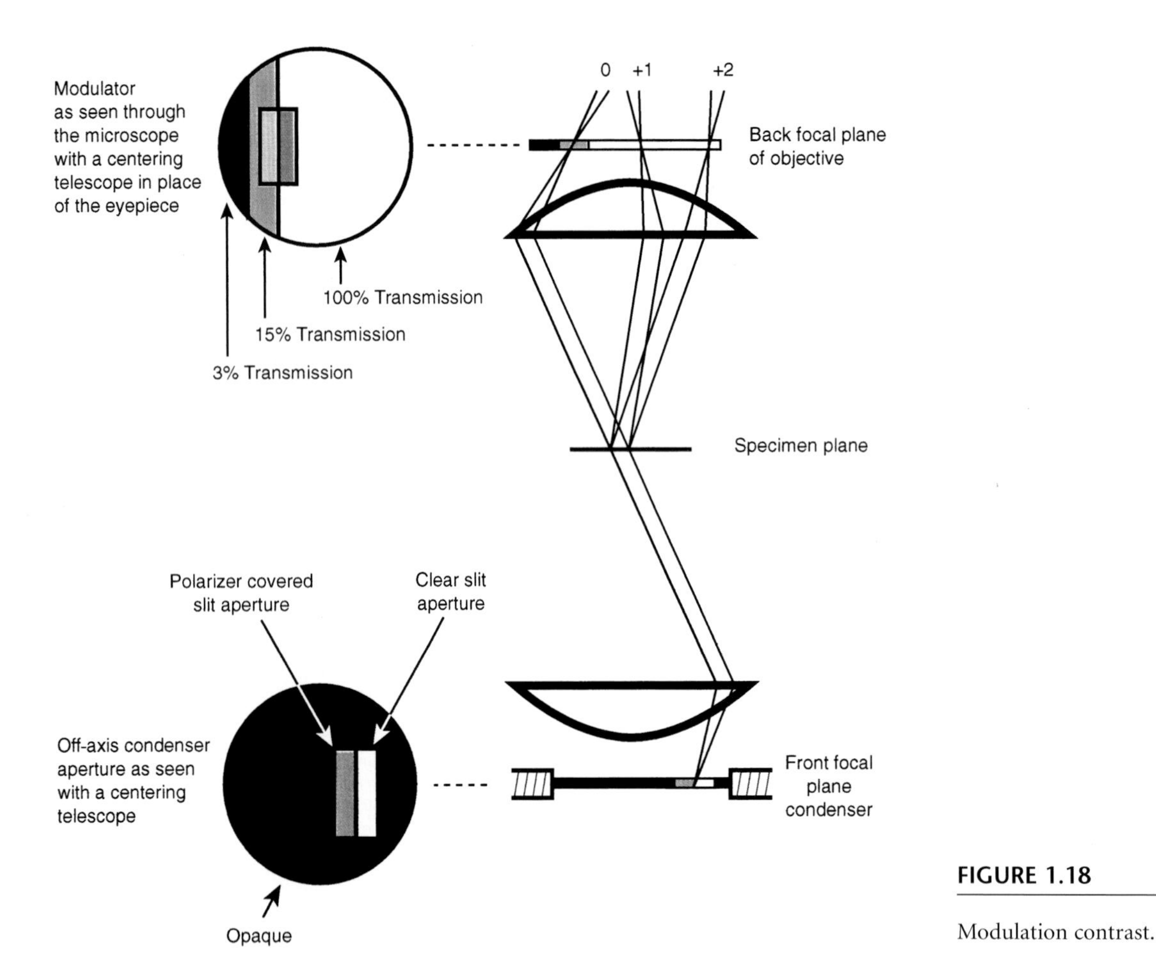

FIGURE 1.18

Modulation contrast.

passing through this slit can be regulated, permitting oblique BF to be mixed with MC. (After aligning the microscope for Köhler illumination, alignment for MC requires a centering or phase telescope to orient and position the condenser slit precisely over the modulator.)

Varel Contrast

One of the more recent and innovative advances in contrast enhancement is Varel contrast. This technique is similar to MC, but it employs an annular modulator in the objective and a corresponding partial annulus in the condenser. The advantage of this method is that it makes alignment simpler and is readily interchangeable with phase contrast. This provides added flexibility with respect to making different types of specimen structures visible. This technique is also very useful for in vitro fertilization and other microinjection techniques.

DARK-FIELD MICROSCOPY

Dark-field microscopy (DF) can render high resolution and excellent detection of particles or features far below the resolution limits of the light microscope. With DF optics, bacteria, viruses, particles in suspension, and cytoskeletal filaments in vitro such as actin can be detected. DF has also been used to enhance information in autoradiographs of cells and tissues (Fig. 1.19). The cleaner the microscope, the more structural detail can be detected. Dirt and dust destroy a good DF image.

In DF, no direct light enters the objective (see Fig. 1.20A, B). The source waves pass by; the back-

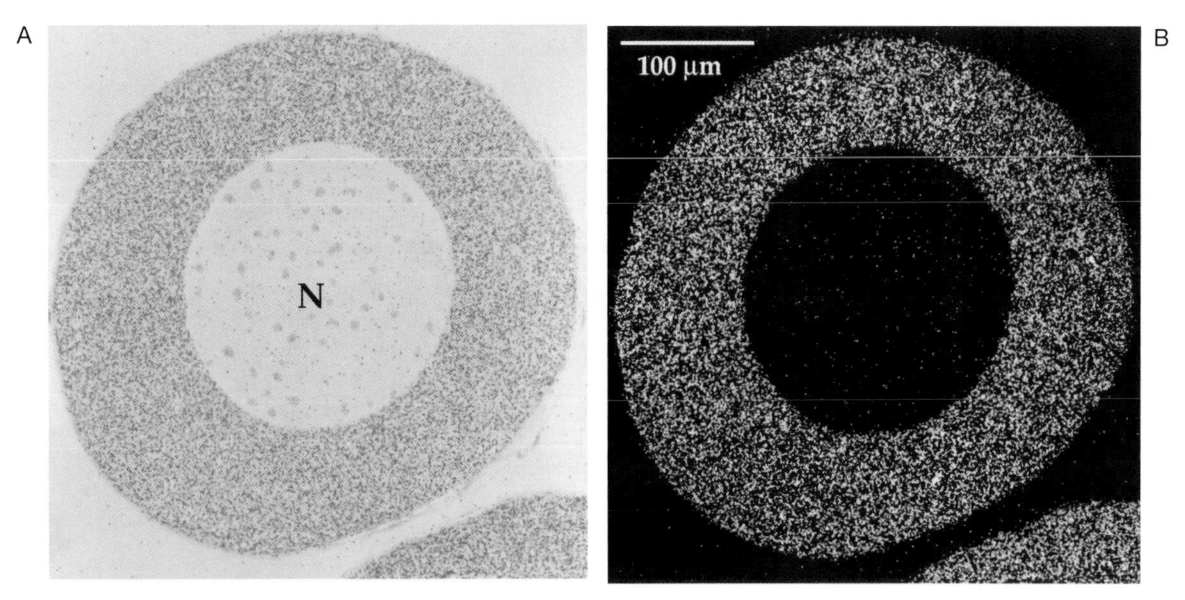

FIGURE 1.19

(*A*) A 1-μm section of a *Xenopus* oocyte (Dumont stage 1) hybridized with ^{3}H-labeled RNA complementary to one strand of the histone gene clone XlhlC. The nucleus (N) is not detectably labeled above background, whereas the cytoplasm is uniformly and heavily labeled. Probe-specific activity 5 x 10^7 d.p.m./mm. Exposure 3 weeks. Bright-field illumination. Bar, 100 μm. (*B*) Same preparation viewed in dark field to accentuate the silver grains. (Photo provided by J.G. Gall, Carnegie Institution of Washington.) (Reprinted, with permission, from Jamrich et al. 1984 [©Macmillan Publishers Ltd.].)

ground is dark. Only where specimen features diffract, refract, reflect, or disperse the incoming, highly oblique wavefronts do they become visible. Only higher orders of diffracted light participate in image formation, resulting in contrast reversal. Specimen features appear bright on a dark background.

Special dark-field condensers are required, either dry or, for objectives with high NAs, with an immersible front lens. The condenser NA must always be higher than that of the objective. High-NA objectives therefore require an aperture iris to set the NA below that of the condenser. Simple dark-field for low-power objectives (NA < 0.5) can be obtained using the highest-NA (largest) phase ring in a phase-contrast condenser (see below).

PHASE CONTRAST

Phase contrast (Ph), originally developed by the Dutch physicist Frits Zernicke, is based on Abbé's theory of image formation. It is a widely used method in studies of live cells or unstained fixed material.

The vast majority of living cells are phase objects. In cells that are phase objects, absorption differences among cellular structures are minimal and differences in refractive index or thickness cannot be detected, because they generate insufficient phase shifts for achieving full constructive or destructive interference of light rays. Furthermore, the intensity or amplitude of the diffracted light is so low that it is totally overwhelmed by the nondiffracted source waves. This is the reason most live cells are barely visible in a well-aligned BF optical system.

In the Zernicke Ph microscope, an annulus (phase annulus) in the condenser aperture (front focal plane) generates a hollow cone of zero-order illumination that is projected onto the back focal plane of the objective (see Fig. 1.21A). A matching phase ring in the objective performs two functions: (1) It attenuates or absorbs the nondiffracted zero-order light by 70–80% and (2) it shifts its

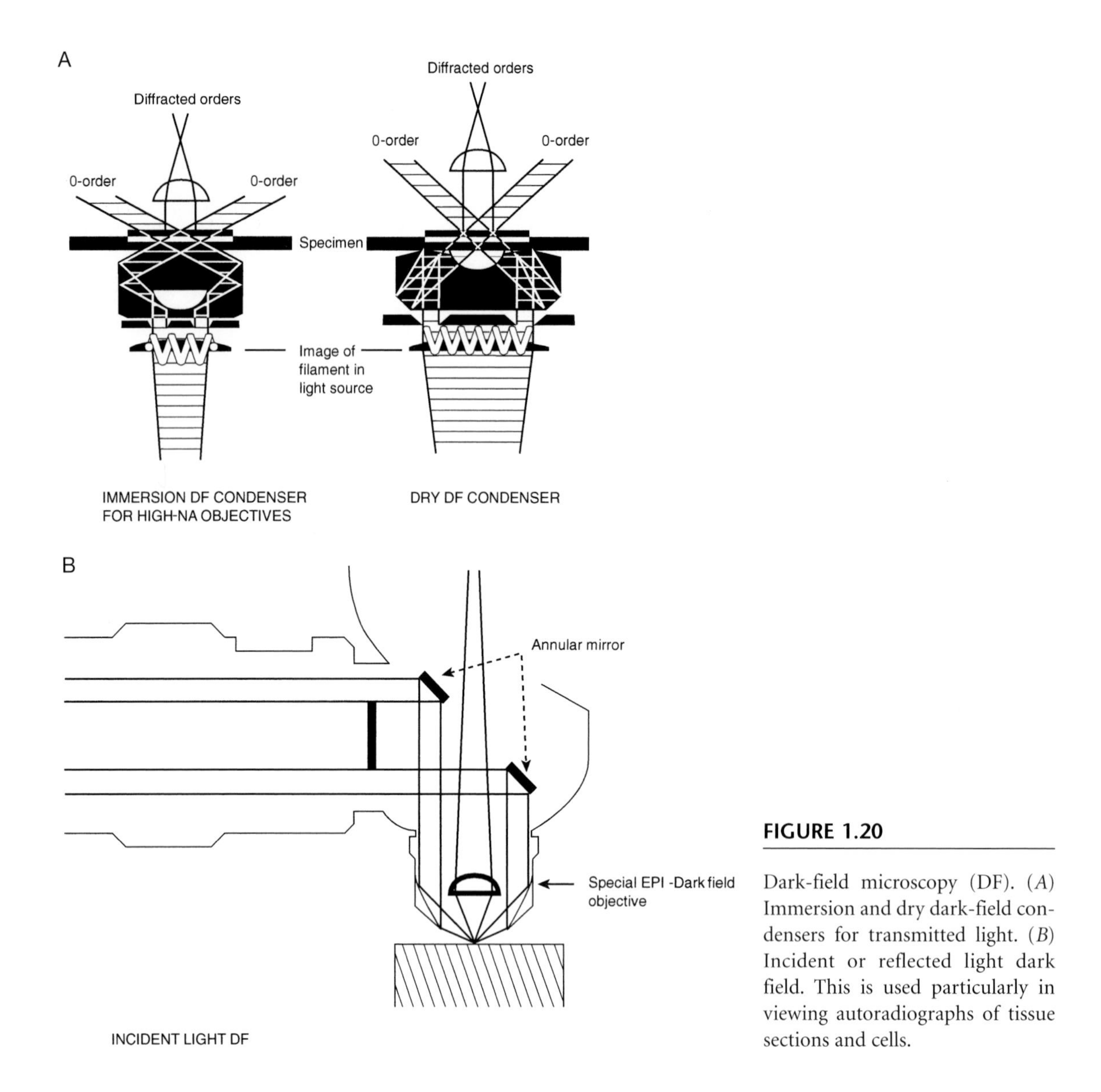

FIGURE 1.20

Dark-field microscopy (DF). (*A*) Immersion and dry dark-field condensers for transmitted light. (*B*) Incident or reflected light dark field. This is used particularly in viewing autoradiographs of tissue sections and cells.

phase position by one quarter of the wavelength ($\lambda/4$) of the light. This arrangement alters the amplitude and phase relationships of diffracted versus nondiffracted wavefronts to achieve better interference, thereby enhancing contrast.

To optimize structures with Ph optics, the specimen must be thin (<5 μm) to prevent excessive halation ("halo" formation due to refractive gradients passing zero-order light outside the Ph ring). The formation of halos in Ph optics makes it very difficult to determine the location of the boundaries or the edges of cell structures. The resolution is somewhat reduced by the aperture of the matching phase annuli in the condenser and the rings in the objectives. For very thin specimens, such as ultrathin sections prepared for transmission electron microscopy (TEM), special, high-absorption phase objectives are available. Phase contrast produces specific gray levels that distinguish areas of different specific refractive indices (n) times thickness (t). Positive phase-contrast regions of higher refractive index (n) within the specimen usually appear darker (e.g., mitochondria appear dark gray on lighter background cytoplasm and nucleoli appear even darker within the nucleoplasm). Very large n or t differences, however, can cause a contrast reversal.

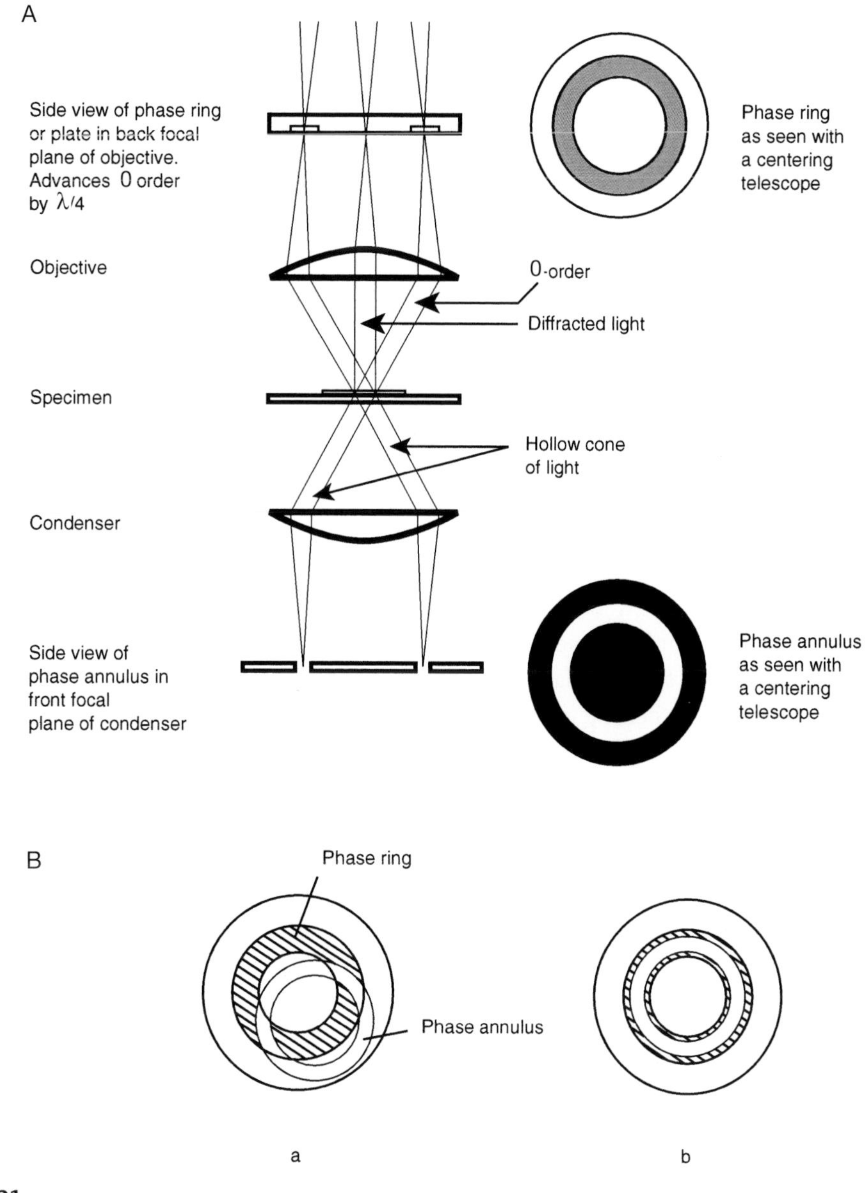

FIGURE 1.21

Phase contrast. (*A*) Light path in a typical phase contrast microscope, showing phase (Ph) ring and condenser (Ph) annulus. (*B*) Proper alignment of phase annulus in condenser with phase ring (*b*). Improper alignment of phase annulus with phase ring (*a*).

Köhler alignment also requires centering or exact superimposition of the matched phase ring and annulus in the condenser and *objective* (Fig. 1.21A). This superimposition is accomplished through the use of a telescope lens that can focus at the back focal plane of the objective. In this fashion, both the condenser annulus and the objective phase ring can be seen simultaneously (see Fig. 1.21B). The position of the phase ring in the objective is fixed. The phase annulus of the condenser, however, is moveable through the use of adjusting screws built into the turret condenser. This permits the observer to superimpose the two rings as seen in Figure 1.21B. Note that the bright-phase annulus of the condenser fits into the dark-phase ring of the objective; an example of a misaligned phase ring and annulus is shown in Figure 1.21B,a, and the corresponding image with an aligned ring and annulus is depicted in Figure 1.21B,b.

POLARIZED-LIGHT MICROSCOPY

Muscle, nerve, bone tissue, fibers, or any fiber-like structure can be contrasted effectively in polarized light, sometimes with the help of compensators for increased sensitivity. Furthermore, the quantitative capabilities of the polarizing microscope have made it an indispensable instrument not only for the crystallographer, but also for cell biologists studying cytoskeletal systems (e.g., the mitotic spindle).

Polarized light (Pol) has become an essential tool to detect or enhance the contrast of birefringent or anisotropic structures within cells (e.g., the A bands of skeletal muscle myofibrils). These structures have some degree of systematic molecular structure or preferential molecular orientation that results in different refractive indices in two orthogonal directions. Depending on their orientation between two "crossed" polarizers, they will in effect rotate plane polarized light (delivered by the first polarizer to the specimen) by specific amounts or generate elliptically or circularly polarized light, which then can pass through the second polarizer and be detected.

Polarizing filters or prisms are used to convert unpolarized light to plane polarized light that oscillates in only one plane (see Fig. 1.22A). The most basic polarizing microscope has such a polarizing element, the "polarizer" somewhere below the condenser and a second polarizing element, called the "analyzer," behind the objective. Both elements are usually in crossed or orthogonal positions. For example, the polarizer E-W and the analyzer N-S generate a dark background (Fig. 1.22B). The efficiency of the polarizers can be determined by comparing their parallel versus crossed positions; this can then be expressed as an extinction factor that can reach 10^{-6}, but is usually greatly reduced when employing high-NA optics or optics that contain glass elements that are internally strained. This problem can be corrected to some extent through the use of so-called "rectified optics" (Inoue 1959).

Striking, intense interference colors can be obtained when a birefringent specimen with its two orthogonal oscillation directions of differing refractive index (n_γ and n_α) is positioned diagonal (at an angle of 45°) to the polarizer and analyzer (Fig. 1.22C). If the product of thickness $d \times (n_\gamma - n_\alpha)$ results in a path difference or retardation of 1λ between the two directions, the wavelength to which λ is assigned is canceled by the analyzer, and the residual interference color is a mixture of the visible spectrum with one region missing. For most biological birefringent structures, the path differences generated are much less than 1λ. Under these conditions, subcellular birefringent structures appear either bright or dark depending on their orientation between the crossed polarizer and analyzer.

Polarized light microscopy is also a quantitative tool. Path differences (retardation values) can be measured with wave plates and compensators, elements that produce known path differences (e.g., for biological materials usually $\lambda/20-\lambda/30$). Using such devices, accurate determinations of optical axes (the direction in which plane polarized light is not altered by the object) and the directions for higher or lower refractive index can also be identified (e.g., the sign of birefringence of the microtubule bundles [spindle fibers] in the mitotic apparatus).

As a full-fledged quantitative tool, the optimal polarizing microscope should employ strain-free optics and it should be equipped with a centerable rotating stage, centerable objectives, and receptacles in the microscope stand for the introduction of wave plates and compensators.

DIFFERENTIAL INTERFERENCE CONTRAST MICROSCOPY

Because of its wide dynamic range of contrast control and its performance at the highest NAs, DIC (differential interference contrast) microscopy has become a widely used technique for the study of unstained fixed cells or tissues, as well as for studying dynamic events in vivo at the limits of resolution of the light microscope. DIC can be generated using either the Smith or Nomarski systems (see Fig. 1.23).

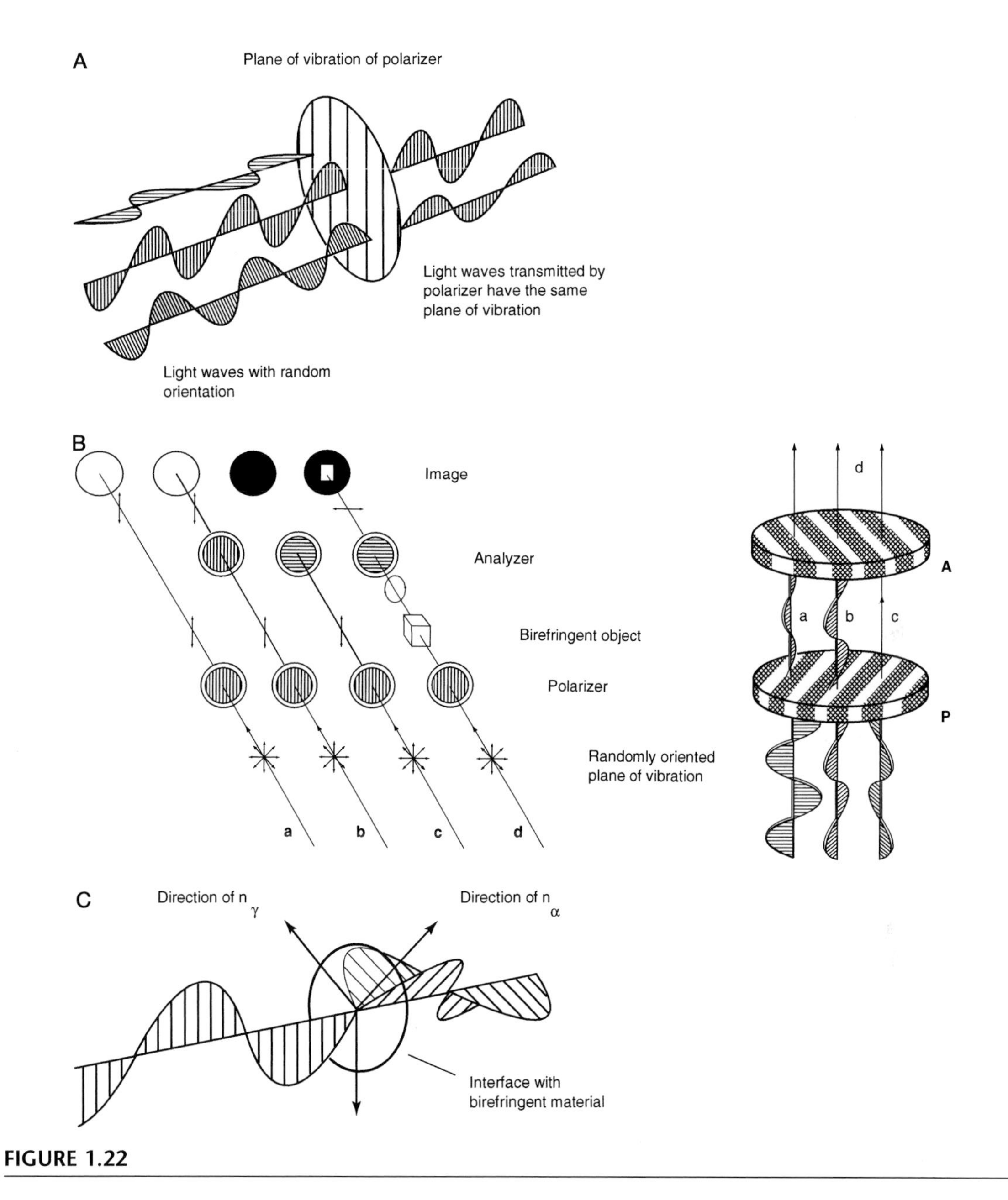

FIGURE 1.22

Polarized-light microscopy. (*A*) Random light waves incident on a polarizer are converted to light vibrating in one plane. (*B*) Polarizer only (a), parallel polarizer and analyzer (b), crossed polarizer and analyzer (c), and the birefringent specimen (d). (*C*) Splitting of linearly polarized light into two orthogonal vibration directions by birefringent structures (e.g., muscle fibers or mitotic spindles).

A simple polarizing microscope with crossed polarizer and analyzer is equipped with so-called Wollaston prisms (two quartz prisms cemented together with their optical axes at 90° to each other), one at or near the back focal plane of the objective and, in transmitted light, a second one in the focal plane of the condenser (see Fig. 1.23A, B). The γ (slow) and α (fast) directions of the prism are oriented at 45° to polarizer and analyzer. The incoming plane polarized beam (dashed line in Fig. 1.23) is split into two equal, perpendicularly oscillating components (dotted and cross-hatched lines) as it enters the first prism. At the inclined interface between the two prisms, slow and fast directions become reversed, and refraction in opposing directions occurs for the two beams

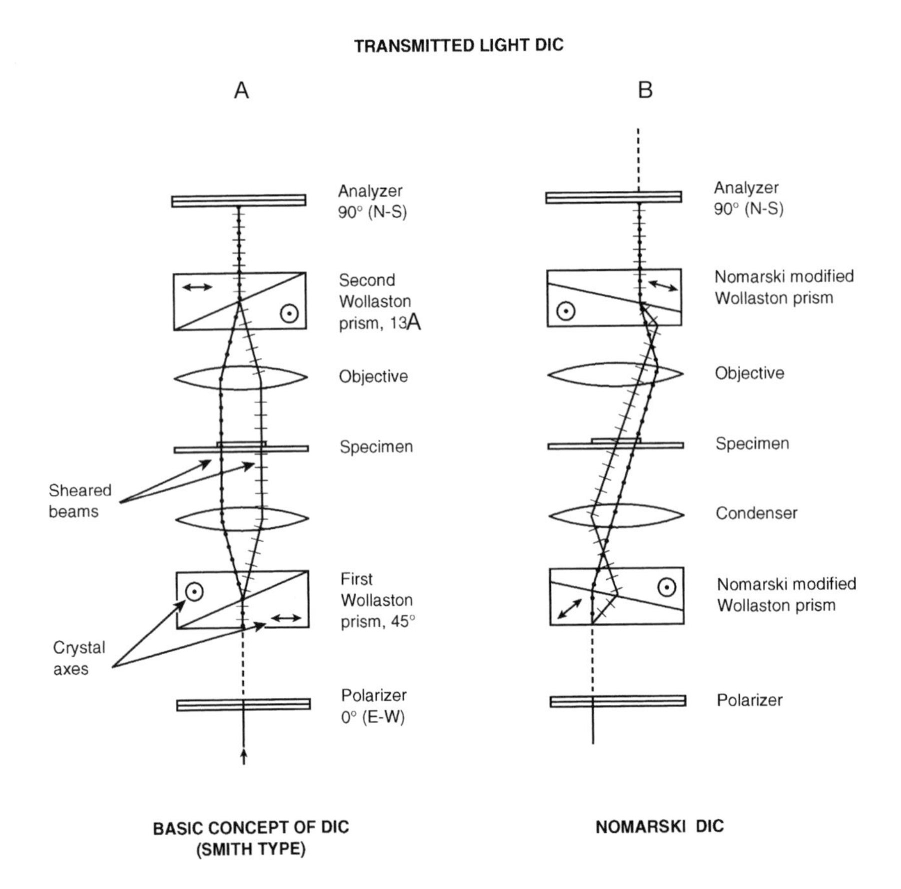

FIGURE 1.23

Differential interference contrast (DIC) microscopy. (*A*) Transmitted light DIC (Smith type). (*B*) Nomarski DIC in transmitted light.

that exit the second prism "sheared" from each other. Their phase relationship is a function of where the shear in the Wollaston prisms has taken place. With the inclined shearing surface in the front focal plane of the condenser (transmitted light), the two diverging beams oscillating perpendicularly to each other are made parallel, traverse the specimen, and enter the objective, which converges the two beams to meet at its back focal plane. Here, the second Wollaston recombines both beams to a common single beam, still consisting of two perpendicularly oscillating components at 45° to the analyzer. Equal amounts of light from both components are delivered to the analyzer, which brings them into its vibration direction, where both can now interfere with each other. The relative phase shift between both components is a function of refractive index or thickness gradients in the specimen and of the relative lateral position of one of the Wollastons.

The amount of shear selected is usually right at the resolution limit of a given objective. For the highest resolution, one sometimes goes below the shear; for more contrast, often higher shear is chosen. With the "shear direction" at 45°, feature gradients such as thickness and refractive index differences within the specimen should be aligned in the same direction. This makes a rotating stage for specimen orientation desirable. Opposing specimen gradients generate opposing phase shifts between the two sheared wavefronts, and opposing gray levels (constructive interference on one side of the structure and destructive interference at the other) result, giving the impression of a third dimension.

For best DIC images, make sure that the polarizer and analyzer are crossed for maximum darkness. This should be checked without the Wollaston prisms in place. With either condenser or objective prism in place, a diagonal interference fringe appears in the back focal plane of the objective (which can be observed by pulling out the eyepiece or using a centering telescope). This fringe must disappear with both prisms in position. The contrast control darkens the objective pupil evenly. This contrast control can either be on the objective Wollaston prism (by movement across the optical axis) or be adjusted using a "de-Senarmont compensator," consisting of a $\lambda/4$ plate in a special fixed orientation combined with a rotating analyzer or polarizer. This permits more precise and more reproducible settings, particularly when using video-enhanced contrast (VEC).

DIC (according to Smith) places the objective's Wollaston prism right at its back focal plane, an area often not accessible, particularly in high-magnification objectives. Nomarski redesigned this double-quartz prism such that the optical axis of the first quartz is tilted with respect to the incoming beam. Shear now takes place at the air–quartz interface (compare A and B in Fig. 1.23). Refraction at the interface between both prisms causes both sheared beams to converge with their crossover outside the prism (Fig. 1.23B). This crossover can now be placed at the back focal plane of the most highly corrected objectives while the actual Nomarski prism remains behind the lens mount in a separate receptacle, from where it can be easily removed, allowing the objective to be used for other techniques as well.

The major benefit of DIC is its ability to contrast optimally any specific specimen gradient (see Fig. 1.24). Since refractive index and thickness gradients both generate phase shifts between the sheared wavefronts, the interpretation of DIC images must take this into account. Furthermore, DIC allows the use of full objective and condenser apertures for the highest resolution and the shal-

A

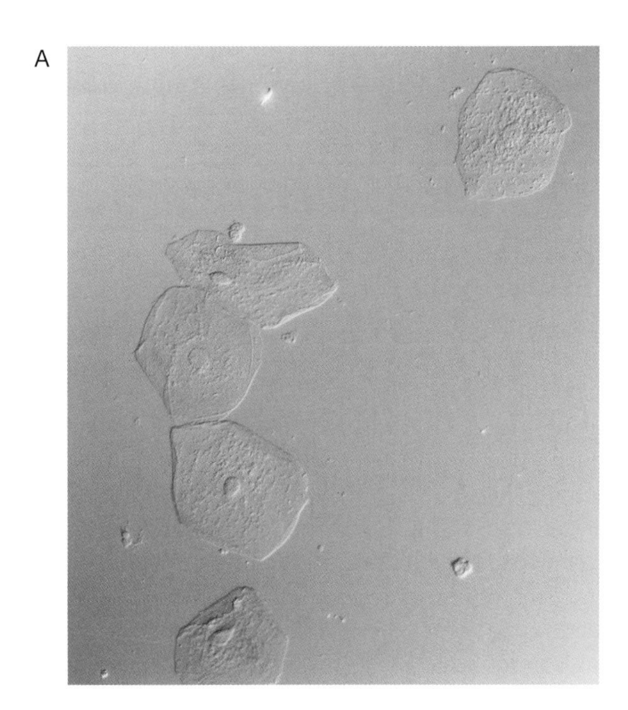

B

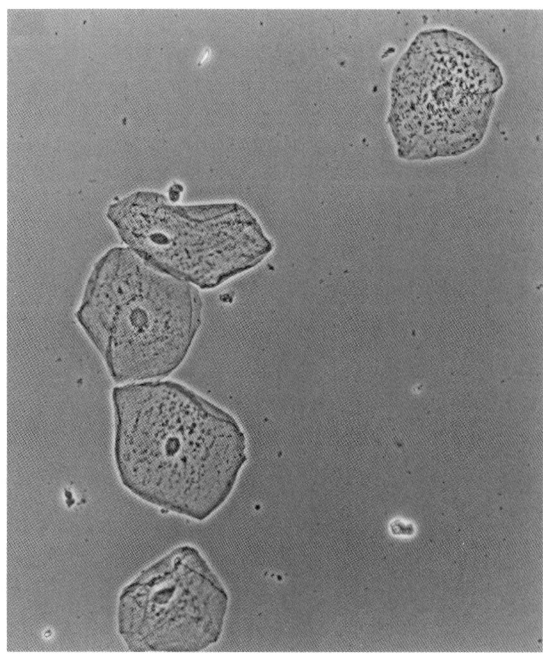

FIGURE 1.24

Differential interference contrast (DIC). (*A*) The same field of human cheek epithelial cells as seen in Fig. 1.15, but at a lower magnification. Note the bas-relief effect seen with DIC and the sharp cell borders. (*B*) In contrast, note the bright "halo" seen at the edge of each cell when viewed by phase contrast. This halo makes it much more difficult to define an edge in phase contrast relative to DIC. (Photos provided by R.D. Goldman, Northwestern University.)

lowest depth of field. Optimal contrast is generated in the thin slice of best focus with relatively little influence of the layers above and below this slice. The thickness of this optical slice varies according to the NAs of the condenser/objective combination.

CAUTION: Because DIC uses polarized light, plastic, anisotropic dishes or coverslips can seriously reduce or completely void its performance.

REFLECTION INTERFERENCE MICROSCOPY

Reflection interference microscopy (RIM) allows observation of the interference pattern generated by the thin film between a cell surface and a substrate (e.g., cover glass). Wavefronts reflected from the substrate interfere with those reflected from the cell surface. The latter experience a phase shift of $\lambda/2$ resulting in destructive interference (dark areas) wherever the gap between cell and substrate approaches zero. Therefore, cell-substrate adhesions, such as focal contacts and hemidesmosomes, can be effectively contrasted and their dynamic properties can be studied in vivo. Working at lower-illumination apertures and using an interference filter for monochromatic light, interference patterns between two cell surfaces can also be generated and studied (Izzard and Lochner 1980).

RIM utilizes a microscope equipped for reflected light Köhler illumination (Fig. 1.3B); preferably, a mercury arc lamp with a line selection filter of 546 nm is used. Cell-substrate adhesions, however, can be contrasted in white light as well. A standard 50% T/R plane glass reflector is used to illuminate the specimen as shown in Figure 1.3. To improve the signal-to-noise ratio at the low reflectivities of cell and substrate (<4%), the field diaphragm may be set to less than the full field diameter. The field diaphragm further serves as a focusing aid as it comes into sharp focus only when the cover glass is in focus. Special so-called antiflex objectives further enhance signal-to-noise ratios by eliminating all residual internal scatter in the objective using a crossed polarizer and analyzer in the microscope while a $\lambda/4$ plate in front of the objective in effect rotates the signal waves from the specimen $2 \times 45°$ for free passage through the analyzer. Furthermore, RIM can be greatly enhanced by using video image processing.

FLUORESCENCE MICROSCOPY

A wide range of specimens absorb light radiation, become "excited," and then reradiate or emit light. When the emission continues for some time after excitation, the process is called phosphorescence. Emission that ceases almost instantly after excitation ($\sim 10^{-7}$ sec) is called fluorescence. Both forms of emitted radiation, along with chemical reactions that cause light emission, comprise photoluminescence, an area of ever-growing interest to biomedical researchers.

Many organic and some inorganic substances (e.g., drugs, vitamins, and chlorophyll) display autofluorescence or primary fluorescence. When irradiated with light of a specific spectral region, they emit radiation of a longer wavelength (lower energy). Stokes observed this in fluorspar, and the spectral relationship he described between excitation and emission became Stokes' law. The study of primary fluorescence dates back to the early part of this century with the investigations of Köhler, Reichert, and others. Fluorescence microscopy garnered serious attention only in the 1950s, after Coons and Kaplan successfully developed indirect immunofluorescence for specific staining of tissues and cells with fluorescent compounds (e.g., FITC [fluorescein isothiocyanate]) conjugated to an antibody, thereby detecting and localizing specific antigens within cells.

Labeling tissues and cellular components with proteins (e.g., antibodies) and nucleic acids derivatized directly or indirectly with specific fluorochromes has become one of the most important methods for both research and clinical diagnostic studies. The range of fluorescence markers or labels for every conceivable cellular organelle, protein, and nucleic acid component is steadily growing. For example, techniques ranging from cellular pH detection or dynamic events involving Ca^{++} fluxes in vivo to fluorescence in situ hybridization (FISH; see Chapers 15–18) for localizing specific genes in situ have revolutionized many aspects of cell biology and diagnostic pathology. It should be noted that both confocal microscopy (see Chapter 2) and two-photon or multiphoton microscopy (see Chapter 3) also employ fluorescence. The following factors make fluorescence an important tool in cell and molecular biology.

- *Specificity:* Fluorescent molecules or fluorochromes have specific absorption and emission wavelengths.
- *Sensitivity:* Under optimum conditions, a single fluorochrome molecule can be detected.
- *Spectral sensitivity:* Excitation and/or emission wavelengths of certain fluorochromes can reflect physiological, physical, and/or chemical changes (e.g., ratio imaging).

Obtaining High-Quality Fluorescence Images

In addition to the appropriate light source and the proper optics, selection of the best filters is essential. The exciting radiation is intense and covers the whole field, and only small portions of the field emit relatively weak fluorescence signals. The exciting, shorter-wavelength light must be blocked, after traversing the specimen, permitting only the longer-wavelength fluorescence to pass to the eye or detector. This is accomplished with two filters whose transmission curves optimize the excitation and emission wavelengths, for a given fluorochrome, without overlapping areas of transmission. The primary, or excitation, filter is used to select wavelengths from the light source corresponding to those in the excitation spectrum of the fluorochrome and to reflect or absorb other wavelengths (Fig. 1.25A). This primary filter is placed in the illumination beam. The secondary, or barrier, filter separates unabsorbed excitation light from fluorescence emission light (most excitation light is in fact not absorbed by the fluorochrome) and is placed in the observation path (see Fig. 1.25B).

Transmitted Fluorescence versus Epifluorescence Microscopy

Fluorescence was initially studied with the transmitted light microscope. For best a priori separation of emitted from exciting radiation, usually a dark-field condenser was used (see discussion of dark-field microscopy above). Presently, transmitted light fluorescence is limited to low-magnification objectives (<10×), where the relatively higher condenser aperture provides more intense excitation energy and offsets the generally lower-objective aperture with its limited light gathering power.

The most widely used contemporary fluorescence microscopes are equipped with incident or epiillumination and incorporate the use of a dichroic mirror—or chromatic beam splitter—to assist in the separation of the fluorescence emission light from unabsorbed reflected excitation light. The surfaces of dichroic mirrors have thin coatings that facilitate the reflection of shorter wavelengths and transmission of longer wavelengths (see Fig. 1.25).

The advent of highly efficient chromatic beam splitters or dichroic reflectors has made epifluorescence the method of choice for most applications. The advantages are obvious.

- The exciting radiation travels in a path opposite to that of the observed fluorescence. This minimizes the effects of the absorption of light due to specimen thickness.
- The full-objective aperture provides excitation and collects emitted fluorescence for maximum brightness.
- The objective becomes the condenser assuring perfect alignment between both.
- Epifluorescence can be readily combined with other transmitted light techniques such as phase or DIC. It is often critical to find specimen areas of interest with these latter techniques prior to

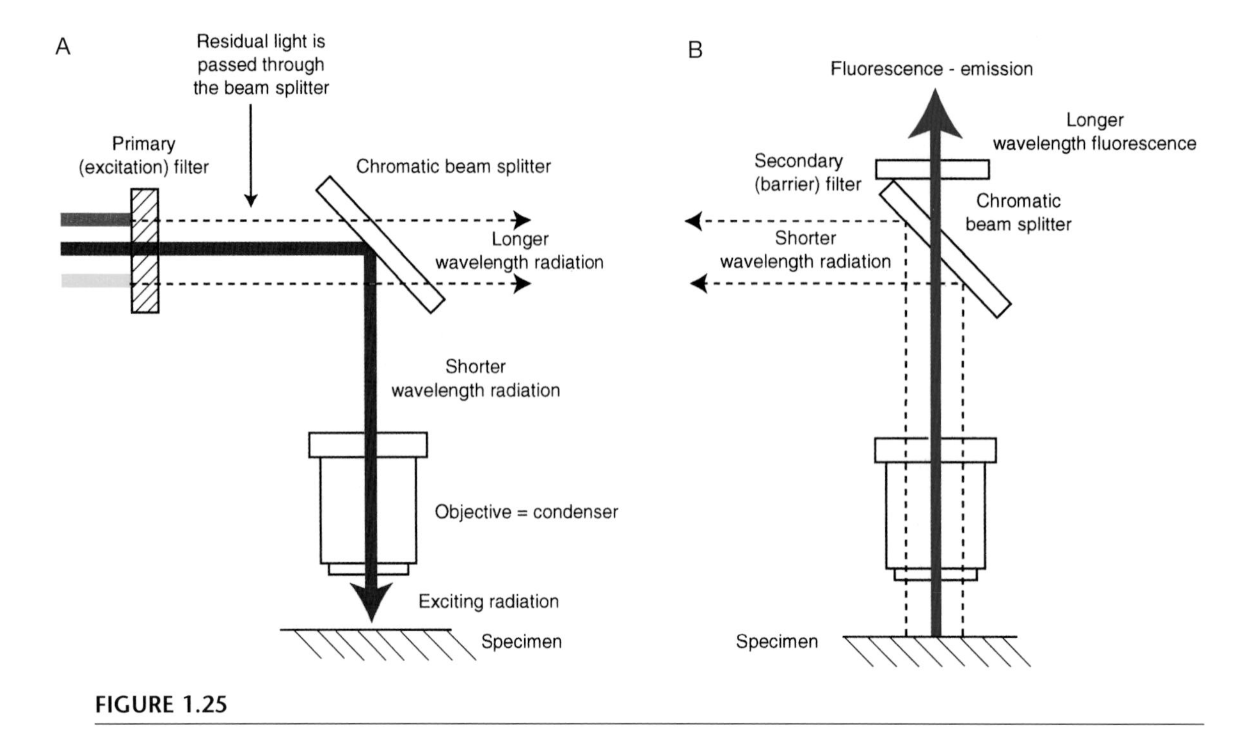

FIGURE 1.25

Filter components of the epifluorescence microscope. The blue excitation light and the green light emitted from the specimen are typical for the fluorochrome fluorescein.

subjecting the specimen to excitation in order to reduce quenching or fading of fluorescence intensity.

- Only the objective needs to be immersed in oil for studies using the highest NA.

Figure 1.26 shows typical transmission curves for some of the exciter and barrier filters. Broadband (dyed glass filters), band-pass (BP), short-pass (KP), and long-pass (LP) interference filters are designed and combined to provide the highest possible transmission for the excitation (absorption) wavelength range of a given fluorochrome (exciter filter) and the highest transmission for the fluorochrome's emitted fluorescence light (barrier filter). A minimum of overlap of the transmission curves of exciter and barrier filters (cross talk) assure the best contrast (signal to noise [*S*/*N*]) in the image. The spectral characteristics of a typical chromatic beam splitter are shown in Figure 1.27. Filter combinations are available for a wide range of fluorochromes useful in cell biological studies (see Appendix 1).

Exciter filters, chromatic beam splitters, and barrier filters are usually mounted in either sliders, filter blocks, or external filter wheels for quick insertion into the light path for a range of different fluorochromes.

CAUTION: Chromatic beam splitters have exposed thin layers of soft interference coatings that should not be touched or wiped!

Light Sources for Fluorescence

Choosing the best light source is primarily dictated by the absorption (excitation) peak of the fluorochrome to be used (see Appendix 1). In addition, however, many filter sets contain exciter filters specifically designed for the intense emission lines of mercury arc lamps at wavelengths of 365, 405, 436, and 546 nm and therefore mandate the use of a mercury burner (see below). Other consider-

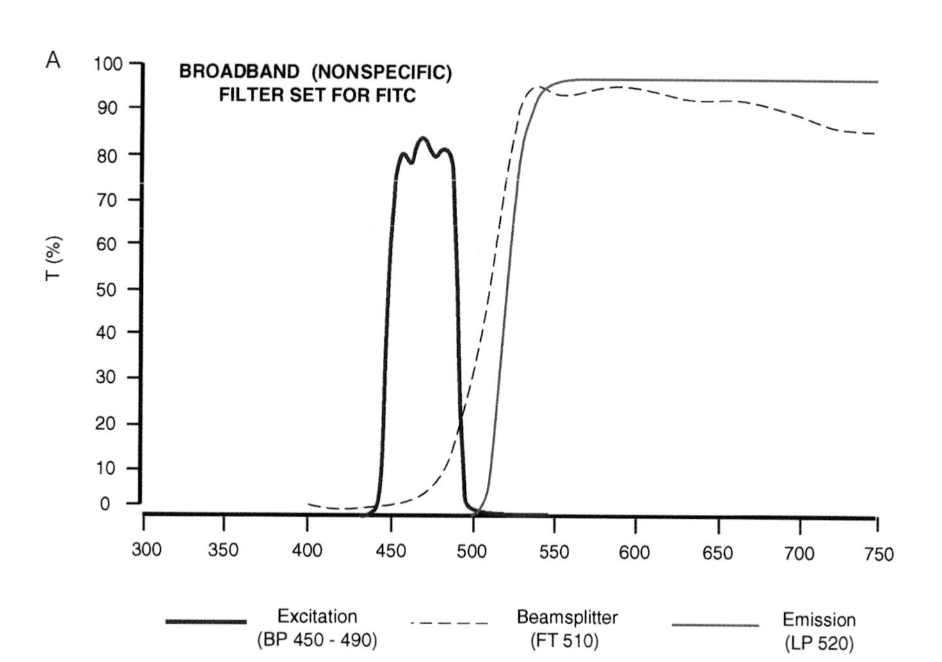

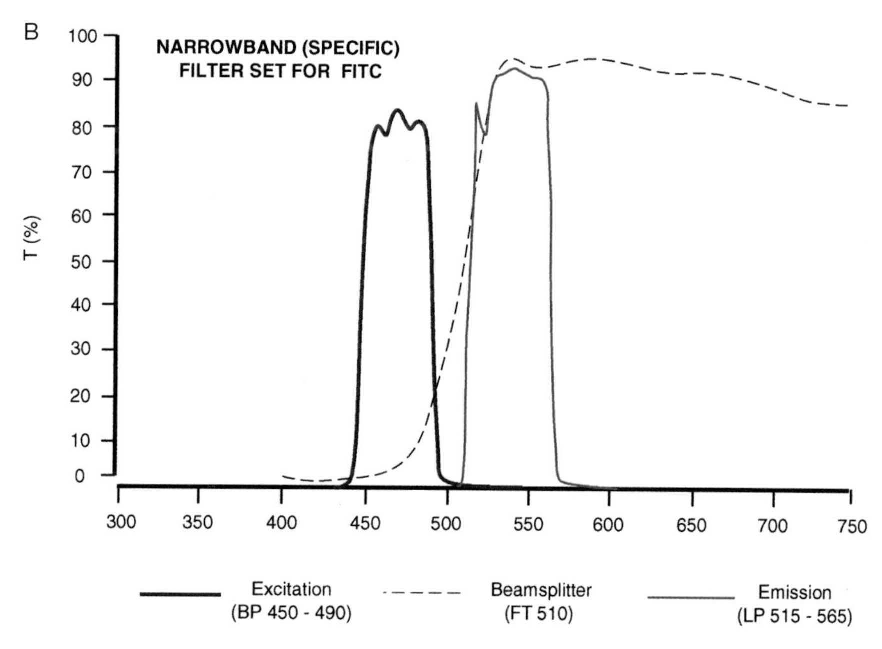

FIGURE 1.26

Transmission curves for some typical exciter and barrier filters. (*A*) Broadband (nonspecific) filter set for FITC. (*B*) Narrowband (specific) filter set for FITC.

ations are based on the fact that some fluorophores have absorption peaks in spectral regions where a xenon lamp has relatively more output (FITC at 480 nm) or where even a tungsten halogen lamp may suffice (e.g., with the use of low-light-level video) and may actually minimize fading of fluorescence, often encountered with FITC and mercury lamps. For further information, refer to the relative spectral intensity curves under light sources for the microscope covered below.

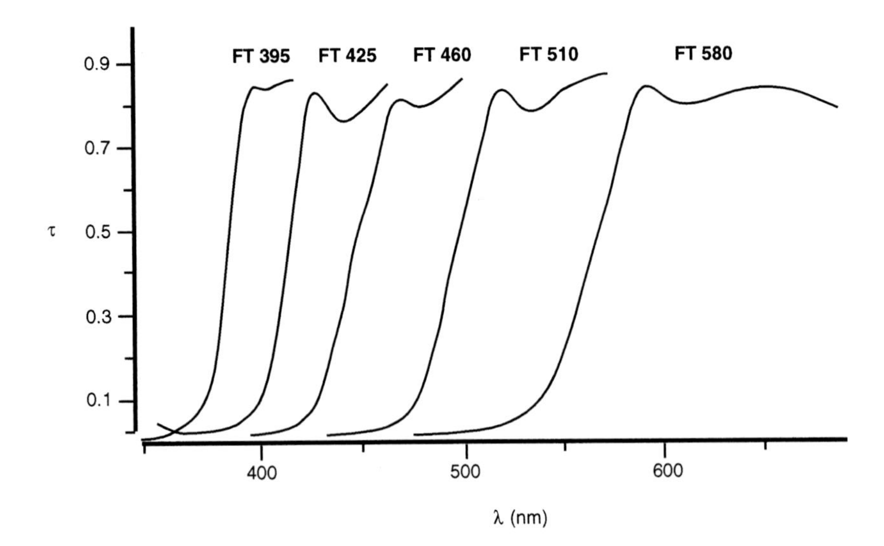

FIGURE 1.27

Transmission of typical chromatic beam splitters. "Tau" (τ) represents transmission.

Optimal Alignment of the Light Source

Aligning the light source optimally in an epifluorescence microscope, particularly a DC source such as the HBO 100, is critical for a uniformly illuminated field. Follow the manufacturer's instructions carefully to achieve maximal illumination. Below is a typical procedure for mercury/xenon lamp exchange and alignment.

1 Remove the old lamp by following instructions outlined in the illuminator manual.

   Never touch a new bulb with your hands! *If you do touch the bulb, gently wipe off the fingerprints with 100% ethanol. If this precaution is not taken, the bulb will become etched when turned on because of the intense heat output of these light sources.*

   Remember, when placing the arc lamp into the socket, try to line up the heat sink so that it is parallel to the front of the socket. Because the lamp housing is compact, it will be easier to position the socket into the housing with the heat sink lined up!

2 After placing the arc lamp and the socket into the housing, turn on the power, and open any shutters and iris diaphragms in the light path. Use an open position in the nosepiece to allow the light to be projected down to the stage area. Place a white piece of paper on the stage and move the epireflector/filter holder to a position that reduces light intensity to a comfortable level.

   This basic method for lamp alignment is also used for transmitted light optical systems.

3 Adjust the lamp collector focusing control so that the real and mirror images of the lamp are sharply focused. The mirror image is obtained with the reflecting mirror mounted behind the lamp. Adjust the up and down control of the real image (move the socket up and down) until both the real and mirror images are visible on the paper.

4 Adjust the focus/size of the mirror image using the mirror position control screws so that the focus and size of the mirror image match that of the real image.

5 Fine-tune the adjustment screws so that the real image is located close to center.

6 Once the real image is positioned, move the mirror image using the left and right controls and the up and down controls so that the mirror image is located symmetrical to the real image.

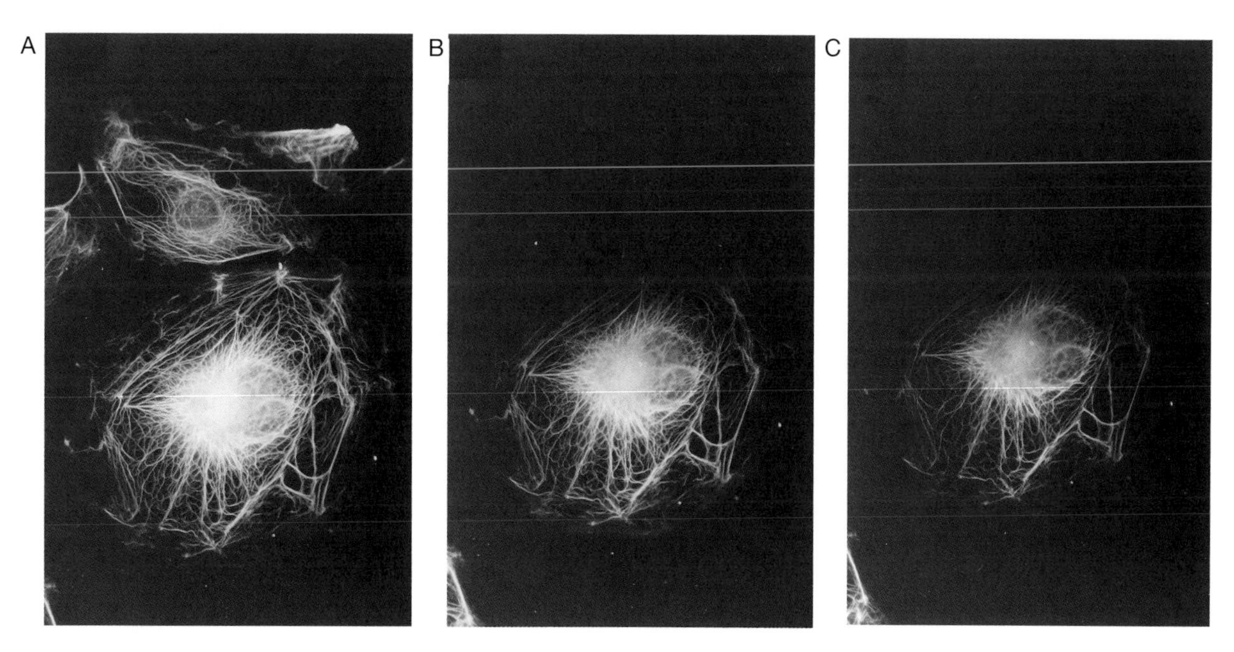

FIGURE 1.28

These epifluorescence photomicrographs show the effects of improper alignment of the light source in the same field of view. This is one of the most commonly encountered problems in epifluorescence, and the results are loss of information and resolution. In addition, exposure times for recording images are greatly increased when the light is off center (see *B* and *C*). (*A*) A micrograph of a field containing several PtK2 cells prepared for indirect immunofluorescence with a keratin antibody. The light source is well aligned and the field is uniformly illuminated. (*B*) Light source off center. Note loss of information at the top of the field of view. (*C*) Light source even more off center. (Photos provided by R.D. Goldman, Northwestern University.)

7 After inserting the objective of choice, use the collector lens focus control (Fig. 1.2B) until uniform field illumination is achieved. Different objectives require slightly different adjustments to the collector focus control for even field illumination.

Figure 1.28 illustrates proper alignment of the light source in epifluorescence microscopy. Allow a newly installed bulb to burn for at least 2 hours before turning it off. Always remember that the total lifetime of a bulb is the product of the total hours it is in use as well as the number of times it has been turned on and off. Hence, be conscious of the future use of the bulb during the course of a day, before turning the bulb off. If the bulb has been turned off and it is needed again, wait for 20 minutes before turning it back on.

With AC-operated lamps (HBO50), real and mirror images should be side by side. With DC-operated lamps (HBO100), real and mirror images are above and below the center.

Optics for Fluorescence Microscopy

The brightness of the fluorescence image is a function of the quantum efficiency and concentration of the fluorophore, the spectral brightness of the source, the transmission through the optical system used for both excitation and emission wavelengths, the total magnification, and, most significantly, the NA of the objective and, in the case of transmitted light fluorescence microscopy, the NA of the condenser.

The relative image brightness (B) in transmitted light fluorescence changes by the ratio

$$B = \frac{\mathrm{NA}^2_{\mathrm{Obj.}} \times \mathrm{NA}^2_{\mathrm{Cond.}}}{\mathrm{Tot.\ Mag.}^2}$$

In incident light fluorescence this becomes

$$B = \frac{NA^4_{Obj.}}{Mag.^2}$$

The brightest fluorescence image is obtained with an objective of relatively low magnification and high NA combined with an eyepiece or transfer optics of the lowest magnification needed to detect the resolved detail.

For excitation in the visible range, most objectives have good transmission properties. The type of objective chosen depends on aberration corrections, NA, and the magnification required for any given experiment. For near-UV excitation down to 365 nm, objectives such as Plan Neofluars have been specially antireflection-coated to provide high-throughput transmission. In addition, some Plan-apochromats (63/1.4 and 100/1.4) are designed for this spectral region. For deeper UV excitation (≤340 nm), special Fluar(fluor)-lenses have been designed, again combining high NA with relatively low magnification (see Appendix 1, Table A1.3). Figure 1.29 represents the differences in resolution observed at higher NAs and magnifications.

A

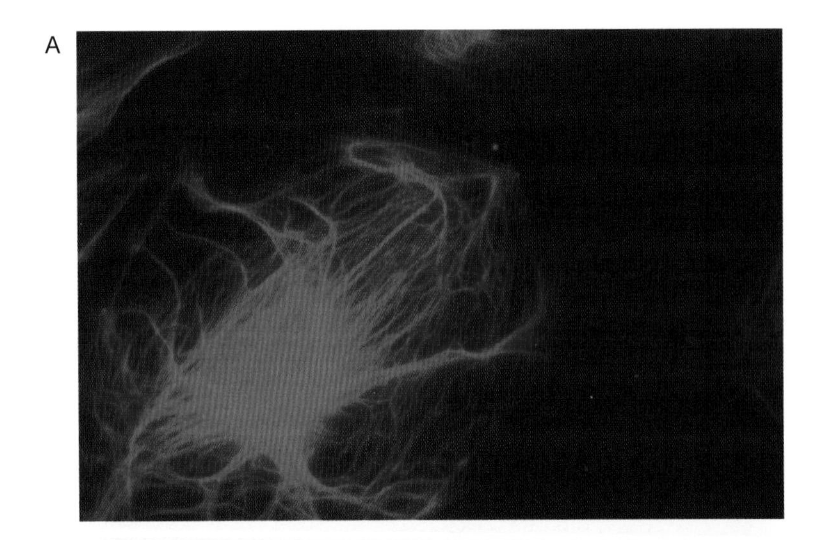

B

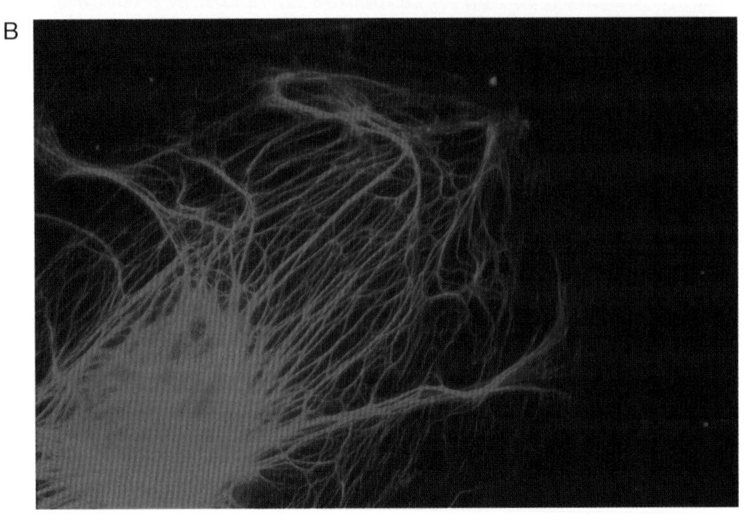

FIGURE 1.29

This is a region of a PtK2 cell prepared for indirect immunofluorescence with keratin antibody (see Chapter 7). Same region of the same cell is viewed by epifluorescence with (*A*) a 40×, 0.9 NA oil immersion lens and (*B*) a 63x, 1.4 NA oil immersion lens. This demonstrates the differences in resolution as seen in the detailed fibrillar structure at higher NAs and magnifications. (Photos provided by R.D. Goldman, Northwestern University.)

Potential Causes of Poor Fluorescence Contrast

- Autofluorescence in mounting medium (see Chapter 4).
- Autofluorescence in immersion oil: Some oils have residual autofluorescence, and a special water or glycerin immersion objective may render superior results. Multi-immersion Plan Neofluars are particularly advantageous.
- Poor filter combination: "bleed-through" from exciter to barrier filter or pinholes in exciter filter. The coatings of interference filters are prone to this latter defect. To alleviate this problem, a second exciter filter can be used.
- Thick specimens with fluorescence generated above and below the focal plane may become "washed out" and require confocal observation or deconvolution (see Chapter 2).
- Specimens in aqueous media and located at some depth behind the cover glass may look fuzzy because of spherical aberration. A high-NA oil immersion lens is not designed to look through water! A special water immersion objective with a correction collar would render crisp and sharp images despite its lower NA as necessitated by the water.

Multiple Fluorescence Markers

So-called multiparameter fluorescence (e.g., double- or triple-label immunofluorescence) can be carried out with filter sets that are used to visualize directly up to four fluorochromes simultaneously (see Chapter 6). However, more efficient separation is achieved by successive recording of each fluorochrome with separate filter sets. This would also permit better balancing of the brightness in each channel.

Another potential problem arises when the same cell is viewed with multiple fluorescent tags. The fluorescence images obtained through the different filter sets may not be in exact *x*/*y* registration because of either placement of the barrier filter in the converging imaging beam path or a filter that is non-plane-parallel. Infinity optics directly offer an advantage in this regard (see above). Furthermore, special plane-parallel filter sets are available for obtaining precise registration.

LIGHT SOURCES AND FILTERS FOR THE LIGHT MICROSCOPE

Light Sources

An ideal light source for the microscope is one that evenly, uniformly, and brightly illuminates the full aperture of the condenser or, in reflected light, the objective pupil with a spectral intensity distribution that best matches that of the eye, film, or other detector. In the early days of light microscopy, carbon arc lamps fulfilled some of these requirements, but these were cumbersome to use and lacked stability. The sun was even a better light source, with obvious limitations.

These shortcomings of the early light sources for microscopy have been largely resolved by today's low-voltage *halogen bulbs.* Low voltage allows a closely woven (tungsten) filament to glow without arcing while the halogen gas in the glass/quartz envelope prevents the tungsten from coating the envelope internally, which would slowly reduce the output. Tungsten halogen bulbs (6 V or 12 V) are now the source in most modern microscopes and can have 10-, 20-, or 100-W power consumption and somewhat corresponding outputs. The power consumption is not directly related to the amount of light that is transmitted through the microscope. This depends on the shape of the filament, the collector lens aperture (see Fig. 1.2B) and focal length, and many other factors.

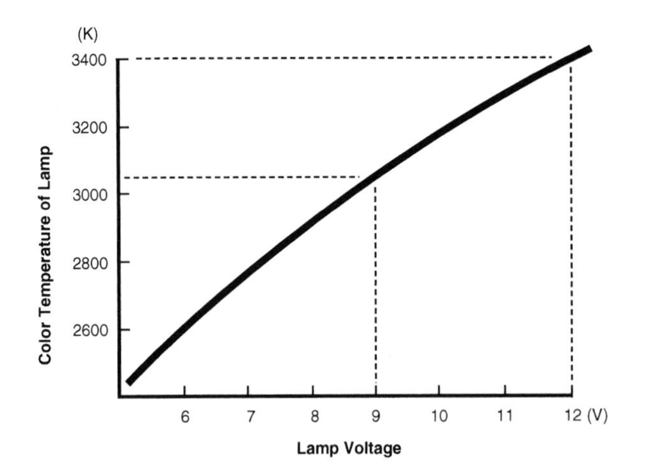

FIGURE 1.30

Relationship of voltage to color temperature.

The lifetime of today's bulbs can range from 100 to 2000 hours and is, of course, largely a function of the basic design, the rated voltage, and the treatment of the bulb during use. Most modern power supplies prevent switch-on surges and gently raise the voltage from zero, helping to prolong the life of the bulb. The emitted radiation at the rated voltage has a spectral intensity distribution or color temperature of between 3000°K and 3200°K. The relationship between voltage and color temperature is shown in Figure 1.30. To achieve daylight illumination (5500°K), a blue filter is usually furnished.

For some of the microscope techniques that require either added output, such as high-magnification DIC or polarized light, or specific spectral characteristics as required for fluorescence, high-pressure gas discharge lamps such as mercury (HBO) or xenon (XBO) "burners" are available. These range from 50 W to 100 W, have different arc dimensions, and can develop exceptionally high luminous densities, often in a very small arc (0.25 × 0.25 mm for an HBO 100). Under these circumstances, proper focus and alignment are particularly critical (for further details, see the discussion on fluorescence microscopy above). To ionize the internal gas, discharge lamps are ignited with high-voltage pulses provided by their power supply. They reach their full intensity after several minutes and then operate at high internal pressure (10–15 atm) and must be used in a safe housing that provides adequate cooling. Follow the manufacturer's instructions on safety, installation, and alignment closely! Figure 1.31 shows the relative spectral output for a range of different light sources.

Where the intensity or high luminous density of a 100-W HBO is needed along with uniform filling of the large condenser aperture, as in high-resolution DIC, a so-called Ellis scrambler (after Gordon Ellis, University of Pennsylvania) makes an ideal interface between source and microscope: A quartz fiber approximately 1 m long and 1 mm in diameter receives the nonuniform arc at its entry through special collector optics. Giving the fiber a loop, the exit of the fiber becomes a round uniform source of high intensity that through additional optics can be enlarged to fill the condenser's aperture evenly. Needless to say, the alignment of fiber entry to source and fiber exit to condenser is critical if the full benefit of such a system is to be realized.

Filters

A diverse array of filters are used in microscopy; some major types are listed below.

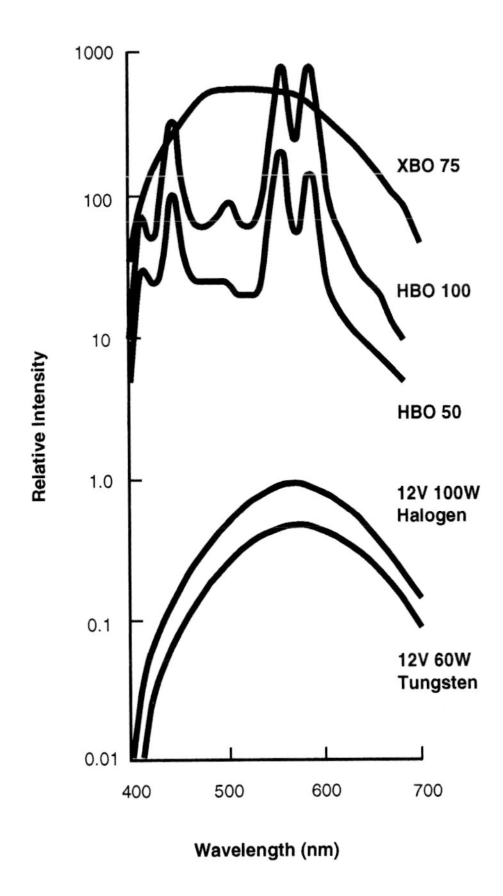

FIGURE 1.31

Relative spectral intensity of various light sources with heat reflection filter in light path.

Neutral-Density Filters

For light attenuation without a change in color temperature, *neutral-density* (ND) filters are used. These are usually metal-coated glass filters with uniform absorption over the whole visible spectrum. They can have transmission values from 50% to 1.5% and allow for comfortable viewing under bright light conditions or serve to optimize light levels for video microscopy or photomicroscopy, without changing the color temperature.

Heat Filters

Excessive heat radiation from tungsten halogen as well as gas discharge lamps needs to be removed by *heat filters.* Either KG-1 glass filters or so-called Calflex heat reflection filters or both will effectively protect not only the specimen, but other sensitive elements such as polarizers and other filters.

Color-balancing Filters

To optimally color balance the microscope's illumination to the spectral sensitivity of film, the eye, or other detectors, *color-balancing filters* are available. The following filters shift overall color temperature:

Wratten type 80: tungsten to daylight
Wratten type 81: reduces color temperature (red)
Wratten type 82: increases color temperature (blue), and is available in many different strengths

Color-compensating Filters

Filters that enhance or attenuate light levels in given spectral regions are called *color-compensating* (CC) filters and also come in different degrees of absorption. The following are some examples of these filters:

CC Y (yellowish)
CC M (magenta)
CC B (bluish)
CC R (reddish)

Spectral Filters

Black and white photomicrography, video microscopy, or any microscope technique that enhances contrast by interference can be further enhanced with *spectral filters.* These filters select a narrow band or broad band from the full visible spectral range. Broadband filters are made from stained glass, whereas narrowband filters are usually interference filters that can be designed for any bandwidth at any spectral region. Three major benefits are derived from the use of these filters.

- Chromatic aberrations in the optics are largely eliminated, for example, by the use of a green filter. This is particularly advantageous with Achromats and Planachromats that have residual chromatic aberrations.
- The contrast generated with techniques like phase or DIC can be enhanced by using a green filter. This limits the wavelength range of the illumination to the optimal design parameters of the optical system.
- The contrast in weakly stained specimens can be improved. For example, a faint reddish stain in a histological preparation will appear much darker when illuminated with a blue filter, and vice versa.

Such spectral filters are needed whenever measurements of path differences in polarized light microscopy or other types of interferometry are performed.

Special Fluorescence Filters

Special filters for fluorescence microscopy are described in some detail in the discussion on fluorescence above and in Appendix 1. Whenever possible, these filters are placed in the illumination path of the microscope somewhere between the lamp collector and condenser. Their optical quality may not always be such that the best image quality is retained when placing them behind the objective.

PHOTOMICROGRAPHY: PHOTODOCUMENTATION THROUGH THE MICROSCOPE

Special microscope cameras are ideal for photomicrography (documentation on film). Two designs are available.

- Direct projection of the microscope's intermediate image to the film plane via, for example, a 1.5× projector lens.
- Utilizing the microscope eyepiece (for best image quality, a special photoeyepiece) and converging the infinity-focused imaging beam for sharp focus in the film plane with a special camera objective. Under these conditions, the system is insensitive to slight differences in mounting dis-

tance of the camera above the eyepiece and has the flexibility of using different eyepieces for the desired magnification on film.

This discussion is limited to hardware and its proper adaptation and describes the key steps for taking a high-quality photomicrograph. Excluded are discussions of the multitude of film materials available, as well as darkroom and film-processing techniques. Microscope cameras can range in film format from 35 mm (24 × 36-mm) to 4 × 5-inch sheets. The latter are often useful for an instant Polaroid print.

The body of the microscope camera contains a vibration-isolated central shutter behind the camera objective and a photosensor (either photodiode or photomultiplier) to measure available light for manual or automatic exposure. This type of camera accepts either a 35-mm film cassette or 4 × 5-inch sheet film. Exposure can be determined via either integrated (30% or more of format) or spot readings (1–3%). A separate control unit allows for fully automatic or manual/time exposure and offers functions such as ISO settings for any film sensitivity, 2–3 stops of over- or underexposure for automatic "bracketing" in full or 1/3 stops, double exposure, wind, and reciprocity failure compensation. (The film's exposure index holds true only in a limited exposure range. During long exposures, the sensitivity decreases and the exposure needs to be extended accordingly.) Furthermore, all important data are displayed.

For 35-mm film, an alternate and more economical system is the single lens reflex (SLR) camera. Properly adapted, precisely parfocalized, and solidly mounted, it can render excellent results. For SLR, the following should be considered:

- Focus through the camera viewfinder is not adequate for photomicrography. The film plane must be parfocal with the viewing eyepieces of the microscope equipped with focusing reticules.
- The focal plane shutter of SLR cameras causes vibration. To minimize this problem on fixed-stained specimens, short exposure times (<1 sec) should be extended with ND filters to >1 second.

Optical adapters for most SLR cameras (without lens) are available as well as adapters for SLRs equipped with a lens. In the latter case, the image, even with the so-called wide-field/high-eyepoint eyepieces, may show some vignetting (loss of field) at the corners.

Transfer Factors and Magnification on Film

Magnification of the intermediate image × transfer factor = magnification on film. For 35-mm film, a generally useful transfer factor of 2.5× selects a 17-mm diagonal of the intermediate image field to cover the film format. This is a range in which most optics are well corrected for flat-field and off-axis aberrations. For 4 × 5- inch format cameras, the transfer factor usually becomes 10×.

Achieving Best Focus on Film

Achieving good focus is more critical at lower objective magnifications, where the depth of field (in the object) is greater, but the depth of focus (toward the image) is more shallow. For optimal results use the following procedure.

1. With the specimen out of focus and with relaxed eyes, adjust the focusing eyepiece until double lines in the focusing reticule are sharp. (Use your glasses if you have astigmatism.) For critical focus, use a magnifying telescope of 2.5× or 3×, adjusted to infinity over the eyepiece and fine-tune its adjustment.

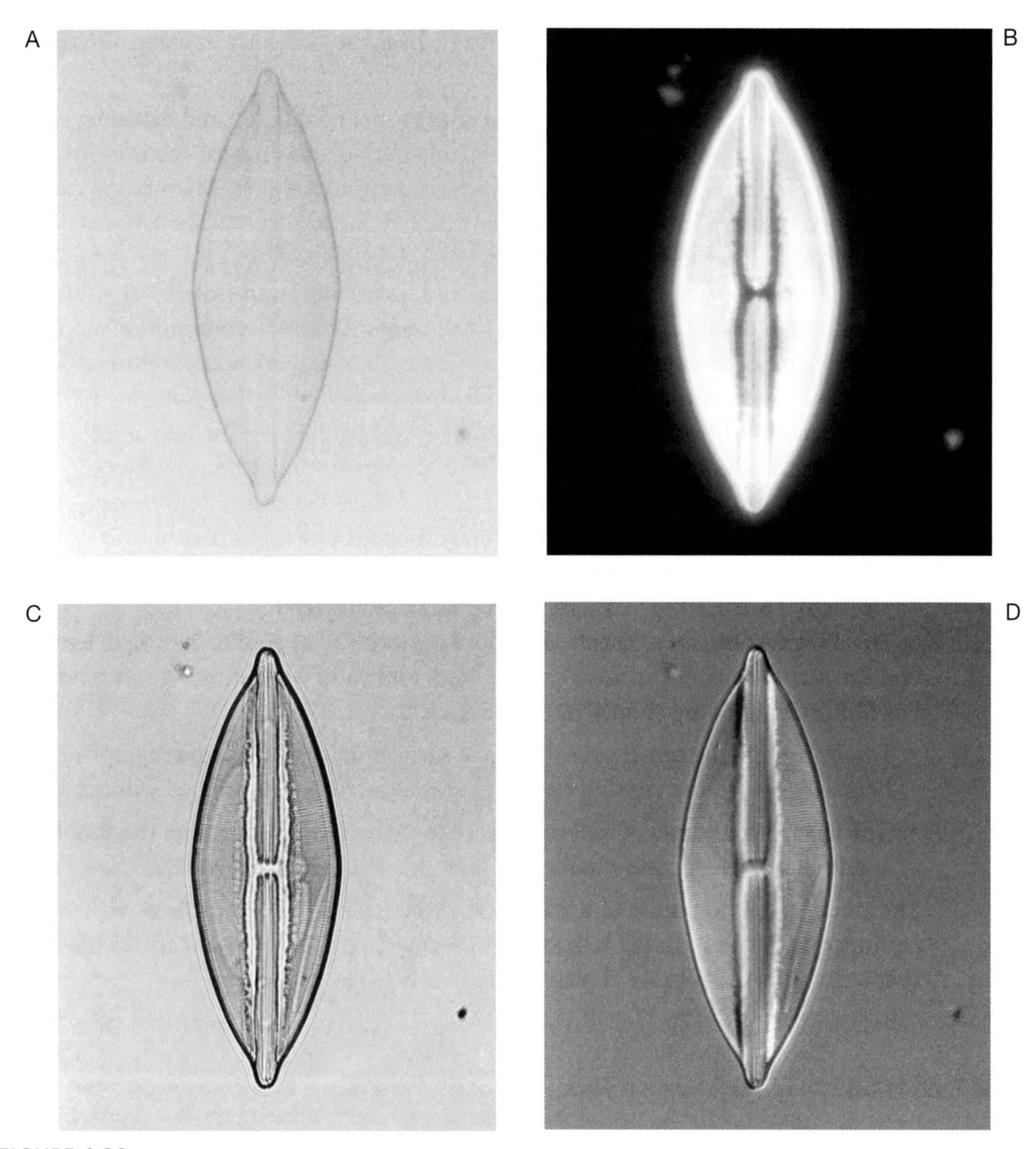

FIGURE 1.32

Diatoms are frequently used as test specimens for light microscopy. This is an unfixed, unstained preparation of a diatom that demonstrates the images obtained with bright-field (*A*), dark-field (*B*), phase-contrast (*C*), and DIC (*D*) microscope systems. These photos were taken on 35-mm black and white film and printed in a darkroom according to classic techniques. (Photos provided by R.D. Goldman, Northwestern University.)

2 Focus the microscope until the specimen is in sharp focus together with the double cross of the focusing reticule. Move your eyes either vertically or horizontally. Sharpest focus is attained when there is no parallax (shift) between image and crossline.

For best photomicrography in black and white or color, the choice of the best film depends much on the task at hand and on personal preferences. The choice of the best optics and the correct filter (see above) is essential to obtain a photographic document of the specimen that clearly records everything you have seen.

Figure 1.32 represents and compares the structure of a diatom using different optical systems. These photos were taken on 35-mm black and white film.

REFERENCES

General Microscopy

Born M. and Wolf E. 1970. *Principles of optics.* Pergamon Press, Oxford.

Bradbury S., Evennett P.J., Haselmann H., and Piller H. 1989. *Dictionary of light microscopy.* Oxford University Press, Oxford.

Herman B. and Jacobsen K. 1990. *Optical microscopy for biology.* Wiley-Liss, New York.

Lacey A.J. 1989. *Light microscopy in biology. A practical approach.* IRL Press, Oxford.

Pawley J. 1989. *Handbook of biological confocal microscopy.* Plenum Press, New York.

Pluta M. 1988. *Advanced light microscopy,* vol. I. Elsevier, Amsterdam.

———. 1989. *Advanced light microscopy,* vol. II. Elsevier, Amsterdam.

———. 1992. *Advanced light microscopy,* vol. III. Elsevier, Amsterdam.

Spencer, M. 1982. *Fundamentals of light microscopy.* Cambridge University Press, London.

Oblique Illumination Techniques

Ellis G.W. 1981. Edge enhancement of phase phenomena, U.S. Patent no. 4255014.

Hoffman R. 1975. The modulation contrast microscope. *Nature* **254:** 586–588.

Kachar B. 1985. Asymmetric illumination contrast. *Science* **227:** 766–768.

Spector D.L., Goldman, R.D., snd Leinwand L.A. 1998. *Cells: A laboratory manual.* Cold Spring Harbor Laboratory, Cold Spring Harbor, New York.

Phase-Contrast Microscopy

Francon M. 1962. *Progress in microscopy,* Chap. II, pp. 64–93. Row, Peterson, Evanston, Illinois.

Ross K.F.A. 1967. *Phase contrast and interference microscopy for cell biologists.* Edward Arnold, London.

Zernicke F. 1942. Phase contrast: A new method for the microscopic observation of transparent objects. *Physica* **9:** 686–693.

———. 1955. How I discovered phase contrast. *Science* **121:** 345–349.

Dark-Field and Polarized-Light Microscopy

Jamrich M., Mahon K.A., Gavis E.R., and Gall J.G. 1984. Histone RNA in amphibian oocytes visualized by in situ hybridization to methacrylate-embedded tissue sections. *EMBO J.* **3:** 1939–1943.

Patzelt W.J. 1985. *Polarized light microscopy: Principles, instruments, applications.* E. Leitz, Wetzlar, Germany.

Shurcliffe W.A. 1962. *Polarized light.* Harvard University Press, Cambridge, Massachusetts.

Shurcliffe W.A. and Ballard S.S. 1964. *Polarized light.* Van Nostrand, Princeton, New Jersey.

Swindell W. 1975. *Polarized light: Benchmark papers in optics.* Halsted Press, New York.

Differential Interference Contrast

Allen R.D., David G.B., and Nomarski G. 1969. The Zeiss-Nomarski differential interference equipment for transmitted light microscopy. *Wiss. Mikrosk. Mikrosk. Tech.* **69:** 193–221.

Francon M. 1962. *Progress in microscopy.* Row, Peterson, Evanston, Illinois.

Lang W. 1979. *Nomarski differential interference contrast microscopy.* Reprint MA IX/79. Carl Zeiss, Oberkochen, Germany.

Padawer J. 1968. The Nomarski interference contrast microscope. *J. R. Microsc. Soc.* **88:** 305–349.

Reflection Interference Microscopy

Beck K. and Bereiter-Hahn J. 1981. Evaluation of reflection interference contrast images of living cells. *Microsc. Acta* **84:** 153–178.

Gingell D. and Todd J. 1979. Interference reflection microscopy: A quantitative theory for image interpretation. *Biophys. J.* **26:** 507–526.

Izzard C.S. and Lochner L.R. 1976. Cell to substrate contacts in living fibroblasts. *J. Cell Sci.* **21:** 129.

Ploem J.S., ed. 1975. Reflection contrast microscopy as a tool for investigation of the attachment of living cells to a glass surface, In *Mononuclear phagocytes in immunity,* pp. 405–421. Blackwell, Oxford.

Fluorescence Microscopy

Bright G.R. 1993. Multiparameter imaging of cellular function. In *Fluorescent probes for biology function of living cells—A practical guide* (eds. W.T. Mason and G. Rolf), pp. 204–215. Academic Press, New York.

Herman B. and Lemasters J.J. 1993. *Optical microscopy: Emerging methods and applications.* Academic Press, San Diego.

Taylor D.L. and Wang Y.L. 1989. *Fluorescence microscopy of living cells in culture,* Part A, vol. 29 and Part B, vol. 30. Academic Press, San Diego.

Waggoner A.S., DeBiasio R., Bright G.R., Ernst L.A., Conrad P., Galbraith W., and Taylor D.L. 1989. Multiple spectral parameter microscopy. *Methods Cell Biol.* **30:** 449–478.

Photomicrography

Delly J.G. 1980. *Photography through the microscope.* Kodak Publication P-2.

Loveland R.P. 1981. *Photomicrography: A comprehensive treatise.* Wiley, New York.

Video Microscopy

Allen R.D. and Allen N.S. 1983. Video enhanced microscopy with a computer frame memory. *J. Micros.* **129:** 3–17.

Allen R.D., Allen N.S., and Travis J.L. 1981. Video enhanced contrast, differential interference contrast microscopy. *J. Cell Motil.* **1:** 298–302.

Inoué S. 1981. Video image processing greatly enhances contrast quality and speed in polarization microscopy. *J. Cell Biol.* **89:** 346–356.

———. 1986. *Video microscopy.* Plenum Press, New York.

———. 1987. Video microscopy of living cells and dynamic molecular assemblies. *Appl. Optics* **26:** 3219–3225.

———. 1988. Progress in video microscopy. *Cell Motil. Cytoskeleton* **10:** 13–17.

Schotten, D. 1993. *Electronic light microscopy.* Wiley-Liss, New York.

Weiss D.G., Maile W., and Wick R. 1992. Video microscopy. In *Light microscopy in biology* (ed. S.J. Lacey), pp. 221–278. IRL Press, Oxford.

CHAPTER 2

Confocal Microscopy, Deconvolution, and Structured Illumination Methods

John M. Murray

University of Pennsylvania School of Medicine, Philadelphia, Pennsylvania

INTRODUCTION

When a thick specimen is viewed through a conventional microscope, the sum of a sharp image of an in-focus region is seen, plus blurred images of all of the out-of-focus regions. The depth of field (i.e., the distance between the top and bottom of the in-focus region at a fixed setting of the focus knob) is less than 1 µm for the high numerical aperture (NA) objective lenses that are used for fluorescence microscopy. Thus, even when viewing a specimen as thin as 5 µm, 80% of the light may be coming from out-of-focus regions. The result is a low-contrast image composed of an intensely bright, but very blurred background on which is superimposed the much dimmer in-focus information.

In this chapter, "thick" and "thin" refer to the thickness of the fluorescent sample; overall specimen thickness per se does not increase the background. As the overall specimen thickness increases beyond 5–10 µm, however, two other factors begin to degrade the image quality. When the illumination or imaging path intersects regions of widely different refractive index such as small granules or organelles, their curved surfaces act as microlenses to deflect the light in random directions. The consequence of multiple deflections may be to distort the light path enough to introduce aberrations into the image or even to scatter the light completely out of the field of view.

One way to eliminate the high background, scattering, and aberrations is to slice the thick specimen into many thin sections. This approach requires fixation, dehydration, and embedding, so it has limited application to the microscopy of living cells, but several other methods do work well with living samples. These methods can be grouped into three classes: primarily "optical" (e.g., confocal microscopy, multiphoton microscopy), primarily "computational" (e.g., deconvolution techniques), and mixed (e.g., structured illumination) approaches. These techniques make it possible to see details within thick specimens (e.g., the interiors of cells within living tissue) by optical sectioning, without the artifacts associated with physically sectioning the specimen.

How does one decide which method to use? All of these methods address problems encountered in imaging thick specimens. For routine qualitative observation of relatively thin specimens (<3 µm), it is probably much quicker and less frustrating to use conventional ("wide-field") microscopy. There are a few situations, however, in which the benefits of these more complex methods are important enough, even for a thin specimen, to warrant the extra cost, inconvenience, and time.

The most common application to thin specimens is when intrinsic contrast is very low, so that any loss of contrast, even the minimal decrease due to a small amount of out-of-focus light, complicates interpretation of the data. In this situation, all of these methods can usually improve contrast for any sample thicker than ~1 µm. Another common application is to enable accurate measurement of the amount of a fluorescent component present in a cell, a task in which deconvolution methods excel. Finally, in the relatively rare case in which even a modest enhancement of resolution would change the interpretation of the data, then confocal, deconvolution, and some of the structured illumination methods are capable of delivering a small improvement over a conventional microscope.

For thicker objects that produce a moderate amount of out-of-focus light (typically 5–30 µm), any of the methods discussed here (and multiphoton microscopy, discussed in Chapter 3) should give a dramatically better result than a conventional microscope. When the sample is living (i.e., photobleaching and phototoxicity constrain exposure) and the signal is weak or the contrast is low, methods that must use photomultipliers for detection (e.g., point-scanning methods, confocal or multiphoton) have a severe handicap compared to methods that can use charge-coupled device (CCD) cameras (e.g., deconvolution, disk-scanning confocal microscopy, and structured illumination), because of the much higher quantum efficiency of CCDs. With very thick specimens that produce an overwhelming amount of out-of-focus light, only point-scanning (confocal or multiphoton) microscopy will give a satisfactory result.

How much is a "moderate" amount of out-of-focus light? Typically, the image seen through a conventional microscope will be too blurred to discern details, but one will be able to locate the region of interest and at least roughly set the focus level. If the view through a conventional microscope is virtually featureless and gives no landmarks for choosing the appropriate area or for setting the focus, two methods—point-scanning confocal or multiphoton microscopy—can be used to produce extremely useful images from terrible specimens. From very thick specimens, however, it is not realistic to expect a final image quality comparable to the best that a conventional microscope produces with a thin specimen, for reasons that will be considered in this chapter.

DECONVOLUTION METHODS

The goal of deconvolution is to improve the images of thick objects by computationally removing the out-of-focus blur. The strategy is to calculate the structure of a hypothetical object that could have produced the observed, partially focused image. The calculation is based on fundamental optical principles, in particular, a quantitative understanding of the effects of defocus, and may also take into account prior information or guesses about the specimen. The method commonly employed is to refine iteratively an initial guess about the true object until the estimated image (i.e., the estimated object appropriately blurred by the effects of defocus) corresponds to the actual observed image.

Optical Principles

Successful application of these techniques requires an appreciation of how an image is formed by a microscope and what happens to an image when the lens is defocused. For this purpose, it is helpful to introduce the twin concepts of *point spread function* (PSF) and *contrast transfer function* (CTF). Both of these concepts describe the relationship between a real object and the image that is formed of it by an optical system. The PSF describes this relationship in terms of the image of a very small object, effectively a single point. Although the microscope can "see" objects that are vanishingly small, the limited resolution of a microscope prevents the image from accurately portraying their size, no matter what magnification is applied.

An illustration of this phenomenon is given in Figure 2.1. Note that below a certain object size, images of every object appear the same. Increasing the magnification does not help; the image can

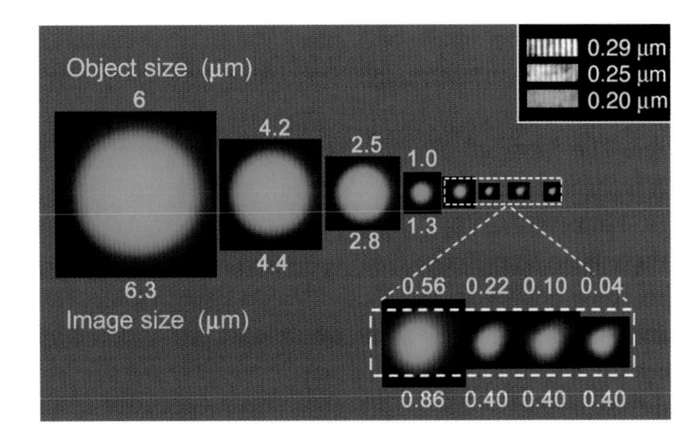

FIGURE 2.1

Epifluorescence microscope images of eight beads of known, decreasing size. The images of the fluorescent beads are shown in green; all are adjusted to have the same maximum brightness. The actual brightness varies by ~1000-fold. The true diameter of each bead is given above its image, and the diameter measured from the image is listed below it. The four smallest beads are shown at two different magnifications. The apparent diameters measured in the image are slightly larger than the true diameters, and the apparent diameters are the same for the three smallest beads, even though their true diameters differ by more than fivefold. These three images reveal the PSF of this microscope. The images in the upper right corner, obtained with the same microscope setup, show the appearance of three gratings with spacings of 0.29, 0.25, and 0.20 μm between the white bars. Note that the contrast between the black and white bars decreases as the spacing gets smaller. In the actual object, the grating contrast is the same for all three scales. The bead images in the upper row are displayed at the same magnification as the gratings.

be made larger, but not sharper. This limiting image is called an "Airy disk," after the British astronomer G.B. Airy who first recognized its significance in 1834 (also see Chapter 1). Notice that for the microscope used for this figure, the Airy disk is not quite a perfect circle. The three smallest beads all appear to be slightly elliptical, with a weaker tail extending toward the upper right, although electron microscopy shows that the beads are nearly perfect spheres. As Airy was the first to point out, the shape of this limiting image provides no information about the shape of the object—the Airy disk is an intrinsic property of the optical system itself. The Airy disk shows that the image of a point-like object is not a single point, but is spread into a fuzzy disk, hence the name "point spread function." The PSF is often used as a means of quantitatively characterizing the performance of an imaging system.

The grating images in Figure 2.1 show that the size of the PSF sets the resolving power of an optical system. An optical system forms an image by substituting its PSF for every point in the object, and then sums all of these PSFs to make the image. The width of the PSF determines how far apart two points in the object must be to avoid being smeared together in the image. If the PSF of the optical system is broad, two points must be rather well separated to prevent overlap of the two corresponding PSFs in the image. If their PSFs overlap extensively, the two points will appear to be just a single point, smeared into an indistinct average like the lines in the image of the 0.2-μm grating (Fig. 2.1, upper right box).

Suppose that the experiment in Figure 2.1 included not an individual 0.04-μm bead, but rather a pair of 0.04-μm beads separated by 0.08 μm, twice their own diameter. This pair would still be a smaller "object" than the single 0.22-μm bead, and their joint image would have the same shape as any of the three limiting images in Figure 2.1. In other words, 0.08 μm is well below the resolution of this microscope. It is important to realize, however, that the Airy disk for the pair of beads would be twice as bright as the Airy disk from a single bead, i.e., the imaging process is a *linear* operation.

The total intensity in an image of A and B is exactly the sum of the total intensity in an image of A plus the total intensity in an image of B.

A single number, such as the Rayleigh criterion, is often quoted for the resolving power of a microscope. The Rayleigh criterion is the radius of the Airy disk, given numerically by $0.6\lambda/NA$, where NA is the numerical aperture of the objective lens and λ is the wavelength of light that forms the image. However, using a single number for the resolving power is somewhat misleading, because there is no sharp cutoff. As the size of small details approaches the resolution of the imaging system, these details do not suddenly disappear. Instead, their contrast in the image becomes a smaller and smaller fraction of their true contrast in the object, until finally the image contrast approaches the size of the random fluctuations due to noise, and they then become "invisible." The images of the gratings in Figure 2.1 show this progressive loss of contrast with decreasing size.

A complete description of the resolving power of an optical system thus requires information about the variation in the ratio of image contrast to object contrast as a function of size. The PSF contains this information, but it is revealed more clearly in its twin, the CTF. This function describes the extent to which contrast variations in the object are faithfully replicated in the image. Perfect contrast transfer means that image contrast equals object contrast. The CTF is usually expressed as a ratio, so that perfect contrast transfer means that the CTF has a value of 1.0 (in reality, the CTF is always less than 1.0).

It is reasonable to expect that information about some features of the object might be transferred into the image more faithfully than others. For instance, the image may be a nearly perfect representation of the large-scale features of the object, but contain much less information about the very smallest details. This will always be true for images obtained from an optical microscope, because one cannot see clearly those details of the object that are small compared to the wavelength of light (i.e., details on the scale of the PSF). Thus, the CTF is a function of the size of the feature being observed (Fig. 2.2, left). Normally, the CTF is shown in graphical form, plotting the ratio of image contrast to object contrast (vertical axis) against the reciprocal of size (i.e., spatial frequency). The CTF is simply a different representation of the same information that is given by the PSF of an optical system. Mathematically, the CTF is the Fourier transform of the PSF, and vice versa.

When we speak of image resolution, we are therefore making a statement about the *ratio* of image-to-object contrast at small spacings. There is always an interaction between image contrast, image signal-to-noise ratio, and image resolution, even though it is sometimes convenient to think of these as independent properties. What is normally referred to as *visibility* is determined by all three of these parameters, as well as properties of the display system and the observer. Keep in mind that much useful information can often be extracted concerning features that, by eye, are "invisible."

The example of a typical, good microscope CTF and PSF shown on the left side of Figure 2.2 represents the case in which the specimen is thin and lies exactly in the focal plane of the objective lens. In fact, the CTF and PSF are three-dimensional (3D) functions. Their third dimension is revealed by comparing image to object when the object is displaced vertically from the focal plane of the lens. As the focus changes, concentric rings appear in the PSF, and the CTF develops ripples and in some regions becomes negative. For features in the size range corresponding to these negative oscillations of the CTF, dark parts of the object will appear bright in the image and vice versa (Fig. 2.2, right). As the degree of defocus increases, the CTF becomes increasingly oscillatory, with the contrast reversals affecting ever larger features in the image.

The image of a small fluorescent bead (i.e., the PSF) develops concentric rings as the lens moves away from focus (Fig. 2.2, top right). Changing the focus of the lens means that the 3D PSF is viewed at different levels along the optical axis. A vertical slice of the complete 3D PSF, viewed from the side, is shown in Figure 2.3 (Born and Wolf 1999).

It surprises most people that microscopes can produce such wildly "incorrect" images as depict-

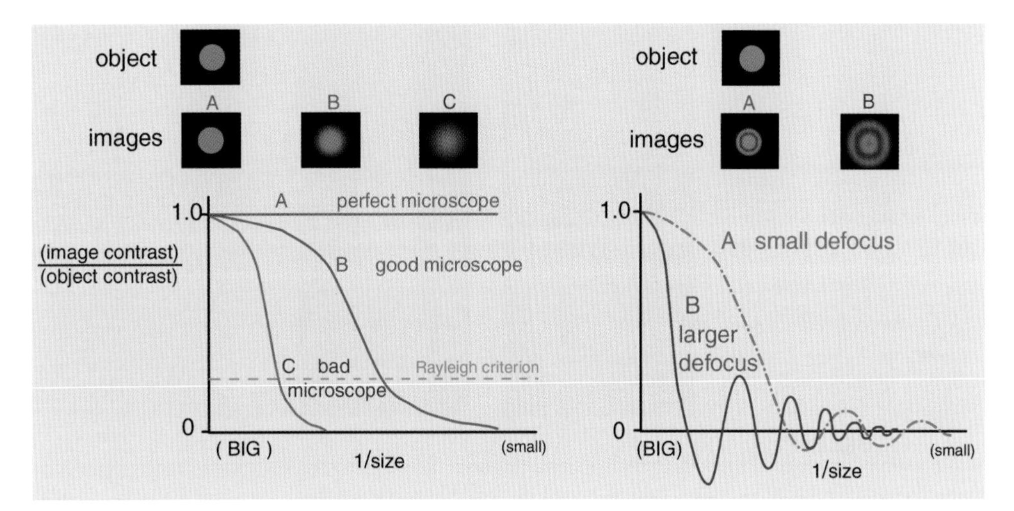

FIGURE 2.2

(*Left*) Schematic representation of some contrast transfer functions, and the corresponding PSFs (object-image pairs for a very small object). (*A*) A perfect (impossible!) microscope; (*B*) a typical, good microscope; (*C*) a poor or improperly used microscope. The dashed line lies at a relative contrast of 25%, corresponding to the Rayleigh criterion for the resolving power of a microscope. (*Right*) Images of a tiny object from a microscope at two different values of defocus reveal the 3D nature of the PSF and CTF. Objects in some size ranges appear to change from black to white or vice versa as the focus level changes between the indicated values. An example of this contrast reversal in a transmitted light image is shown in Fig. 2.4.

ed in Figures 2.2 (right) and 2.3. A striking example is shown in Figure 2.4, but the same effect can easily be observed on almost any specimen. In bright field, a small high-contrast feature such as a small dust particle or a scratch demonstrates the contrast reversal phenomenon clearly. If a good dry or oil immersion lens is used to carefully focus up and down by a small distance on either side of the correct focus, the particle will oscillate from bright to dark and back again. If the focus can be controlled carefully enough, the position can be found, halfway between a bright and a dark oscillation, where the particle becomes practically invisible (i.e., CTF ~ 0 for details of this size).

To reiterate, a microscope substitutes its 3D PSF for each point in the 3D object, and then sums all of those innumerable PSFs to give the final 3D image. The mathematical operation called *con-*

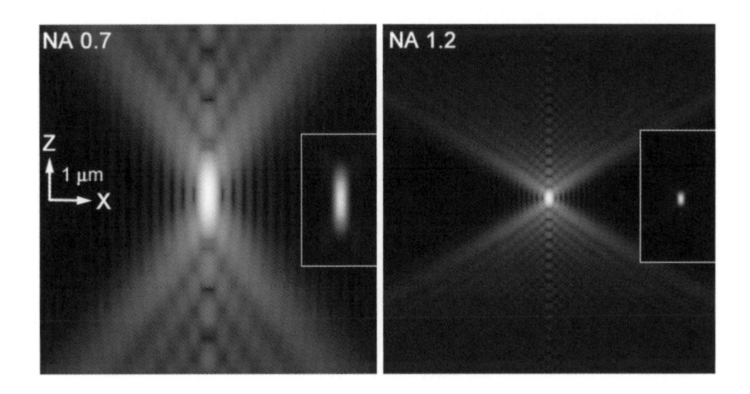

FIGURE 2.3

Vertical sections through the 3D PSFs calculated for lenses of two different numerical apertures (NAs). The contrast has been greatly enhanced to show the weak side lobes (which are the rings of the Airy disk, viewed edge on). The small insets show the PSFs at their true contrast level. Note that vertical smearing (proportional to NA^2) is much more strongly affected by the lens NA than is lateral smearing (proportional to NA).

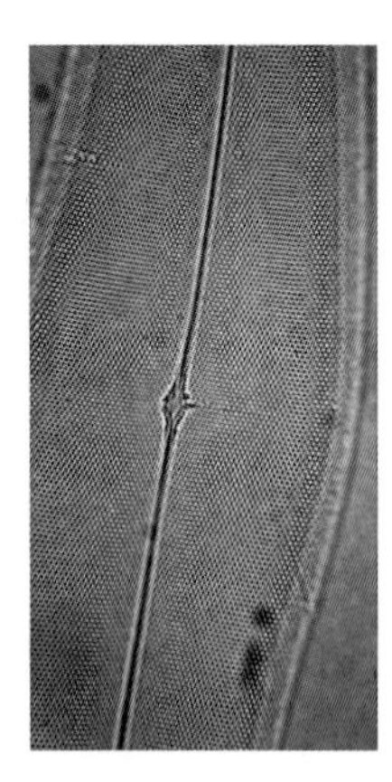

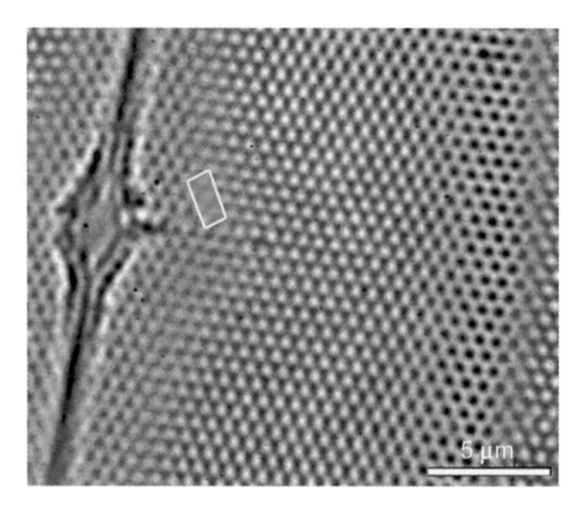

FIGURE 2.4

Contrast reversal because of defocus. (*Left*) A bright-field image of a diatom taken through a conventional microscope using a 60X, 1.4 NA lens. (*Right*) An enlargement of a small portion of the image on the left, showing contrast reversals because of changing amounts of defocus. The diatom shell is curved, being thinner at the edges. As a result, the distance from the lens to the surface of the diatom varies, i.e., the view includes a range of defocus values. Over this range, the CTF changes sign several times: From left to right, the holes change from black to white, to black again, and finally to white on the right-hand edge. The white rectangle indicates a narrow band midway between a black and a white hole region where the contrast of the holes is low, i.e., the CTF is nearly zero for structures of this size at this value of defocus.

volution precisely describes this "substitute and sum" process. The distribution of intensities in the 3D image is the result of *convolving* the object intensity distribution with a 3D PSF. The 3D PSF is an intrinsic property of the optical system and does not depend on the object.

Deconvolving Wide-Field Microscope Images

The 3D PSF (or equivalently, the 3D CTF) contains all of the information needed to predict the appearance of a known object when viewed through the corresponding optical system, for any choice of focus. For the microscopist, however, the appearance (i.e., the image) of the object is known, and the real structure is unknown. It is in principle possible to go "backward" in a one-step calculation from the observed appearance to the actual structure using the mathematical procedure called *deconvolution*. In practice, for realistic signal-to-noise ratios, a much better approach is to perform this calculation in a multistep, iterative process.

To illustrate this procedure, imagine that a 3D object consists of a stack of discrete, two-dimensional (2D) planes. Normally, the imaging data will also be a stack of 2D image planes, collected by changing the fine focus of the microscope by a small increment between successive images. Consider plane number 5 of the observed stack of images in Figure 2.5. When this image was recorded, what the detector "saw" was the sum of an in-focus view of object plane 5, plus a view of object plane 6 blurred by one increment of defocus, plus object plane 4 blurred by one increment of defocus in the opposite direction, plus planes 7 and 3 blurred by plus and minus two increments of defocus respectively, and so forth. To deconvolve the image data, an initial guess is made at the real structure of the object in planes 1–4 and 6–9. Using the known CTF appropriate for each plane's defocus, these initial estimates are blurred and added together to estimate the contribution from out-of-focus blur to the *observed* plane 5. The sum of blurred object planes 1–4 and 6–9 is subtracted from the image data for plane 5, and the result is an estimate for the in-focus appearance of the object in plane number 5. Repeating these steps for all nine planes generates an improved estimate of the object. The entire sequence of operations on all nine planes is repeated in a loop until the object estimate no longer changes significantly (Fig. 2.6).

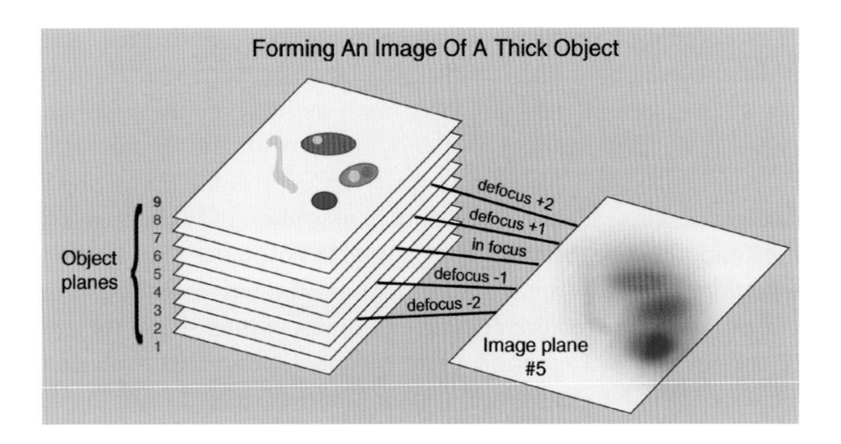

FIGURE 2.5

A simplified conception of how one plane in the microscopic image of a thick object is formed. The process can be thought of as adding the in-focus image of one object plane to the images of neighboring object planes viewed at different amounts of defocus. For clarity, the process is illustrated for only two neighbors on each side of the in-focus plane. In reality, each image plane receives contributions from all object planes.

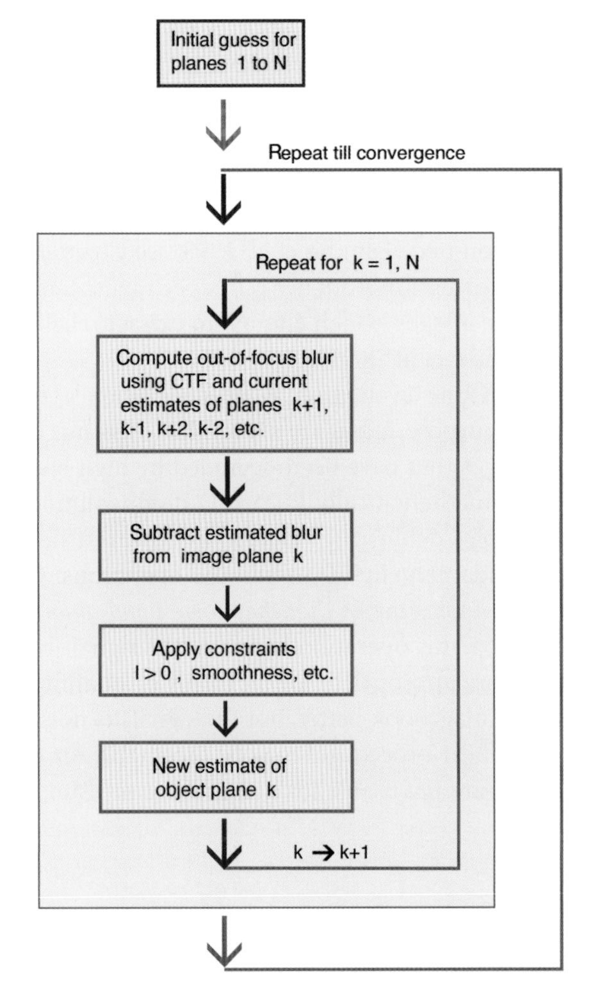

FIGURE 2.6

Flowchart for the constrained iterative deconvolution algorithm.

This description of iterative deconvolution is a simplified rendering of one of the early methods for deconvolving light microscope (LM) images (Castleman 1979; Agard 1984; Agard et al. 1989). Several other computational approaches have been reported (Erhardt et al. 1985; Fay et al. 1989; Holmes 1992; Carrington et al. 1995), and each of the commercially available imaging packages incorporates its own additional proprietary features for improving the speed and accuracy of the convergence. These features include the application of various constraints at each cycle of the iteration. Smoothness constraints can be used to enforce the physical impossibility of seeing intensity fluctuations on a scale much smaller than the known resolution of the microscope. Another possible constraint is the requirement that all values of intensity in the object must be greater than zero. One commercial package employs the so-called "blind-deconvolution" approach (Holmes 1992), in which the 3D PSF of the optical system is also estimated from the data to be deconvolved, instead of being experimentally measured from a separate 3D image of a small bead or computed from a theoretical model (Hopkins 1955; Stokseth 1969; Born and Wolf 1999).

Constrained iterative deconvolution is a computationally intensive task. It demands significant computer power and requires 10–20 minutes for a typical 3D data set even on today's fast processors. Why would one choose this approach rather than the quicker and, at least on the surface, simpler approach of confocal microscopy? The resolution achieved by the two methods is comparable, and with samples that are not excessively thick, the two methods are approximately equal in their ability to remove the out-of-focus background light that degrades contrast. Where deconvolution plus wide-field microscopy is clearly superior to confocal microscopy is in the quality of the image data, measured as signal-to-noise ratio.

The reasons for the difference are discussed later in the section on Interpreting the Results, but the point to note here is that with living samples, attaining a higher signal-to-noise ratio becomes a top priority. If examining only fixed cells by, for example, labeling with antibodies, then quantitative measurements of the fluorescence intensity are not usually worthwhile because of the huge uncertainties inherent in immunocytochemical detection. However, it is now possible to see protein molecules in living cells. Suddenly, the types of questions asked have changed beyond recognition because the molecules can be directly counted (Femino et al. 1998; see Chapters 34 and 35 of Goldman and Spector 2005). The yield of information from live cell experiments can be enormously increased if the signal-to-noise ratio in the images is high enough to extract reliable quantitative estimates of fluorophore distribution (Swedlow et al. 2002).

There are occasions, particularly if real-time evaluation is needed, when a rough contrast enhancement procedure is useful. For this purpose, there are several less-rigorous computational methods for enhancing the contrast in images that have been degraded by high background. It is very important to distinguish between the mathematically linear 3D deconvolution as described above and the several varieties of fast simple "deblurring" algorithms (nearest neighbor, multineighbor, unsharp masking), which are fundamentally 2D, nonlinear operations. *Only the linear operation of 3D deconvolution restores image intensities so that they correspond quantitatively to the intensity distribution in the object.* In doing so, this operation increases the signal-to-noise ratio of the data (Fig. 2.7), and the output images are appropriate for all forms of quantitative measurement. Deblurring algorithms can make the image look better, but the signal-to-noise ratio is often degraded. The output data from these nonlinear procedures may be acceptable for distance measurements on high contrast objects, but they are *not* usable for quantitative measurements of fluorophore distribution.

Practical Aspects and Tips for Generating Reliable Images

Modern objective lenses are nearly perfect when used under the conditions for which they were designed. Living cells, however, with their structural variation and optical inhomogeneities (e.g., the

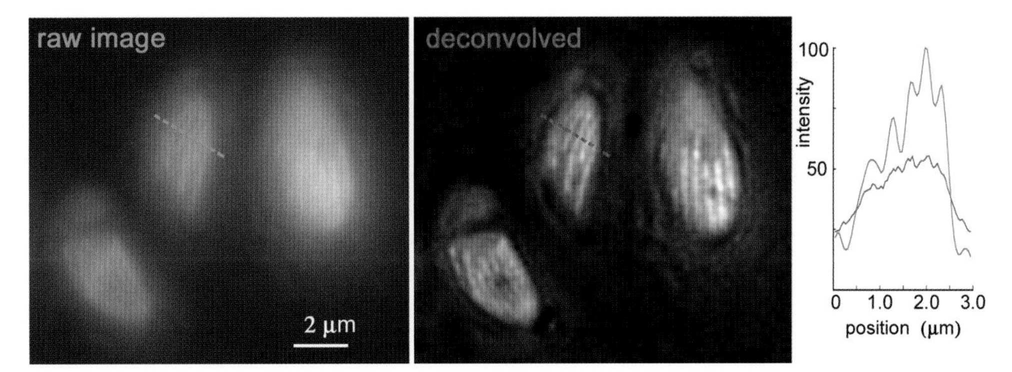

FIGURE 2.7

Deconvolution of wide-field microscope images of the parasite *Toxoplasma gondii* expressing YFP-tubulin (Swedlow et al. 2002). (*Left*) A focal plane near the plasma membrane from a 3D stack of raw images of YFP fluorescence in several living parasites. (*Middle*) The same focal plane after processing of the 3D stack by constrained iterative deconvolution. Microtubules are clearly visible as bright striations. (*Right*) A plot of the change in intensity along the dashed lines in the raw data and in the data after deconvolution. Deconvolution greatly improves the signal-to-noise ratio by restoring the out-of-focus light to its proper location. (Specimen kindly provided by Dr. Ke Hu, University of Pennsylvania. Images and deconvolution courtesy of Paul Goodwin, Applied Precision Inc.)

refractile blobs in the phase-contrast images of Fig. 2.8) are far from the optical engineer's ideal object. Furthermore, when the goal is live cell imaging, a beautiful image of a dead cell loses to a mediocre image of a happy cell; when choices have to be made, optics is compromised to improve the cell's environment, rather than the other way around. As a consequence, it is necessary to be able to recognize the more common forms of optical aberration (Cagnet 1962; Agard et al. 1989; McNally et al. 1994; Keller 1995) that afflict live cell imaging and to do what one can to compensate (Hell and Stelzer 1995; see the sections on Confocal Microscopy and on Interpreting the Results).

Spherical aberration is the manifestation of a difference in focal position of paraxial rays compared to peripheral rays. It is usually induced by the presence of material with the "wrong" (i.e., not anticipated by the optical engineer) refractive index between the lens surface and the focal plane. For instance, using an oil immersion lens to image a sample immersed in water will cause serious spherical aberration unless the sample is within a few micrometers of the coverslip. Use of a water immersion lens avoids this particular problem, but currently these lenses are often afflicted with other more complex aberrations that interfere with deconvolution. To some extent, proper choice of immersion oil can compensate for the induced spherical aberration (Hiraoka et al. 1990). This compensation, however, is adequate only over a limited range of depths within the water and at the expense of introducing some *chromatic aberration* (Scalettar et al. 1996), because the dispersion of water is quite different from that of glass or immersion oil. Chromatic aberration is a manifestation of a difference in focal position and in magnification for light of different wavelengths. It causes a lateral and axial shift in apparent position of objects of different color, obviously a significant problem when trying to determine colocalization of different fluorophores. The apparent shift becomes worse with increased distance from the optic axis (also see Chapter 1).

The success of the iterative deconvolution procedure depends critically on the accuracy of the image data and of the PSF. Computer software and hardware are the easy inexpensive components of a quantitative 3D imaging system. The difficult expensive parts are the optical, mechanical, and electronic components that are required to collect image data that are precise and artifact-free and have a high signal-to-noise ratio. To ensure the reliability of the raw data, the imaging conditions must be painstakingly optimized (Wallace et al. 2001), and there are several important preprocessing steps that must be performed to correct various artifacts typically present in 3D epifluorescence image data (Hiraoka et al. 1990; Scalettar et al. 1996; McNally et al. 1999; Markham and Conchello 2001). For a superb guide to the practical aspects of deconvolving light microscope images, see Wallace et al. (2001).

Limitations

As with any other technique, limitations to the use of deconvolution exist: Some specimens are unsuitable, and some suitable specimens challenge the currently available computational methods. One straightforward limitation, not unique to deconvolution methods, is the need for a certain minimal signal-to-noise ratio in the input data. All of the algorithms have the potential for amplifying noise. If the input signal is too noisy (i.e., the noise is large compared to the contrast between signal and background), the algorithms will fail and the output will be meaningless. Effectively, this limits deconvolution methods to specimens of "moderate" thickness (Swedlow et al. 2002; see Introduction).

Other limitations are not intrinsic to the method itself, but are imposed by limited computational resources. For instance, the algorithms assume that the PSF of the optical system is the same for all points in the field of view ("shift-invariance"), because the computations required to take account of a spatially variant PSF are not feasible for most applications (McNally et al. 1994). It is easy to show experimentally that this assumption is routinely violated in images of typical large (i.e., nonyeast) eukaryotic cells. Incorporating a measurement of local optical inhomogeneities based on differential interference contrast (DIC) imaging into an algorithm that allows for space-variant deconvolution is one promising approach to this problem (Kam et al. 2001). Another assumption that is incorporated into most algorithms (e.g., in smoothing filters that are used to constrain the intermediate calculations) is that the data are "stationary" in a statistical sense, which would require that the power spectrum be the same for every small region of the image. This is far from true for any biological sample, particularly when the image records fluorophore distribution. Again, this assumption is not a necessary feature of the restoration algorithms (Castleman 1996), but a simplification to reduce the computational load. The practical consequence of violating this assumption is that small features can become unstable after a number of iterations and suddenly disappear from the calculated result even though they may be present in the raw data. To avoid being misled by this artifact, sensible controls must be designed and the experiments repeated using independent methods and different conditions.

Notwithstanding these minor difficulties, deconvolution of images from a wide-field microscope is a tremendously powerful technique that can yield information unobtainable in any other way (Swedlow et al. 2002). It is an invaluable tool that is becoming increasingly important with the rapid progress in methods for visualizing gene products in living cells.

CONFOCAL MICROSCOPY

The goal of this method is to improve imaging of thick objects by physically removing the out-of-focus light before the final image is formed (Minsky 1961; Petráň et al. 1968; Brakenhoff et al. 1979; Carlsson et al. 1985; Amos et al. 1987). The method takes advantage of differences in the optical paths followed by in-focus and out-of-focus light, selectively blocking the latter while allowing the former to pass to the detector.

Optical Principles

Confocal microscopes differ from conventional (wide-field) microscopes because they do not "see" out-of-focus objects. In a confocal microscope, most of the out-of-focus light is excluded from the

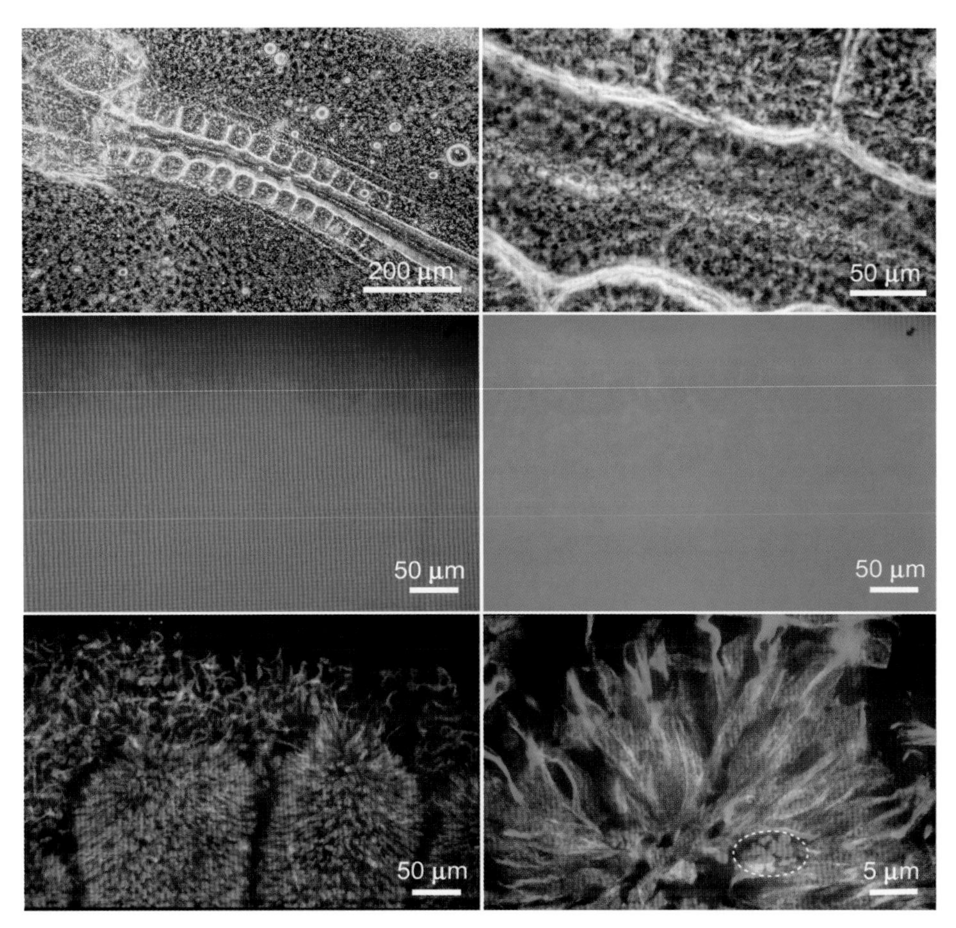

FIGURE 2.8

Images of a thick fluorescent specimen from a confocal and a conventional microscope. The sample is a chick embryo labeled with propidium iodide and antibody against the carboxy-terminal glutamic acid form of α-tubulin (fluorescein isothiocyanate [FITC] label). (*Top left*) Low-magnification, wide-field, phase-contrast image of the entire embryo. The sample is ~0.5 mm thick and contains a high density of refractile globules that scatter light efficiently. (*Top right*) Phase-contrast image at the same magnification as the fluorescence images. (*Middle row*) Conventional epifluorescence images showing (*left*) propidium iodide and (*right*) glu-tubulin distribution. The large amount of out-of-focus light severely reduces contrast. (*Bottom left*) Optical section obtained by confocal microscopy of exactly the same field and focal plane as the middle row. (*Bottom right*) Higher-magnification confocal view of a portion of the same field. Mitotic nuclei with condensed chromatin can be readily identified. (*Dotted white ellipse*) Bundles of microtubules are also seen. The mitotic spindle in these cells is formed predominantly of the tyrosinylated form of α-tubulin, and hence is not seen. (Sample kindly provided by Dr. Camille DiLullo, Philadelphia College of Osteopathic Medicine, Philadelphia.)

final image, greatly increasing the contrast and hence the visibility of fine details in the specimen. Figure 2.8 shows a comparison of images of a thick specimen viewed by both wide-field and confocal microscopes. Figures 2.9 and 2.10 schematically illustrate how a confocal microscope works. On the left of Figure 2.9 is a wide-field microscope. A light source, in conjunction with a condenser, distributes light uniformly across the area of the specimen under observation. The diagram illustrates the paths followed by light arising from the specimen, passing up through the objective lens and eventually reaching a detector (e.g., film, video camera, or retina). Three paths are shown, corresponding to light arising from three locations in the sample. The first location is in the center of the field of view and in the focal plane of the objective lens. The heavy dashed lines in Figure 2.9 are the limits of the bundle of light rays that contribute to the image from this point. Similarly, the lighter dashed lines mark the rays from a second point in the same plane but displaced horizontal-

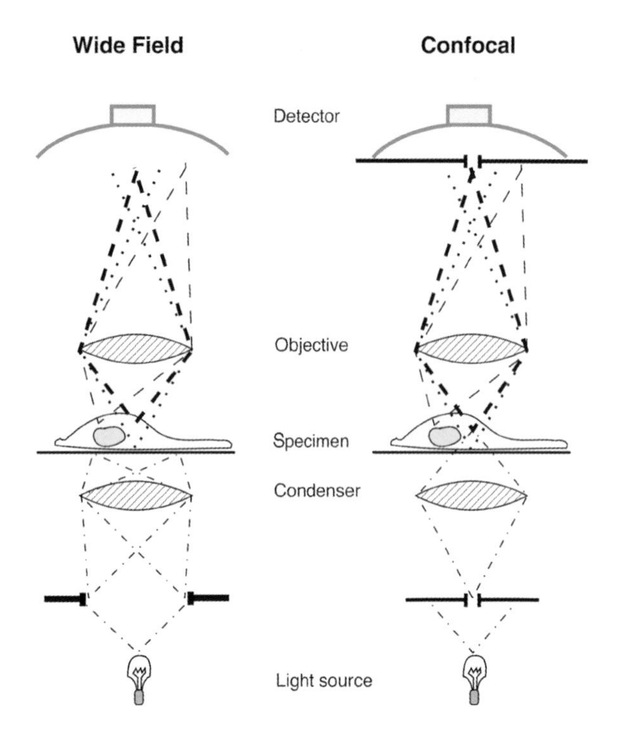

FIGURE 2.9

Schematic illustration of the operating principle of the confocal microscope. (*Left*) A conventional, or *wide-field*, microscope. The specimen is illuminated over an extended region by a light source and condenser. Light rays arising from three points in the specimen are shown. The dashed lines emanate from two points in the focal plane, one centrally located (*darker dashed lines*), the other off-axis (*lighter dashed lines*). The third point is on-axis but located below the plane of focus (*dotted lines*); it gives a blurred image at the detector. The detector forms an image from the sum of all the simultaneously arriving light rays. (*Right*) A *confocal* microscope. Two pinhole apertures have been introduced. The upper aperture allows the focused light rays from only the on-axis, in-focus point of the specimen to pass to the detector. The lower aperture restricts the illumination so that it is focused on the point seen by the upper pinhole aperture.

ly from the first point. Finally, light represented by the dotted lines is coming from a third point located below the first point (i.e., from an out-of-focus plane). This light contributes to the blurred background.

The right side of Figure 2.9 shows how the background is eliminated simply by adding a pinhole aperture to the wide-field microscope. Note that behind the objective lens, all of the light rays are brought together at a crossover point, the location of the intermediate image plane of the microscope. Normally, the microscope oculars are focused on this plane to form the final, fully magnified image. The location of this crossover plane along the vertical axis of the microscope is different for different light rays, depending on the distance of the corresponding point in the specimen from the front of the objective lens. The crossover point for light rays from the illustrated out-of-focus plane (dotted lines) is below that for rays from the in-focus plane (dashed lines). As illustrated, a pinhole aperture at the correct height will pass the converged rays from the in-focus point, but block nearly all of the dispersed rays from points higher or lower than the focal plane. (The geometry is slightly different in so-called "parallel beam" confocal systems, but the principle is identical [Amos et al. 1987; Shao et al. 1991].) Out-of-focus points therefore contribute little to the final image and are essentially invisible. A side effect of the pinhole aperture is that most of the in-focus points also become invisible; only the rays from the central spot are allowed to pass through the aperture.

Because all of the specimen will be invisible except for the tiny spot imaged through the pinhole aperture, there is no need to illuminate an extended area. There are three good reasons for restricting the incoming light to the minimum necessary area. First, light going to other parts of the specimen will be scattered, and inevitably some of it will leak through the pinhole aperture, degrading the contrast in the image. Second, all of the illuminated area will be subject to photobleaching. Third, restricting the illumination to a single, focused point gives a dramatic improvement in the discrimination against points above and below focus; i.e., it enhances the vertical resolution.

The reason for this enhancement is as follows. If the incoming illumination is focused sharply to a point in the focal plane, then regions above or below this focal point will receive dispersed, much less intense, illumination. In fact, the intensity of illumination falls off as the square of the axial dis-

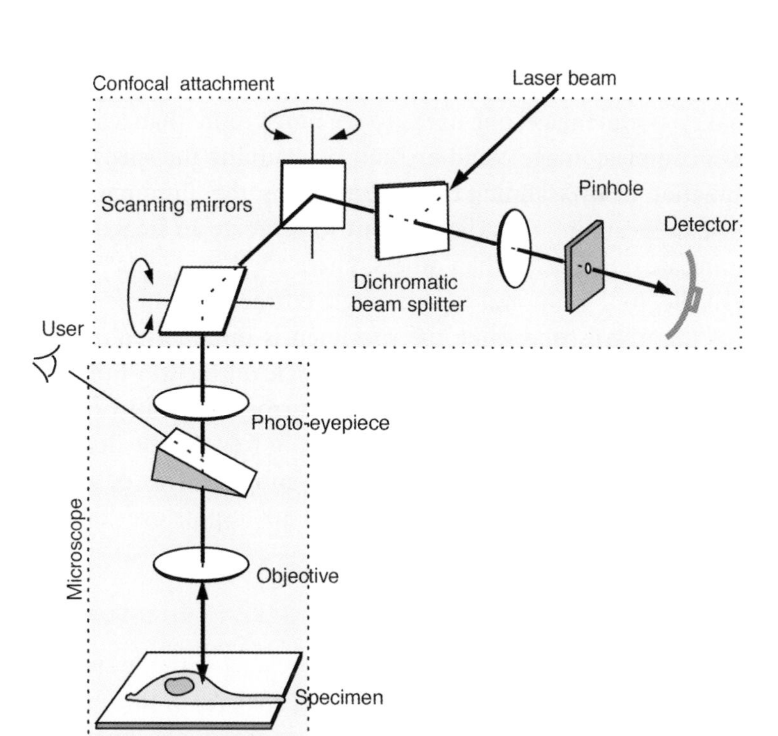

FIGURE 2.10

A typical laser-scanning confocal microscope. The instrument consists of a conventional fluorescence microscope (enclosed in the *lower shaded rectangle*) attached to a confocal scanning unit (*upper shaded rectangle*) comprising a pair of scanning mirrors, a laser, some wavelength-selective filters, a pinhole aperture, and a photomultiplier detector. The laser illumination is directed down the phototube of the microscope, having been deflected by the rapidly oscillating scanning mirrors so that it sweeps across the specimen in a raster pattern. Fluorescent light emitted by the sample passes back up through the phototube, is "descanned" by the scanning mirrors, and passes through the dichromatic beam splitter (which removes any reflected laser light) to the pinhole aperture. Light originating from the focal plane passes through the pinhole to the detector, but all other light is blocked. For reflectance imaging, the dichromatic beam splitter is replaced by a half-silvered mirror. A sliding prism allows visual (nonconfocal) observation through the usual binocular eyepieces, using the normal microscope lamps for illumination.

tance from the focal plane (i.e., the intensity within the cone of illumination is inversely proportional to the cross-sectional *area* of the cone). Thus, when using this type of focused spot illumination in combination with the pinhole-blocked detector not only will the pinhole aperture reject most of the light from out-of-focus planes, but the light emitted from those planes will be less than it would have been with wide-field illumination. By exactly the same reasoning, the lateral resolution of the microscope will also be enhanced if a focused spot of illumination is used. These two modifications, limiting the area "seen" by the detector and the area illuminated by the light source, are the key ingredients of a confocal microscope. A confocal microscope is simply a light microscope in which both the field of view of the objective lens and the region of illumination have been restricted to a single point in the same focal (*con*focal) plane (Wilson and Sheppard 1984).

To gain the optical sectioning capability of the confocal microscope, other aspects of the microscope's performance have been sacrificed. Field of view has been traded for increased axial resolution. The pinhole aperture effectively excludes light from out-of-focus planes, but it also restricts the field of view laterally to a spot the size of the demagnified pinhole. Thus, to gain the advantages conferred by the confocal pinhole, one must give up the convenience of acquiring an image from an extended area in parallel. The confocal image must be built sequentially by scanning one or more spots over the specimen until the region of interest has been covered (Fig. 2.10).

Instruments

To make a useful image, one needs to see much more than a single tiny spot of the sample. In principle, a complete image could be built by scanning the specimen to and fro under a fixed spot of illumination or by scanning the objective lens, the illumination, or the pinhole itself. In practice, because the scanning must be very fast to generate an image in an acceptable time, some types of scanning are much easier than others. There are three major types of confocal microscopes currently available. In the simplest type, a single diffraction-limited spot is held stationary on the optical axis of the microscope while the specimen is moved (*specimen scanning*). In the *beam-scanning* instruments, a laser beam is focused to a single diffraction-limited spot or line that is deflected over the stationary specimen using oscillating mirrors or acousto-optical deflectors. This type of microscope employs a single, fixed pinhole or slit in front of the detector. In the *pinhole scanning* instruments, a disk (called a Nipkow disk) containing an array of more than 10^4 pinholes rotates in the illumination path, sweeping approximately 1000 spots over the specimen simultaneously.

Specimen Scanning

There are important optical advantages associated with the stationary beam of the specimen-scanning type of instruments (Brakenhoff et al. 1979). All of the imaging takes place exactly on the optical axis, which minimizes many of the lens aberrations that plague the beam-scanning instruments. Alignment of the fixed optical path is also greatly simplified compared to systems with moving optical components. The primary disadvantage is that the specimen must be moved, along with specimen holder, chambers, and, for living cells, liquid culture medium. The total mass of moving material is much larger than in a beam-scanning instrument, creating many opportunities for vibration and loss of positional accuracy. For the scan to be completed within a reasonable time, the mechanical accelerations must be large, and only certain specimens are suitable. Even with the lightest specimens, the time resolution of specimen-scanning instruments is much worse than for other types of confocal microscopes and can often be problematic for living samples. Sweeping a beam of light over a stationary specimen can be done much more rapidly than moving a specimen under a stationary light beam. On the other hand, when the illumination and imaging light travel separate paths, as in all forms of transmitted light imaging (e.g., bright field, phase contrast, DIC), only the stationary-beam, specimen-scanning instruments are truly confocal. The beam-scanning instruments are confocal only when the objective lens also serves as the condenser (e.g., epi-illumination fluorescence or reflection modes), for reasons that will be explained below.

Beam Scanning: Single-Spot Mode

In beam-scanning confocal microscopes (Fig. 2.11), the illumination is scanned while the specimen is held stationary (Carlsson et al. 1985). In the single-spot mode, a small (diffraction-limited) spot is swept over the specimen by means of a rapidly oscillating mirror interposed between the light source and the condenser lens (which is also the objective lens in epifluorescence mode). Because a useful image often consists of 10^5–10^6 pixels, the dwell time for each pixel must be kept very short to accumulate a useful image in a reasonable length of time. The need for fast scanning places stringent demands on the source of illumination, because the number of photons collected per pixel also decreases as the dwell time is shortened. To collect a 512 x 512-pixel image in 1 second, the scanning spot of light can dwell on each point for 4 μsec at most. During this time, as many photons as possible must be collected so that the statistical noise in the image is minimized. For this reason, a very intense source of light is needed, which in most instances means a small laser.

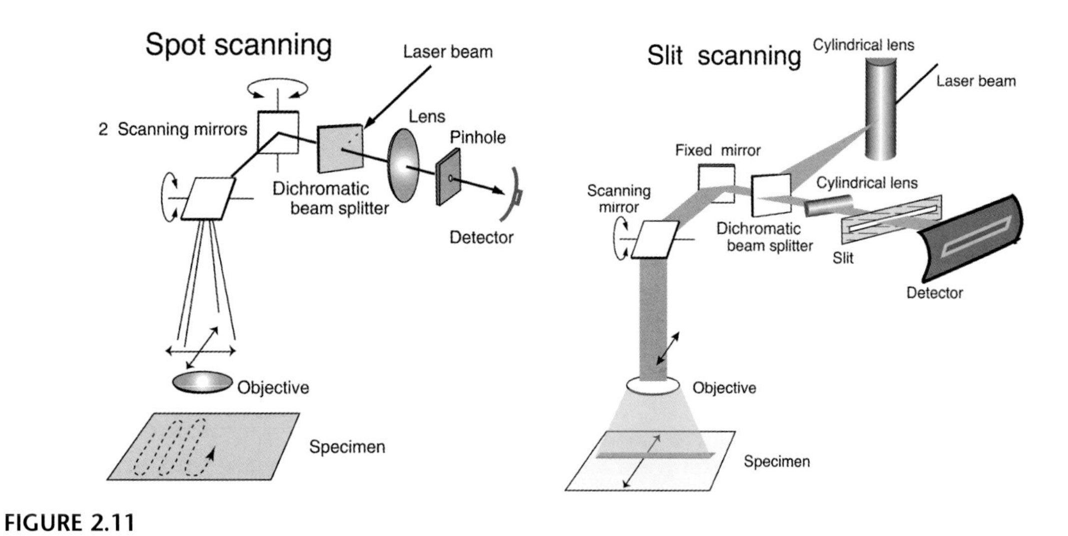

FIGURE 2.11

Comparison of spot-scanning and line (slit)-scanning modes of confocal microscopy.

Beam Scanning: Line Mode

A second form of beam-scanning microscope (Fig. 2.11, right) has been developed that is, strictly speaking, only partially confocal. "Slit-scanning" instruments replace the round pinhole aperture behind the objective lens with a long, very narrow slit (Lichtman et al. 1989; Amos and White 1995). The illumination is shaped into a single narrow line, focused to the same line on the specimen that is seen by the slit aperture. Scanning is necessary in only one direction, because the slit is long enough to admit light from points all the way across the field of view. The resolution and contrast in the image are no longer completely isotropic, but, in practice, the image quality is almost equal to that achieved with spot-scanning instruments for some types of sample, and the images are collected in a fraction of the time.

Pinhole (Tandem) Scanning

The renaissance in optical microscopy that has been under way for the last two decades can in some respects be traced to the stir caused in the biological research community by the appearance of a remarkable new type of microscope, a homemade tandem scanning confocal, and its equally remarkable Czech inventor, Mojmír Petráň (Egger and Petráň 1967; Petráň et al. 1968; Egger et al. 1969; an earlier, undeveloped microscope was reported by Minsky 1961, 1988). This small device, carried by Dr. Petráň in his coat pocket, provided the first glimpse of the promise of confocal microscopy. It was a nightmare to align, but the images were astonishing and inspired the development of more user-friendly instruments (Boyde et al. 1983; Petráň et al. 1986; Xiao and Kino 1987).

In tandem-scanning microscopes, the spot of illumination and the detector pinhole move over the field of view in tandem. The same pinhole is used for forming the spot of illumination on the input light path and blocking out-of-focus light on the return path. Multiple spots are formed and imaged in parallel by means of an array of pinholes arranged in an Archimedean spiral on a rotating disk (Fig. 2.12).

Modern tandem-scanning microscopes are enormously improved from early days, and they offer two important advantages compared to spot-scanning methods: First, the detector is a CCD

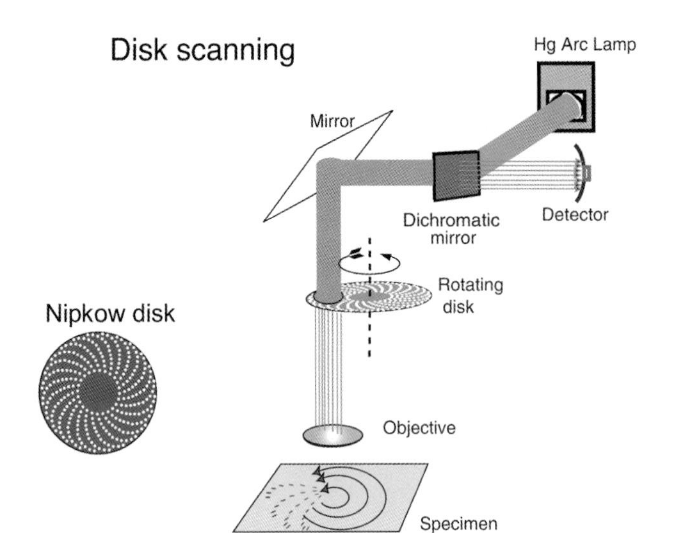

FIGURE 2.12

Tandem (disk)-scanning mode of confocal microscopy. These instruments make use of an array of pinhole apertures on a rotating disk.

(60-70% quantum efficiency) instead of a photomultiplier (10-20% quantum efficiency), and, second, for some bright specimens, direct visual observation in real time is possible. At any one instant, only a few percent of the field of view is illuminated (1000 pinholes of the ~25,000 on the disk), but the disk rotates fast enough that the moving spots fuse into a seemingly uniform image. Tandem-scanning instruments can in principle use broad-spectrum light sources such as high-pressure arc lamps for illumination. This would allow a wider selection of fluorescent probes and would eliminate one source of noise that commonly contaminates images from laser-based instruments. In practice, severe loss of illumination intensity at the disk with conventional light sources limits their use to reflected light imaging rather than fluorescence.

To avoid interference and maintain confocality, adjacent pinholes on the disk must be separated from each other by a distance that is large compared to their diameter. For example, if they are positioned 10 diameters apart, only 1% of the incoming illumination will pass through the disk. To circumvent this inefficiency, one commercially available design uses a second disk, consisting of an array of tens of thousands of microlenses, positioned above the "pinhole" disk. Each lens collects light and focuses it on one pinhole (which is actually a small transparent area etched in the opaque coating of a clear plate), thus increasing the effective aperture of each pinhole. With this design, ~25% of the illumination light actually passes through the disk. Nevertheless, coherent laser illumination is still necessary, because with conventional illumination, the focused spots from the lenslet array are much larger than the pinholes and light throughput is severely reduced (Watson et al. 2002). A completely different design has been described that has even better efficiency than the dual-disk lenslet array, in which a random mask of opaque and transparent patches, with 50% overall transmittance, replaces the array of pinholes (Juskaitis et al. 1996). Another promising approach that has been demonstrated (Hanley et al. 1999, 2000; Heintzmann et al. 2001), but is not yet commercially available, is the replacement of the spinning disk with a stationary array of approximately 10^6 individually addressable micromirrors on a chip (digital micromirror device [DMD], Texas Instruments).

Imaging Modes

Confocal microscopes can form an image using several different sources of optical information from the specimen. A "reflectance" image can be formed by using the light that is scattered (backscatter) from the specimen in a backward direction, i.e., back along the path of the incoming

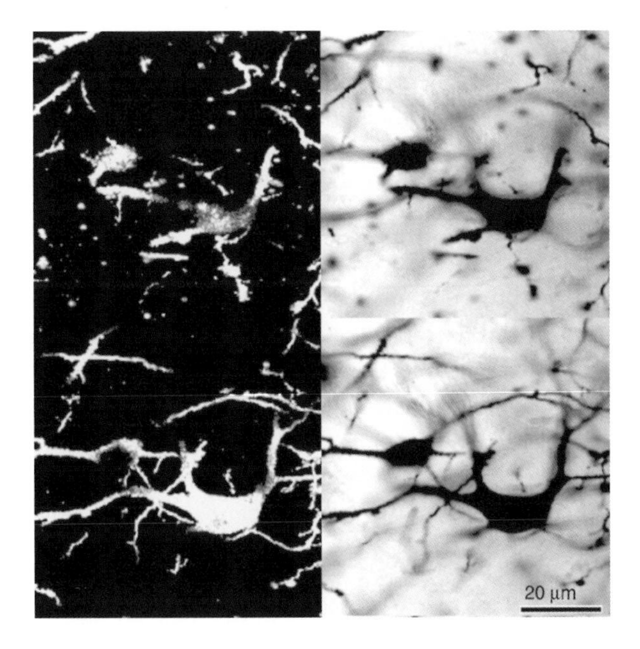

FIGURE 2.13

Confocal reflectance and nonconfocal transmitted light images of Golgi-stained neurons. (*Left*) Two optical sections made by using reflectance mode imaging in a laser-scanning confocal microscope. The silver precipitate gives a very bright backscattered image. (*Right*) The corresponding nonconfocal transmitted light (bright-field) images.

epi-illumination. This light has the same wavelength as the original illumination. Colloidal gold labels are easily visualized in reflectance mode. The insoluble precipitates formed by the Golgi stain procedure for neurons (Fig. 2.13) or horseradish peroxidase (HRP) oxidation of substrates such as diaminobenzidine also give bright backscatter images.

The confocal microscope also can easily be configured to detect the interference pattern between light reflected from cell membranes and that reflected from the underlying substrate (Sato et al. 1990). This technique, known as interference reflection contrast microscopy (IRM) (Izzard and Lochner 1976; DePasquale and Izzard 1987, 1991), works well with intact living cells (Fig. 2.14).

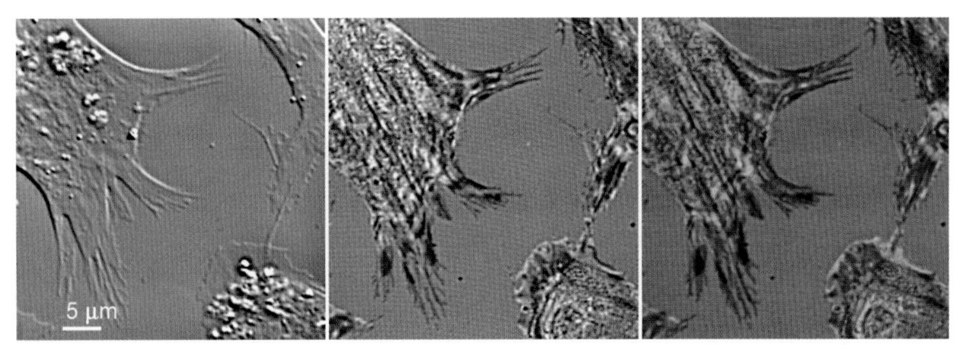

FIGURE 2.14

Confocal interference reflectance contrast imaging (IRM). Transmitted light DIC (*left*), single-wavelength (488-nm) IRM (*middle*), and dual-wavelength (488-, 633-nm) IRM (*right*) images of living fibroblasts from dissociated embryonic quail heart tissue growing on a glass substrate. In the IRM images, the contrast arises from interference between the laser light reflected from two surfaces (e.g., basal cell membrane and glass substrate). The image was acquired at full aperture with a 63x, 1.2 NA water immersion lens, but the high degree of coherence of the illuminating laser light nevertheless causes the higher-order fringes (e.g., from interference between reflections off the apical and basal cell membranes) to have much higher contrast than with conventional illumination (Izzard and Lochner 1976; Sato et al. 1990). The dual-wavelength IRM image is quite useful for discriminating between the zero-order (*black*) fringes, which occur at the same location for both wavelengths, and first- or higher-order (*green, yellow, red*) interference bands, which occur at different positions for the two laser lines. (Specimen kindly provided by Dr. J. Sanger, University of Pennsylvania.)

Fluorescence emitted by the specimen is the most common source of optical information used to generate confocal microscope images. In this case, the image-forming light has a different wavelength from the illumination, so the fluorescence and reflectance signals can be separated using dichromatic beam splitters as in conventional epifluorescence microscopes. A particularly valuable feature of commercial instruments is the ability to acquire simultaneous, perfectly registered images from multiple different fluorescent labels with (in some instruments) independent control of the trade-off between sensitivity and resolution in each channel.

As well as being able to form an image from the emitted fluorescent light that is collected by the objective lens, most beam-scanning confocals have the useful ability to simultaneously collect illuminating light that passes through the specimen and thus acquire a transmitted (e.g., bright-field, phase-contrast, or DIC) image in parallel with the backscatter or epifluorescence image. The quality of these scanned, transmitted images is usually higher than could be obtained using a conventional wide-field microscope, and they should be in perfect register with the simultaneously acquired fluorescence or reflectance confocal image. It is important to realize, however, that the transmitted light image is *not* confocal, because there is no pinhole between the specimen and the transmitted light detector. The crucial difference between the transmitted and epifluorescent light is that only the latter encounters the scanning mirrors en route to the detector. Epifluorescent light from every point in the scanned field of view is reflected from the mirrors back along exactly the same path that the incoming laser illumination traversed: In other words, the epifluorescent light is *descanned* by the scanning mirrors, and thus forms a stationary beam that can pass through a fixed pinhole. The transmitted light is not descanned, and thus is *not* stationary anywhere along its path. In principle, a second set of mirrors could be introduced below the condenser, synchronized with the mirrors on the input side, to descan the exiting transmitted light. In practice, technical difficulties of this and other approaches to descanning (Goldstein et al. 1990; Art et al. 1991; Dixon et al. 1991; Dixon and Cogswell 1995) have blocked commercialization of a spot- or tandem-scanning instrument that is truly confocal in transmitted light.

Lasers and Fluorescent Labels

The total power required for imaging typical specimens is quite modest (~0.1 mW) when compared to the power of commonly available lasers, but the intensity at the focal spot can be enormous (MW/cm^2). It is important to use the minimum power necessary to acquire each image, which usually means reducing the beam intensity 10–100-fold by using neutral density filters or an acousto-optic modulator in the illumination path. At very low laser power, the strength of the emitted fluorescence will increase directly in proportion to increases in the intensity of the illumination. However, as the power is increased, the emitted light will reach a constant value, which occurs when the strength of illumination is high enough to excite every fluorescent molecule in the illuminated spot. Increases of illumination power beyond this point result in no further increase in emission from the fluorescent molecules in the focal plane. Away from the plane of focus, where the illumination is less intense, increases in laser power will continue to excite more and more fluorescent molecules. This is an undesirable effect, because the out-of-focus light does not contribute to the image, and the excited molecules are subject to photobleaching. In this respect, multiphoton imaging (see Chapter 3) can have an advantage over confocal imaging, because the light intensity is not high enough to generate multiphoton absorption events outside the focal spot. Some fluorophores, however, photobleach faster with multiphoton than with single-photon excitation (Patterson and Piston 2000).

Each type of laser emits light at a set of characteristic wavelengths, so the type of laser available determines the fluorophores that can be imaged. Table 2.1 shows the wavelengths available from

Table 2.1 Visible laser and mercury arc emission wavelengths

Source	Emission Wavelengths (nm)												
Argon	{351}	{364}	458	466	477	488	496	502	514				
Krypton	{337}	{356}	468	476	482	521	531	568	647				
Argon/krypton	488	568	647										
Helium/neon	543	594	604	612	629	633	1152						
Helium/cadmium	325	442	534	539	636								
Solid state	???	405	457	532	635	650	670	685	694	750	780	810	???
Mercury arc	313	334	365	405	436	546	577						

Unbracketed values are the lines available from the commonly used low-power (5–50-mW), air-cooled lasers. New, solid-state lasers are introduced frequently; other wavelengths will probably be available soon. Many more lines, including those in the UV ranges that are listed in brackets, are available from the large, high-power (1–5-W), water-cooled versions of the gas lasers. In contrast to the lasers, the mercury arc lamp emits at all wavelengths between its major peaks, at a level of 5–10% of the peak intensity.

some of the more common lasers and the major peaks in the spectrum from a mercury arc lamp. It is important to remember that the peaks in the arc lamp spectrum are much broader than the spectral lines from the lasers and, in addition, significant emission (5–10% of peak intensity) occurs at all wavelengths between the peaks. Thus, the range of fluorophores that can be excited by mercury arc illumination is much broader than that for any single laser. Table 2.2 shows the wavelengths for peak excitation and the range over which the excitation efficiency is at least 25% of the peak for some commonly used fluorophores.

Simultaneous Imaging of Multiple Labels

The distribution of wavelengths available from the light source becomes especially important when two or more fluorophores must be imaged in the same specimen. In general, one can expect problems with cross talk between two channels ("bleed-through") when the emission ranges of two fluorophores overlap significantly and one of them is much more strongly excited than the other. Table 2.2 lists the range of wavelengths over which emission intensity is at least 5% of the maximum for each fluorophore. This range can be used together with the range of excitation wavelengths to predict when problems are likely to arise. For example, a pair of "rhodamine-class" and "fluorescein-class" labels (with spectra similar to rhodamine and fluorescein) is one popular combination for double-labeling experiments in conventional epifluorescence microscopy, using mercury arc illumination at 495 nm and 546 nm. However, pairs of these fluorophores often give unsatisfactory results in confocal microscopes that use only an argon or argon/krypton laser, because these lasers do not emit appropriate wavelengths for efficient excitation of the rhodamine-class dye. Some instruments attempt to use the 488-nm and 514-nm lines of the argon laser to excite the fluorescein class (typical peak excitation at 490 nm) and rhodamine class (excitation optimum ~ 550 nm). Only the first fluorophore is efficiently excited at 488 nm, but it has an extended long wavelength "tail" of emission that completely overlaps the emission spectra of the second dye. At 514 nm, both dyes are excited equally (~20% of maximum excitation). This combination of spectral properties and laser excitation wavelengths thus leads to severe problems with bleed-through.

The problem is solved by using different fluorophores and/or different lasers. For example, the 488-nm argon and 543-nm green helium/neon laser lines work well with these two dyes, or the fluorescein-class dye can be used, plus a longer-wavelength ("Texas Red class") dye using the 488-nm argon and the 567-nm line of the argon/krypton laser. As one increases the number of fluorophores used simultaneously, the probability of significant cross talk approaches 100%. There are several approaches to this problem, and new methods are introduced regularly. One approach is to give up simultaneous data acquisition and instead employ sequential scans with single laser lines.

Table 2.2 Excitation and emission ranges for common fluorophores

Fluorophore	Ex_{max}	Em_{max}	25% Ex	5% Em
For labeling nucleic acids				
Acridine orange	500	526	450–520	496–631
7-Amino-actinomycin D	540	655	460–600	570 to >780
Chromomycin A3	430	550	370–470	450–700
DAPI	359	461	310–395	385–600
Ethidium homodimer	528	617	450–570	550–750
Hoechst 33258	346	460	315–438	390–610
Propidium iodide	536	620	470–580	560 to >750
Quinacrine	420	490	360–460	435–600
TOTO-1	514	533	470–535	515–650
YOYO-1	491	509	450–515	475–615
For labeling membranes/organelles				
di-I	540	565	495–570	540–680
di-OC6 (3)	478	496	445–505	480–610
Rhodamine 123	505	534	470–530	460–610
For labeling ions				
BCECF	505	530	425–550	495–605
fluo-3	506	526	455–540	480–680
indo-1	339	405,490	260–370	370–580
For labeling protein (intrinsic fluorescence)				
ECFP	426	476	350–475	440–600
EGFP	491	512	430–514	470–600
EYFP	515	532	478–540	507–630
Tetrameric RFP (DsRed)	558	583	473–578	550–700
Monomeric RFP (mRFP1)	584	607	516–604	556–732
For labeling protein (covalent labels)				
Alexa Fluor 488	495	519	450–520	460–610
Alexa Fluor 546	556	573	510–580	540–680
Alexa Fluor 568	578	603	520–610	560–720
Alexa Fluor 594	590	617	530–625	570 to >750
Alexa Fluor 633	632	647	560–660	610–750
AMCA	343	442	<250 to 376	390–546
Bodipy Fl	503	512	450–515	485–580
Cascade Blue	395	420	340–420	400–510
Cy-3	555	570	495–575	540–660
Cy-5	650	670	585–680	635–750
Fluorescein (FITC), Cy2	490	525	450–520	460–610
Lucifer yellow	430	540	380–470	470–660
NBD	460	534	420–510	480–630
Rhodamine	555	570	495–580	540–680
Texas Red	590	615	540–610	570–720

The wavelengths for maximum excitation (ex) and emission (em) are listed. The last two columns give the range of wavelengths over which excitation efficiency is ≥25% and emission intensity is >5% of the maximum values. For more information, spectra, and an excellent discussion of applications of these and other fluorophores, the Molecular Probes CD-ROM catalog or Web site (http://www.probes.com/) is invaluable. mRFP1 is from Campbell et al. (2002).

This works for combinations of fluorophores that have overlapping emission spectra but can be individually excited by different laser lines. It does not solve the problem when both emission and excitation spectra overlap extensively. An alternative is to greatly increase the spectral resolution of the detection apparatus (using a dispersive element such as a prism, grating, or acousto-optic deflector) so that arbitrary wavelength bands of emission can be collected (see Chapter 3). Although it is sometimes possible to choose a narrow range of wavelengths that gives acceptable discrimination between fluorophores with partially overlapping excitation/emission spectra, the detected signal becomes weaker as the detection window is narrowed. Additional criteria can also be introduced for "gating" the fluorescence output. For instance, fluorophores that have similar emission and excitation spectra may nevertheless have quite different fluorescent lifetimes. Very short pulsed excitation and time-gated detection synchronized to the laser pulses then allow discrimination based on fluorescent lifetime (Gadella et al. 1993; Cole et al. 2001). Fluorescence polarization (or "hole-burning" plus time-gated polarization-sensitive detection) offers additional possibilities for discrimination (Massoumian et al. 2003).

Inevitably, the number of signals that need to be detected separately will exceed the capability of the detection system to discriminate, so ultimately, the problem of fluorophore cross talk will have to be addressed by postacquisition processing. In this approach, multiple images contaminated with cross talk among different fluorophores are collected, each with a different combination of excitation/emission wavelengths. Reference "images" are also collected from pure samples of each fluorophore individually, using the same set of excitation/emission combinations. The contributions of each fluorophore are then determined by solving the appropriate system of simultaneous linear equations for each pixel of the image. Applications of this computational method to karyotyping routinely discriminate among more than 20 different "colors" of fluorescent label (e.g., the SKY system from Applied Spectral Imaging) (Schrock et al. 1996; Ried et al. 1997).

Specimen Preparation

For a comprehensive discussion of preparing specimens for live cell imaging, refer to Chapter 17 of Goldman and Spector (2005). A few points that are particularly important for confocal microscopy are mentioned here. Confocal microscopy is compatible with any of the conventional specimen preparation methods, including imaging of unprepared living tissue. Modest-resolution images of material down to a depth of about 0.2 mm below the surface can be obtained from many tissues if the working distance of the objective is large enough. Thicker slices can often be examined completely if they are mounted between two thin coverslips and imaged from both sides. For the highest-resolution work, spherical aberration introduced by the sample limits the maximum depth to about 0.2 mm. However, severe attenuation of both the incoming laser illumination and exiting fluorescence emission, because of scattering by local inhomogeneities in the refractive index of the sample, often limits the quality of confocal images deeper than 0.05 mm. Attenuation is sometimes less for multiphoton microscopy, because of the longer wavelength used for illumination and the fact that a detector pinhole is unnecessary; thus, multiphoton microscopy has an advantage over confocal for deep imaging in some tissues (see Chapter 3). When mounting thick specimens, care must be taken to avoid compressing them while at the same time minimizing the distance between the coverslip and the specimen. It is also important to use small coverslips when possible. Large coverslips flex with each motion of the objective lens, causing fluid displacements and specimen motion.

Time-lapse imaging of living samples for extended periods at 37°C poses several problems for confocal microscopy. The cycling of the heater for the sample chamber results in vertical movements that shift the plane of focus. Because immersion lenses are usually required, the sample chamber will be strongly thermally coupled to the objective lens and will require a separate objective lens heater, whose cycling also changes the focal plane. In a typical microscope room, the cycling of the laboratory air conditioner/heater system causes additional focal shifts. With well-designed chambers and lens heaters, and a modern laboratory building with good environmental stability, these thermally induced shifts are small enough to be barely noticeable under visual observation. Even under these ideal conditions, however, the shifts are still about tenfold larger than the vertical resolution of the confocal microscope. To compensate for these shifts in unattended time-lapse imaging, the vertical extent of each 3D stack must be increased both above and below the specimen by the size of the thermal shift. For instance, to be certain of completely capturing a 5-μm-thick cell, images might have to be taken over a 15-μm span, tripling the total exposure and making it very difficult to do extended time-lapse imaging. A third problem arises when using water immersion lenses. At 37°C, the water between the sample chamber and the lens quickly evaporates, and it is impossible to replace it without interrupting the time-lapse study and moving the specimen.

Specifications provided by a microscope manufacturer reveal the fundamental problem. According to Zeiss, the overall thermal response for their Axiovert inverted microscope system, including lenses, stage, and focus drive, is 10 μm of focal plane shift per 1°C temperature change. Therefore, to reduce the focal shifts to below the vertical resolution of the confocal microscope, the temperature of the microscope must be stabilized to better than ± 0.04°C. That is not a realistic goal when the room temperature fluctuates by 1°C or 2°C and the difference between the room temperature and the sample/lens temperature is 15–20°C. Moving the microscope to a "warmroom" held at 37°C is one solution, but the humidity must be very high to reduce the evaporation rate of the immersion water. A more user-friendly solution is to enclose the entire microscope except for the mercury arc lamp in a box held at 37°C and 80–90% humidity (Fig. 2.15).

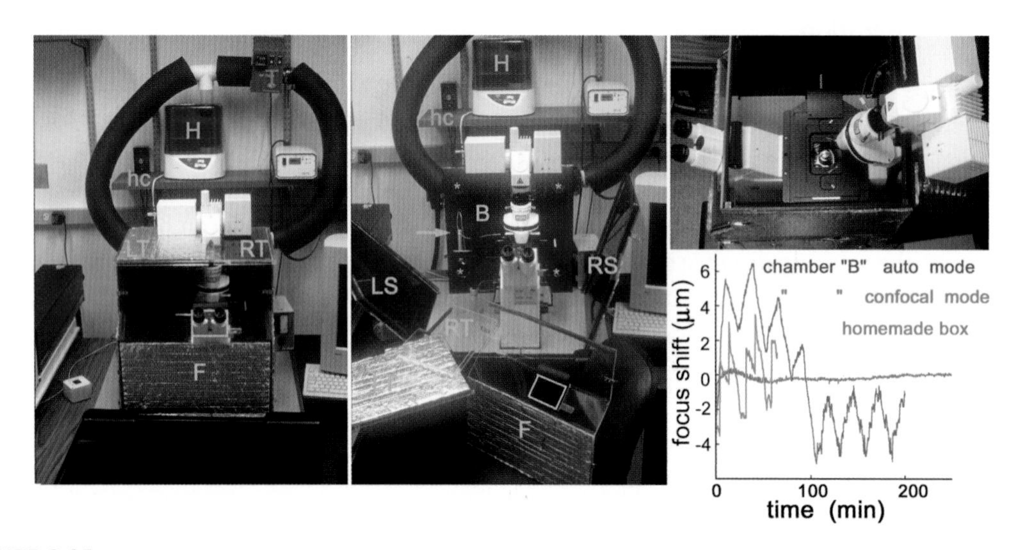

FIGURE 2.15

"Homemade" confocal microscope environmental chamber for live cell imaging. (*Left*) Front view of the temperature- and humidity-controlled box that encloses the entire inverted microscope except for the mercury arc and tungsten lamps. Orange letters indicate the separate pieces of the box (see description below). (H) Ultrasonic humidifier (Vicks); (hc) relative humidity controller (RHCN-3A, Omega Engineering); (T) heater-air circulator (Air-Therm, World Precision Instruments). (*Middle*) The disassembled pieces of the box, which are made of a 6-mm acrylic sheet, that fit snugly together and are locked into place with clasps. The front piece (F) has a rectangular aperture through which the eyepieces protrude (*orange rectangle*). The front piece and right half of the top (RT) are clear acrylic. The left and right sides (LS, RS), left half of top (LT), and back panel (B) are opaque, black acrylic covered with reflective thermal insulating "bubble wrap." The back panel and floor plate remain permanently mounted on the microscope. The remainder of the box disassembles for use at room temperature. (*Upper right*) Top view from the right side, with the top of the box removed. The condenser lamp housing post of the microscope tilts backward for access to the stage, pushing a swiveling panel that is set into the back panel of the box. (*Lower right*) Graphs showing focal-plane shifts with different sample chambers. Blue and green lines represent shifts with a commercial sample chamber and objective lens heater in "auto" and "confocal" modes, respectively. The red line represents the shifts with the homemade box. After an initial equilibration period, the focal plane with the homemade box is stable to within ~0.2 μm.

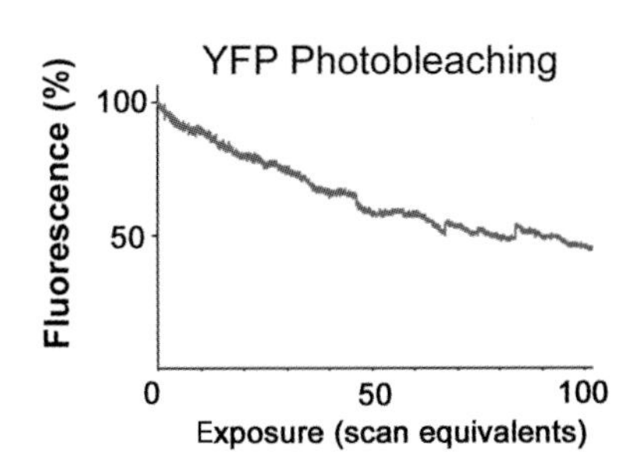

FIGURE 2.16

Loss of fluorescence of yellow fluorescent protein (YFP) because of photobleaching. Fluorescence from a YFP fusion protein expressed in a living cell (a nondiffusible cytoskeletal component) was recorded using 514-nm light in a laser-scanning confocal microscope (Zeiss LSM 510). The illumination intensity, dwell time per pixel, and photomultiplier gain were adjusted to give a high-quality image using a pinhole diameter equivalent to ~1 Airy disk. The average fluorescence intensity in a small region was then recorded over the course of repeated scans. The fluorescence decreased by ~1/2 after 100 scans.

Photobleaching and Phototoxicity

These two phenomena are in most cases actually the same process viewed from two different perspectives. When the emphasis is on accurate measurement of the 3D distribution of a fluorophore, then the primary concern is with photobleaching. When the emphasis is on observing the fluorophore distribution in a physiologically "normal" state, then phototoxicity will be the foremost concern. As a general rule, for quick observations of living cells at a single time point, photobleaching is the relevant phenomenon, and it is usually a surmountable problem. For repeated observations of the same cell, phototoxicity is *always* a major problem and usually necessitates accepting compromises that limit image quality.

Modern confocal microscopes can acquire a high-quality digital image with much lower illumination than is necessary for visual observation of the same sample. For example, Figure 2.16 shows the photobleaching of yellow fluorescent protein (YFP) in a living cell observed by confocal microscopy. In this specimen, as is often the case, the error caused by photobleaching is small for a single image, or even a moderately large 3D stack of images. Long before photobleaching makes the intensity measurement inaccurate, however, phototoxicity will have made the experiment irrelevant (refer to Chapter 17 of Goldman and Spector 2005). To keep cell damage to a minimum, careful attention must be paid to optimizing the microscopy. The goal is to extract as much information as possible from the limited number of photons that the cell will tolerate before phototoxicity becomes unacceptable.

When the sample is not too thick, the much larger quantum efficiency of CCD cameras compared to photomultipliers gives disk-scanning confocals, wide-field deconvolution, and structured illumination methods an important advantage over point-scanning confocal microscopes. For very thick, living specimens for which photobleaching is a serious problem, multiphoton illumination may provide some sample-dependent improvement (Patterson and Piston 2000; see Chapter 3).

Deconvolution of Confocal Images

Although the confocal pinhole is said to remove out-of-focus light, this removal is never perfect, and confocal images always show residual effects of the ripples in the 3D PSF caused by defocus. An example is shown in Figure 2.17, a confocal reflectance image of the same diatom imaged by wide-field microscopy in Figure 2.4. Notice that the phase reversals are seen in the confocal image; the holes change from black to white across the curved diatom surface. This result suggests that confocal images might benefit from deconvolving the 3D PSF, and indeed such a benefit has been reported (Shaw and Rawlins 1991; Cox and Sheppard 1995; van der Voort and Strasters 1995; Verveer et al. 1999; Boutet de Monvel et al. 2001). However, there are several considerations that might make investigators wary of the results of deconvolution, and several more reasons why they might conclude it is not worth the effort.

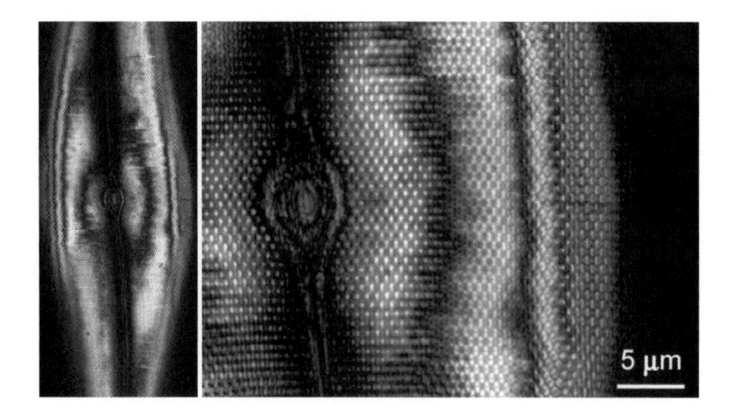

FIGURE 2.17

Contrast reversal because of defocus in a confocal image. (*Left*) A reflectance image of a diatom taken with a confocal microscope (Olympus Fluoview 300) using a 60x, 1.4 NA lens. (*Right*) An enlargement of a small portion of the image on the left, showing contrast reversals because of changing amounts of defocus. Compare with Fig. 2.4.

The first challenge is to determine the correct 3D PSF for a confocal microscope. In theory, it can be computed (Wilson and Sheppard 1984), but in practice, the computation is inaccurate, as it is for conventional wide-field microscopy with high-NA objectives. Can one then experimentally determine a 3D PSF, as is routinely done for wide-field microscopes? Unfortunately, the images from confocal microscopes often suffer from "patterned noise" artifacts (Fig. 2.18). Some of the patterns repeat with a periodicity that is a significant fraction of the entire field of view, which means that they are not represented in the images of tiny beads that are used for PSF measurements. Effectively, the PSF varies across the field of view, contrary to the assumptions of typical deconvolution algorithms. A second concern is the poor signal-to-noise ratio of confocal images (Figs. 2.18 and 2.19). This makes it quite difficult to measure the (local) 3D PSF to the accuracy required for reliable deconvolution and greatly exacerbates the tendency of deconvolution to amplify noise in the raw data. These concerns apply even to images of thin specimens. If a specimen is thick enough that a confocal microscope must be used (instead of using wide-field plus deconvolution), then the 3D PSF is certain to be seriously degraded by spherical and chromatic aberration (see Fig. 2.23), and this distortion will change dramatically between the top and bottom of the 3D data stack. In this situation, deconvolution is unlikely to give a correct result.

Finally, it could reasonably be argued that deconvolution comes too late to correct the most important defects of confocal images. The great benefit of deconvolving wide-field images is that the signal-to-noise ratio is enhanced because at least some of the out-of-focus light can be restored to the in-focus plane, thus increasing its total information content. In a confocal image, however, virtually all of the out-of-focus light is blocked by the pinhole before reaching the detector, and thus cannot be retrieved by deconvolution. This removes much of the motivation for using deconvolution methods with confocal images.

Practical Aspects and Tips for Generating Reliable Images

The currently available confocal microscopes are rather delicate, unstable instruments. Typically, they are controlled by complex computer programs that are prone to unexplained crashes and failed operations or missing functions. For these reasons and more, acquiring high-quality confocal images that are a faithful representation of the sample is a slow, often frustrating process. Even with a complete novice at the controls, an image of some kind will usually appear on the screen,

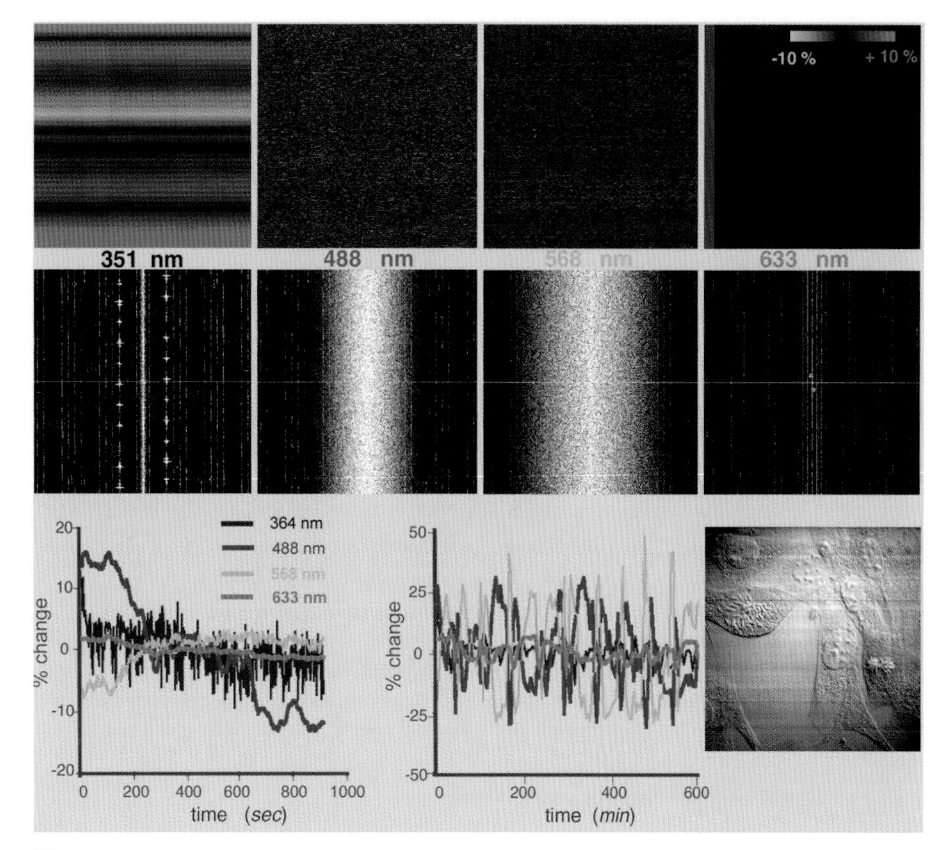

FIGURE 2.18

Fluctuations in illumination intensity and polarization in a confocal microscope (Zeiss LSM510). Mean-corrected images of a uniform specimen (*top row*) and their power spectra (*middle row*) acquired with four different laser lines. The images would ideally be of uniform intensity, thus entirely black after subtraction of the mean value, but instead show artifactual intensity fluctuations. Deviations from the mean value of up to ±10% are color-coded according to the scale in the top right corner. The power spectra show that each laser contributes patterned noise with a complex mixture of periodicities. Contrast of the power spectra has been enhanced to make the weaker features visible in the print. Some of the patterns have repeat lengths that are much longer than a single scan line. (*Bottom row*) The average intensity in a 100 x 100-pixel image of a uniform, stable fluorescent sample is plotted for a series of 1000 images acquired at 1-sec intervals (*left*) or 600 images at 1-min intervals (*middle*). The effect of these artifacts is to severely degrade the image signal-to-noise ratio (see Fig. 2.19). The horizontal stripes in the DIC image at the lower right are caused by random changes in the plane of polarization that accompany the fluctuations in intensity of the illumination.

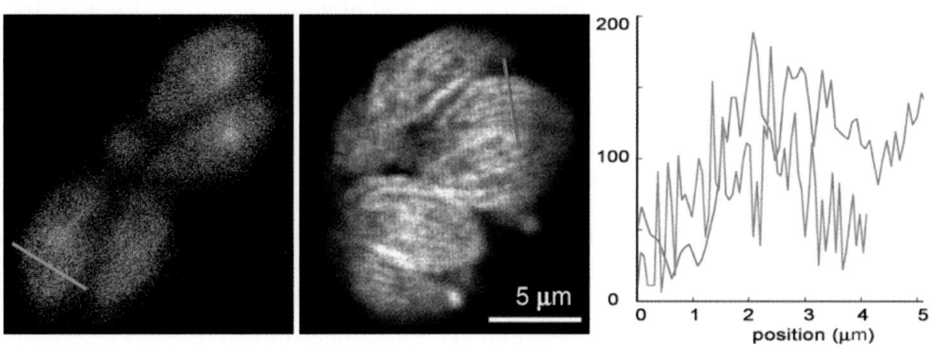

FIGURE 2.19

Confocal images of transgenic *T. gondii* expressing a YFP–α-tubulin fusion protein. Microtubules near the cell surface are included in these single optical sections. (*Left*) A typical confocal image; (*middle*) a superior image (Olympus confocal, much less intensity fluctuation), the best ever recorded from this specimen. Compare the signal-to-noise ratio in this image with the wide-field image of Fig. 2.7. (*Right*) The intensity profile along the red and blue lines in the images.

but distinguishing image from garbage takes time and considerable care regardless of how much experience the operator has.

Below are listed some guidelines that may help in adjusting the microscope parameters to obtain interpretable confocal images (see also the Troubleshooting Guide). For the preliminary adjustment of imaging parameters, choose an area of the specimen that is roughly equivalent to the area that will be recorded but is not the best area. The chosen area will be rendered unusable during the setup phase.

1 Choose the appropriate combination of laser, dichromatic mirror (beam splitter), and emission filter (see the subsection on Simultaneous Imaging of Multiple Labels).

2 Decide what pixel spacing is appropriate for collecting the information needed from this particular sample, and set the magnification or electronic zoom factor accordingly.

   Do not oversample (the pixel spacing should be only slightly smaller than the Nyquist criterion; i.e., slightly less than one half of the spatial resolution required for the experiment). Use the highest-NA objective available. For thick samples mounted in aqueous media, a water immersion objective with correction collar set to minimize spherical aberration is the best choice (see the section on Interpreting the Results). For thinner samples, an oil immersion lens may be acceptable, but the refractive index of the immersion oil must be carefully selected to minimize the spherical aberration for each specimen.

3 Estimate the imaging parameters.

   a Set the pinhole initially to ~1 Airy disk diameter.

   b Set the laser to the minimum power that gives a decent signal at maximum gain on the detector.

4 Find the linear range of the detector system. Use a pseudocolor lookup table (LUT) that highlights underflow (intensity = 0) and overflow (intensity = 255 for 8-bit or 4096 for 12-bit detection) in color, but is gray scale in between the two.

   a With the laser off or set to zero power, scan at the speed to be used for the specimen, and adjust the "offset" (dark current compensator) so that the recorded image intensity is minimized but there are no pixels at zero intensity.

   b Find a region of the specimen that is likely to be the brightest. With the laser on, decrease the detector gain until the recorded intensity in the brightest region of the image is safely below the saturation value (e.g., 200 of a maximum 255 for an 8-bit system).

5 Find the linear range for the specimen. Check that the recorded fluorescence emission increases linearly with an increase in laser power up to at least twice the laser power that will be used for imaging.

   If the emission does not increase in proportion to laser power (i.e., ground state depletion is occurring), temporal resolution (work with lower laser power and longer scan times), spatial resolution (work with lower laser power and increased pinhole diameter or pixel size), or both must be sacrificed.

6 Verify that settings are below the instantaneous damage threshold. With the scan speed and laser power set at the preliminary values determined in Steps 2–5, monitor the image intensity in a small area of the specimen over the course of numerous repeated scans.

   One would like to be able to scan dozens of times before the cumulative photobleaching reaches 50%. If the fluorescence is bleaching too much, sacrifice either temporal resolution (use longer intervals between scans in a time-lapse series), spatial resolution (increase the pinhole diameter and spacing between optical sec-

tions; increase the xy *pixel size), or both. If the fluorescence is not bleaching measurably, the experiment is going to be quick, or photodamage is not a concern, then decrease the photomultiplier gain (which will decrease the noise), decrease the pinhole size (which may improve contrast and resolution), and increase the laser power to maintain maximum intensities just below saturation.*

7 Iteratively readjust the parameters according to Steps 4–6 until the image signal-to-noise ratio is optimized.

TROUBLESHOOTING GUIDE

This guide is primarily for confocal microscopes, but the same general principles apply to all methods. For troubleshooting deconvolution, see McNally et al. (1994, 1999) and particularly Wallace et al. (2001).

When poor images are obtained, the first question to be answered is whether the problem is with the specimen or with the equipment. An enormous amount of frustration and wasted time can be avoided if standard samples are available that can be used to compare the system performance at the moment with its performance in the past (i.e., on a day when good images were obtained). Four simple specimens are useful for this purpose: a resolution test target such as used for Figure 2.20; an optically flat mirror or bare glass slide; small beads, 0.2-0.5 μm in diameter and labeled with multiple fluorophores (e.g., Molecular Probes TetraSpeck) that are excited by all of the laser lines and detected through all of the filter sets on the instrument; and a solution of fluorescent dyes covering similarly broad excitation and emission spectra (e.g., a mixture of DAPI, Alexa Fluor 488, Alexa Fluor 594, and Alexa Fluor 633). The beads should be spread into a film on a coverslip, allowed to dry, and mounted on a thin layer of antifade solution or optical cement (e.g., Epo-Tek #301). To make the fourth standard specimen, a generous layer of the solution of fluorescent dyes should be sealed under a coverslip.

On a day when the equipment seems to be in good working order, collect and store a 2D image of the resolution test target at optimum focus, a 3D stack of images of the fluorescent beads, and an image of the fluorescent dye solution a few micrometers underneath the coverslip. Use the smallest pinhole. Collect similar images for all laser lines and all filter/detector channel combinations. Also collect an "*X-Z*" scan of the beads as in Figure 2.22. For these measurements, place any gain, sensitivity, background, dark level, or other adjustments in manual mode. Experiment to find settings of these parameters that give a zero intensity reading in the absence of illumination and a peak intensity reading that is just below saturation for the in-focus illuminated beads (see above section on Practical Aspects and Tips for Generating Reliable Images). Carefully record these settings along with the objective lens used and the parameters relevant to illumination and signal intensity (laser tube current, neutral density filters, acousto-optic tunable filter [AOTF] settings, beam splitters, pixel spacing, pixel integration time, etc.). These measurements serve as a calibration that can be repeated later when the performance of the system becomes questionable. Some common problems affecting confocal microscope systems are listed below.

Symptom: Image intensity is decreased over the whole field of view (at all magnifications).
Likely causes: If the image is at first bright but then gets dimmer, the problem is either photobleaching or drift of the focus level. Check for focus drift by collecting a reflectance image of a mirror or bare glass surface using the smallest possible pinhole. (This image is very sensitive to focus level.) If the image is always dim, then there is probably a misalignment of the confocal pinhole or of the internal mirrors of the laser. If the mirrors of the laser are misaligned, some wavelengths will be affected more strongly than others. For instance, the argon/krypton and krypton lasers are very prone to loss of their yellow (568-nm) and red (647-nm) lines while retaining the blue (488- or 476-nm) line.

Symptom: Image intensity is decreased over the periphery of the field of view (more pronounced at lower magnification).
Likely causes: If the effect is seen in both fluorescence and reflectance images, then it is caused by either a misalignment (most systems) or an intrinsic design problem (older Bio-Rad systems). If the effect is much more pronounced in fluorescent images than in reflectance images, chromatic aberration is indicated.

Symptom: Resolution is poor.
Likely causes: If the problem is apparent in both thin (e.g., fluorescent beads) and thick samples, then the fault is probably in the alignment of the confocal optics (but first make sure that the objective lens is clean!). Verify that the laser beam is correctly centered on the axis of the objective lens and that the entire back aperture of the lens is filled with incoming light. If the problem is restricted to thick samples, then spherical aberration is probably the culprit. The newer water immersion objectives with a correction collar greatly ameliorate this problem for thick samples in aqueous media, but with the drawback of slightly decreased resolution for optically ideal specimens (i.e., very thin samples with a refractive index the same as glass and positioned immediately adjacent to the coverslip).

Symptom: Focusing for maximum brightness does not give the sharpest image. The image can be made bright or sharp, but not both.
Likely causes: The system is not *con*focal. The focal plane for the illumination system does not coincide with the focal plane for the imaging system. For visible wavelength illumination and imaging, the pinhole or an intermediate lens that focuses the light on the pinhole is probably misaligned. If the illumination or imaging wavelengths are ultraviolet or far red, then a misaligned collimator lens is the likely culprit.

Symptom: Alternating stripes of higher and lower intensities appear in the image.
Likely causes: Mechanical vibration, defects on the scanning mirrors, or an electronic oscillation in the laser or the detector circuits are the likely causes. To decide among these possibilities, collect an image of the fluorescent dye test sample with the largest available pinhole aperture. This image is very insensitive to vibration but is still sensitive to electronic oscillations and mirror defects. Mirror defects cause a fixed pattern of bright–dark stripes that does not change between images. Most electronic oscillations (and mechanical vibrations) give a different pattern with each image. Confocal systems from some manufacturers use single-mode polarization-preserving fiber-optic coupling of the laser to the scan head. Some of these systems are very prone to fluctuating illumination intensity and polarization angle. Painstaking rotational alignment of the fiber polarization axis with the laser polarization axis mitigates the effect, but within a few days, the fiber drifts out of alignment and the stripes reappear.

Symptom: A circular bright spot or a set of rings appears at a fixed point in every image.
Likely cause: A reflection of the laser beam off an internal glass surface is being detected. If fluorescence images are being collected, then an inappropriate set of filters is being used (reflected laser light is getting through). Most systems now include a quarter-wave plate and polarizer combination to minimize the problem in reflectance mode imaging. One of these elements probably has been rotated.

Symptom: Images from different fluorophores are misregistered.
Likely cause: Displacement of a very-short-wavelength (e.g., DAPI) channel or very-long-wavelength (e.g., CY5 or Alexa 633) channel from the middle-wavelength channels is usually due to misalignment of a ultraviolet/visible or infrared/visible collimator lens (a necessary component of the illumination path that corrects for the small residual chromatic aberration present in all currently

available objective lenses). Be aware that some manufacturers' service personnel will not check collimator lens alignment unless specifically asked, so a recent inspection by unsupervised service personnel provides no assurance that the system is correctly aligned.

STRUCTURED ILLUMINATION METHODS

The goal of these techniques is to improve the images of thick objects by a combination of optical and computational manipulations. By a wonderfully simple manipulation, the blurring caused by defocus can be turned into an effective tool for separating in-focus from out-of-focus light, when the light is right (i.e., structured).

Investigators normally strive to achieve completely uniform illumination (i.e., completely "unstructured") across the entire field of view, so that variations in intensity across the image arise solely from variations in the structure of the object. Contrary to what one might expect, superimposing artifactual intensity fluctuations across the image by using carefully patterned nonuniform illumination can actually *increase* the amount of information about the object that is stored in the image.

Development of new techniques for using structured illumination to enhance microscope performance is proceeding rapidly and in many different directions (Bailey et al. 1993; Neil et al. 1997, 1998, 2000; Wilson et al. 1998; Gustafsson et al. 1999; Hanley et al. 1999, 2000; Gustafsson 2000; Cole et al. 2001; Heintzmann et al. 2001; Dubois et al. 2002; for an excellent survey, see Gustafsson 1999). At this time, most of the techniques have been demonstrated only on very thin specimens and are available only in a few specialized laboratories. In this chapter, we describe one method (Neil et al. 1997, 2000) that does work quite well with thick specimens and is now commercially available from at least two vendors.

Optical Principles

The basic principle is shown in Figure 2.20, which illustrates in a different way the same information contained in Figures 2.1– 2.3. Figure 2.20 shows an image of a resolution test target, a series of gratings of different spacing, which was tilted by ~1.5° before being photographed. The tilt, which is from right to left in the image, displaces the sample from the focal plane by an amount that increases, in opposite directions, away from the center of the field. Notice that the 0.33-µm grating

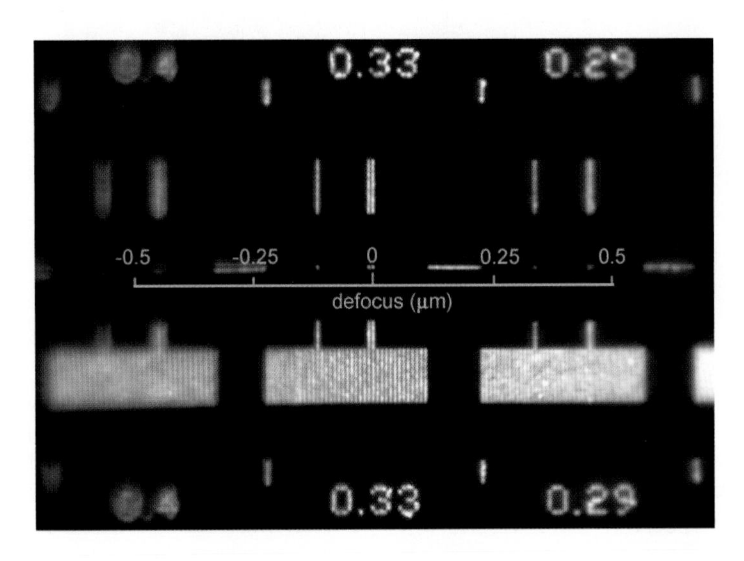

FIGURE 2.20

Image of a resolution test target that was tilted by ~1.5° on the microscope stage. The spacing of the gratings in micrometers is marked on the specimen. Because of the tilt, the defocus varies across the image as indicated by the orange scale. The specimen-to-lens distance increases from right to left. Note that the black and white bars of the 0.29-µm grating are fairly distinct at the left end but smeared to an average gray on the right. 60X, 1.4 NA objective, 546 nm illumination.

is sharp and well resolved in the center, but becomes quite blurred toward the left. The left side of the 0.29-µm grating is just resolved at a defocus of 0.25 µm, but the right side is completely smeared out when the defocus increases to 0.5 µm. How can this "problem" be turned into a "solution?"

Imagine the use of a grid of evenly spaced lines as a mask through which the specimen is illuminated (Fig. 2.21). The grid is placed in the light path of the microscope, with appropriate lenses to bring its shadow on the specimen into sharp focus in exactly the focal plane of the objective lens. If the specimen is thin, the result is simply an image of the in-focus specimen crossed by a set of sharply demarcated shadows where the illumination has been interrupted by the mask. If the specimen is thick, then superimposed on this in-focus image will be a blurred image of the other planes of the specimen (see Figs. 2.5 and 2.21). However, the contribution from the out-of-focus planes of the specimen will *not* be modulated by the lines of the grating; away from the in-focus plane, the shadow of the grating quickly becomes smeared and the illumination is uniform, at a level that is the average of the light and dark bars. Note in Figure 2.20 that a defocus of 0.5 µm is sufficient to completely eliminate the 0.29-µm modulation.

A simple algebraic combination of three images of the object illuminated through a grating shifted by exactly one third of its period between each image yields an in-focus image uncontaminated by out-of-focus blur (Neil et al. 1997): $I_{\text{in-focus}} = [(I_1 - I_2)^2 + (I_1 - I_3)^2 + (I_2 - I_3)^2]^{1/2}$. The simple sum of the three raw images $(I_1 + I_2 + I_3)$ is identical to the normal wide-field image with no grating. The best optical sectioning is obtained when the period of the grating is slightly larger than the diameter of the Airy disk (Neil et al. 1997).

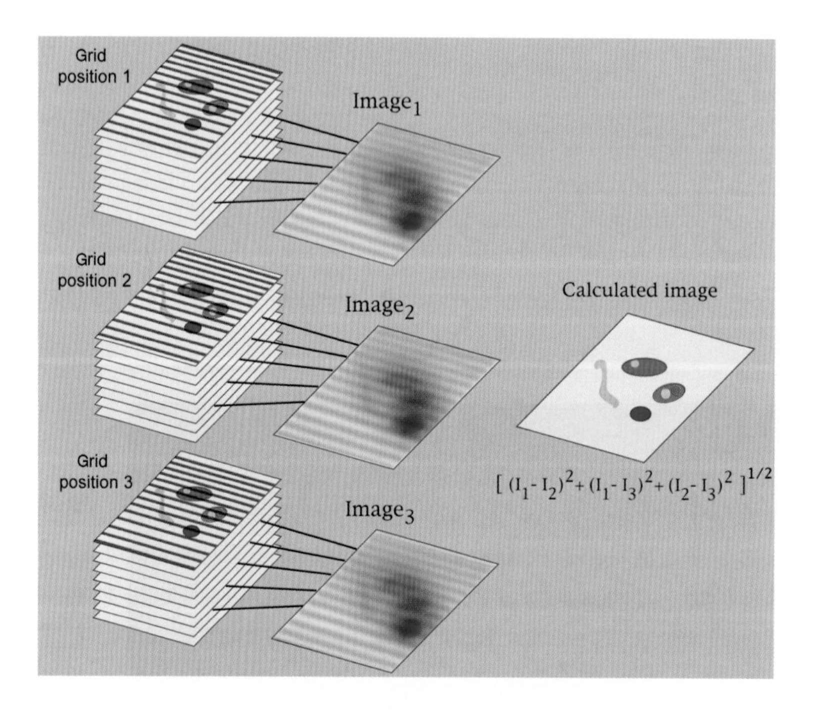

FIGURE 2.21

Structured illumination employed for optical sectioning. The light source is masked by a regular grid, which casts a pattern of stripes on the in-focus object plane. Three images of the thick object are acquired, shifting the grid by 1/3 repeat between each acquisition. Each image is the sum of contributions from the in-focus plane, which is shadowed by sharply defined stripes, plus blurred out-of-focus planes without distinct stripes. The square root of the sum of the squared difference images removes the out-of-focus blurred light, giving the in-focus plane.

The grating can be shifted rapidly and precisely (e.g., by a piezoelectric element), and the calculation is very fast. The rate-limiting step for optical sectioning by this method is thus the acquisition of the three images. Compared to wide-field plus deconvolution, the maximum acquisition rate would be threefold slower, but the finished product, the in-focus optical sections, is available immediately instead of after 15–20 minutes as for the deconvolution computation. The total exposure is slightly greater than for wide-field plus deconvolution, because the bars of the grating are typically not completely opaque. The *z*-axis resolution is comparable to what is achieved by either confocal or deconvolution methods.

The first commercial versions of this methodology have only recently been introduced (http://www.thales-optem.com/optigrid, http://www.zeiss.de/us/micro/home.nsf), so very few investigators have experience with it and consequently its limitations are not yet fully known. One would expect that the computation would fail for extremely thick samples, when the background out-of-focus light becomes overwhelming. In that case, the contrast of the grid lines in the raw data will be very low, and the difference between I_1, I_2, and I_3 would become comparable to the noise level. For a discussion of some of the technical problems that need to be addressed to make the technique useful in practice, see Cole et al. (2001). The simplicity of the hardware component means that it is inexpensive and can be user-installed as an add-on to virtually any modern microscope. It is safe to predict that this and other variants of structured illumination microscopy will quickly supplant expensive confocal microscopes for many applications on modestly thick specimens, particularly living samples.

INTERPRETING THE RESULTS

The Meaning of "Optical Section"

Although the images produced by confocal microscopy, deconvolution, and structured illumination methods are referred to as optical "sections," they differ from true sections in that their top and bottom edges are not sharply defined. In a physical section that has been cut by a knife in a microtome, there is no ambiguity about which section contains each point of the original object, at least not at the resolution of the light microscope. A specified point in the cell was either included in one particular microtome section or it was not, but there is no intermediate state. An optical section, however, includes some locations fully (i.e., present at their true intensity), and other locations above and below at less than their true intensity. There is no sharp cutoff that demarcates what is included in the optical section and what is excluded. Instead, there is a continuous decrease in the ratio of image intensity to object intensity for locations further and further away from the midpoint (i.e., the CTF falls off steadily with distance from the focal plane; see Figs. 2.2 and 2.3).

A measurement that is commonly used as the analog of section "thickness" for optical sections is the width at half-amplitude (full-width at half-maximum, FWHM) of the curve that describes the relative intensity of points at different distances from the midpoint of the section. This curve can be measured by collecting a series of closely spaced optical sections (0.1 μm for the highest-NA lenses) of a small, bright object. A small fluorescent bead is a good specimen for this measurement, but any bright object that is small compared to the expected FWHM can be used. A plot of the total intensity of the image of the object in each optical section should give a curve similar to those in Figure 2.22, the axial intensity profile of the 3D PSF for the optical system (equivalent to a vertical line through the center of Fig. 2.3). Confocal microscopy, deconvolved wide-field images, or structured illumination microscopy should give a PSF with an axial FWHM $\leq$ 0.6 μm for the highest-NA objective lenses. Much smaller values of FWHM have been obtained, but not with samples and equipment that are realistic for use in live cell imaging experiments (Bailey et al. 1993; Hell et al. 1997; Gustafsson et al. 1999).

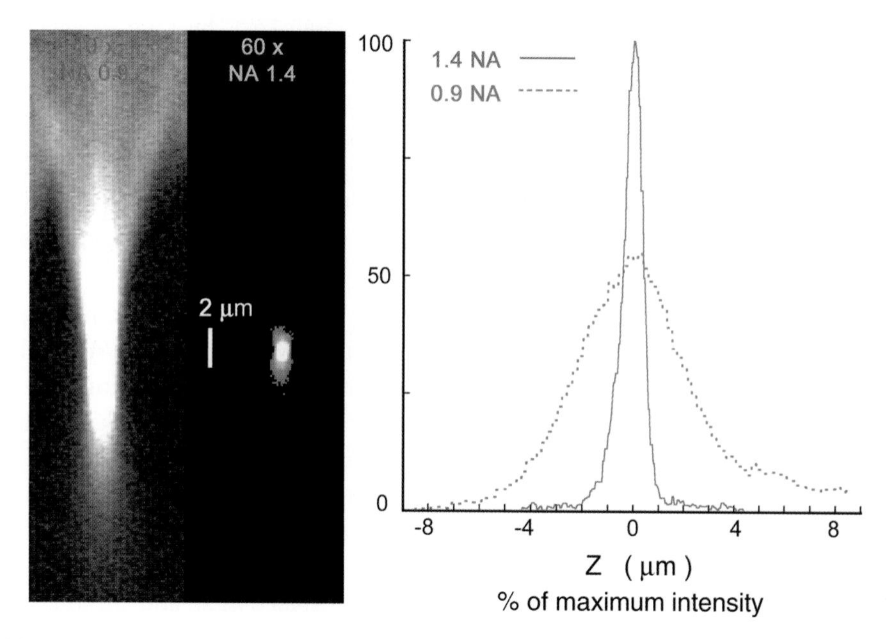

FIGURE 2.22

X-Z scans of 0.9-µm-diameter fluorescent beads and measured axial intensity profiles. The left side of the image shows a scan using a 40x, 0.9 NA objective lens. The right half is a scan of a different sample using a 60x, 1.4 NA lens. Intensity for the two images has been adjusted to the same maximum value; in reality, the 40x, 0.9 NA image is much dimmer. The round bead appears much more elongated with the lower-NA objective for two reasons. Decreasing the NA affects vertical resolution much more severely than lateral resolution. The left scan is also markedly asymmetrical above and below focus, indicative of spherical aberration, which causes further elongation. The right scan (of a different sample) also shows a small amount of spherical aberration. Bar, 2 µm. The graph shows the measured axial intensity profiles for these two situations. In these curves, spherical aberration is manifest as asymmetry of the profile to the right and left of the peak (e.g., 40x curve beyond 5 µm).

Spherical Aberration

The FWHM of the vertical PSF decreases with the square of the NA of the objective lens. For confocal systems, the width also decreases with decreasing pinhole size. For all microscopes, the vertical PSF is very sensitive to the presence of spherical aberration (Fig. 2.22). Unfortunately, a certain amount of spherical aberration must often be accepted when examining living specimens (Fig. 2.23). The highest-NA oil immersion objectives are designed for work with a specimen that is located immediately beneath a coverslip connected to the lens by immersion oil. Images of thick specimens, for which these conditions are not uniformly possible, are increasingly degraded by spherical aberration as the focal plane is lowered. The problem is greatly ameliorated by using the newer, long-working-distance water immersion objectives. These objectives are designed to be used *with* a coverslip, in contrast to older designs (e.g., those used by electrophysiologists for patch-clamp studies). They are very expensive, but with thick specimens immersed in water, they perform better than standard, very-short-working-distance, high-NA oil immersion objectives. The PSF of these new water immersion lenses is not yet as good as with the best high-NA oil objectives, so their performance in deconvolution is limited.

As illustrated in Figure 2.22, spherical aberration leads to an asymmetrical response to defocus; the image looks different when defocused by the same amount in opposite directions from the in-focus plane. When present, significant spherical aberration is readily visible. A convenient way to check for it is to find a very small, very bright "dot" of fluorescence and observe its appearance as the lens is defocused by a small amount (a few micrometers) in either direction. A typical observa-

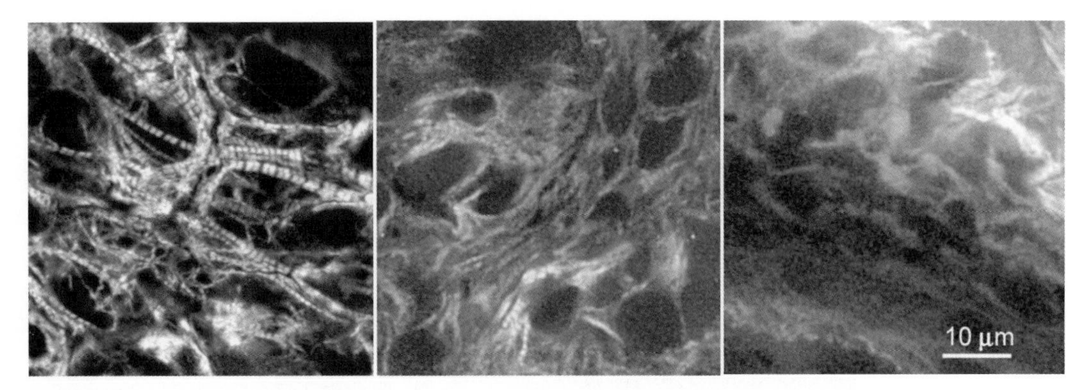

FIGURE 2.23

Confocal optical sections of thick tissue. The developing heart in a chick embryo, labeled with fluorescent antibody against cardiac myosin, was imaged with a 60x, 1.4 NA oil immersion lens. (*Left*) The first optical section of a 3D stack of 160 sections spaced at 0.5-μm increments. The first section was ~20 μm below the coverslip. (*Middle*) Section 30; (*right*) section 130. Spherical aberration increasingly degrades the resolution, so that the 2.2-μm myofibrillar striations, clearly visible in section 1, are barely resolved in section 130. (Sample kindly provided by Howard Holtzer, University of Pennsylvania.)

tion is the appearance of bright rings on one side of focus and general fuzziness without rings on the other side. With the water immersion lenses, this asymmetry can often be eliminated over an extended range of focal planes in a thick specimen by careful adjustment of the correction collar. With oil immersion lenses, one can choose an immersion oil with a refractive index that minimizes the spherical aberration at the depth in the specimen where the major interesting features lie, but optical performance in other focal planes will be degraded (Fig. 2.23).

Chromatic Aberration

With modern, highly corrected objective lenses and an ideal sample, all wavelengths of light in the visible range should be focused to the same point with an accuracy of better than 0.3 μm (Keller 1995). However, if the mixture of refractive indices in the sample deviates from the design parameters of the objective lens enough to cause noticeable spherical aberration, then chromatic aberration is also likely to be induced. The lens design is specific not only for a particular arrangement of refractive indices, but also for the way in which those refractive indices vary with wavelength (dispersion). Sample dispersion often does not match design specifications, and the result of this mismatch is a shift in focal position according to wavelength. Figure 2.24 shows this behavior in a series of optical sections of a single bead. In this case, the chromatic aberration is caused not by the sample, but by misalignment of collimator lenses that are supposed to correct the confocal system for residual chromatic aberration in the objective. This correction allows the system to be used with UV, infrared, and visible light (also see Chapter 1).

As the composite image at the top of Figure 2.24 shows, chromatic aberration is a serious problem in determining colocalization of different fluorescent molecules. In an extended sample, complicated artifactual shifts are observed because the direction and amount of apparent shift between different colors varies with position in the field of view. Interpretation of slight differences in localization demands careful control experiments to rule out chromatic aberration.

Chromatic aberration is a particular concern with confocal microscopes (Hell and Stelzer 1995; Keller 1995). If the focal spot for the illumination wavelength is not in the same position as the focal spot seen through the pinhole aperture, then the system is not *con*focal. In this situation, image intensity is severely decreased.

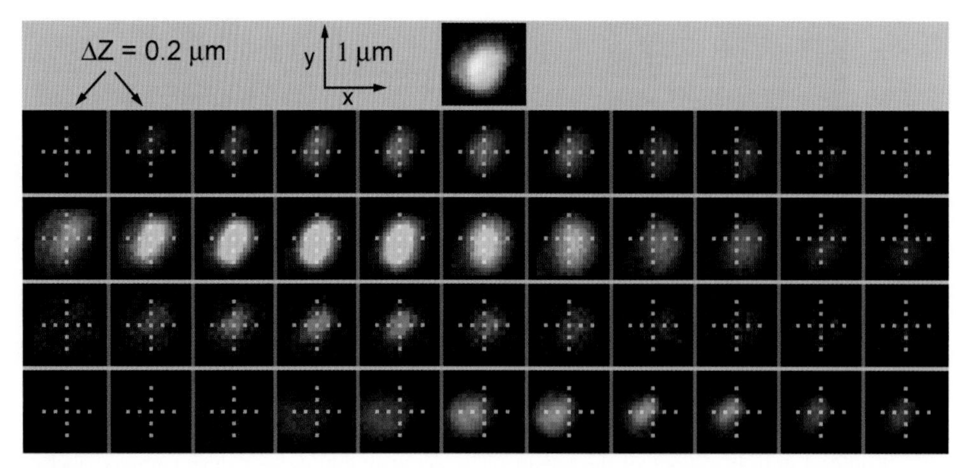

FIGURE 2.24

Chromatic aberration in confocal images. Eleven optical sections, spaced at 0.2-μm intervals, of a single, 0.2-μm-diameter bead labeled with four different fluorophores. The dotted cross in each section marks the *x-y* location of the center of the bead in the green image at best focus. Note that the other three wavelength bands are shifted in *x, y,* and *z* with respect to the green. The yellow and blue images also reveal some spherical aberration. A composite color image for the sixth optical section is shown at the top. (*Red*) Excitation (ex) 633 nm, emission (em) > 650 nm; (*yellow*) ex 568 nm, em 585–615 nm; (*green*) ex 488 nm, em 505–600 nm; (*blue*) ex 364 nm, em 385–470 nm. 63x, 1.2 NA water immersion lens.

Signal-to-Noise Ratio

Live cell imaging is considerably more demanding than observation of fixed samples. The motivation to invest the extra time and effort is often the need to observe *changes* in the distribution or amount of some fluorescent molecule. To be confident that an observed change is real, rather than the result of random fluctuations in the image intensity, the investigator must consider the signal-to-noise ratio (S/N) in the image.

Suppose that an experiment requires detecting changes of 25% or more in the local concentration of a fluorescent molecule. As an example, consider the cells shown in Figure 2.7. These cells will be used to follow the incorporation or removal of YFP-tubulin at the growing and shrinking ends of microtubules. Because this specimen is thin enough that the background from out-of-focus light is not overwhelming, it is appropriate for any of the microscopic techniques described in this chapter.

The S/N must be large enough to be confident that a 25% change in fluorescence at the end of a microtubule is not caused by random noise. To draw this conclusion with 95% confidence, the expected fractional change in signal (i.e., $0.25S$) must be greater than twice the standard deviation (i.e., greater than twice the noise, $2N$). Thus, S/N must be at least 8 for this measurement to succeed. If the required precision is expressed as a fractional change P (i.e., 25% precision in the measurement of S means $P = 0.25$), then for 95% confidence, S/N must be greater than $2/P$.

To decide if this experiment is feasible, a rough estimate of the signal and its standard deviation (s) is needed. First, an area of a typical sample that is similar to the region of interest must be located. Two images must be collected in rapid succession from this area without changing anything between exposures. Photobleaching must be avoided. In the absence of noise, these two images should be identical, so the difference between them can be used to estimate the noise level in a typical measurement. In fact, the standard deviation of this difference image is $\sqrt{2}$ times the standard deviation of a single pixel in the original images. The noise in the measurement of microtubule fluorescence will be larger than s because the background must be subtracted. If the ratio of background to total intensity is b, then the standard deviation in one pixel of the background-corrected microtubule fluorescence image will be a factor larger than the standard deviation in the raw intensity measurement. In Figure 2.7, $b \approx 0.95$ before deconvolution and 0.8 after deconvolution. Finally, the size of the target area must be taken into account. If the target includes n pixels and the average fluorescence per pixel of microtubule after background correction is denoted by F, then

$$S/N \cong \frac{F}{s\sqrt{(1 + b)/2n}}$$

For the experiment in Figure 2.7, F was ~50 (of 4096 maximum for the 12-bit CCD) averaged over a 5 x 5-pixel box, s was ~12, and b was 0.95, so $S/N \approx 21$. It can be concluded that the images collected under these conditions are indeed good enough to detect the hoped-for change in YFP-tubulin incorporation.

Now suppose one contemplated doing this experiment with a laser-scanning confocal microscope instead of a wide-field plus deconvolution. Doing this simple calculation beforehand would save a lot of time and frustration, because one would find that the experiment cannot be done with an ordinary confocal! Typical numbers (Zeiss LSM510) are roughly $F = 40$ (of 255 maximum for 8-bit detection), $s = 30$, and $b = 0.5$, which for the same size target give an $S/N \approx 4$ (compare Figs. 2.7 and 2.19).

Point-scanning confocal microscope images are much noisier than wide-field images for several reasons. Photomultipliers, the usual detectors on point-scanning confocal systems, are about fourfold less efficient than a good CCD camera (quantum efficiency [QE] ~15% vs. ~60%). Thus, for the same exposure (photobleaching, phototoxicity), the wide-field image would be formed from four times as many photons and have twice the S/N as the confocal image. In addition, many confocal microscopes add a large amount of unnecessary noise to the image, generated by electronic artifacts in the detector circuitry (Fig. 2.18) and random fluctuations in the illumination intensity. Figure 2.25 shows a direct comparison of the noise in the illumination of a laser-scanning con-

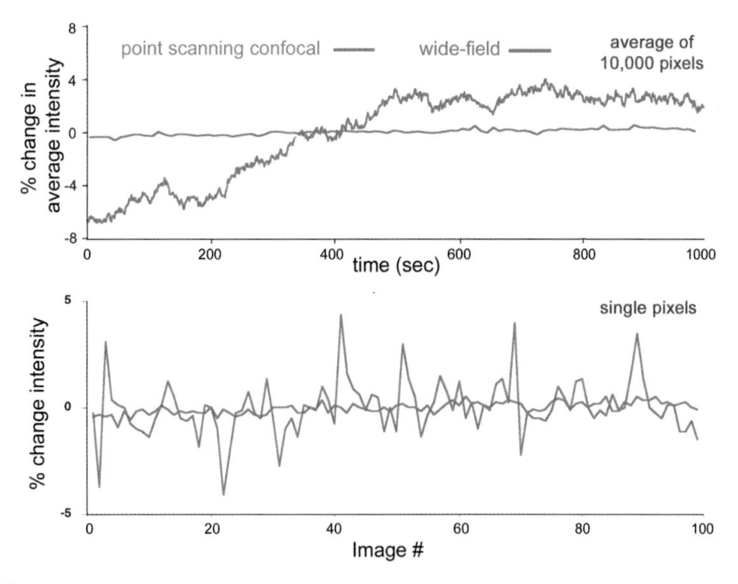

FIGURE 2.25

Fluctuations in illumination intensity in confocal versus wide-field microscope. (*Top*) The average intensity in a 100 x 100-pixel image of a uniform, stable, fluorescent sample is plotted for a series of 1000 images acquired at 1-sec intervals, for a point-scanning confocal and a wide-field microscope with CCD detector. (*Bottom*) The intensity from a single pixel in a sequence of 100 successive images of the same specimen is plotted.

focal and a wide-field microscope. It should be emphasized that this extra noise in the confocal is entirely unnecessary and is simply a matter of poor design in most of the commercially available instruments.

Disk-scanning confocals, which use CCD detectors, should have an *S*/*N* at least twofold better than point-scanning instruments for the same exposure levels. The noise level in images from the structured illumination method described above will be increased by the processing steps needed to calculate the in-focus image, but model calculations suggest that this effect will be less than a factor of 2. Thus, the structured illumination technique also has the potential for giving an image *S*/*N* substantially higher than the commercially available point-scanning confocals.

Why Confocal?

After this comparison of signal-to-noise ratios with a thin sample, it is perhaps worth restating the message of the Introduction to this chapter: Each of the microscopic techniques described here has its own realm within which it reigns unchallenged, and outside of which it must take second place to other approaches. When specimen thickness is modest, the *S*/*N* of the raw data acquired with a wide-field microscope is high enough to allow excellent contrast enhancement by restoration methods such as deconvolution or structured illumination. However, when the sample is very thick, out-of-focus background reduces the contrast, and only point-scanning confocal or multiphoton techniques are useful. As a counterpoint to the *S*/*N* comparison for a thin sample, Figure 2.26 shows a comparison between point-scanning confocal microscopy and wide-field microscopy plus deconvolution with a sample that is definitely in the confocal realm.

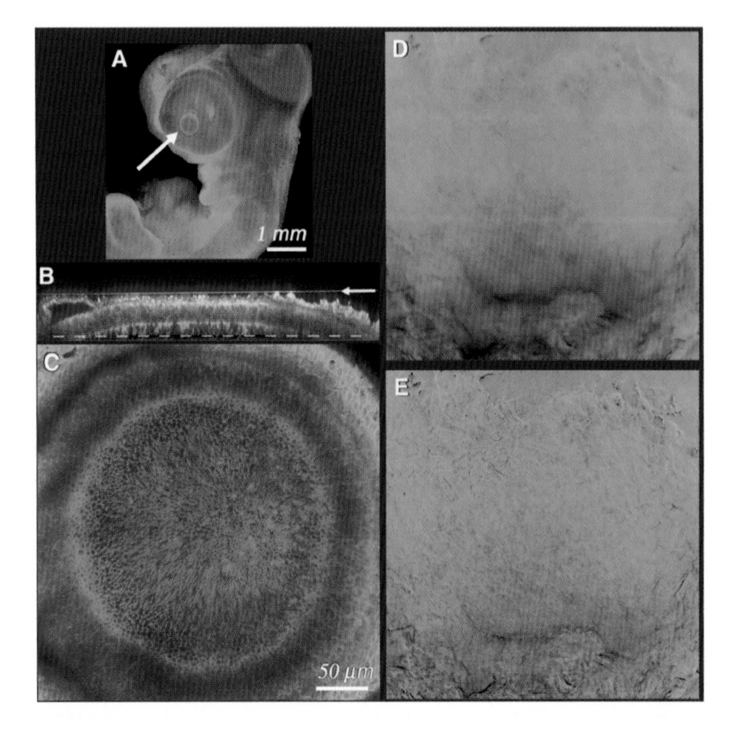

FIGURE 2.26

Comparison of laser-scanning confocal and wide-field microscope performance in imaging a thick tissue (Swedlow et al. 2002). A 5-day-old quail embryo stained with Alexa488-phalloidin (*green*) and DAPI (*blue*) was imaged by laser-scanning confocal (*A–C*) and wide-field microscopy (*D*). (*A*) A low-magnification survey image of the entire embryo. The head is at the top of the figure. The arrow points to the developing eye, the region shown at higher magnification in *B–E*. (*B*) "*X-Z*" section (parallel to optical axis) showing position of coverslip (*arrow*) and location of the focal plane shown in *C* (*dashed line*). (*C*) A single optical section ~50 µm below the exterior surface of the embryonic eye. Concentrations of actin at cell cortices are visible. (*D*) The same embryo imaged by wide-field microscopy and recorded with a CCD camera (Alexa488 only). The image shown is one of 60 recorded from a series of optical sections. (*E*) The same optical section as in *D* after restoration by deconvolution. No cellular details are visible in wide-field images of this thick tissue. Bars: (*A*) 1 mm; (*C*) 50 µm; (*B,D,E*) same scale as *C*. (Images *D* and *E*, courtesy of Jason Swedlow, University of Dundee.)

Three-Dimensional Reconstruction

The output from confocal microscopy, deconvolution, or structured illumination methods is typically a stack of optical sections, collected by changing the focus by a constant amount between each image acquisition. If the goal is to collect enough information to permit a 3D reconstruction of a specimen at the highest resolution possible, then adjacent optical sections must be spaced at increments of roughly half the FWHM of the vertical PSF. The resolution in the vertical direction will always be worse than in the horizontal direction, by ~3x at high NA. Thus, it is common in 3D reconstructions, particularly from confocal microscopes, for objects to appear elongated in the vertical direction (see Fig. 2.22). The software supplied for deconvolution often compensates for this effect, and a similar compensation (essentially a one-dimensional edge sharpening filter) is helpful on confocal images.

Display of 3D image stacks in a form that efficiently and faithfully transmits the data to the viewer is a challenging task. Simplified representations of 3D image data are essential for display purposes. Software abounds for "3D rendering" or various other forms of displaying 3D intensity distributions, and the output can be visually striking and quite persuasive. However, injudicious use of 3D rendering software can also produce misleading representations that exaggerate the contrast (hence the reliability of the segmentation between different regions of the image) and the resolution (rendering fuzzy intensity gradients as sharp boundaries). As always, caveat emptor!

ADDITIONAL READING

Davidson M. 2003. Molecular expressions: Images from the microscope http://micro.magnet.fsu.edu/index.html.

Inoué S. and Spring K.R. 1997. *Video microscopy: The fundamentals.* Plenum Press, New York.

Taylor C.A. 1978. *Images: A unified view of diffraction and image formation with all kinds of radiation.* Wykeham Publications, New York.

World Wide Web Virtual Library: Microscopy http://www.ou.edu/research/electron/www-vl.

REFERENCES

Agard D.A. 1984. Optical sectioning microscopy: Cellular architecture in three dimensions. *Annu. Rev. Biophys. Bioeng.* **13:** 191–219.

Agard D.A., Hiraoka Y., Shaw P., and Sedat J.W. 1989. Fluorescence microscopy in three dimensions. *Methods Cell Biol.* **30:** 353–377.

Amos W.B. and White J.G. 1995. Direct view confocal imaging systems using a slit aperture. In *Handbook of biological confocal microscopy* (ed. J.B. Pawley), pp. 403–415. Plenum Press, New York.Amos W.B., White J.G., and Fordham M. 1987. Use of confocal imaging in the study of biological structures. *Appl. Optics* **26:** 3239–3243.

Art J.J., Goodman M.B., and Schwartz E.A. 1991. Simultaneous fluorescent and transmission laser scanning confocal microscopy. *Biophys. J.* **59:** 155a.

Bailey B., Farkas D.L., Taylor D.L., and Lanni F. 1993. Enhancement of axial resolution in fluorescence microscopy by standing-wave excitation. *Nature* **366:** 44–48.

Born M. and Wolf E. 1999. *Principles of optics: Electromagnetic theory of propagation, interference and diffraction of light.* Cambridge University Press, London.

Boutet de Monvel J., Le Calvez S., and Ulfendahl M. 2001. Image restoration for confocal microscopy: Improving the limits of deconvolution, with application to the visualization of the mammalian hearing organ. *Biophys. J.* **80:** 2455–2470.

Boyde A., Petráň M., and Hadravsky M. 1983. Tandem scanning reflected light microscopy of internal features in whole bone and tooth samples. *J. Microsc.* **132:** 1–7.

Brakenhoff G.J., Blom P., and Barends P. 1979. Confocal scanning light-microscopy with high aperture immersion lenses. *J. Microsc.* **117:** 219–232.

Cagnet M. 1962. *Atlas optischer Erscheinungen (Atlas of optical phenomena).* Springer, Berlin.

Campbell R.E., Tour O., Palmer A.E., Steinbach P.A., Baird G.S., Zacharias D.A., and Tsien R.Y. 2002. A monomeric red fluorescent protein. *Proc. Natl. Acad. Sci.* **99:** 7877–7882.

Carlsson K., Danielsson P.E., Lenz R., Liljeborg A., Majlof L., and Aslund N. 1985. Three-dimensional microscopy using a confocal scanning laser microscope. *Optics Lett.* **10:** 53–55.

Carrington W.A., Lynch R.M., Moore E.D., Isenberg G., Fogarty K.E., and Fay F.S. 1995. Superresolution three-dimensional images of fluorescence in cells with minimal light exposure. *Science* **268:** 1483–1487.

Castleman K.R. 1979. *Digital image processing.* Prentice-Hall, Englewood Cliffs, New Jersey.

———. 1996. *Digital image processing.* Prentice Hall, Englewood Cliffs, New Jersey.

Cole M.J., Siegel J., Webb S.E., Jones R., Dowling K., Dayel M.J., Parsons-Karavassilis D., French P.M., Lever M.J., Sucharov L.O., et al. 2001. Time-domain whole-field fluorescence lifetime imaging with optical sectioning. *J. Microsc.* **203:** 246–257.

Cox G. and Sheppard C.J.R. 1995. Effects of image deconvolution on optical sectioning in conventional and confocal microscopes. *Bioimaging* **1:** 82–95.

DePasquale J.A. and Izzard C.S. 1987. Evidence for an actin-containing cytoplasmic precursor of the focal contact and the timing of incorporation of vinculin at the focal contact. *J. Cell Biol.* **105:** 2803–2809.

———. 1991. Accumulation of talin in nodes at the edge of the lamellipodium and separate incorporation into adhesion plaques at focal contacts in fibroblasts. *J. Cell Biol.* **113:** 1351–1359.

Dixon A.E. and Cogswell C. 1995. Confocal microscopy with transmitted light. In *Handbook of biological confocal microscopy* (ed. J.B. Pawley), pp. 479–490. Plenum Press, New York.

Dixon A.E., Damaskinos S., and Atkinson M.R. 1991. A scanning confocal microscope for transmission and reflection imaging. *Nature* **351:** 551–553.

Dubois A., Vabre L., Boccara A.C., and Beaurepaire E. 2002. High-resolution full-field optical coherence tomography with a Linnik microscope. *Appl. Optics* **41:** 805–812.

Egger M.D. and Petráň M. 1967. New reflected light microscope for viewing unstained brain and ganglion cells. *Science* **157:** 305–307.

Egger M.D., Gezari W., Davidovits P., Hadravsky M., and Petráň M. 1969. Observation of nerve fibers in incident light. *Experientia* **25:** 1225–1226.

Erhardt A., Zinser G., Komitowski D., and Bille J. 1985. Reconstructing 3-D light-microscopic images by digital image-processing. *Appl. Optics* **24:** 194–200.

Fay F.S., Carrington W., and Fogarty K.E. 1989. Three-dimensional molecular distribution in single cells analysed using the digital imaging microscope. *J. Microsc.* **153:** 133–149.

Femino A.M., Fay F.S., Fogarty K., and Singer R.H. 1998. Visualization of single RNA transcripts in situ. *Science* **280:** 585–590.

Gadella T., Jovin T.M., and Clegg R.M. 1993. Fluorescence lifetime imaging microscopy (FLIM): Spatial resolution of microstructures on the nanosecond time scale. *Biophys. Chem.* **48:** 221–239.

Goldman R.D. and Spector D.L., eds. 2005. *Live cell imaging: A laboratory manual.* Cold Spring Harbor Laboratory, Cold Spring Harbor, New York.

Goldstein S.R., Hubin T., Rosenthal S., and Washburn C. 1990. A confocal video-rate laser-beam scanning reflected-light microscope with no moving parts. *J. Microsc.* **157:** 29–38.

Gustafsson M.G. 1999. Extended resolution fluorescence microscopy. *Curr. Opin. Struct. Biol.* **9:** 627–634.

———. 2000. Surpassing the lateral resolution limit by a factor of two using structured illumination microscopy. *J. Microsc.* **198:** 82–87.

Gustafsson M.G., Agard D.A., and Sedat J.W. 1999. I5M: 3D widefield light microscopy with better than 100 nm axial resolution. *J. Microsc.* **195:** 10–16.

Hanley Q.S., Verveer P.J., Arndt-Jovin D.J., and Jovin T.M. 2000. Three-dimensional spectral imaging by Hadamard transform spectroscopy in a programmable array microscope. *J. Microsc.* **197:** 5–14.

Hanley Q.S., Verveer P.J., Gemkow M.J., Arndt-Jovin D., and Jovin T.M. 1999. An optical sectioning programmable array microscope implemented with a digital micromirror device. *J. Microsc.* **196:** 317–331.

Heintzmann R., Hanley Q.S., Arndt-Jovin D., and Jovin T.M. 2001. A dual path programmable array microscope (PAM): Simultaneous acquisition of conjugate and non-conjugate images. *J. Microsc.* **204:** 119–135.

Hell S.W. and Stelzer E.H. 1995. Lens aberrations in confocal fluorescence microscopy. In *Handbook of biological confocal microscopy* (ed. J. Pawley), pp. 347–354. Plenum, New York.

Hell S.W., Schrader M., and van der Voort H.T. 1997. Far-field fluorescence microscopy with three-dimensional resolution in the 100-nm range. *J. Microsc.* **187:** 1–7.

Hiraoka Y., Sedat J.W., and Agard D.A. 1990. Determination of three-dimensional imaging properties of a light microscope system. Partial confocal behavior in epifluorescence microscopy. *Biophys. J.* **57:** 325–333.

Holmes T.J. 1992. Blind deconvolution of quantum-limited incoherent imagery: Maximum-likelihood approach. *J. Opt. Soc. Am. A* **9:** 1052–1061.

Hopkins H.H. 1955. The frequency response of a defocused optical system. *Proc. R. Soc. Lond. Ser. A* **231:** 91–103.

Izzard C.S. and Lochner L.R. 1976. Cell-to-substrate contacts in living fibroblasts: An interference reflexion study with an evaluation of the technique. *J. Cell Sci.* **21:** 129–159.

Juskaitis R., Wilson T., Neil M.A., and Kozubek M. 1996. Efficient real-time confocal microscopy with white light sources. *Nature* **383:** 804–806.

Kam Z., Hanser B., Gustafsson M.G., Agard D.A., and Sedat J.W. 2001. Computational adaptive optics for live three-dimensional biological imaging. *Proc. Natl. Acad. Sci.* **98:** 3790–3795.

Keller H.E. 1995. Objective lenses for confocal

microscopy. In *Handbook of biological confocal microscopy* (ed. J.B. Pawley), pp. 111–126. Plenum Press, New York.

Lichtman J.W., Sunderland W.J., and Wilkinson R.S. 1989. High-resolution imaging of synaptic structure with a simple confocal microscope. *New Biol.* **1:** 75–82.

Markham J. and Conchello J.A. 2001. Artefacts in restored images due to intensity loss in three-dimensional fluorescence microscopy. *J. Microsc.* **204:** 93–98.

Massoumian F., Juskaitis R., Neil M.A., and Wilson T. 2003. Quantitative polarized light microscopy. *J. Microsc.* **209:** 13–22.

McNally J.G., Karpova T., Cooper J., and Conchello J.A. 1999. Three-dimensional imaging by deconvolution microscopy. *Methods* **19:** 373–385.

McNally J.G., Preza C., Conchello J.A., and Thomas L.J. 1994. Artifacts in computational optical-sectioning microscopy. *J. Opt. Soc. Am. A* **11:** 1056–1067.

Minsky M. 1961. "Microscopy apparatus." U.S. Patent #3,013,467.

———. 1988. Memoir on inventing the confocal scanning microscope. *Scanning* **10:** 128–138.

Neil M.A., Juskaitis R., and Wilson T. 1997. Method of obtaining optical sectioning by using structured light in a conventional microscope. *Optics Lett.* **22:** 1905–1907.

———. 1998. Real time 3D fluorescence microscopy by two beam interference illumination. *Optics Commun.* **153:** 1–4.

Neil M.A., Squire A., Juskaitis R., Bastiaens P.I., and Wilson T. 2000. Wide-field optically sectioning fluorescence microscopy with laser illumination. *J. Microsc.* **197:** 1–4.

Patterson G.H. and Piston D.W. 2000. Photobleaching in two-photon excitation microscopy. *Biophys. J.* **78:** 2159–2162.

Petráň M., Hadravsky M., Benes J., and Boyde A. 1986. In vivo microscopy using the tandem scanning microscope. *Ann. N.Y. Acad. Sci.* **483:** 440–447.

Petráň M., Hadravsky M., Egger M.D., and Galambos R. 1968. Tandem scanning reflected light microscope. *J. Opt. Soc. Am. A* **58:** 661–664.

Ried T., Koehler M., Padilla-Nash H., and Schrock E. 1997. Chromosome analysis by spectral karyotyping. In *Cells: A laboratory manual. Subcellular localization of genes and their products* (ed. D.L. Spector et al.), vol. 3, pp. 113.1–113.9. Cold Spring Harbor Laboratory Press, Cold Spring Harbor, New York.

Sato M., Sardana M.K., Grasser W.A., Garsky V.M., Murray J.M., and Gould R.J. 1990. Echistatin is a potent inhibitor of bone resorption in culture. *J. Cell Biol.* **111:** 1713–1723.

Scalettar B.A., Swedlow J.R., Sedat J.W., and Agard D.A. 1996. Dispersion, aberration and deconvolution in multi-wavelength fluorescence images. *J. Microsc.* **182:** 50–60.

Schrock E., du Manoir S., Veldman T., Schoell B., Wienberg J., Ferguson-Smith M.A., Ning Y., Ledbetter D.H., Bar-Am I., Soenksen D., et al. 1996. Multicolor spectral karyotyping of human chromosomes. *Science* **273:** 494–497.

Shao Z.F., Baumann O., and Somlyo A.P. 1991. Axial resolution of confocal microscopes with parallel-beam detection. *J. Microsc.* **164:** 13–19.

Shaw P.J. and Rawlins D.J. 1991. The point-spread function of a confocal microscope—Its measurement and use in deconvolution of 3-D data. *J. Microsc.* **163:** 151–165.

Stokseth P.A. 1969. Properties of a defocused optical system. *J. Opt. Soc. Am.* **59:** 1314.

Swedlow J.R., Hu K., Andrews P.D., Roos D.S., and Murray J.M. 2002. Measuring tubulin content in *Toxoplasma gondii:* A comparison of laser-scanning confocal and wide-field fluorescence microscopy. *Proc. Natl. Acad. Sci.* **99:** 2014–2019.

van der Voort H.T.M. and Strasters K.C. 1995. Restoration of confocal images for quantitative image analysis. *J. Microsc.* **178:** 165–181.

Verveer P.J., Gemkow M.J., and Jovin T.M. 1999. A comparison of image restoration approaches applied to three-dimensional confocal and wide-field fluorescence microscopy. *J. Microsc.* **193:** 50–61.

Wallace W., Schaefer L.H., and Swedlow J.R. 2001. A workingperson's guide to deconvolution in light microscopy. *BioTechniques* **31:** 1076–1082.

Watson T.F., Juskaitis R., and Wilson T. 2002. New imaging modes for lenslet-array tandem scanning microscopes. *J. Microsc.* **205:** 209–212.

Wilson T. and Sheppard C.J.R. 1984. *Theory and practice of scanning optical microscopy.* Academic Press, London.

Wilson T., Neil M.A., and Juskaitis R. 1998. Real-time three-dimensional imaging of macroscopic structures. *J. Microsc.* **191:** 116–118.

Xiao G.O. and Kino G.S. 1987. A real-time scanning optical microscope. *SPIE Scan. Imag. Tech.* **809:** 107–113.

CHAPTER 3

Multiphoton and Multispectral Laser-scanning Microscopy

Mary E. Dickinson

Department of Molecular Physiology and Biophysics, Baylor College of Medicine, Houston, Texas

INTRODUCTION

Fluorescence microscopy has become an invaluable and common tool for research scientists in a wide variety of fields. Consistent developments in fluorescent probes and proteins, better imaging technology, and more robust protocols for live cell imaging have made fluorescence imaging one of the most versatile tools available to researchers. This chapter focuses on two advanced forms of fluorescence imaging that increase the information that can be obtained from biological samples: multiphoton laser-scanning microscopy (MPLSM) and multispectral laser-scanning microscopy (MSLSM). In a broad sense, these techniques have advantages over standard confocal laser-scanning microscopy by enhancing the ability to collect data in multiple dimensions. MPLSM improves the ability both to collect data from deep in the sample (along the *z* axis) and to obtain more data over time with less lethality, whereas MSLSM reveals the spectra of markers in biological samples and enhances the color separation that is possible, enabling more markers to be viewed simultaneously. Live imaging of whole, intact animals is possible using imaging techniques such as magnetic resonance imaging (MRI), but the lack of resolution and the availability of contrast agents make it difficult to study molecular and cellular events. Both of the techniques featured in this chapter offer submicron resolution for depths of less than a millimeter and benefit greatly from the availability of a large number of fluorescent probes to examine specific cellular and biochemical events. This chapter is devoted to the practical considerations that affect the success of imaging experiments and is designed as a starting point for scientists who are considering the use of these techniques in their own research.

BASIC PRINCIPLES OF MULTIPHOTON LASER-SCANNING MICROSCOPY

MPLSM emerged as a technique in the early 1990s (Denk et al. 1990), although multiphoton excitation has been predicted since the 1930s (Goppert-Mayer 1931). In MPLSM, fluorescent molecules are excited by the quasi-simultaneous absorption of two or more near-infrared photons. Multiphoton, also referred to as nonlinear excitation, has a quadratic dependence, producing excitation only in a small focal volume; thus, out-of-focus fluorescence does not contribute to the acquired image, reducing background, and photodamage outside the plane of focus is greatly reduced. In practical terms, MPLSM makes it possible to acquire images with a high signal-to-noise ratio by using a wavelength that is less harmful to live cells. The use of near-infrared light makes it possible to image deeper in the specimen, owing to less scatter and absorption of the incident light. However, the efficiency of multiphoton excitation depends on criteria that differ from those favoring single-photon excitation events.

In single-photon excitation, a fluorescent molecule or fluorochrome (also called a chromophore) absorbs a high-energy photon within a certain wavelength range and then, within nanoseconds, releases a photon of longer wavelength (lower energy). The absorption of a photon results in the excitation of the molecule, by displacing an electron from the ground state to an excited state. For single-photon excitation, the excitation is directly proportional to the incident photon flux of the source, since each photon has an equal probability of exciting a molecule in the ground state. As the molecule relaxes back to the ground state, some energy is lost through nonradiative exchange (heat or vibration within the molecule), but the rest is shed as a photon of light. The energy loss accounts for the Stokes' shift (or redshift) seen between the excitation and the emission wavelength. Multiphoton excitation of the fluorochrome is induced by the combined effect of two or more lower-energy near-infrared photons and can be achieved by two photons of the same or different wavelengths, but with a single laser source, two photons of approximately the same wavelength are used. The probability of two-photon excitation is proportional to the intensity squared (I^2), because a quasisimultaneous absorption of two photons is necessary. It follows that for three-photon excitation, the probability of three-photon absorption is the intensity cubed (I^3). The emission characteristics of the excited fluorochrome are unaffected by the different absorption processes (Fig. 3.1).

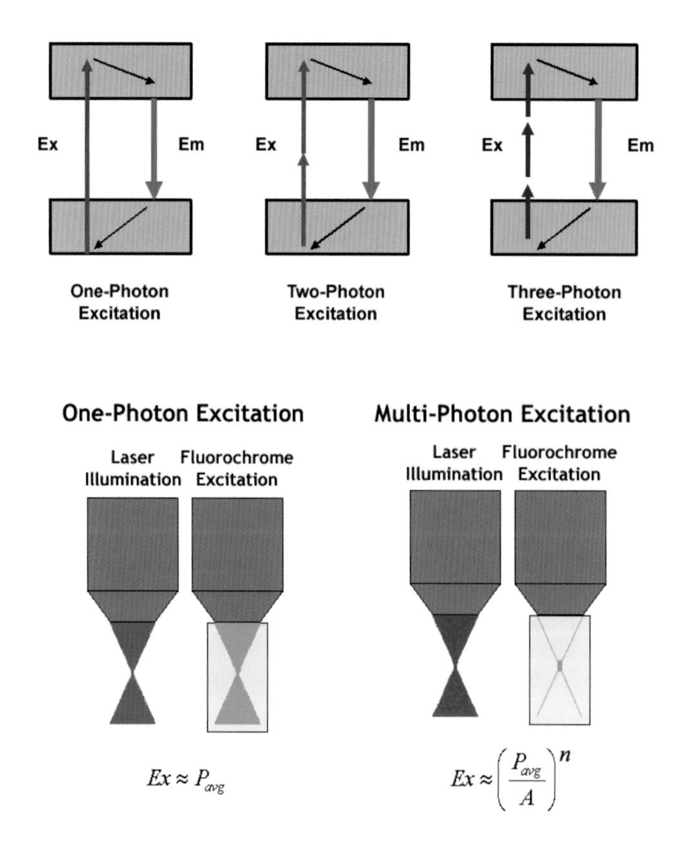

FIGURE 3.1

Principles of multiphoton excitation. (*Top*) Three diagrams showing theoretical differences between single-, two-, and three-photon excitation with respect to the energy levels of the fluorochrome. (*Bottom*) Diagram showing the difference between one- and multiphoton excitation in the sample. For single-photon excitation, fluorochrome excitation is directly proportional to the photon flux of the incident light, whereas for two-photon excitation, excitation depends on the square of the intensity of the incident light and for three-photon excitation, on the intensity cubed. Thus, multiphoton excitation is limited to the focal plane.

To improve the efficiency of multiphoton excitation, ultrafast lasers emitting short pulses of light at a rapid frequency are used (for a review, see Wise 2000). Using these lasers, very large peak intensities can be achieved in a repeated pulse train to sustain fluorophore excitation exclusively at the focal plane where the density of the photon flux is sufficiently high. Although several laser sources have been utilized, Titanium:Sapphire (Ti:Sapphire) lasers have become the most common light source for this application. Further developments in laser technology now include full automation and software control of Ti:Sapphire lasers, resulting in turnkey laser sources for multiphoton microscopy that are convenient to use and provide enough power over a large enough wavelength range to be useful for many applications (Coherent-Inc.; Spectra-Physics).

If we examine the parameters that affect the probability of two-photon excitation, we find that excitation probability relies on (1) how efficiently a given molecule is excited via a nonlinear event, which is a property of the fluorochrome itself, and (2) the density of the photon flux or, in other words, how well the photons at the plane of focus are concentrated, which depends on the output of the laser, in addition to how well the incident beam is focused. Two-photon excitation probability can be expressed as

$$n_a \propto \delta \left(\frac{P_{\text{avg}}^2}{\tau f^2} \right) \left(\pi \frac{\text{NA}^2}{hc\lambda} \right)^2$$

where the n_a is the probability of excitation, δ is the excitation cross section of a dye, P_{avg} is the average power of the incident beam at the sample, τ is the duration of the pulse or pulse width, f is the repetition frequency of the pulse train from the laser, NA is the numerical aperture of the lens, h is Planck's constant, c is the speed of light, and λ is the wavelength (from Denk et al. 1990). Given this, multiphoton excitation is favored when using molecules with large cross sections, high peak power excitation sources, and high-NA objective lenses. Discussed below are how these parameters can be optimized to produce the best quality images revealing the most information.

Optimizing Multiphoton Excitation in Biological Samples

Many factors contribute to the success of biological imaging experiments, as evidenced by the many chapters in this book. Summarized below are some of the common areas that have the greatest effect on success for multiphoton microscopy.

Choosing a Dye

Although multiphoton excitation appears to be conceptually simple, two factors make choosing the best dye for the experiment a complicated choice. First, it is difficult to predict whether a molecule will efficiently absorb two lower-energy photons simultaneously (Xu et al. 1987, 1996; Xu 2000). Drastic differences in multiphoton absorption between different molecules have been identified, and it is difficult to predict, either by the structure of a molecule or by its single-photon properties, how efficiently a dye will absorb simultaneous low-energy photons, although some theories are emerging (Albota et al. 1998; Rumi et al. 2000). Second, the wavelengths for maximum multiphoton excitation are difficult to predict (Xu 2000). Deriving the multiphoton excitation wavelength maximum is not as simple as doubling the single-photon excitation wavelength maximum, although in some cases, this can be a good place to start. The absorption efficiency at a given wavelength must be measured and is reflected in the multiphoton cross section (usually referred to as δ) for a given fluorochrome (for a review, see Xu 2000). In most cases, this value is determined for a given dye using a specially constructed multiphoton fluorimetry system. Cross sections for some dyes have been characterized (Xu et al. 1987, 1996; Xu 2000; Bestvater et al. 2002; W. Zipfel, unpubl.; http://microscopy.bio-rad.com/products/multiphoton/Radiance2100MP/mpspectra.htm). These data are generally gathered from solutions of dyes in a cuvette and quantified against a known standard. However, the biological environment may affect the absorbance spectrum of some dyes. Recently, the multiphoton excitation peak has been determined directly in biological samples, providing the ability to optimize multiphoton excitation on the microscope under the same conditions in which the experiment will be performed (Dickinson et al. 2003). Despite some of the difficulties in predicting the best dyes for MPLSM experiments, many common dyes and fluorescent proteins are suitable. Figure 3.2 shows MPLSM images of several common fluorochromes that emit at various wavelengths throughout the visible spectrum.

The Laser Source

Since the density of the photon flux has an important role in excitation efficiency, both the performance of the laser and the optics of the system are important considerations in optimizing efficiency of excitation. As mentioned above, ultrafast Ti:Sapphire lasers have become the most popular laser source for MPLSM, and several commercial sources for Ti:Sapphire lasers are available. For most applications, a broadly tunable source is desired because different fluorochromes with different excitation peaks may be used for different applications. The laser systems currently available to microscope users can be tuned either manually across a broad range from 690 nm to approximately 1000 nm (such as with the Coherent MIRA 900 or the Spectra-Physics Tsunami) or be tuned via software-controlled motorized mirrors (such as with the Coherent Chameleon or the Spectra-Physics Mai Tai), with only small sacrifices to the tuning range and power output. Most dyes can be excited by two-photon absorption between 710 nm and 950 nm. The maximum power output over the tuning range is a function of the fluorescence spectra of the Ti:Sapphire crystal and the optics within the cavity. For the Ti:Sapphire cavity to lase, the crystal is pumped by a 532-nm doubled vanadate pump source. Typically, 5–10-W pump sources are used, and as more pump power is added, more Ti:Sapphire output is seen. The Ti:Sapphire lasers currently marketed can produce a large amount (well over 2 W) of average power at the peak of the tuning curve. Typically, the peak of the Ti:Sapphire output is near 780–800 nm (a wavelength range that might be used to optimally excite fluorescein or rhodamine using two-photon absorbance), but overall power output of these lasers usually falls to lower values at the wings of the tuning curve, toward 700 nm and past 950 nm. For the manually tunable lasers, it is possible to improve the power output of the laser at the wings by installing optics in the cavity that are optimized for specific wavelengths; however, above 930 nm, significant water absorption bands exist, thus limiting power output. Purging the laser cavity with nitrogen or by recirculating dry air will provide the best performance and mode-locking stability. In addition to their other advantages, software tunable lasers are constructed with sealed and dry cavities for maxium performance throughout the tuning range.

FIGURE 3.2

MPLSM images of common fluorochromes. (*A*) Hoechst 33342, NIH 3T3 cells, 740 nm excitation; (*B*) YoPro-1, NIH 3T3 cells, 800 nm excitation; (*C*) cyan fluorescent protein targeted to the nucleus by histone 2B fusion, NIH 3T3 cells, 800 nm excitation; (*D*) DiO, NIH 3T3 cells, 800 nm excitation; (*E*) green fluorescent protein (sample provided by Carolyn Smith, NIH), 850 nm excitation; (*F*) Oregon green phalloidin, NIH 3T3, 850 nm excitation; (*G*) Sytox green, NIH 3T3 cells, 950 nm excitation; (*H*) ToPro-1, NIH 3T3 cells, 800 nm excitation; (*I*) DiA, NIH 3T3 cells, 800 nm excitation; (*J*) yellow fluorescent protein, NIH 3T3 cells, 950 nm excitation; (*K*) rhodamine phalloidin, NIH 3T3 cells, excited, 850 nm excitation; (*L*) DiI, NIH 3T3 cells, 930 nm excitation. Excitation wavelengths are not suggested to be optimal; they just happened to be convenient. (All dyes were purchased from Molecular Probes, OR; fluorescent proteins were obtained from Clontech, CA.)

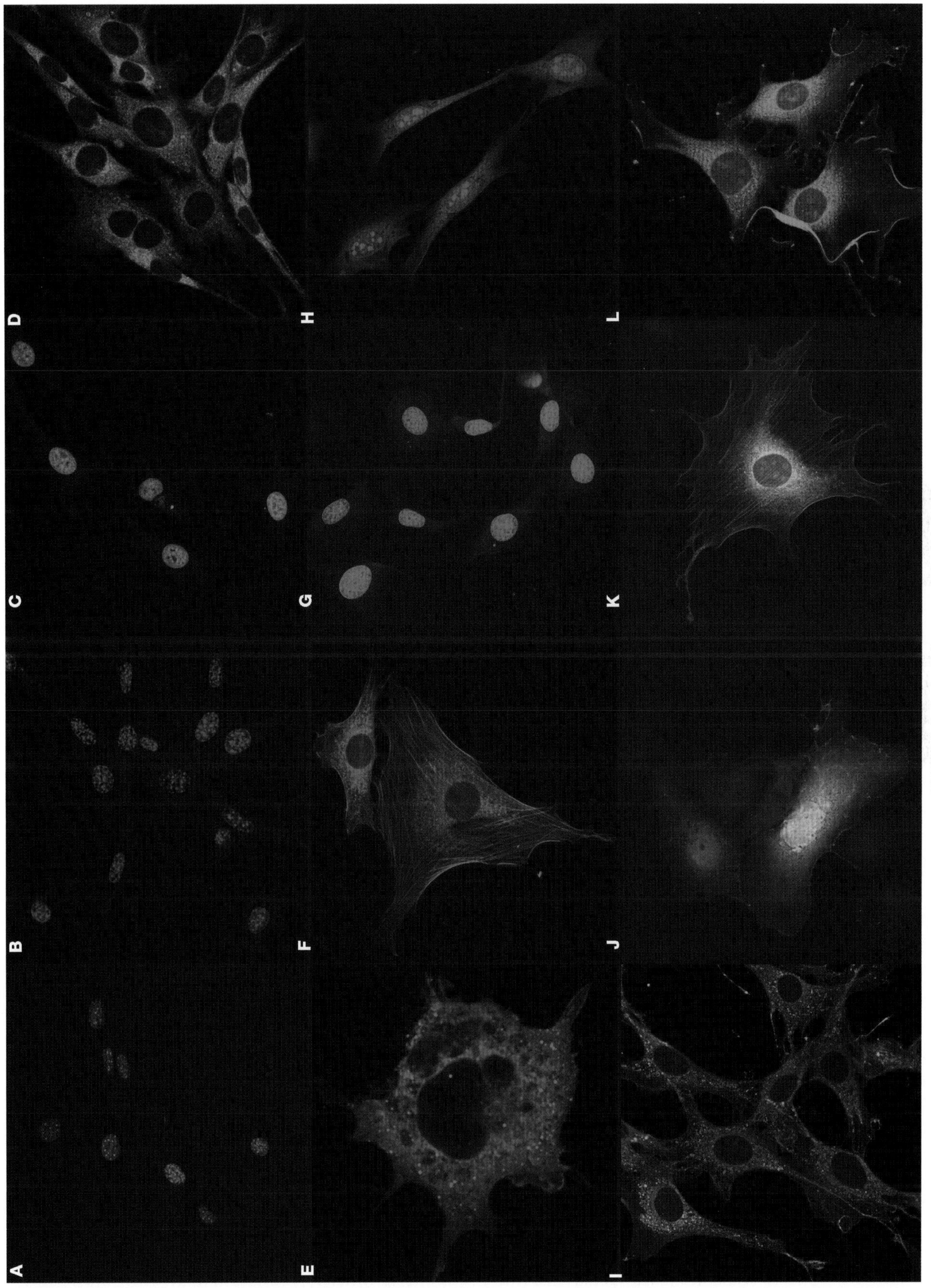

FIGURE 3.2 (*See facing page for legend.*)

In addition to the power output and the wavelength, two other parameters also affect the efficiency of fluorochrome excitation: the frequency of the pulse train (abbreviated as *f* above) and the duration of the pulse (often referred to as the pulse width or τ). The frequency of the pulse train is a fixed value and depends on the geometry of the laser cavity. For most Ti:Sapphire lasers, this range is between 70 MHz and 100 MHz. Thus, the duration between pulses is in the range of 10 ns and fits well with the lifetime of most fluorochromes (Denk et al. 1995).

The pulse width is an important and often elusive parameter. According to the equation given above, in the discussion on Basic Principles of MPLSM, the shorter the pulse width, the greater the peak power and the greater the likelihood of an excitation event. That said, it is also true that the shorter the pulse width exiting the laser, the more the pulse will be broadened as it encounters glass on the way to the specimen. When the pulse width becomes broader, more average power is needed to achieve the same excitation efficiency. Thus, if power is limiting or if the specimen is sensitive to the level of average power, optimizing pulse width can be important.

Pulse broadening is due to an effect termed group velocity dispersion (GVD). GVD is encountered when pulses of light encounter normal dispersive media such as glass (for a review, see Wolleschensky et al. 2002). Pulses, centered at a particular wavelength, are produced as the result of constructive and destructive interference of several wavelengths lasing at once. The more frequencies or colors that are combined, the larger the bandwidth and the shorter the pulse width. Photons of different wavelengths in the pulse do not pass through glass at equal rates, lengthening the pulse duration, so less GVD is seen when there is less bandwidth to the pulse (less difference in the wavelengths). For instance, consider the difference in pulse broadening between an 80- and 180-fs pulse traveling through a single objective lens (assume the objective lens has a dispersion parameter of 2000 fs^2). The 80-fs pulse will stretch to ~105 fs, whereas the 180-fs pulse will not be stretched at all (181 fs) as it passes through the same lens. Considering all the glass that is encountered in an entire laser-scanning microscope system, it can be expected that an 80-fs pulse will be somewhere between 290 fs and 450 fs at the sample, whereas a 180 fs pulse will be between 220 fs and 270 fs at the sample. It is possible to compensate for the dispersive effects of the optics in the microscope system, or even compensate for fiber delivery (Helmchen et al. 2002; Wolleschensky et al. 2002); however, precompensation systems add to alignment complexities and can often result in limiting the amount of available power at the sample.

Measuring Bandwidth and Pulse Width

Manually tunable ultrafast lasers allow the user to optimize the length of the pulse exiting the cavity by adjusting the prism pair that compensates for GVD within the laser cavity. Information about the duration of the pulse exiting the laser head can be inferred by measuring the bandwidth. This can be done quite easily by using a spectrometer (IST-Rees; Ocean-Optics). A pulse duration of 150–200 fs generally corresponds to an approximately 6–8-nm bandwidth, depending on the particular laser. Although this provides a useful benchmark, spectrometers do not provide a direct measure of pulsewidth and are not designed to take measurements at the sample plane of the microscope where multiphoton excitation efficiency is most critical.

Directly measuring the pulsewidth of ultrafast laser output requires a device called an autocorrelator. Most autocorrelators take advantage of a nonlinear effect called second harmonic generation (SHG) in which light of twice the input energy (half the wavelength) is generated when light of sufficiently high intensity (large peak power) is focused in a crystal (termed a doubling crystal) or other such media. Autocorrelators measure pulse width by splitting the input light into two paths of equal intensity, such as in a Michelson interferometer. A variation in the two paths is introduced, thereby producing a slight delay between pulses. The two paths are focused onto a doubling crystal, and a detector is used to measure the intensity of the SHG signal generated by the input. The intensity of the SHG signal as a function of the delay in the pulse gives the pulse width.

In the past, it has been difficult to measure the pulse width at the sample on an MPLSM system. Although it is possible to measure the pulse width produced by the laser, as described above, a measure of the pulse width at the sample is needed because of the effects of GVD. Recently, an autocorrelator has been developed for this purpose that is quite easy to use (APE, Berlin, Germany). The Carpe autocorrelator makes two measurements, one from the laser path outside the microscope and then another using an external sensor that is placed on the microscope stage. Thus, it is possible to determine the amount of pulse broadening due to dispersion in the microscope and to optimize the excitation efficiency related to pulse width. With such an autocorrelator, the laser output can be optimized to provide the shortest pulse width at the sample, producing the best excitation efficiency for a given fluorochrome.

Choosing Optics and Objective Lenses for MPLSM

For nonlinear microscopy, objective lenses should be optimized for the following parameters.

- *Long working distance with a high NA:* Excitation efficiency increases as the NA of the objective lens increases. However, many long-working-distance objective lenses have a low NA. Therefore, it is best to choose an objective lens with the highest NA for the working distance needed. NA has an effect not only on excitation efficiency, but also on collection efficiency, so NA is extremely important. Low-magnification, high-NA lenses can be better for collecting an emission signal from deep within the sample (Oheim et al. 2001), and often offer more working distance than higher-magnification lenses with a similar NA.
- *High transmission in the NIR and the visible-wavelength range:* Many objective lenses used for biomedical microscopy are corrected for the UV/visible range, and the efficiency of transmission of these lenses often decreases significantly in the near-infrared (NIR) range. Thus, both the laser output and the transmission of the objective can combine to lower the power available to the sample at longer wavelengths. Further on in this chapter, we will see that this effect is significant when characterizing the multiphoton excitation spectra of fluorochromes using the microscope. If excitation energy at longer wavelengths is limiting, lenses corrected for better transmission in the infrared may be worth considering. For instance, a comparison of the transmission of the Zeiss 40x Achroplan lens with that of the Zeiss 40x IR-Achroplan lens at 900 nm demonstrates almost 25% higher transmission using the infrared-corrected version of this lens.
- *Limited pulse broadening:* Objective lenses should be minimally dispersive to reduce the chance that short pulses will be lengthened en route to the sample, which will again reduce the peak intensity. In addition to the group velocity dispersion, chromatic aberration of lenses leads to pulse distortions. Specifically, a radius-dependent group delay is introduced (Kempe and Rudolph 1993; Netz et al. 2000). Therefore, different radial portions of the beam across the pupil of the objective lens arrive at different times at the focal region and cause a temporal broadening of the pulse, resulting in lower peak intensity in the focal region. This effect is also referred to as propagation time difference (PTD). Known dispersion and PTD values for some lenses can be found in Wolleschensky et al. (2002).
- *Limiting or correcting for focal plane mismatch:* Chromatic aberrations can also affect the ability to both excite and collect photons at the same focal plane, because of the mismatch in focal distance between two such different wavelengths (see Wokosin and Girkin 2002 and references therein). Correcting for this mismatch properly can ensure not only that visible and NIR beams

are focused at the same plane, but also that the greatest collection efficiency for the visible emission signal will be realized. Corrections for chromatic aberrations can be made either by using a collimator to correct for the mismatch by focusing the NIR beam at the visible plane or by using a wider collection lens or nondescanned detection to recover photons that are not focused back through to the detector. Although correcting for the collection can mitigate the loss of collection signal, proper collimation of the incident beams will ensure that images produced by the simultaneous use of visible and NIR beams can be overlayed and that photochemistry performed by the NIR beam can be imaged accurately using excitation by the visible lasers.

Optimizing Laser Alignment into the Microscope

Poor alignment of the laser into the microscope can have a very profound effect on excitation efficiency. The following protocol describes a sequence for optimizing alignment of the laser into the scanhead and for ensuring the best overlay between the visible and NIR excitation beams.

Aligning the NIR Laser to the Optical Path of the Microscope Using a VIS Laser

1. Place a mirror slide (part 453001-9062, Carl Zeiss, Inc.) on the stage and focus on the slide using a low-magnification (10x) lens.

2. Configure the system to reflect one of the visible laser lines down to the sample by choosing the appropriate primary beam-splitter selection.

3. While scanning, make sure that the visible laser line is reflected back out of the scan head. Use the focus knob if necessary to change the spot size.

4. The visible and NIR beams can be walked together using the routing mirrors (see Fig. 3.3). Place a piece of lens paper in the path close to where the NIR beam enters the scan head. Adjust the mirror farthest from the scan head using both the *x* and *y* knobs or adjustment screws to overlay the visible and NIR beams.

5. Hold the lens paper in the path closest to the laser. Align the overlay of the two beams by adjusting the routing mirror closest to the scan head.

6. Repeat Steps 4 and 5 until no further improvements can be made.

7. Switch to an open position in the objective turret. Place a piece of lens paper on the stage to determine whether the NIR spot is centered. If it is not centered, adjust the mirror closest to the scan head to center the beam.

   Many lasers have an elliptical beam shape and may not fill the entire aperture.

8. Use a slide with a thin mirrored grid (part 474028-0001, Test Grid Specimen for LSM, Carl Zeiss, Inc.) to ensure that the NIR and visible lasers are aligned well to one another. Place the grid slide on the stage and focus on the grid pattern using the objective lens of choice.

9. Set up an imaging configuration that allows separate channels for the reflected laser light from a visible line and the NIR line. Use a very low amount of power from the Ti:Sapphire laser—too much power will burn the slide. Confirm that these channels are correct by switching off one line at a time; the corresponding image should disappear.

External alignment of the Titanium:Sapphire laser into the microscope scan head

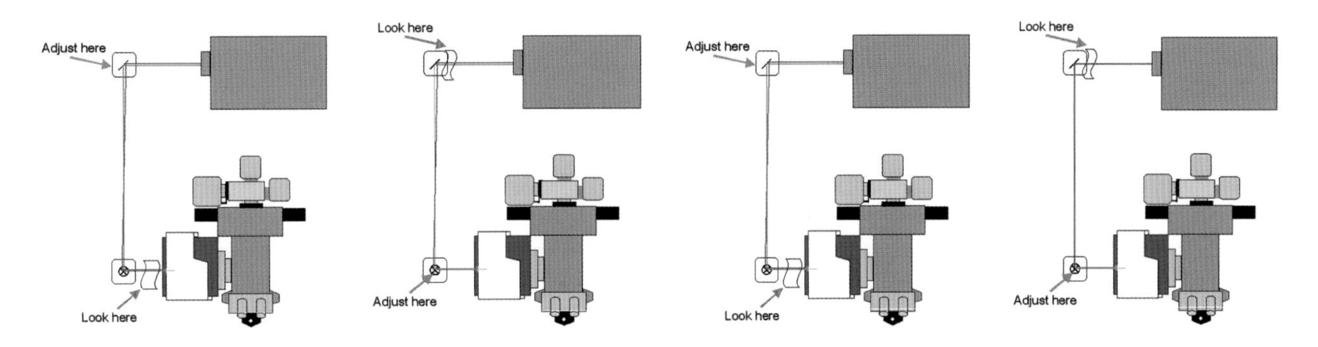

Alignment of the Titanium:Sapphire laser to the optical path of the microscope using a reflective grid slide

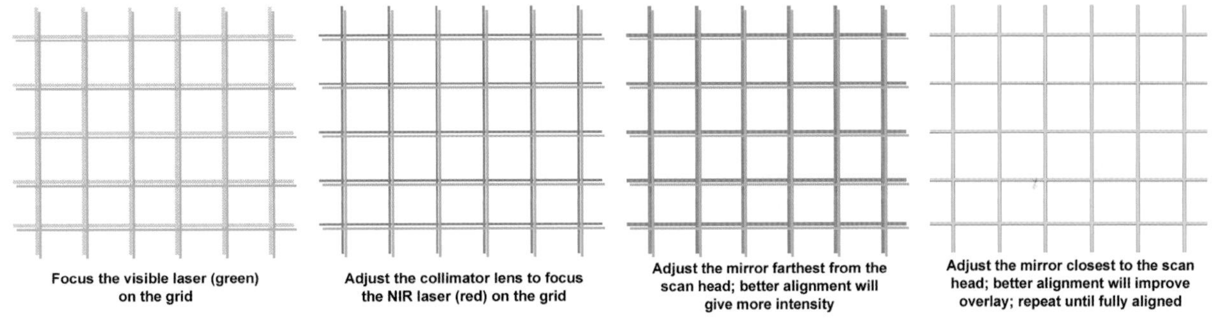

FIGURE 3.3

Alignment of an external Ti:Sapphire laser to the microscope using visible laser alignment as a guide (see text).

10 While imaging, adjust the focus of the visible reflection and then adjust the collimation lens if there is one available. Optimizing this lens ensures that the visible and NIR excitation planes are parfocal (Fig. 3.3).

11 To begin, peak the mirror farthest from the scan head. Better alignment is indicated by an increase in intensity (Fig. 3.3). Next adjust the alignment of the two beams on the grid using the mirror closest to the scan head (Fig. 3.3).

12 Repeat the sequence until no further improvements can be made (Fig. 3.3).

Detecting Photons: Descanned versus Nondescanned

Nondescanned detectors (NDDs) are used to improve the collection of emission photons, especially from deep within the sample. In a confocal microscope, pinholes and detectors are aligned with respect to the incident light so that the position of the pinhole accurately rejects light that is out of the plane of focus. For multiphoton microscopy, excitation is limited to the focal plane so a pinhole is not required. In fact, since all fluorescence that is produced is a result of excitation at the focal plane, collecting scattered photons improves the signal in the image. NDDs, placed very near to the emission signal at the sample, can be used to collect scattered photons that might be lost before reaching the detectors within the scan head. For highly scattering samples, this approach can greatly improve light collection, improving the signal-to-noise ratio within the image and increasing the detection efficiency when deep imaging is needed. Because of the increased ability to collect scattered light, many users immediately notice that stray light from fluorescent bulbs or monitors in the room can contribute a significant amount of noise. Therefore, very dark conditions are recommended when using NDDs.

Although a single NDD can offer significant improvement, signal from multiple NDDs can be added. Because fluorescence emission occurs in a radial fashion from the focal point, NDDs can be placed above and below the sample; the signal from the detectors in the epipath (through the objective lens) and the forward path (toward the condenser) can be summed to offer more sensitivity if proper collection optics are used. High-NA condensers can often be useful for collecting the light scattered in the forward direction. In addition to MPLSM, other nonlinear imaging modalities, such as SHG imaging (Campagnola et al. 2001; Millard et al. 2003) or coherent anti-Stokes Raman scattering (CARS) imaging (Muller et al. 2000; Cheng et al. 2002; Nan et al. 2003) often make use of nondescanned detection. For these methods, forward detection can be more sensitive than epidetection.

Live Cell Imaging

MPLSM is currently being applied to a large variety of experiments and many imaging protocols have evolved to look at different samples. For instance, Figure 3.4 shows vital labeling of a chick embryo neural tube labeled with the lipophilic dye DiA. With MPLSM it is possible to image all the way through the neural tube in this speciman. The popularity of multiphoton imaging, in part, is due to the availability of robust culture methods and the creativity of many microscopists who have devised clever ways of preparing living samples for light microscopy. Some of these protocols are reviewed in Goldman and Spector (2005), or in recent reviews (Helmchen and Denk 2002; Hadjantonakis et al. 2003; Ragan et al. 2003). One area in which MPLSM has been well utilized is intravital imaging (see Chapters 22 and 23 in Goldman and Spector 2005). MPLSM has been used to follow cellular processes, often for days, weeks, or months, in whole rodents. Neurons in the brain (Helmchen et al. 1999; Svoboda et al. 1999; Trachtenberg et al. 2002; Chaigneau et al. 2003; D'Amore et al. 2003; Stosiek et al. 2003) and neuromuscular junctions (Keller-Peck et al. 2001; Walsh and

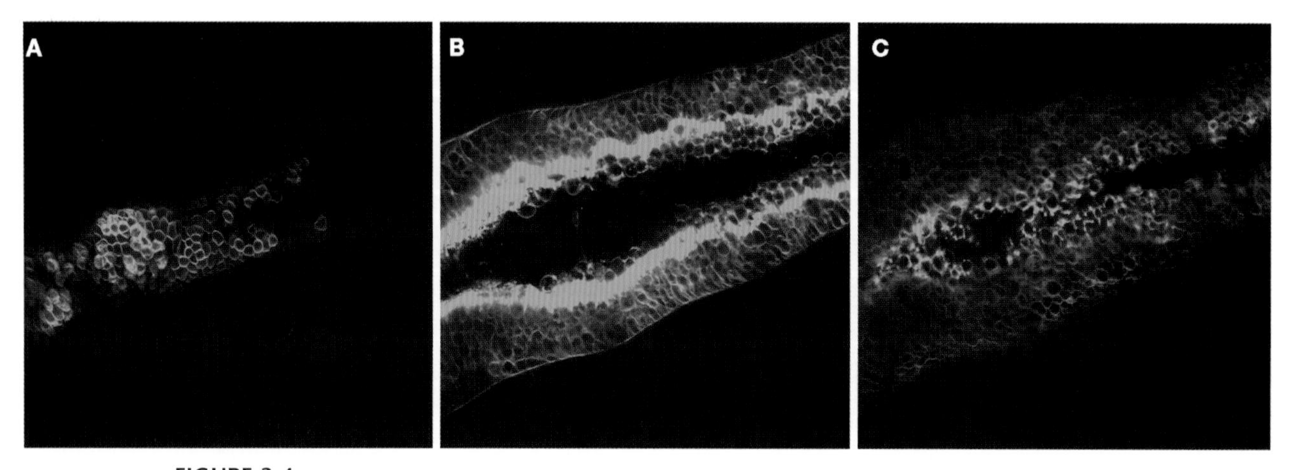

FIGURE 3.4

DiA-injected chick neural tube. An HH (Hamburger Hamilton) stage 10 embryonic chick caudal neural tube was injected with DiA/EtOH solution (0.005 μg/ml solution; Molecular Probes, OR). (*A*) Dorsal-most neural tube, 15 μm from the overlying ectoderm; (*B*) 80 μm deep into the same sample; (*C*) 101 μm deep in the same sample to the very ventral neural tube. Individual images are 325.7 μm^2.

Lichtman 2003) have been studied, as well as cells in the immune system (Miller et al. 2002, 2003; Padera et al. 2002), in implanted tumors (Brown et al. 2001; Wang et al. 2002), and in the kidney (Dunn et al. 2002, 2003). These methods utilize minimally invasive procedures and provide evidence about cell behaviors that can be gleaned only by watching cells in their native environment.

Intravital imaging methods are increasing as technology becomes available. Many advances have been made toward engineering fiber-optic multiphoton microscopes (Helmchen et al. 2001; Helmchen 2002; Helmchen and Denk 2002; Jung and Schnitzer 2003). Recently, a fiber-optic, head-mounted MPLSM system has been used to image from the brain as the rodent subject is free to move about in its environment (Helmchen et al. 2001). Such methods make long-term observations possible and provide direct evidence of the cellular and molecular consequences of animals interacting with their environment. In addition to having an impact on basic research, fiber-optic methods may enable the use of MPLSM in surgical and clinical procedures.

MULTISPECTRAL LASER-SCANNING MICROSCOPY

During the last several years, there has been an increased interest in multicolor fluorescence experiments. The development of new dyes and fluorescent proteins has made it possible to label a wide range of cell types and to generate specific protein tags in live animals (for a review, see Hadjantonakis et al. 2003). This has increased the need for more advanced methods to discriminate between different dyes and fluorescent markers simultaneously. In response to this need, multispectral imaging has become an important tool for laser-scanning microscopists. Although the first multispectral imaging systems were developed for wide-field microscopy, this chapter focuses on the methods used in laser-scanning microscopy, as this is a rapidly growing field on its own. (For further information about wide-field multispectral imaging, see Eng et al. 1989; Wiegmann et al. 1993; Morris et al. 1994; Schrock et al. 1996; Zangaro et al. 1996; Mooney et al. 1997; Richmond et al. 1997; Wachman et al. 1997; Garini et al. 1999; Ornberg et al. 1999; Zeng et al. 1999; Levenson and Hoyt 2000; Tsurui et al. 2000; Farkas 2001; Ford et al. 2001; Yang et al. 2003.)

Traditional Approaches for Fluorochrome Separation

When more than one fluorochrome is used, fluorescence cross talk can make it difficult to interpret the results of an experiment. This is particularly true if colocalization experiments are performed or if quantitative measurements are made, e.g., for fluorescence resonance energy transfer (FRET) (see Chapter 8 in Goldman and Spector 2005) and for fluorescence recovery after photobleaching (FRAP) (see Chapter 7 in Goldman and Spector 2005). Several strategies can be used to eliminate or reduce cross talk.

Traditional Filter Methods

In laser-scanning microscopy (see Chapter 2), laser illumination is used as the excitation source, and coated glass is commonly used to separate emission signals. Typically, a beam splitter splits light into two paths—one transmitted, one reflected—and then a band-pass filter further refines the spectral composition of the light collected by the detector, such as a photomultiplier tube (PMT). The dichroic and band-pass filter combination perform two functions: to limit excitation light from reaching the detector and to collect as much emission signal as possible for one fluorochrome, without allowing the emission signal from the other fluorochrome(s) to contaminate the signal. Thus, separation of fluorescent signals depends on the characteristics of the filters that are used. For closely overlapping spectra, the narrower the band-pass filter, the better the spectral separation; in this case, however, fewer photons are collected.

Multitracking

In cases where there is overlap in the excitation spectra of two fluorochromes, laser illumination can be carefully controlled during the scan cycle to reduce or avoid cross talk. This approach, called multitracking, allows excitation and emission collection of each fluorochrome separately and is accomplished by the fast switching of laser lines, either by line or by frame. For instance, for two dyes such as fluorescein and rhodamine, the 488-nm line used to excite fluorescein would be turned on as the laser scans across the line, collecting fluorescein emission using a fluorescein filter in front of a detector; then, on the return, the 488-nm would be turned off and the 543-nm line used to excite rhodamine would be turned on. As the 543-nm line scans across the sample, rhodamine signal is acquired using a rhodamine band-pass filter. This avoids exciting both dyes at once and can significantly reduce spectral cross talk, particularly when the choice of emission filters is limited. The fast switching of the laser lines is conferred through an acousto-optic tunable filter (AOTF) or similar device that has a very rapid response time.

Quantitative Cross-Talk Correction Using Ratiometric Analysis

For quantitative experiments, fluorescence cross talk and background signal can be particularly problematic. In these instances, correction methods based on the ratio of intensities can be used to quantify the fluorescence from each of the probes in a sample or to estimate changes in a ratiometric indicator (for a review, see Helmchen 2000). By measuring the amount of cross talk present when using single-labeled samples, it is possible to estimate the amount of signal bleed-through that is likely to occur when fluorochrome pairs are imaged. Often, two different excitation wavelengths and/or two different channels for emission collection are used to determine the amount of signal obtained when each fluorochrome is excited. Once the ratio of signal is determined for each dye, or for a single dye with and without an indicator present, the pattern can be applied to the experimental situation.

Correction methods based on a ratio of fluorescence intensities have been used to determine FRET efficiencies (for a review, see Berney and Danuser 2003). FRET occurs via a dipole–dipole interaction between fluorochromes such that excitation of the donor fluorochrome results in emission of an acceptor fluorochrome through a nonradiative exchange of energy from one molecule to the other (see Chapter 8 in Goldman and Spector 2005). Because FRET efficiency falls off as the distance increases between fluorochromes, FRET is a valid way to examine protein–protein interactions (Herman et al. 2001; Periasamy 2001; Periasamy et al. 2001). For FRET to occur, there must be overlap between the donor emission and acceptor excitation spectra. Often this means that there is cross talk between acceptor and donor emission in addition to coexcitation of both the donor and the acceptor. In many cases, it is not possible to separate dyes using narrow-band-pass filters or multitracking, and correction methods are needed to determine donor and acceptor fluorescence. This method has been used successfully in many cases for cross-talk elimination and FRET measurements, but methods such as these are complicated by the concentration of different components, background fluorescence, the spectral response of the detector, and differences between detectors and noise, making it cumbersome to acquire and process multiple images and difficult to control for all variables (for a review, see Berney and Danuser 2003). In addition, this method may require optimized filters for each set of fluorochromes, and, although these correction schemes are generally successful for two fluorochromes, they are not very effective for more than two labels.

Multispectral Laser-scanning Microscopy: Image Acquisition

Careful choices of fluorescent labels, narrow filters, multitracking and ratio imaging can reduce cross talk in a large number of cases; however, there are some instances when spectral overlap cannot be eliminated by these strategies. Despite the efforts of dye and filter manufacturers, there are often cases when it is not possible or feasible to choose nonoverlapping probes for certain experiments. This is particularly true when multiple fluorescent protein variants are used in the same experiment (see Chapter 1 in Goldman and Spector 2005). In addition, naturally occurring fluorescence or fluorescence produced by fixatives or other treatments may mask the signal of some fluorochromes and can bleed through to many channels, making these difficult to eliminate. In addition, when a large number of probes are used, spectral cross talk is inevitable. In these instances, multispectral imaging can be used to achieve cross talk elimination. Multispectral imaging has been used extensively in remote sensing analysis, which involves studying how light waves are emitted and scattered off the surface of the earth and within the atmosphere. Characterizing the spectral frequencies that are present allows the different structures and terrains to be identified. Similar approaches are now being used to determine fluorochrome identity.

Multispectral imaging is a method used to collect spectral information from samples with a mixed set of fluorochromes by collecting a series of images in which spectral information is known for each pixel (Levenson and Hoyt 2000; Dickinson et al. 2001, 2003; Farkas 2001; Lansford et al. 2001; Hiraoka et al. 2002; Zimmermann et al. 2003). Spectral information in an image can be used to determine the identity of the dye and shows how dye fluorescence is affected by environmental changes. To determine the spectral content of each pixel within a field of view, images are collected at a series of wavelength bands (x, y, λ). Images acquired in a wavelength series, similar to a time series or a z series, are called a Lambda stack, image cube, or spectral cube (Fig. 3.5). Figure 3.5 shows an example of a Lambda stack of a group of 3T3 cells that express either cyan fluorescent protein (CFP) or green fluorescent protein (GFP) in the nucleus. The spectrum of any field of view can be revealed by graphing how the average intensity changes with respect to wavelength. In this

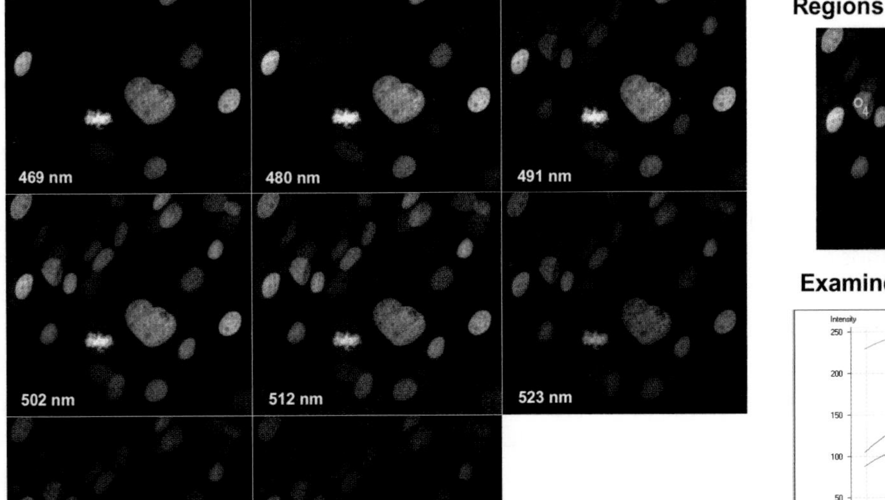

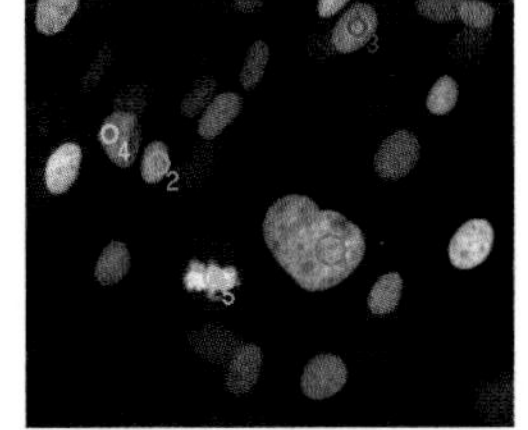

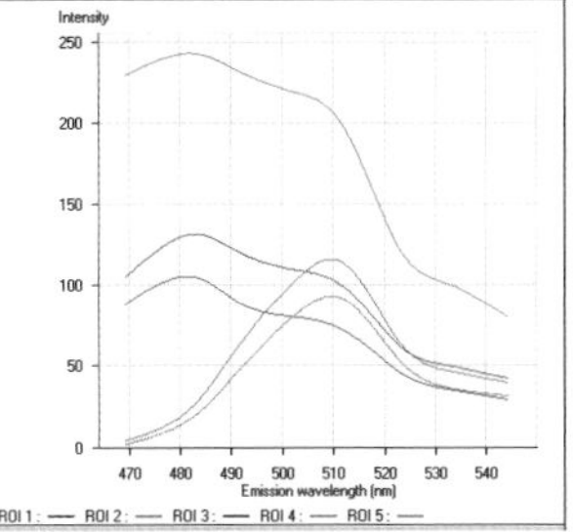

FIGURE 3.5

Spectral analysis of a mixed group of cells. Lambda stack images of individual cells that express either CFP or GFP targeted to the nucleus. Selecting a region of interest for each nucleus reveals the spectrum of fluorescent protein emission.

example, simply graphing the spectra of individual nuclei makes it obvious which nuclei have which dye, but for more complex samples with spatial and spectral overlap of signals, dye identity can be determined using mathematical approaches (discussed later in the chapter).

To acquire a Lambda stack, the microscope system must be capable of collecting images at different emission or excitation wavelengths. In contrast to the push broom approach employed in wide-field microscopy and remote sensing (for a review, see Farkas 2001), where spectral information is collected a line at a time, in laser-scanning microscopy, acquiring a Lambda stack is more like sweeping each floor tile, separating the yield into different bins for each tile. Different methods of acquiring data at wavelength bands can be used to generate a Lambda stack. These are summarized below.

Acquiring Image Stacks at Variable Emission Wavelengths

Initial attempts to add spectra detection to laser-scanning microscopes (LSMs) involved coupling a spectrometer to the microscope to characterize the wavelength content of the emission signal (Hanley et al. 1998; Favard et al. 1999; Haralampus-Grynaviski et al. 2000). In most instances, a CCD (charge-coupled device) camera was used as a detector, limiting the sensitivity and resolution of the collected signal. Nevertheless, these successful implementations led the way to the development of PMT-based spectral detectors that can be built in to the descanned path of the LSM.

Multiple filters placed in front of a PMT offer a simple approach for building an imaging spectrometer. The same detector is used, and different band-pass filters or different long-pass and short-pass filter pairs are rotated in front of it to acquire bands of equal bandwidth successively. This approach is now being used on some confocal and multiphoton laser-scanning systems (Bio-Rad; Konig et al. 2000). The larger the number of filters, the greater the flexibility of the band pass collected and the spectral range. This method benefits from having defined, reproducible bands, but a major drawback to the filter-based approach is that repetitive scanning is necessary as an image is collected at each wavelength increment. This process can be slow and increases the chance that photobleaching will occur during acquisition. Tunable filter devices (liquid crystal tunable filters, acousto-optic tunable filters, etc.) can be used for greater flexibility, but the poor transmission efficiency of some filters, due in part to polarization, can reduce the signal to noise (Farkas 2001; Lansford et al. 2001).

Greater demand for multispectral strategies for live cell imaging has pushed for the development of rapid acquisition schemes that limit the amount of times the field of view is exposed to excitation light. A new device, called the META detector (Carl Zeiss), uses a multichannel detector to collect bands of emission signals that have been separated using a diffraction grating (Dickinson et al. 2001, 2002; Haraguchi et al. 2002). Using this device, multiple bands of data can be acquired in parallel instead of in series, greatly reducing the amount of time needed to collect a Lambda stack. The grating allows for the precise register of 10.7-nm bands onto each channel of a 32-channel multi-anode detector, and a separate image can be made from each channel that is activated. In a single scan, eight images, corresponding to eight different wavelength bands, can be acquired simultaneously to produce a Lambda stack that covers 85.6 nm of the visible spectrum. If a larger part of the spectrum must be sampled, additional scans can be made or adjacent channels can be binned to double, triple, or quadruple the width of the band detected. The limitation to the number of images collected in a Lambda stack relates only to the rate of signal processing. Future improvements in this area may make it possible to image the entire spectrum at 10-nm resolution in a single scan. In addition, with the use of a multichannel detector and a grating, the spectral registration to different bands is highly reproducible. Because of this, acquired spectra are consistent between experiments and resemble spectral information provided by other sources (Beechem et al.

2003). A similar approach utilizing a 16-channel PMT detector and a grating has shown similar capabilities (Buehler et al. 2003).

Another approach to acquiring multiple bands in parallel utilizes a prism to separate emission wavelengths and movable gates that allow a specific band of light to pass to a detector (Leica Microsystems Inc.), much like a prism monochrometer. Mirrored gates allow what is not detected along the primary path to be reflected to additional detectors to acquire multiple image bands at once. In this case, the number of scans in the Lambda stack is limited by the number of detectors available on the system, usually three to four, regardless of the bandwidth of the signal in each image, and care must be taken to calibrate the input from multiple detectors.

In addition to Lambda stack acquisition, with the implemention of methods for variable band-pass detection, LSMs now have a greater capacity than ever before for flexible emission collection. In some cases, fluorochrome separation fails because the best band-pass filter for separation is not present on the system. Some approaches outlined above offer the ability to freely configure the wavelength range of detection, making it possible to design custom band-pass settings for the fluorochromes in use. With the META detector, for example, this is accomplished by activating a series of channels in the multichannel array, and the signal collected by those channels is binned to make an image. The wavelength range depends on the position of the channel in the detector, and the width of the band pass, which can be as small as 10.7 nm, depends on the number of detector channels that are active. Eight images, all with different band-pass configurations, can be collected. For prism-gate systems, the width of the gate specifies the width of the band collected, and the position with respect to the prism determines the wavelength content. The number of different images depends on the number of detectors. For systems using filter pairs, choices are limited to the filters that are present, but this certainly extends the range of possibilities beyond single band-pass filters. All of these approaches offer a tremendous advantage for avoiding cross talk and provide the flexibility necessary to keep up with the growing number of available fluorescent conjugates and proteins.

Multiphoton, Multispectral Analysis

Multiphoton excitation produces fluorescence with the same spectral emission properties as single-photon excitation (for reviews, see Denk et al.1995; Xu 2000). Any of the methods discussed above can be used to acquire MPLSM/MSLSM Lambda stacks. However, as discussed earlier in the chapter, NDDs can be used to improve the detection efficiency, especially during deep tissue imaging. Thus, it would be advantageous to develop methods for spectral separation that can be used to acquire nondescanned photons.

In addition to characterizing fluorchrome emission, spectral information can also be collected by relating fluorescence intensity to varying excitation wavelength. MPLSM utilizes a continuously tunable Ti:Sapphire laser for excitation, making it possible to alter the excitation wavelength to collect an excitation Lambda stack. Recently, the author's laboratory has utilized the software-controlled, tunable Chameleon laser to acquire Lambda stacks with variable multiphoton excitation (Dickinson et al. 2003). Because this approach requires that only the excitation wavelength be varied, signal can be collected with any PMT including an NDD. By varying the excitation signal, multispectral nondescanned detection is possible, providing a powerful tool for eliminating autofluorescence and separating overlapping fluorochromes deep within tissues. This method can be used for eliminating spectral cross talk and also enables the user to determine how the tuning curve of the laser, objective transmission, and environment of the fluorochrome effect dye excitation. By determining excitation curves directly on the microscope, users can quickly determine the ideal wavelength for the most efficient excitation of a single fluorochrome or labels used in combination. Interestingly, some fluorochromes that have very closely overlapping excitation spectra with single-photon excitation, such as CFP and YFP, or Alexa 488 and Sytox Green, display significantly less overlap in their multiphoton excitation curves (Dickinson et al. 2003; W. Zipfel, unpubl.), and for triple fluorochrome applications, this may result in more robust signal separation (Dickinson et al. 2003). Collecting excitation Lambda stacks can be slow since it requires sequential scans, but for some fluorochromes, only two or three scans may be necessary for cross-talk elimination.

Mathematical Approaches to Fluorochrome Separation

As mentioned above, fluorochromes have unique spectral signatures. Once a Lambda stack or spectral cube is collected, these signatures can be used to assign the proper dye identities to individual pixels. This process has been called emission or excitation fingerprinting, depending on how the Lambda stack is collected. Two methods for determining the contribution of different fluorochromes to the image are discussed here. The first, called linear unmixing, is a very reliable method for determining dye identity using known reference spectra for the fluorochromes present in the sample, and the second, principle component analysis, can help to identify unknown fluorochromes that contribute to the signals in the image.

Linear Unmixing

Simply put, linear unmixing is a way of matching the spectral variations in the Lambda stack with the known spectral variations for the fluorochrome labels that are used. The fluorescence intensity in a pixel in which fluorochromes are colocalized and have overlapping emission spectra will be the sum of the intensity of each label (Lakowicz 1999). Thus, the pixel is said to be linearly mixed where the sum spectrum, $S(\lambda)$ equals the proportion or weight, A, of each individual spectrum $R(\lambda)$: $S(\lambda) = A_1R_1(\lambda) + A_2R_2(\lambda) + A_3R_3(\lambda) + \cdots$, etc.

Consider the example in Figure 3.6. The orange spectrum is the sum of the two-component spectra shown in pink and blue according to the proportions listed on the graph. In fact, for every possible combination of dye concentrations, the sum spectrum can be predicted. Therefore, if we wanted to determine the spectral content of each pixel in an image, we could match the shape of the sum spectrum from each pixel to a known sum spectrum in a library, much like matching a fingerprint in a database. For instance, if the shape matched the sum spectrum in the first panel of Figure 3.6, then we would conclude that the two dyes are evenly mixed in that pixel, whereas if it matched the orange spectrum in the second pixel, we would conclude that the pixel contained 75% blue and 25% pink, and so on.

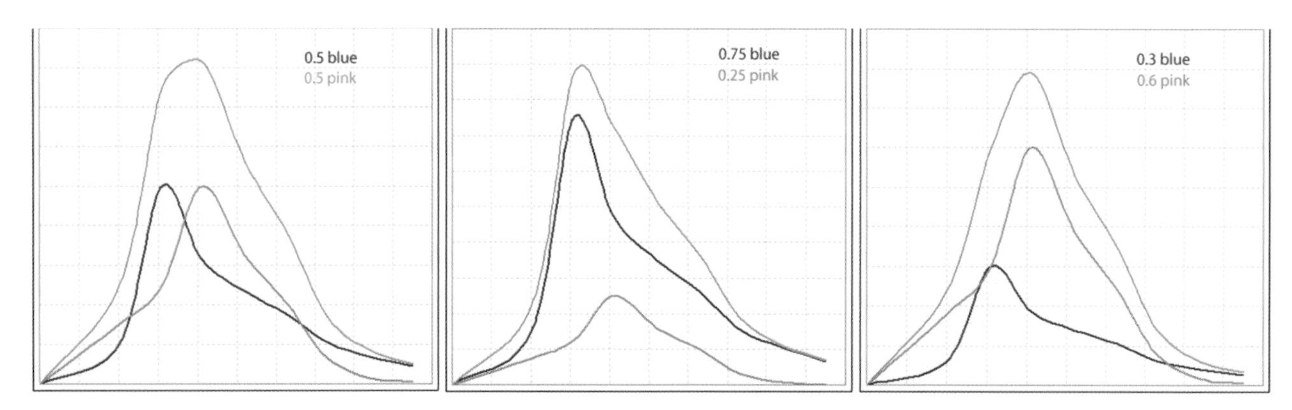

FIGURE 3.6

Theoretical linear mixing. Emission signal from two different components yield different sum spectra when signals overlap. With known components, it is possible to predict the sum spectra of any concentration of components.

Linear unmixing algorithms are used to very quickly solve the problem described above. A matrix of values representing the sum spectra is matched against a library of predicted spectra according to whatever best-fit parameters are imposed. Once the weight of each spectral component is determined, the Lambda stack can be separated into individual images for each fluorescent label. The intensity of each pixel in the unmixed images is the total collected intensity of the pixel multiplied by the proportion of spectra in the pixel.

Practical Considerations for Linear Unmixing

The success of linear unmixing depends on using reference spectra that faithfully represent the spectra in the sample to ensure the best fit. Thus, controls must be imaged under the same conditions as the mixed sample, pixel saturation should be avoided, and Lambda stack acquisition must be reliable. The optical components of each microscope system will impose some bias, making it impossible to use spectra from other sources as reference spectra. In fact, the best reference spectra are those that are acquired in exactly the same way the Lambda stack is acquired (same objective, offset, amplifier gain, wavelength range, dichroics, etc.) or those chosen from nonoverlapping regions within the Lambda stack itself. In addition, it is often helpful to consider any background signal from mounting media or laser light reflections as a known reference spectra in order to properly separate background from specific fluorochrome signals. Spectral signals that do not match those in the spectral library are considered residual and are not assigned. If a residual image is part of the output, it is possible to see how well the reference spectra fit the acquired spectra. It is often the case that the residual signal comes from pixels that are saturated or from background in the sample.

Linear unmixing can be used to separate fluorescent signals that have subtle differences in emission spectra (Dickinson et al. 2001). To determine the limits of spectral separation using linear unmixing and the META detector, we performed a pairwise comparison of seven commonly used dyes with peak emissions from 509 nm to 531 nm (Dickinson et al. 2002). The results of these comparisons showed that many of these dye combinations could be separated using linear unmixing, despite the similarity in the spectra. In fact, dyes with as little as 4–5 nm of separation in the spectra could be unmixed if the fluorescence signals were balanced. Dyes such as enhanced GFP and FITC (fluorescein isothiocyanate) or Alexa 488, which have 7 nm of difference, could be unmixed reliably, even if very different levels of signal were present (Fig. 3.7). These experiments illustrate the power of multispectral imaging for eliminating cross talk. The result of linear unmixing for a sample labeled with Sytox Green and fluorescein is shown in Figure 3.8.

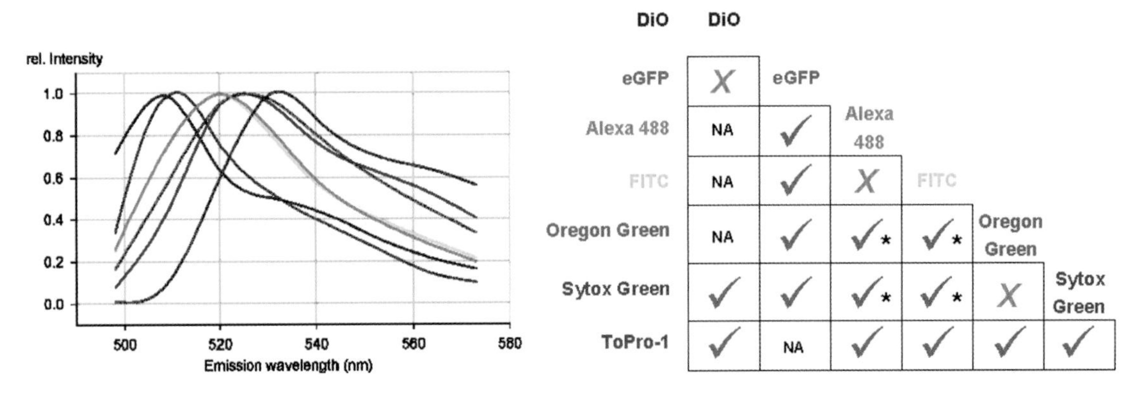

FIGURE 3.7

Summary of data from a pairwise comparison of the linear unmixing of green dyes. Spectra are shown to the left of individual fluorochromes listed on the right. Signals that could be unmixed are indicated by a check mark, whereas those that could not be unmixed are indicated by an *X*.

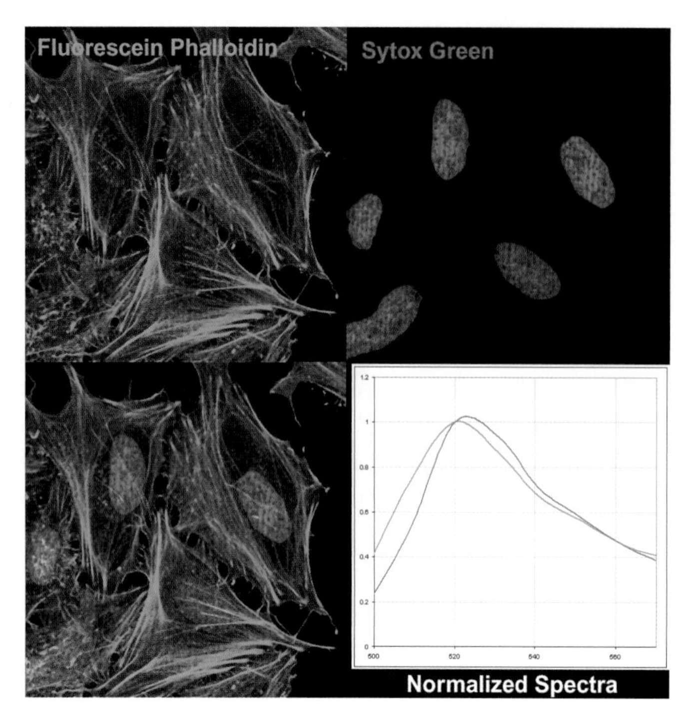

FIGURE 3.8

Linear unmixing of HeLa cells labeled with fluorescein phalloidin (actin; *green*) and Sytox green (nuclei; *red*). Linear unmixing produces two individual images for each dye shown at the top. The overlay of the individual images is shown at the bottom with the emission spectra of the two dyes. This figure was acquired first as both a *z* stack and a Lambda stack. A maximum intensity projection of the *z* stack was used for linear unmixing.

It is often asked how many bands of data or how many images in a Lambda stack are needed for accurate spectral separation. In some cases, particularly in the case of two components, a Lambda stack with images at only two wavelengths may be needed; however, the two-wavelength bands that are collected must be representative of the spectral differences between the dyes. Although they may provide accurate separation, acquiring data from throughout the emission spectra will increase the signal that is collected and may result in more sensitive detection. If samples contain more than one dye, more than two channels are needed to resolve all the components properly.

Principal Component Analysis

Although the best way to properly assign the identity of pixels in a Lambda stack is to use linear unmixing based on known components, it is also possible to identify individual component spectra in a Lambda stack or spectral cube using principal component analysis (PCA). This type of analysis is performed routinely in a large number of fields from astronomy to music to find patterns in data sets. The main goal of PCA is to reduce a complicated series of overlapping spectra into a series of individual peaks. This is done by mapping spectral intensities onto a set of orthogonal axes in order to reveal variances. PCA works best when spectra are spatially distinct in the sample so that individual spectral components are abundant. Once individual components are detected, they can be used for linear unmixing in order to produce images containing separate components.

Applications of MSLSM

MSLSM has only recently been introduced as an option for commercial systems for laser-scanning microscopy, yet several groups have already applied these methods to FRET analysis (Haraguchi et al. 2002; Zimmermann et al. 2003; Nashmi et al. 2004), live cell imaging of fluorescent protein variants (Haraguchi et al. 2002; Bertera et al. 2003), and elimination of autofluorescence (Dickinson et al. 2003). All of these instances represent key advantages of multispectral imaging over other methods, as cross-talk elimination without linear unmixing is extremely difficult if not impossible. In addition, the ease and speed of Lambda stack acquisition and linear unmixing make it an attractive option for many researchers.

CONCLUSIONS

Improvements in imaging technology have made it possible to use multidimensional methods for fluorescence imaging and quantitative analysis in biological samples. The methods described here add to the flexible strategies that are available to researchers interested in answering complex questions at the level of cells and molecules. The more tools available to isolate individual components and then study them in the context of other related components and processes, the greater the chance that our interpretations will reflect what really happens in biological systems.

ACKNOWLEDGMENTS

I thank (alphabetically) Greg Bearman, Scott Fraser, Zbi Iwinski, Richard Levenson, Eva Simbuerger, Sebastian Tille, and Ralf Wolleschensky for discussions and contributions to this chapter.

REFERENCES

Albota M., Beljonne D., Bredas J.L., Ehrlich J.E., Fu J.Y., Heikal A.A., Hess S.E., Kogej T., Levin M.D., Marder S.R., et al. 1998. Design of organic molecules with large two-photon absorption cross sections. *Science* **281:** 1653–1656.

APE (Berlin, Germany). http://www.ape-berlin.com.

Beechem J.M., Haugland R., Janes M., Clements I., Kilgore J., Salisbury J., and Ignatius M. 2003. Advanced spectroscopy of fluorescent indicators/dyes imaged in-situ with wavelength-resolved confocal laser-scanning microscopy (CLSM). *Biophys. J.* **84:** 587a.

Berney C. and Danuser G. 2003. FRET or No FRET: A quantitative comparison. *Biophys. J.* **84:** 3992–4010.

Bertera S., Geng X., Tawadrous Z., Bottino R., Balamurugan A.N., Rudert W.A., Drain P., Watkins S.C., and Trucco M. 2003. Body window-enabled in vivo multicolor imaging of transplanted mouse islets expressing and insulin-Timer fusion protein. *BioTechniques* **35:** 718–722.

Bestvater F., Spiess E., Stobrawa G., Hacker M., Feurer T., Porwol T., Berchner-Pfannschmidt U., Wotzlaw C., and Acker H. 2002. Two-photon fluorescence absorption and emission spectra of dyes relevant for cell imaging. *J. Microsc.* **208:** 108–115.

Bio-Rad, Inc. (Hercules, California). http://www. biorad. com.

Brown E.B., Campbell R.B., Tsuzuki Y., Xu L., Carmeliet P., Fukumura D., and Jain R.K. 2001. In vivo measurement of gene expression, angiogenesis and physiological function in tumors using multiphoton laser scanning microscopy. *Nat. Med.* **7:** 864–868.

Buehler C., Kim K.H., Greuter U., Schlumpf N., and So P.T.C. 2003. Multi-color two-photon scanning microscopy using a 16-channel photomultiplier. In *Progress in biomedical optics and imaging. Multiphoton microscopy in the biomedical sciences* (ed. A. Periasamy and P.T.C. So), vol. 426, pp. 217–230. Proceedings of the International Society for Optical Engineering (SPIE), Bellingham, Washington.

Campagnola P.J., Clark H.A., Mohler W.A., Lewis A., and Loew L.M. 2001. Second-harmonic imaging microscopy of living cells. *J. Biomed. Opt.* **6:** 277–286.

Carl Zeiss (Jena, Germany). http:// www.zeiss.com.

Chaigneau E., Oheim M., Audinat E., and Charpak S. 2003. Two-photon imaging of capillary blood flow in olfactory bulb glomeruli. *Proc. Natl. Acad. Sci.* **100:** 13081–13086.

Cheng J.X., Jia Y.K., Zheng G., and Xie X.S. 2002. Laser-scanning coherent anti-Stokes Raman scattering microscopy and applications to cell biology. *Biophys. J.* **83:** 502–509.

Coherent Inc. (Santa Clara, California). http://www. coherentinc.com.

D'Amore J.D., Kajdasz S.T., McLellan M.E., Bacskai B.J., Stern E.A., and Hyman B.T. 2003. In vivo multiphoton imaging of a transgenic mouse model of Alzheimer disease reveals marked thioflavine-S-associated alterations in neurite trajectories. *J. Neuropathol. Exp. Neurol.* **62:** 137–145.

Denk W., Piston D.W., and Webb W.W. 1995. Two-photon

molecular excitation in laser-scanning microscopy. In *Handbook of biological confocal microscopy* (ed. J.B. Pawley), pp. 445–458. Plenum Press, New York.

Denk W., Strickler J.H., and Webb W.W. 1990. Two-photon laser scanning fluorescence microscopy. *Science* **248:** 73–76.

Dickinson M.E., Bearman G., Tilie S., Lansford R., and Fraser S.E. 2001. Multi-spectral imaging and linear unmixing add a whole new dimension to laser scanning fluorescence microscopy. *BioTechniques* **31:** 1272, 1274–1276, 1278.

Dickinson M.E., Simbuerger E., Zimmermann B., Waters C.W., and Fraser S.E. 2003. Multiphoton excitation spectra in biological samples. *J. Biomed. Opt.* **8:** 329–338.

Dickinson M.E., Waters C.W., Bearman G., Wolleschensky R., Tilie S., and Fraser S.E. 2002. Sensitive imaging of spectrally overlapping fluorochromes using the LSM 510 META. In *Progress in biomedical optics and imaging. Multiphoton microscopy in the biomedical sciences* (ed. A. Periasamy and P.T.C. So), vol. 426, pp. 123–136. Proceedings of the International Society for Optical Engineering (SPIE), Bellingham, Washington.

Dunn K.W., Sandoval R.M., and Molitoris B.A. 2003. Intravital imaging of the kidney using multiparameter multiphoton microscopy. *Nephron Exp. Nephrol.* **94:** e7–11.

Dunn K.W., Sandoval R.M., Kelly K.J., Dagher P.C., Tanner G.A., Atkinson S.J., Bacallao R.L., and Molitoris B.A. 2002. Functional studies of the kidney of living animals using multicolor two-photon microscopy. *Am. J. Physiol. Cell. Physiol.* **283:** C905–C916.

Eng J., Lynch R.M., and Balaban R.S. 1989. Nicotinamide adenine dinucleotide fluorescence spectroscopy and imaging of isolated cardiac myocytes. *Biophys. J.* **55:** 621–630.

Farkas D.L. 2001. Spectral microscopy for quantitative cell and tissue imaging. In *Methods in cellular imaging* (ed. A. Periasamy), pp. 345–361. Published for the American Physiological Society by Oxford University Press, New York.

Favard C., Valisa P., Egret-Charlier M., Sharonov S., Herben C., Manfait M., Da Silva E., and Vigny P. 1999. A new UV-visible confocal laser scanning microspectrofluorometer designed for spectral cellular imaging. *Biospectroscopy* **5:** 101–115.

Ford B.K., Volin C.E., Murphy S.M., Lynch R.M., and Descour M.R. 2001. Computed tomography-based spectral imaging for fluorescence microscopy. *Biophys. J.* **80:** 986–993.

Garini Y., Gil A., Bar-Am I., Cabib D., and Katzir N. 1999. Signal to noise analysis of multiple color fluorescence imaging microscopy. *Cytometry* **35:** 214–226.

Goldman R.D. and Spector D.L., eds. 2005. *Live cell imaging: A laboratory manual.* Cold Spring Harbor Laboratory Press, Cold Spring Harbor, New York.

Goppert-Mayer M. 1931. Ueber Elementarakte mit zwei Quantenspruengen. *Ann. Phys. (Leipzig)* **9:** 273–295.

Hadjantonakis A.K., Dickinson M.E., Fraser S.E., and Papaioannou V.E. 2003. Technicolour transgenics: Imaging tools for functional genomics in the mouse. *Nat. Rev. Genet.* **4:** 613–625.

Hanley Q.S., Verveer P.J., and Jovin T.M. 1998. Optical sectioning fluorescence spectroscopy in a programmable array microscope. *Appl. Spectrosc.* **52:** 783–789.

Haraguchi T., Shimi T., Koujin T., Hashiguchi N., and Hiraoka Y. 2002. Spectral imaging fluorescence microscopy. *Genes Cells* **7:** 881–887.

Haralampus-Grynaviski N.M., Stimson M.J., and Simon J.D. 2000. Design and applications of a rapid-scan spectrally resolved fluorescence microscopy. *Appl. Spectrosc.* **54:** 1727–1733.

Helmchen F. 2000. Calibration of fluorescent calcium indicators. In *Imaging neurons: A laboratory manual* (ed. R. Yuste et al.), pp. 32.1–32.9. Cold Spring Harbor Laboratory Press, Cold Spring Harbor, New York.

———. 2002. Miniaturization of fluorescence microscopes using fibre optics. *Exp. Physiol.* **87:** 737–745.

Helmchen F. and Denk W. 2002. New developments in multiphoton microscopy. *Curr. Opin. Neurobiol.* **12:** 593–601.

Helmchen F., Tank D.W., and Denk W. 2002. Enhanced two-photon excitation through optical fiber by single-mode propagation in a large core. *Appl. Opt. 41:* 2930–2934.

Helmchen F., Fee M.S., Tank D.W., and Denk W. 2001. A miniature head-mounted two-photon microscope. High-resolution brain imaging in freely moving animals. *Neuron* **31:** 903–912.

Helmchen F., Svoboda K., Denk W., and Tank D.W. 1999. In vivo dendritic calcium dynamics in deep-layer cortical pyramidal neurons. *Nat. Neurosci.* **2:** 989–996.

Herman B., Gordon G., Mahajan N., and Centonze V. 2001. Measurement of fluorescence resonance energy transfer in the optical microscope. In *Methods in cellular imaging* (ed. A. Periasamy), pp. 257–272. Published for the American Physiological Society by Oxford University Press, New York.

Hiraoka Y., Shimi T., and Haraguchi T. 2002. Multispectral imaging fluorescence microscopy for living cells. *Cell. Struct. Funct.* **27:** 367–374.

IST Rees (Horseheads, New York). http://www.istspectech.com.

Jung J.C. and Schnitzer M.J. 2003. Multiphoton endoscopy. *Opt. Lett.* **28:** 902–904.

Keller-Peck C.R., Walsh M.K., Gan W.B., Feng G., Sanes J.R., and Lichtman J.W. 2001. Asynchronous synapse elimination in neonatal motor units: Studies using GFP transgenic mice. *Neuron* **31:** 381–394.

Kempe M. and Rudolph W. 1993. Femtosecond pulses in the focal region of lenses. *Phys. Rev. A* **48:** 4721–4729.

Konig K., Riemann I., Fischer P., and Halbhuber K.J. 2000. Multiplex FISH and three-dimensional DNA imaging with near infrared femtosecond laser pulses. *Histochem.Cell Biol.* **114:** 337–345.

Lakowicz J.R. 1999. *Principles of fluorescence spectroscopy.* Kluwer Academic/Plenum, New York.

Lansford R., Bearman G., and Fraser S.E. 2001. Resolution

of multiple green fluorescent protein color variants and dyes using two-photon microscopy and imaging spectroscopy. *J. Biomed. Opt.* **6:** 311–318.

Leica Microsystems Inc. (Bannookburn, Illinois). http://www.leica.com.

Levenson R.M. and Hoyt C.C. 2000. Spectral imaging and microscopy. *Am. Lab.* **32:** 26–34.

Millard A.C., Campagnola P.J., Mohler W., Lewis A., and Loew L.M. 2003. Second harmonic imaging microscopy. *Methods Enzymol.* **361:** 47–69.

Miller M.J., Wei S.H., Cahalan M.D., and Parker I. 2003. Autonomous T cell trafficking examined in vivo with intravital two-photon microscopy. *Proc. Natl. Acad. Sci.* **100:** 2604–2609.

Miller M.J., Wei S.H., Parker I., and Cahalan M.D. 2002. Two-photon imaging of lymphocyte motility and antigen response in intact lymph node. *Science* **296:** 1869–1873.

Mooney J.M., Vickers V.E., An M., and Brodzik A.K. 1997. High-throughput hyperspectral infrared camera. *J. Opt. Soc. Am.* **14:** 2951–2961.

Morris H.R., Hoyt C.C., and Treado P.J. 1994. Imaging spectrometers for fluorescence and raman microscopy—Acoustooptic and liquid-crystal tunable filters. *Appl. Spectrosc.* **48:** 857–866.

Muller M., Squier J., De Lange C.A., and Brakenhoff G.J. 2000. CARS microscopy with folded BoxCARS phase-matching. *J. Microsc.* **197:** 150–158.

Nan X., Cheng J.X., and Xie X.S. 2003. Vibrational imaging of lipid droplets in live fibroblast cells with coherent anti-stokes raman scattering microscopy. *J. Lipid Res.* **44:** 2202–2208.

Nashmi R., Dickinson M.E., McKinney S., Jareb M., Labarca C., Fraser S.E., and Lester H.A. 2004. Effects of localization, trafficking, and nicotine-induced upregulation in clonal mammalian cells and in cultured midbrain neurons. *J. Neurosci.* **23:** 11554–11567.

Netz R., Feurer T., Wolleschensky R., and Sauerbray R. 2000. Measurement of the pulse-front distortion in high numerical aperture lenses. *Appl. Phys.* **B70.**

Ocean Optics (Dunedin, Florida). http://www.oceanoptics.com.

Oheim M., Beaurepaire E., Chaigneau E., Mertz J., and Charpak S. 2001. Two-photon microscopy in brain tissue: Parameters influencing the imaging depth. *J. Neurosci. Methods* **111:** 29–37.

Ornberg R.L., Woerner B.M., and Edwards D.A. 1999. Analysis of stained objects in histological sections by spectral imaging and differential absorption. *J. Histochem. Cytochem.* **47:** 1307–1314.

Padera T.P., Stoll B.R., So P.T., and Jain R.K. 2002. Conventional and high-speed intravital multiphoton laser scanning microscopy of microvasculature, lymphatics, and leukocyte-endothelial interactions. *Mol. Imaging 1:* 9–15.

Periasamy A. 2001. Fluorescence resonance energy transfer microscopy: A mini review. *J. Biomed. Opt.* **6:** 287–291.

Periasamy A., Elangovan M., Wallrabe H., Barroso M., Demas J.N., Brautigan D.L., and Day R.N. 2001. Wide-field, confocal, two-photon, and lifetime resonance energy transfer imaging microscopy. In *Methods in cellular imaging* (ed. A. Periasamy), pp. 295–308. Published for the American Physiological Society by Oxford University Press, New York.

Ragan T.M., Huang H., and So P.T. 2003. In vivo and ex vivo tissue applications of two-photon microscopy. *Methods Enzymol.* **361:** 481–505.

Richmond K.N., Burnite S., and Lynch R.M. 1997. Oxygen sensitivity of mitochondrial metabolic state in isolated skeletal and cardiac myocytes. *Am. J. Physiol.* **273:** C1613–C1622.

Rumi M., Ehrlich J.E., Heikal A.A., Perry J.W., Barlow S., Hu Z.Y., McCord-Maughon D., Parker T.C., Rockel H., Thayumanavan S., et al. 2000. Structure-property relationships for two-photon absorbing chromophores: Bis-donor diphenylpolyene and bis(styryl)benzene derivatives. *J. Am. Chem. Soc.* **122:** 9500–9510.

Schrock E., du Manoir S., Veldman T., Schoell B., Wienberg J., Ferguson-Smith M.A., Ning Y., Ledbetter D.H., Bar-Am I., Soenksen D., et al. 1996. Multicolor spectral karyotyping of human chromosomes. *Science* **273:** 494–497.

Spectra Physics (Mountain View, California). http://www.spectraphysics.com/.

Stosiek C., Garaschuk O., Holthoff K., and Konnerth A. 2003. In vivo two-photon calcium imaging of neuronal networks. *Proc. Natl. Acad. Sci.* **100:** 7319–7324.

Svoboda K., Helmchen F., Denk W., and Tank D.W. 1999. Spread of dendritic excitation in layer 2/3 pyramidal neurons in rat barrel cortex in vivo. *Nat. Neurosci.* **2:** 65–73.

Trachtenberg J.T., Chen B.E., Knott G.W., Feng G., Sanes J.R., Welker E., and Svoboda K. 2002. Long-term in vivo imaging of experience-dependent synaptic plasticity in adult cortex. *Nature* **420:** 788–794.

Tsurui H., Nishimura H., Hattori S., Hirose S., Okumura K., and Shirai T. 2000. Seven-color fluorescence imaging of tissue samples based on Fourier spectroscopy and singular value decomposition. *J. Histochem. Cytochem.* **48:** 653–662.

Wachman E.S., Niu W., and Farkas D.L. 1997. AOTF microscope for imaging with increased speed and spectral versatility. *Biophys. J.* **73:** 1215–1222.

Walsh M.K. and Lichtman J.W. 2003. In vivo time-lapse imaging of synaptic takeover associated with naturally occurring synapse elimination. *Neuron* **37:** 67–73.

Wang W., Wyckoff J.B., Frohlich V.C., Oleynikov Y., Huttelmaier S., Zavadil J., Cermak L., Bottinger E.P., Singer R.H., White J.G., et al. 2002. Single cell behavior in metastatic primary mammary tumors correlated with gene expression patterns revealed by molecular profiling. *Cancer Res.* **62:** 6278–6288.

Wiegmann T.B., Welling L.W., Beatty D.M., Howard D.E., Vamos S., and Morris S.J. 1993. Simultaneous imaging of intracellular [Ca^{2+}] and pH in single MDCK and glomerular epithelial cells. *Am. J. Physiol.* **265:** C1184–C1190.

Wise F.W. 2000. Lasers for multiphoton microscopy. In *Imaging neurons: A laboratory manual* (ed. R. Yuste et al.), pp. 18.1–18.9. Cold Spring Harbor Laboratory Press, Cold Spring Harbor, New York.

Wokosin D.L. and Girkin J.M. 2002. Practical multiphoton microscopy. In *Confocal and two-photon microscopy: Foundations, applications, and advances* (ed. A. Diaspro), pp. 207–235. Wiley-Liss, New York.

Wolleschensky R., Dickinson M.E., and Fraser S.E. 2002. Group-velocity dispersion and fiber delivery in multiphoton laser scanning microscopy. In *Confocal and two-photon microscopy: Foundations, applications, and advances* (ed. A. Diaspro), pp. 171–189. Wiley-Liss, New York.

Xu C. 2000. Two-photon cross sections of indicators. In *Imaging neurons: A laboratory manual* (ed. R. Yuste et al.), pp. 19.1–19.9. Cold Spring Harbor Laboratory Press, Cold Spring Harbor, New York.

Xu C., Zipfel W., Shear J.B., Williams R.M., and Webb W.W. 1996. Multiphoton fluorescence excitation: New spectral windows for biological nonlinear microscopy. *Proc. Natl. Acad. Sci.* **93:** 10763–10768.

Xu Y.W., Zhang J.R., Deng Y.M., Hui L.K., Jiang S.P., and Lian S.H. 1987. Fluorescence of proteins induced by two-photon absorption. *J. Photochem. Photobiol. B* **1:** 223–227.

Yang V.X., Muller P.J., Herman P., and Wilson B.C. 2003. A multispectral fluorescence imaging system: Design and initial clinical tests in intra-operative Photofrin-photodynamic therapy of brain tumors. *Lasers. Surg. Med.* **32:** 224–232.

Zangaro R.A., Silveira L., Manoharan R., Zonios G., Itzkan I., Dasari R.R., VanDam J., and Feld M.S. 1996. Rapid multiexcitation fluorescence spectroscopy system for in vivo tissue diagnosis. *Appl. Opt.* **35:** 5211–5219.

Zeng H.S., Weiss A., MacAulay C., and Cline R.W. 1999. System for fast measurements of in vivo fluorescence spectra of the gastrointestinal tract at multiple excitation wavelengths. *Appl. Opt.* **38:** 7157–7158.

Zimmermann T., Rietdorf J., and Pepperkok R. 2003. Spectral imaging and its applications in live cell microscopy. *FEBS Lett.* **546:** 87–92.

CHAPTER 4

Preparation of Cells and Tissues for Fluorescence Microscopy

Andrew H. Fischer,[1] Kenneth A. Jacobson,[2] Jack Rose,[3] and Rolf Zeller[4]

[1]*Emory University Hospital, Atlanta, Georgia*
[2]*University of North Carolina, Chapel Hill, North Carolina*
[3]*Yale University School of Medicine, New Haven, Connecticut*
[4]*University of Basel, Basel, Switzerland*

INTRODUCTION

Fluorescence microscopy is one of the most widely used approaches for localizing proteins and subcellular compartments at the light microscopic level. The strategies described in the following eight chapters take advantage of the sensitivity and specificity of nonimmunological as well as immunologically based fluorescent probes for revealing structure–function relationships. A selected group of protocols is provided to serve as representative examples; many others may be found in the literature.

Nonimmunological fluorescent probes may be used to directly label specific subcellular components and macromolecules such as the Golgi apparatus or components of the nuclear pore complex (for a complete discussion, see Chapter 5). Typical fluorescent groups associated with probes used in these studies are rhodamine, fluorescein, BODIPY, NBD, and the carbocyanines (available from Molecular Probes). Many of these probes are directly taken up into living cells and are incorporated and concentrated in specific organelles that can then be examined using the fluorescence microscope. A protocol for mounting live cells for microscopic examination is provided below.

Immunofluorescence, another widely used application of fluorescence microscopy, involves the use of antibodies, obtained by conventional methods of antibody generation and purification (see Harlow and Lane 1988) or derived from the serum of individuals with a variety of autoimmune disorders, to localize a particular protein or other antigen. Direct or one-step immunofluorescence involves the conjugation of a fluorophore (e.g., fluorescein isothiocyanate [FITC]) directly to the primary antibody, thereby making the localization a one-step procedure. Indirect immunofluorescence, which is used somewhat more frequently, involves the initial binding of the primary antibody to the antigen in a fixed cell. Subsequently, a fluorescently labeled secondary antibody is incubated with the specimen to form a fluorescent "sandwich" at the site of the target. The secondary antibody is chosen to react with the host species in which the primary antibody was raised. For example, if the primary antibody was raised in a mouse, the secondary antibody may be X-species-anti-mouse (e.g., goat anti-mouse IgG). Secondary antibodies are available as affinity-purified reagents from a variety of companies and can be purchased coupled to many different fluorophores.

Typically, the overall scheme for localization of a cellular protein by indirect immunofluorescence involves:

- Fixation (unless live cells are to be studied)
- Permeabilization of fixed cells to allow penetration of antibodies
- Blocking sites prone to nonspecific interactions
- Labeling the fixed cells/tissues with specific antibodies
- Mounting the sample for microscopic examination

Specific details for fixing and processing samples are given for different systems in the following chapters, along with a discussion of the advantages and applications of the various probes to be used. In general, the choice of conditions for pretreatment and labeling of the sample is dictated by both the nature of the sample and the type of labeling procedure to be used. This chapter presents an overview of the methods for preparing slides and coverslips for fixed and live specimens, a discussion of various approaches to fixation of cells and tissues (with subsequent sectioning of tissue specimens), and a collection of procedures used for mounting live and fixed cells. For extensive protocols on approaches for live cell imaging, see Goldman and Spector (2005).

Preparation of Slides and Coverslips

The following procedures describe various approaches for cleaning slides and coverslips and sterilizing them for cell culture, followed by methods for coating slides and coverslips with a solution that will promote the adhesion of cells or tissues to the glass surface.

CLEANING COVERSLIPS AND SLIDES

It is imperative that the slides and coverslips used in fluorescence procedures be extremely clean. Although coverslips look clean, especially when a new box is first opened, they may have a thin film of grease on them that will not allow tissue culture cells to adhere well and that may interfere with some processing steps in certain protocols. Therefore, coverslips should routinely be washed with acid or base solutions to rid them of this film. Commercial precleaned slides are also likely to be dirty and must be washed prior to use. It is heartbreaking when the final result of an experiment is sitting next to or on top of a piece of dirt.

Primary cells do not attach well to glass slides or coverslips. In these cases, coverslips can be coated with different growth substrates that enhance their adhesion to glass (such as MatriGel or rat tail collagen) or they can be grown on ACLAR plastic (Ted Pella) that is not autofluorescent. However, it should be noted that plastic coverslips are not optimal for most light microscopic preparations and therefore glass should be used whenever possible.

Cleaning Coverslips with Acid

1 Make up 300 ml of two parts **nitric acid** to one part **HCl** in a glass beaker in the hood (solution will turn orange-red).

CAUTION: Nitric acid; HCl; acids and bases that are concentrated (see Appendix 2 for Caution)

2 Place 10 oz of #1.5 coverslips into the acid solution a few at a time so that they are separated and do not break, then allow them to sit for ~2 hours with occasional swirling.

#1.5 coverslips are of a thickness that is compatible with achieving focus and high resolution with most objective lenses. Furthermore, they are less likely to break than #1 coverslips during handling.

3 Decant the acid carefully into a waste receptacle.

4 Wash the coverslips thoroughly in running tap water until the pH of the wash water is back to ~5.5 to 6.0.

5 Store the coverslips in a covered container submerged in 70% ethanol.

6 Flame each coverslip in the tissue culture hood prior to its use.

Cleaning Coverslips with Base

1 Incubate coverslips for 2 hours in 2 N **NaOH**.

2 Rinse extensively in dH_2O.

3 Follow steps 5 and 6 above.

CAUTION: NaOH (see Appendix 2 for Caution)

Cleaning Slides

1 Wash slides (25 x 75-mm-precleaned slides) with a liquid detergent for a few minutes.

2 Rinse slides in H_2O for 30 minutes and dry.

SUBBING OR COATING SLIDES AND COVERSLIPS

In a process called subbing, slides or coverslips are coated with a solution that promotes adhesion of cells or tissue to the surface. Examples of such solutions include gelatin, aminoalkyl **silane**, and poly-L-lysine. Gelatin or aminoalkyl silane is usually used for tissue sections or small organisms, whereas poly-L-lysine is routinely used for cultured cells. After extensive cleaning as outlined above, slides or coverslips are subbed. If coverslips are to be used for cell culture, they should be sterilized uncovered in a laminar flow hood by exposure for 45 minutes to **UV light** after air-drying. Coverslips can be stored at room temperature for several weeks.

Gelatin Subbing

1 Prepare the subbing solution:

a Dissolve gelatin (type Bloom 225; Sigma G 9382) in H_2O at 60°C to make a 0.2% solution.

b Cool the solution to 40°C and add chromium potassium sulfate to 0.02%.

c Cool the solution to 4°C and use immediately, or store at 4°C for several weeks.

2 Place clean slides or coverslips in appropriate racks and immerse into a glass staining dish filled with the above gelatin subbing solution at 4°C for 2 minutes.

Perform the subbing carefully to avoid formation of bubbles on the slide surface.

3 Remove the slide rack from the subbing solution and set it on its side to let the excess solution drain off.

4 Dry slides overnight before use.

Slides and coverslips subbed with gelatin are stable at room temperature for several weeks.

Silanizing Slides

1 Acid-clean the slides in 2 N **HCl** for 5 minutes.

2 Rinse thoroughly in dH_2O.

3 Rinse in **acetone** and air-dry.

4 Immerse slides in a freshly prepared 2% solution of 3-aminopropyltriethoxysilane (aminoalkylsilane) in acetone (Sigma) with agitation for 2 minutes.

5 Rinse slides in distilled dH_2O and air-dry.
Slides may be stored for over 5 years (Nuovo 1997).

CAUTION: Silane; UV light; HCl; acetone (see Appendix 2 for Caution)

Poly-L-lysine Coating

1 Prepare a suitable amount of 500 µg/ml poly-L-lysine (m.w. >150,000; Sigma P 1399) in H_2O.
For some cell types (e.g., for the culture of neurons), a higher concentration of poly-L-lysine (1 mg/ml) may be more appropriate.

2 Coat slides or coverslips by dipping into the poly-L-lysine solution or by applying enough solution to cover the glass surface.

3 Incubate slides/coverslips at room temperature for 10 minutes, and then wash three times with sterile dH_2O.

4 Allow to air-dry.
Poly-L-lysine-coated coverslips can be made more hospitable to cell attachment (and, for example, neurite outgrowth) by the addition of a coating of laminin (2–5 µg/cm^2, GIBCO/BRL). Coated coverslips should be air-dried at least 45 minutes prior to plating. It is not advisable to store laminin-coated coverslips.

Fixation and Permeabilization

Fluorescence microscopy is used to visualize specific cellular components in as native a state and organization as possible. To preserve cellular structure, the specimen is fixed chemically to retain the cells or tissue in a state as near to life as possible by rapidly terminating all enzymatic and other metabolic activities to minimize postfixation changes. Sample fixation is one of the most crucial steps in assuring the accuracy of detection protocols and is therefore decisive in determining the subsequent success or failure of a given experiment. Underfixation of the sample leads to poor morphological preservation and/or loss of signal, whereas overfixation may lead to fixation artifacts, loss of signal, and/or increased nonspecific background signals ("noise"). An ideal fixative should preserve a given antigen in a fashion that reflects the in vivo situation with respect to its distribution (no diffusion or rearrangement). Ideally, cell morphology should be preserved, the antigen of interest should remain accessible to the probe, and the fixation should cause minimal denaturation of the antigen. However, several of these goals are mutually incompatible, and therefore a compromise must be attained.

Glutaraldehyde, formaldehyde, and methanol/acetone are the most commonly used fixatives. Glutaraldehyde, a five-carbon dialdehyde, provides the best preservation of fine structure, and therefore is the fixative of choice for electron microscopy (see Chapter 20) and some immunofluorescence studies (such as microtubule localization (see Chapter 7). However, it is also the harshest of the fixatives, and frequently epitopes are sufficiently altered to prevent binding of the antibody probe. Glutaraldehyde forms a Schiff's base with amino groups on proteins and polymerizes via Schiff's base-catalyzed reactions (Johnson 1985). The ability to polymerize allows glutaraldehyde to form extended cross-links. For immunocytochemistry, 0.01–0.5% glutaraldehyde has been useful in some studies. Two percent glutaraldehyde can penetrate ~ 700 μm into tissue in 1 hour at room temperature. Glutaraldehyde fixation also contributes to nonspecific fluorescence at some excitation wavelengths. This autofluorescence may be partially reduced by treating cells with 1.0–1.5 mg/ml sodium borohydride in PBS, two times, 5 minutes each, after fixation.

Formaldehyde, a one-carbon monoaldehyde, made fresh before use from paraformaldehyde, is widely used and is an excellent general fixative for the localization of most proteins. Formaldehyde does not interfere with epitope recognition to the extent that glutaraldehyde does. In most cases, for fluorescence microscopy, it is the best choice of fixative. Formaldehyde cross-links proteins by forming methylene bridges between reactive groups. Formaldehyde is a less effective cross-linker than glutaraldehyde for several reasons. First, because formaldehyde is a small molecule, many reactive groups in cells escape formation of bridging links. Second, many of the cross-links induced by formaldehyde fixation are reversible, and, thus, tissue fixed in formaldehyde alone often becomes unfixed when exposed to certain buffers during postfixation washing and processing. Because it is a milder fixative, it is extensively used in histochemical and immunocytochemical studies and it is the most effective fixative for nucleic acids (see Chapters 15 and 18). Four percent formaldehyde penetrates ~2 mm in 1 hour at room temperature. In addition, formaldehyde does not contribute significantly to autofluorescence. Cultured cells are usually fixed in 2–4% formaldehyde in PBS, pH 7.4, at 20°C for 15 minutes. *Because most commercially available bulk liquid forms of formaldehyde contain methanol they should not be used in place of preparing formaldehyde fresh from paraformaldehyde.* In some cases, a combination of glutaraldehyde and formaldehyde has been used, because formaldehyde fixes less well but penetrates into tissue quickly to stabilize cellular constituents, whereas glutaraldehyde fixes more thoroughly but penetrates more slowly.

Cold methanol or acetone solutions provide more rapid fixation than the aldehydes and have been used in a variety of studies to examine, for example, components of the cytoskeleton. However, because these fixatives work by precipitating proteins and carbohydrates, they are more likely to alter the localization pattern of some antigens. In addition, because they fix and permeabilize cells at the same time, some of the more soluble antigens may be lost in the preparation protocol. Furthermore, dehydration and fixation occur simultaneously, resulting in possible shrinkage of the samples.

If an intracellular component is to be visualized following aldehyde fixation, it is also necessary to permeabilize the cell, either with detergents or organic solvents that solubilize or extract lipids from the plasma membrane and nuclear envelope and thereby allow probes such as antibodies to gain access to the subcellular structures of interest. Permeabilization is not required for the localization of a cell surface antigen or antigens associated with the extracellular matrix. Some of the more commonly used permeabilization agents include Triton X-100, NP-40, and Brij-58, which solubilize phospholipid membranes. These detergents are most commonly used at concentrations ranging between 0.1% and 0.5% at room temperature or on ice for 5 minutes and are used after fixation. However, some protocols suggest fixing and permeabilizing cells at the same time or permeabilizing cells prior to fixation. Saponin is a detergent that solubilizes cholesterol in the membrane and does less damage to the membrane than the detergents discussed above. However, it must be present during the probe (e.g., antibody) incubation steps and subsequent washes, because the membrane reseals if the saponin is removed from the solution. Saponin has been used in studies where membrane structure is important, and it is commonly used at 0.5% at room temperature. Digitonin is a detergent that selectively permeabilizes the plasma membrane by binding cholesterol and other β-hydroxysterols (Stearns and Ochs 1982). Because of the low level of cholesterol in other membranes, the plasma membrane is selectively permeabilized. Digitonin has been used at a concentration of 50 μg/ml at room temperature for 5 minutes (concentration may need to be optimized for various cell types) and it has been used in fluorescence-based nuclear transport assays.

Methanol or acetone has also been used to permeabilize cells, usually at low temperatures ranging from –20°C to +14°C. These reagents have the advantage of rapid one-step fixation and permeabilization, and they are most useful when studying major protein systems in cells, especially those forming cytoskeletal systems such as actin-containing stress fibers (see Chapter 8) and intermediate filament networks (see Chapter 7). It should be reemphasized, however, that these reagents are harsh and may not be the optimal choice for studies involving more labile cellular antigens.

For a particular application, the best fixation and permeabilization scheme should be determined empirically. The following chapters in this manual provide recommendations and guidelines for fixation and permeabilization for specific organelles or components to be studied. Further discussion of these procedures is provided in Wang et al. (1982). The following section presents protocols for sectioning tissues that may be fixed or unfixed, for staining tissue sections, and for dealing with autofluorescence.

SECTIONING TISSUES

There are three commonly used sectioning methodologies for studying the morphology of tissues at the level of light microscopy: cryosections (frozen sections), paraffin sections, and plastic (or methacrylate) sections. The major advantage of cryosectioning is that the tissue is unfixed, thereby saving time and allowing portions of the frozen tissue to be relatively easily used for immunohistochemistry, genetic studies, or biochemical studies. The disadvantages of cryosectioning are that the quality of the sections is usually inferior to paraffin sections, and certain tissues (e.g., adipose tissue, dense tissues such as wood or chitinous tissue, or calcified materials) are difficult or impossible to section. Plastic is more firm than paraffin, allowing exceptionally thin sections to be cut (1 μm for light microscopy and much thinner for electron microscopy), and allowing very hard tissues including calcified bone to be sectioned. Both paraffin sections and plastic embedding require extensive fixation and processing steps; however, they provide a superior morphology compared with cryosections. Paraffin sections of bone usually require a decalcification step after fixation. The advantages of cryosections over paraffin sections are decreasing as improved techniques are being developed for using paraffin-embedded tissues in genetic studies and immunohistochemistry (Werner et al. 1996; Nuovo 1997).

Sectioning paraffin blocks containing samples requires experience and should be learned from an experienced researcher if possible. Some institutions have service facilities that perform sectioning. If these facilities are used, it will be necessary to provide instructions about the thickness of sections and orientation for mounting on subbed slides (see p. 108). Typically, paraffin blocks containing the specimen are cut into thin (8-μm) tissue sections that are then mounted on subbed slides for further processing and detection. A detailed protocol for embedding in paraffin and sectioning paraffin blocks is presented below.

Cryosections are rapidly and relatively easily prepared prior to fixation, and they provide a good system for visualizing fine details of the cell. Although cryosections are physically less stable than paraffin- or resin-embedded sections, they are generally superior for the preservation of antigenicity and therefore the detection of antigens by microscopy. The preparation of cryosections does not involve the dehydration steps typical of other sectioning methods, and, furthermore, sectioning, labeling, and observation of specimens can usually be carried out in one day. In general, the sample is frozen quickly in either isopentane or liquid nitrogen. (Small samples such as cells and small tissues may be mixed in a slurry of an inert support medium such as OCT compound before freezing). Rapid freezing reduces ice crystal formation and minimizes morphological damage. Frozen sections may be used for a variety of procedures, including immunochemistry, enzymatic detection (for further details, see Hollands 1962), and in situ hybridization. A protocol for cryosectioning is presented below.

Water-soluble methyl methacrylate enables the production of semithin sections that lead to improved resolution for light microscopy. The ability to infiltrate and polymerize both soft and hard tissues at room temperature without prior removal of water results in better preservation and improved tissue morphology, which reduces artifacts. Plastic-embedding medium provides sections with considerably less distortion and shrinkage than comparable paraffin sections. Because the processing procedure does not include harsh organic solvents or heat, the preservation of delicate biological structures and histochemical reaction products leads to improved detection compared with conventional paraffin sections.

The principal disadvantage of methyl methacrylate processing is the initial equipment investment. Some of the cost can be reduced by purchasing a microtome with dual capabilities for paraffin and plastic. Thereafter, the only expense will be the cost of consumables. The majority of reusable materials can be used for both plastic- and paraffin-embedding procedures. An additional difficulty is the toxicity and skin irritating qualities of methyl methacrylate chemicals. Good laboratory handling procedures must be followed at all times, and the use of a fume hood and gloves is recommended.

Cryosectioning

1 Freeze a fresh unfixed tissue sample, up to 2.0 cm in diameter, in Tissue Tek OCT Compound (Sakura Finetek U.S.A. Inc., Torrance, California). Plastic or metal molds of various sizes are sold by many supply companies (such as Fisher, VWR, Shandon-Lipshaw).

The OCT compound is viscous at room temperature and miscible with water, but freezes into a solid support at –20°C. The OCT containing the tissue is frozen onto special metal grids that fit onto the cryostat. Cryostats are essentially a –20° freezer enclosing a microtome.

Certain soft tissues such as brain are optimally frozen in M-1 medium (Shandon-Lipshaw) at –3°C.

Cryostats are expensive, but in many medical centers, hospital pathology laboratories have cryostats that can be rented.

2 Cut sections 5–15-µm thick in the cryostat at –20°C. The temperature of the cutting chamber may have to be adjusted ±5°C for some tissues.

If the tissue frozen in OCT does not cut in a smooth thin sheet, the knife is probably dull.

A camel hair brush is useful to help guide the emerging section over the knife blade.

Watery tissues, fatty tissues, or tissues with variable textures are difficult to section.

3 Transfer the section to a microscope slide. Take a microscope slide at room temperature and touch it to the tissue section. The tissue section then melts onto the slide. To avoid freeze-drying of the tissue section, it must be transferred to the slide within a minute. Poly-L-lysine-coated or silanized slides (as described above) improve the adherence of the section.

Toluidine blue (1–2% in H_2O), hematoxylin and eosin (see p. 117), or any aqueous stain should be used on the first slide to evaluate tissue preservation and orientation.

4 Immediately immerse the slide into an appropriate fixative. See instructions under Fixation and Permeabilization (p. 110) for guidelines on choosing an appropriate fixative.

To maximize the adherence of the section to the slide, some researchers allow the section to air-dry onto the slide at room temperature before fixing the sample. The disadvantage of air-drying the sample is that surface tension forces distort the cells, causing loss of high-resolution detail. Air-drying may also cause some changes in immunostaining results.

5 Cover any unused tissue with a layer of OCT compound to prevent freeze-drying and store the rest of the sample at –70°C.

For long-term storage, a moistened Kimwipe should be added to the container with the block to prevent desiccation (particularly in a frost-free freezer).

Paraffin Sections

The following protocols describe embedding tissues with paraffin, decalcifying fixed calcified tissue for paraffin embedding, and cutting paraffin sections.

Paraffin-embedding Tissue Samples

1 Prepare tissue by cutting into ~2 mm in thickness, and up to 2 cm in length and width. Place tissue into tissue cassettes (e.g., Simport biopsy cassettes, Fisher Scientific). Large cassettes are also available for whole mounts.

Occasionally it is desirable to paraffin-embed very tiny tissue fragments or even cell suspensions. A useful technique is to centrifuge the sample in a glass 15-ml centrifuge tube that has an inner coating of collodion

(EM Science). The centrifuge tube is prepared by adding a small amount of the collodion solution into the tube, swirling the solution to completely coat the inside, then inverting and drying the collodion for 10–15 minutes. The sample is then added to the tube and centrifuged at 500–1000g for 15 minutes. The collodion coating containing the pellet can be pulled from the tube as a thin sac and processed like tissue for paraffin sectioning. Alternatively, small samples can be concentrated and embedded in 2% agar.

Automatic processors for paraffin embedding are also available.

2 Fix tissue by immersion in at least 10 volumes of fixative (typically 10% neutral buffered formalin). Typical fixation time is 2 hours to overnight. Fixation should be standardized for a given procedure, because increasing time in fixative can alter immunoreactivity.

For small pieces of tissue, 1 mm or less in thickness, all incubation times given can be cut in half.

3 Dehydrate tissue in steps by stirring in at least 10 volumes of the following concentration series of alcohols at room temperature:

70%	1 hour
95%	1 hour
95%	1 hour
100%	1 hour
100%	1 hour
100%	1 hour

Ethanol, or mixtures of **methanol** *and isopropyl alcohol (e.g., Flex alcohols, Richard-Allen Scientific, Kalamazoo, Michigan) are most often used. The alcohol is mixed with H_2O to achieve the desired concentration. If all H_2O is not removed from the tissue, then subsequent processing steps will fail.*

CAUTION: Methanol (see Appendix 2 for Caution)

4 "De-fat" the tissue in two changes of at least 10 volumes of **xylene** at room temperature for 1 hour each, with stirring.

CAUTION: Xylene (see Appendix 2 for Caution)

This step is essential to remove all of the alcohol from the previous step and to remove the fat that would otherwise make the paraffin block soft and difficult to cut.

Long-term chronic exposure to xylene can lead to serious health problems. Xylene substitutes are available (Fisher).

5 Infiltrate with paraffin. Melt paraffin (e.g., Surgipath Medical Industries) by heating to 58–60°C. Stir the cassette with the dehydrated, defatted tissue in 10 volumes of melted paraffin. Change the paraffin three times every hour.

Monitor the temperature to be sure it is kept at 58–60°C.

6 Embed the tissue in paraffin. Remove the tissue from the cassette. Use a matchbox as a mold, or purchase a stainless steel or vinyl mold (HistoPrep Base molds, Fisher Scientific). Place a small amount of melted paraffin into the bottom of the box or mold and push the tissue flat against the bottom of the box. Invert the original tissue cassette over the matchbox or mold and pour in paraffin to cover the base of the cassette. Make certain that the base of the cassette is parallel with the tissue, because the base of the tissue cassette will ultimately be held by the microtome. After the paraffin hardens, remove the mold/box. Paraffin tissue blocks can be stored at room temperature with little degradation of immunoreactivity or nucleic acids for long periods of time.

Decalcifying Tissues for Paraffin Embedding

1 Fix the sample with **formaldehyde**. For large pieces of calcified tissue, it is desirable to cut the fresh tissue into 2- to 3-mm-thick slices with a saw. This step improves the speed of fixation and eventual decalcification. It is essential to have well-fixed tissue before decalcification with this procedure because the acids used for decalcification would otherwise damage the tissue.

CAUTION: Formaldehyde (see Appendix 2 for Caution)

2 Rinse tissue for 10 minutes in tap water.

3 Make a decalcifying solution as follows:

88% **formic acid** stock	100 ml
Concentrated HCl	80 ml
dH_2O	820 ml

CAUTION: Formic acid; HCl; acids and bases that are concentrated (see Appendix 2 for Caution)

Other decalcifying solutions with chelators or ion exchange resins may be used (Sheehan and Hrapchak 1980).

4 Immerse the tissue in 100 volumes of decalcifying solution and stir.

Do not mix the solution with formaldehyde, because this produces the carcinogenic compound bis-chloromethyl ether.

5 Monitor the decalcification by trying to flex the tissue. Typically an overnight incubation is required.

Overdecalcification leads to extensive depurination of the DNA, interfering with staining of the nuclei. Decalcification can interfere with immunodetection. Any such effect on immunodetection must be determined empirically for each antigen.

6 Rinse with running cold tap water for 1 hour and proceed to step 3 (dehydration) of the protocol for paraffin embedding.

Cutting Paraffin Sections

1 Prepare a 42–48°C clean dH_2O bath, at least 6 inches in diameter, next to the microtome. Commercial water baths designed for paraffin sectioning are available (Fisher TissuePrep flotation bath, Fisher Scientific). Add 50 mg of gelatin (Fisher Scientific) per liter of H_2O if using unsubbed slides. Gelatin increases the adhesion of the section to glass. Gelatin should not be used if subbed slides are used (e.g., silanized slides).

2 Place the tissue block on ice or cool it to 0–4°C. This makes the paraffin harder, and therefore easier to cut into thin ribbons.

3 Lock the microtome handwheel and get the microtome knife out of the way before loading the paraffin block.

Microtomes may be purchased (Fisher Scientific, VWR, Shandon-Lipshaw). Alternatively, a hospital pathology laboratory may be willing to cut sections for a fee.

4 Mount the tissue cassette bearing the embedded tissue sample into the tissue cassette holder of the microtome. Trim the edges of the paraffin block if necessary to make certain the edge of the knife is parallel with the upper and lower edges of the tissue block, otherwise a ribbon cannot be cut.

5 Advance the microtome knife using coarse adjustments to within ~1 mm of the block. Fine-adjust the knife and/or block position so that the block face is parallel to the sweep of the knife blade. Set the angle of the microtome blade to ~3–8° from the face of the tissue block. This is the minimal angle, measured from the block to the edge of the backside of the knife bevel.

6 Set the section thickness. The lower limit for most paraffin-embedded tissues is ~3 μm in thickness.

7 Face the block (i.e., trim to form a smooth surface), and then prepare a ribbon of tissue sections of the desired thickness.

8 The ribbon can be picked up carefully by hand and floated onto the water bath. Wooden sticks may be used to manipulate the floating sections and stretch the ribbon if needed to remove wrinkles.

If the tissue does not form a ribbon easily, or if the ribbon wrinkles cannot be removed without tearing the tissue, one or more of the following problems are likely:

- The knife is dull.
- The knife angle is improper.
- The tissue is not properly infiltrated with paraffin. Poorly infiltrated tissue shrinks within the wax block, dipping below the surrounding surface. Reprocessing the tissues is not recommended because deterioration of the morphology is almost inevitable and stochastic changes in immunoreactivity are possible. If reprocessing is required, the steps for paraffin embedding must be systematically reversed to bring the tissue back to 100% alcohol, and then the tissue must be defatted as above before reparaffinating.
- The water bath temperature is above or below 42–48°C.
- Tissues that are very dense or have variable textures may be difficult to cut. Increasing the section thickness may help in such cases.
- The tissue may need to be rehydrated for proper sectioning. Soak a Kimwipe in H_2O and apply to the block face for a few minutes.

9 Dip a clean microscope slide under the meniscus of the water bath. Position one or more profiles of the tissue toward the label end of the slide (slides with an opaque end for labeling are available from numerous supply companies). Slowly pull the slide out of the water at an ~45° angle to pick up the section onto the slide. Coated slides may be used to help anchor the tissue section (see above). Convenient coated slides with a built-in spacer that permits capillary action to hold reagents during the reaction are available (Biotech Probe-On Plus, Fisher Scientific).

10 Let the section air-dry at room temperature.

Air-drying overnight will make the section stick more thoroughly to the slide. However, immunoreactivity may decrease if the section is not used within a week or two.

11 Place the slide on its edge and bake at 60°C for 15 minutes to melt the wax.

12 Deparaffinize and rehydrate the section. Before the slides can be reacted with aqueous

reagents, the slide must be immersed successively for ~5 minutes with agitation in xylene, 100% alcohol, 70% alcohol, and 1 minute in buffer.

HEMATOXYLIN AND EOSIN STAINING OF SECTIONS

Hematoxylin and eosin (H & E) stains have been used as a histological stain for at least a century and are still essential for recognizing various tissue types and ultimately for recognizing the morphologic changes that form the basis of contemporary cancer diagnosis. The stain has been unchanged for many years because it works well with a variety of fixatives and is excellent for displaying a broad range of cytoplasmic, nuclear, and extracellular matrix features. Two other stains are also versatile: Giemsa staining, particularly for cells fixed by air-drying, and the Papanicolaou stain, which provides a transparent, highly detailed stain useful for thick cellular samples. For details regarding Giemsa and Papanicolaou stains, the reader may refer to Sheehan and Hrapchak (1980) and Carson (1997).

Hematoxylin has a deep blue-purple color and stains nucleic acids by a complex and incompletely understood reaction. Eosin has a pink color and stains proteins nonspecifically. In a typical tissue, the nuclei are therefore stained blue whereas the cytoplasm and extracellular matrix have varying degrees of pink staining. Well-fixed cells show considerable intranuclear detail. The nuclei show varying cell type- and cancer type-specific patterns of condensation of heterochromatin (hematoxylin staining) that are diagnostically very important. Nucleoli stain with eosin. If abundant polyribosomes are present, the cytoplasm will have a distinct blue cast. The Golgi zone can be tentatively identified by the absence of staining in a region next to the nucleus. Thus, the stain discloses abundant structural information, with specific functional implications.

A limitation of hematoxylin staining is that it is incompatible with immunofluoresence. It is useful, however, to stain one serial paraffin section from a tissue in which immunofluorescence will be performed. Hematoxylin, generally without eosin, is useful as a counterstain for many immunohistochemical or hybridization procedures that use colorimetric substrates (such as alkaline phosphatase or peroxidase).

1. Hydrate the cells or tissue. Use a microscope slide bearing cyrosections or rehydrated tissue sections (see step 12 on p. 116) fixed in either alcohol or an aldehyde-based fixative. Immerse the slide for 30 seconds with agitation in dH_2O.

   A rinse in dH_2O is important; hematoxylin precipitates with salts and buffers. The staining can be performed after immunohistochemical or hybridization reactions with nonfluorescent detection systems.

2. Dip slide into a Coplin jar (Fisher Scientific) containing Mayer's hematoxylin (Sigma) and agitate for 30 seconds. Mayer's hematoxylin is the easiest to use and is compatible with most colorimetric substrates.

3. Rinse in distilled water for 1 minute. The staining intensity can be estimated at this point, and steps 2 and 3 can be repeated if necessary.

4. Stain with eosin solution (Eosin Y, 1% aqueous solution, EM Diagnostic Systems) for 10–30 seconds with agitation.

5. Dehydrate sections with two changes of 95% alcohol and two changes of 100% alcohol for 30 seconds each (ethanol, **methanol**, or Flex alcohols, Richard-Allan Scientific, Kalamazoo, Michigan, are acceptable).

   CAUTION: Methanol (see Appendix 2 for Caution)

Some colorimetric substitutes dissolve in alcohol. If alcohols cannot be used, mount the coverslip with glycerol or other aqueous mounting media.

6 Extract the alcohols with two changes of **xylene**.

CAUTION: Xylene (see Appendix 2 for Caution)

Some colorimetric substitutes dissolve in alcohol. If alcohols cannot be used, mount the coverslip with glycerol or other aqueous mounting media.

7 Add one or two drops of mounting medium (Canada Balsam, Sigma C 1795) and cover with a coverslip.

If using plastic slides, or staining plastic culture dishes, neither xylene nor xylene-based mounting media can be used because they dissolve plastics.

CELLULAR AUTOFLUORESCENCE

Autofluorescence is often a factor limiting detectability of fluorescent probes in cells and tissues. Appropriate unlabeled control slides must be employed to assess the extent of autofluorescence. Autofluorescence can be minimized by spectral discrimination, which encompasses selecting probes and optical filters that maximize the probe fluorescence signal relative to the autofluorescent background (see Appendix 1). Autofluorescence in mammalian cells is principally due to flavin coenzymes (FAD [flavin adenine dinucleotide] and FMN [flavin mononucleotide]: absorption, 450 nm; emission, 515 nm) and reduced pyridine nucleotides (NADH [nicotinamide adenine dinucleotide]: absorption, 340 nm; emission, 460 nm) (Aubin 1979; Benson et al. 1979). In plants, autofluorescence is often due to lignins, which are a source of green autofluorescence and porphyrins, such as chlorophylls, that result in the longer-wavelength red autofluorescence. Therefore, in plants, immunofluorescence can best be performed with a short-wavelength fluorochrome that emits in the blue range (such as coumarin). For fixed cells, especially following glutaraldehyde treatment, autofluorescence can be significantly diminished by washing with 0.1% sodium borohydride in PBS (phosphate-buffered saline) for 30 minutes (Beisker et al. 1987; Bacallao et al. 1995) prior to antibody incubation for indirect immunofluorescence.

Mounting Media

After the specimen is labeled (**see the following chapters for specific labeling and detection protocols**), coverslips containing cells or tissues are mounted onto microscope slides, or slides containing sections are overlaid with a coverslip. A number of recipes for mounting media are presented below, each with particular recommendations. Mounting media are also available from a number of commercial sources (FluorSave, Calbiochem; Slowfade or Prolong, Molecular Probes; Vectashield, Vector). Fading and/or bleaching of a labeled specimen can be a major problem while making observations using fluorescence microscopy. Antifade reagents in the mounting medium can decrease the rate of fading, allowing longer observation times. Many factors influence the fluorescence intensity and bleaching of fluorophores, including the intensity of the excitatory light, the pH, the embedding medium, and the presence of other substances that may quench fluorescence. In general, a mounting medium is made up in a glycerol base with a buffer and antifade reagent. For long-term preservation of specimens, a polyvinyl alcohol mounting medium is better than a glycerol-based medium. Generally, fluorescently labeled slides store well in the cold; slides mounted with paraphenylenediamine, as described below, and kept at –70°C in the dark look as good as new even after 6 months of storage.

MOUNTING MEDIA FOR FIXED CELLS

The rate of fading of fluorescent signals can be retarded by the addition, to mounting media, of certain free-radical scavengers, such as paraphenylenediamine (Johnson and Nogueira Araujo 1981) or *n*-propyl gallate (Giloh and Sedat 1982). A mounting medium with a pH at or above 8.0–8.5 is reported to increase the initial intensity of FITC fluorescence and to reduce its fading. The following are several commonly used mounting media that have been successfully used in many laboratories.

Paraphenylenediamine Mounting Medium (modified from Johnson and Nogueira Araujo 1981)

1. Using a graduated serological pipette, add 9 ml of glycerol (Polysciences, 00084) and 1 ml of 1x PBS, pH 7.4, to a 15-ml Falcon tube. Mix the glycerol and PBS by vortexing, place parafilm around the cap of the Falcon tube, and incubate at 37°C for 20 minutes to remove air bubbles.

2. Wrap a 20-ml glass scintillation vial containing a stir bar with foil and add 10 mg of ***p*-phenylenediamine** (Fisher Scientific, AC13057-5000) and 9.75 ml of the above glycerol/PBS solution.

   CAUTION: *p*-phenylenediamine (see Appendix 2 for Caution)

3. Stir the solution until all of the *p*-phenylenediamine has dissolved. The final color of the medium should be pink/yellow (stirring at room temperature for 4 hours should be sufficient).

4. Add carbonate/bicarbonate buffer to the medium to reach a final pH of 8.0. Check pH with pH paper. A pH of 7 will result in fading of fluorescence (see below for making up buffer).

Carbonate–Bicarbonate Buffer

A. Make up 0.4 M solution of anhydrous sodium carbonate (4.24 g/100 ml).
B. Make up 0.4 M solution of sodium bicarbonate (3.36 g/100 ml).

Take 4 ml of A + 46 ml of B and bring up to 200 ml with dH_2O. The pH should be 9.2.

5 Store, wrapped in aluminum foil in –20°C freezer.

Suppliers are provided for the most critical reagents in this protocol as purity of the compounds is crucial to producing an effective mounting medium. Reagents from other suppliers may also be good.

Triton X-100 quenches the oxidizing antifade agents, so do not use it in final washes before mounting.

Notes

- *p*-phenylenediamine oxidizes readily in air when exposed to light, yielding a fluorescent product that binds to nuclei. If the antifade solution seems to be increasing the background or staining nuclei (particularly noticeable with fluorescein filter sets), it will be necessary to use purer *p*-phenylenediamine. The oxidation products are dark brown, whereas pure *p*-phenylenediamine is nearly white. *p*-phenylenediamine is only slightly soluble in cold water, whereas the oxidation products are readily soluble. Washing the *p*-phenylenediamine powder with about five times its weight of cold water will remove most of the fluorescent material.
- *p*-phenylenediamine containing mounting medium is not compatible with cyanine-conjugated antibodies (i.e., Cy2, Cy3, Cy5).

Glycerol Antifade Mounting Medium

(adapted from Shuman et al. 1989)

Dissolve in 100 ml of glycerol:

n-propyl gallate	5 g
DABCO	0.25 g
***p*-phenylenediamine**	2.5 mg

CAUTION: Glycerol; DABCO; *p*-phenylenediamine (see Appendix 2 for Caution)

Add several pellets of NaOH to bring pH above neutral. Stir thoroughly (>1 day). Store in aliquots at −20°C wrapped in foil.

Notes

- The effectiveness of this glycerol-based antifade solution is greatly diminished by small amounts of H_2O. To minimize residual H_2O in the sample, it is best to first drain off all the buffer before mounting, cover the specimen with this solution, let it sit for 15 minutes or so, then drain it all off and mount in fresh antifade solution.
- Small amounts of residual Triton X-100 (used for permeabilization) and perhaps other detergents convert this antifade solution into a very powerful quenching agent. If the fluorescence

disappears when you apply this solution, remove the coverslip from the slide by flooding with buffer and place it cell side up in a petri dish. Rinse the sample well with excess buffer to remove the detergent, then remount the coverslip with fresh mounting medium.

- Some lectins that bind to surface carbohydrates tend to dissociate in 100% glycerol. In that case, the antifade agents can be dissolved in polyvinylpyrrolidone.
- *p*-phenylenediamine oxidizes readily in air when exposed to light, yielding a fluorescent product that binds to nuclei. If the antifade solution seems to be increasing the background or staining nuclei (particularly noticeable with fluorescein filter sets), it will be necessary to use purer *p*-phenylenediamine. The oxidation products are dark brown, whereas pure *p*-phenylenediamine is nearly white. D-phenylenediamine is only very slightly soluble in cold water, whereas the oxidation products are readily soluble. Washing the D-phenylenediamine powder with about five times its weight of cold water will remove most of the fluorescent material.

n-Propyl Gallate Antifade Medium
(Giloh and Sedat 1982)

1 2% (w/v) *n*-propyl gallate in glycerol.

2 Adjust pH to 8.0.

Gelvatol Mounting Medium

Semipermanent mounts can be made using Gelvatol mounting medium. This medium is not autofluorescent, is viscous, and hardens slowly. Gelvatol is a polyvinyl alcohol-based mountant. Preparations mounted in this way can be stored in the dark at either 4°C or –20°C with little loss of image quality.

1 Prepare the Gelvatol solution by dissolving 0.35 g of Gelvatol (Monsanto, St. Louis, Missouri) in 3 ml of dH_2O (PBS can be substituted for H_2O) and 1.5 ml of glycerol.

2 Heat the solution with stirring in a boiling water bath until the Gelvatol is completely dissolved, and add antifade agents (see above media) as desired.
The mounting medium can be stored at 4°C for long periods (months) with no loss of properties; contamination with mold usually determines the usable shelf life.

MOUNTING OF LIVE CELLS ATTACHED TO COVERSLIPS

1 Carefully remove coverslips from their petri dish with forceps.

2 Wick excess buffer from the coverslip by carefully touching the edges of the coverslip with a piece of Whatman #1 filter paper and gingerly dry the top of the coverslip (the side with no cells). If you lose track of which side of the coverslip the cells are on, mark a corner with a felt-tip marker or India ink pen. Then examine the wet coverslip in a petri dish containing buffer, with a tissue culture or dissecting microscope. Determine whether the mark is in focus with the cells.

3 Place a drop (20 μl) of the mounting medium (see step 2 below) on a clean microscope slide.

4 Gently lower the coverslip onto the mounting medium, cell side down, so that no air bubbles become trapped.

5 Blot away the excess mounting medium with filter paper.

6 Seal the edges of the coverslip with VALAP (live cells, see below) or nail polish (fixed cells).

A Simple Mounting Technique for Observing Live Cells

1 Prepare VALAP.

a Weigh out a mixture of Vaseline, lanolin, and paraffin (1:1:1 w/w/w).

b Melt the mixture over medium low heat.

The wax mixture should spread smoothly and dry quickly on a glass slide. If it hardens too quickly, add more Vaseline and lanolin; if it does not harden quickly enough, add more paraffin.

2 Mount coverslips containing the live cells in PBS(+) as described above. PBS(+) is PBS containing 1.0 mM Ca^{++} and 0.5 mM Mg^{++}, which allows cells to adhere to each other and to the substrate. If live cells are in a medium containing no Ca^{++} or Mg^{++}, they will round up and detach from the coverslip.

Parafilm spacers are sometimes used when mounting coverslips containing live cells on a microscope slide to avoid crushing them. Place narrow strips of Parafilm on a slide spaced so that the coverslip will fit between them with two edges of the coverslip extending onto the Parafilm, which acts as a physical spacer. Instead of Parafilm, small pieces of broken #1 coverslips can be used at the four corners of square coverslips to prevent cell compression.

3 Seal the coverslip around its edges with melted VALAP.

Photobleaching of live samples (such as cells transfected with GFP fusion constructs) can be minimized by including antioxidants in the mounting medium. Oxyrase (Oxyrase Inc., Mansfield, Ohio) is an enzyme additive used to deplete oxygen in order to grow anaerobic bacteria. It has been used at 0.3 units/ml to reduce photodynamic damage during observation of cells (Waterman-Storer et al. 1993). Another approach has been to add ascorbic acid to the mounting medium. This reducing agent is used at 0.1–3.0 mg/ml.

For more long-term observations, sophisticated live cell chambers are available that control temperature on the microscope stage (see Goldman and Spector 2005).

REFERENCES

Aubin J.E. 1979. Autofluorescence of viable cultured mammalian cells. *J. Histochem. Cytochem.* **27:** 36–43.

Bacallao R., Kiai K., and Jesaitis L. 1995. Guiding principles of specimen preservation for confocal fluorescence microscopy. In *Handbook of biological confocal microscopy* (ed. J.B. Pauley). Plenum Press, New York.

Beisker W., Dolbeare F., and Gray J.W. 1987. An improved immunocytochemical procedure for high-sensitivity detection of incorporated bromodeoxyuridine. *Cytometry* **8:** 235–239.

Benson R.C., Meyer R.A., Zaruba M.E., and McKhann G.M. 1979. Cellular autofluorescence—Is it due to flavins? *J. Histochem. Cytochem.* **27:** 44–48.

Carson F.L. 1997. *Histotechnology: A self-instructional text,* 2nd edition. ASCP Press, Chicago.

Giloh H. and Sedat J.W. 1982. Fluorescence microscopy: Reduced photobleaching of rhodamine and fluorescein protein conjugates by *n*-propyl gallate. *Science* **217:** 1252–1255.

Goldman R.D. and Spector D.L. 2005. *Live cell imaging: A laboratory manual.* Cold Spring Harbor Laboratory, Cold Spring Harbor, New York.

Harlow E. and Lane D.L. 1988. *Antibodies: A laboratory manual.* Cold Spring Harbor Laboratory, Cold Spring Harbor, New York.

Hollands B. 1962. Histochemistry and microtomy of fresh-frozen tissue. In *Progress in medical laboratory technique* (ed. F.J. Baker), pp. 112–135. Butterworth, London.

Johnson G.D. and Nogueira Araujo G.M. 1981. A simple method of reducing the fading of immunofluorescence during microscopy. *J. Immunol. Methods* **43:** 349–350.

Johnson T.J. 1987. Glutaraldehyde fixation chemistry. *Eur. J. Cell Biol.* **45:** 160–169.

Nuovo G.J. 1997. *PCR in situ hybridization: Protocols and applications,* 3rd edition. Lippincott-Raven Publishers, Philadelphia.

Sheehan D.C. and Hrapchak B.B. 1980. *Theory and practice of histotechnology,* 2nd edition. Battelle Press, Columbus, Ohio.

Shuman H., Murray J.M., and DiLullo C. 1989. Confocal microscopy: An overview. *BioTechniques* **7:** 154–163.

Stearns M.E. and R.L. Ochs. 1982. A functional in vitro model for studies of intracellular motility in digitonin-permeabilized erythrophores. *J. Cell Biol.* **94:** 727–739.

Wang K., Feramisco J.R., and Ash J.F. 1982. Fluorescent localization of contractile proteins in tissue culture cells. *Methods Enzymol.* **85:** 514–562.

Waterman-Storer C.M., Sanger J.W., and Sanger J.M. 1993. Dynamics of organelles in the mitotic spindles of living cells: Membrane and microtubule interactions. *Cell Motil. Cytoskel.* **26:** 19–39.

Werner M., Von Waseilewski R., and Komminoth P. 1996. Antigen retrieval, signal amplification and intensification in immunohistochemistry. *Histochem. Cell Biol.* **105:** 253–260.

CHAPTER 5

Nonimmunological Fluorescent Labeling of Cellular Structures

Brad Chazotte

Campbell University, Buies Creek, North Carolina

INTRODUCTION

This chapter provides methods to study cellular organization using various non-antibody-based fluorescent probes. Other sources of information include Haugland (1992), Mason (1993), and Wang and Taylor (1989).

Cell Organization

The eukaryotic cell has evolved to compartmentalize its functions and transport various metabolites among cellular compartments. Therefore, in cell biology, the study of organization and structure/function relationships are of great importance. A number of aspects should be considered in studying cellular organization. For purposes of study, one may wish to distinguish between static and dynamic organization. For example, the presence of a plasma membrane is static in the living cell, but the lipid and protein components of the membrane are frequently in dynamic movement, the shape and composition of the membrane itself may also change over time, and phospholipid-based vesicles may be added to or removed from the plasma membrane during such processes as pinocytosis, receptor-mediated endocytosis, and secretion. The type of fluorescent probes used to study the cell may well be dictated by the static or dynamic nature of the organization to be studied as well as the nature of the component to be labeled or mimicked. Finally, a decision must be made as to whether cell-permeant probes can be used or whether the probe will require microinjection or chemical permeabilization of the cell. In some cases, cell permeabilization allows structural analysis but not functional analysis as the cell may no longer be alive.

Labeling Protocols

One extensive source for all types of fluorescent probes and related technical information is Molecular Probes, and their catalog has been used as a general reference for this chapter (Haugland 1992). Typical fluorescent moieties that are used in these studies are rhodamines, fluoresceins, BODIPY, NBD, and the carbocyanines, although this is by no means a complete list. For microscopic imaging techniques, see Chapter 6 in this manual. Note that for single-cell analysis techniques such as microscopic imaging and FRAP (fluorescence recovery after photobleaching), the protocols assume that cells were grown on glass microscope coverslips and immersed in small petri dishes containing culture medium. *At no time in the following protocols should the cells be allowed to dry out.* Bulk techniques, i.e., for cells suspended in solution, will require centrifugation of cells and aspiration of the solutions used in the protocols. Generally, labeling conditions can vary from cell type to cell type, and it may be necessary to vary the protocols for a particular use. In addition, in some of the following sections, separate protocols are given for labeling live or fixed cells.

Two potential problems should always be considered for fluorescence labeling described in this chapter. One potential problem in the study of cells that can negate the inherent advantage of detecting a bright signal against a dark background is autofluorescence. Autofluorescence refers to the fact that cells contain molecules which can fluoresce, e.g., NADH (nicotinamide adenine dinucleotide), FADH (flavine adenine dinucleotide). If the excitation and/or emission wavelengths are similar to that of the chosen fluorescent probe, the probe's signal can be obscured by the native fluorophores of the cell. Autofluorescence is usually dealt with by carefully selecting the excitation and emission wavelengths used, by treatment of fixed cells with reducing agents such as $NaBH_4$ (1% solution for 20 min), and by comparison to unlabeled control slides. Fixation with glutaraldehyde is to be avoided because it can increase interference from cellular autofluorescence, most frequently at wavelengths less than 500 nm. Another potential problem is photobleaching of fluorescence. *Due caution must always be exercised to minimize the intensity and exposure time to incidental light during preparation and handling. Likewise, during observation with a microscope or other instrument, the excitation light should be the minimum necessary to see a sharp, bright, and nonfading image.*

MOUNTING OF LIVE CELLS ON MICROSCOPE SLIDES

1 Prepare a stock supply of VALAP (mixture of Vaseline, lanolin, and paraffin, 1:1:1) to seal coverslips. Weigh the mixture and melt over medium low heat on a hot plate.

   The wax mixture should spread smoothly and dry quickly on a glass slide. If it hardens too quickly, add more Vaseline and lanolin; if it does not harden quickly enough, add more paraffin.

2 After labeling live cells, mount the coverslip in PBS (phosphate-buffered saline) (+). Parafilm spacers are used when mounting coverslips containing live cells on a microscope slide to avoid crushing the live cells. Place narrow strips of parafilm on a slide spaced so that the coverslip will fit between them with two edges of the coverslip extending onto the parafilm, which acts as a physical spacer. Seal the coverslip around its edges with melted VALAP.

PBS (+) is PBS (see Appendix 1) containing 1.0 mM Ca^{++} and 0.5 mM Mg^{++} and allows cells to adhere to each other and to the substrate. If cells are in medium containing no Ca^{++} or Mg^{++}, they will round up and detach from the coverslip.

3 To mount fixed cells, see mounting media in Chapter 4.

Notes

- Fixed cells do not require the physical spacers for mounting.
- Cell culture media containing phenol red, such as DMEM and BME, should be used with caution, because phenol red is fluorescent and may interfere with the fluorescent signal.

LABELING THE COMPONENTS OF THE PLASMA MEMBRANE

The plasma membrane provides the essential, dynamic boundary and a vital selective interface between the cell compartment and its environment (for details, see Alberts et al. 1994). Many plasma membrane proteins are studied most selectively by using immunofluorescence probes, although some quasi-specific chemical labeling is possible, e.g., eosin maleimide conjugation to Band-3 protein of erythrocytes. The bulk of the membrane lipids are phospholipids, although sterols (e.g., cholesterol, glycolipids, and proteolipids) are significant components. Lipid organization and dynamics are frequently studied by reintroduction of fluorescently labeled molecules (e.g., NBD-PE [N-4-nitrobenz-2-oxa-1,3 diazole phosphatidylethanolamine], and BODIPY phosphocholine) or introduction of fluorescent lipid analogs (e.g., DiI, perylene, and pyrene). Certain glycolipids or glycoproteins can be labeled using the plant lectins wheat germ agglutinin or concanavalin A, which bind to certain carbohydrate moieties. Molecules such as DPH (diphenylhexatrene), TMA (trimethylamino)-DPH, and DPH phosphocholine are typically used as reporters on rotational motions in membranes.

Labeling Phospholipids with DiI-C_{16}(3)

Labeling Live Cells

1 Prepare a stock solution (1 mg/ml) of DiI-C_{16}(3) (m.w. 878; Molecular Probes) in absolute ethanol. Store sealed and protect from light at –20°C.

2 Dilute the stock solution 1:500 in PBS(+) for use in labeling.

3 Aspirate cell medium from cells grown on coverslips and rinse three times with PBS(+).
PBS(+) includes the addition of 1.0 mM Ca^{++} and 0.5 mM Mg^{++}. These additions prevent living cells from rounding up and detaching from the glass substratum.

4 Incubate cells at room temperature for 30 seconds with DiI-C_{16}(3) solution.

5 Rinse three times with PBS(+).

6 Mount the coverslip containing cells in PBS(+) as described above.

Notes

- The excitation wavelength maximum for DiI-C_{16}(3) will be near 550 nm and the emission wavelength maximum will be ~565 nm based on methanol as the solvent. Maxima may vary slightly for different carbon chain lengths and especially in different solvents. A rhodamine/Texas Red filter set should be used for observation. For FRAP and confocal microscopy, the 514-nm line of an argon-ion laser may be used for excitation.
- This fluorophore can be used for imaging and especially for FRAP studies. Fluorescence polarization measurements have also been performed using membrane-bound DiIs.
- Other analogs of this carbocyanine dye have different carbon chain lengths (e.g., C_{12}, C_{14}, C_{18}, and C_{22}) that affect the probe's miscibility in the phospholipids of the plasma membrane; e.g., an 18-carbon chain may not mix well in a phase containing primarily 12-carbon chains (Spink et al. 1990). In addition, incorporations should be carried out above the main-phase transition of the membrane, which for most cell membranes is a broad transition below 10°C. Labeling at room temperature or higher should ensure that cell membranes are in the liquid crystalline state without temperature-induced lateral phase separations.
- Enough dye should be present for a strong signal, but not too much to perturb the plasma membrane itself. For FRAP experiments, a probe/phospholipid ratio between 1:1,000 and 1:10,000 is desirable. The plasma membrane composition, the incubation temperature and time, and the dye concentration in the incubating medium all affect the amount of dye incorporated. The use of ethanolic solution and the above concentration is to ensure that the dye *is* intercalated in the bilayer and not adsorbed on the surface or micellized in the medium. Fluorescence polarization measurements have been used to verify that the molecule is incorporated.
- The length of usable observation time will be affected in part by the rate at which the plasma membrane is altered in the normal course of cellular function. DiI tends to locate in the outer leaflet of the plasma membrane and may take several days to appear in cytoplasmic vesicles in unfixed cells. However, in healthy cells labeled and observed at 37°C, significant endocytosis occurs. Fixed cells tend to exhibit more uniform staining since they are permeabilized.
- DiI probes are *not specific* for the plasma membrane and will label most phospholipid membranes provided they can come into contact with the membrane. They are cationic (positively charged) lipophilic probes.

Labeling Phospholipids with NBD-PE

Labeling Live Cells

1. Prepare a stock solution of NBD-PE (Molecular Probes) at a concentration of 1 mg/ml in absolute ethanol. Store sealed and protected from light at –20°C.

2. Label cells as described in the preceding protocol using a 1:500 dilution of the stock probe solution in PBS(+).

Notes

- The excitation wavelength maximum for NBD-PE will be near 460 nm and the emission wavelength maximum will be ~534 nm based on ethanol as a solvent. Maxima may vary slightly for different carbon chain lengths and especially in different solvents. A fluorescein filter set should be used for observation. For FRAP and confocal microscopy, the 488-nm line of an argon-ion laser may be used for excitation.

- The NBD-PE probe is more prone to fading than DiI or rhodamine.
- NBD-PE makes use of an actual phospholipid molecule as opposed to an analog such as DiI, and NBD-PE does not have the net positive charge of DiI.

Labeling the Plasma Membrane with TMA-DPH

Labeling Live Cells

1 Prepare a 1-mM stock solution of TMA-DPH (trimethylamino diphenylhexatriene) in **DMF**. Store sealed and protected from light at –20°C.

CAUTION: DMF (see Appendix 2 for Caution)

2a For conventional fluorescence polarization measurements, use a *bulk suspension* of cells at a concentration of 10^6 cells/ml.

b For microscopy, prepare a labeling solution as a 1000-fold dilution (1 µM) in PBS(+).

3a For the bulk suspension, add enough TMA-DPH for a final concentration of 1 µM and incubate at 37°C for 5 minutes. Centrifuging and rinsing should not be necessary.

b For microscopy, aspirate off the cell medium from cells grown on coverslips, wash three times in PBS(+), and immerse the cells in the labeling solution for 5 minutes at 37°C. Then aspirate the labeling solution and rinse twice with PBS(+). Mount as described above.

Notes

- For fluorescence polarization and lifetime measurements, the excitation wavelength is 360 nm with a 5-nm slit width and the emission wavelength is monitored at 430 nm with a 10-nm slit width. For microscopy and imaging, a UV filter set should be used. See also last note on this list.
- TMA-DPH is closely related to DPH, a probe also used to study the motion dynamics of membranes. However, DPH reports on the phospholipid hydrocarbon (acyl chain) region. With respect to cells, DPH is freely permeable and, in contrast to the positively charged TMA-DPH, labels *all* the cellular membranes. Therefore, TMA-DPH is better suited to study the motional dynamics of the plasma membrane.
- The bulk suspension of cells may be used in fluorescence polarization measurements to obtain information on the anisotropy of the phospholipid headgroup region in the plasma membrane or in fluorescence lifetime measurements of TMA-DPH to report on the local membrane motional dynamics.
- For bulk fluorescence polarization and lifetime measurements, corrections for light-scattering artifacts may be necessary. Control specimens omitting the fluorophore should be prepared simultaneously (Chazotte 1994).
- Recently, both fluorescence polarization and fluorescent lifetime measurements have begun to be adapted to microscopic imaging. They are not readily available but are most promising for future cell studies (Herman and Jacobson 1990; Periasamy and Herman 1994).

Labeling Glycoproteins or Glycolipids with FITC Wheat Germ Agglutinin

Labeling Live Cells

1. Prepare a stock solution of FITC-wheat germ agglutinin (WGA) at a concentration of 2 mg/ml. Store protected from light at 4°C.
2. Prepare a 1:200 dilution in PBS(+) to label the cells.
3. Rinse the cells grown on coverslips three times in PBS(+).
4. Immerse the cell-containing coverslip in a small petri dish containing the labeling solution at 37°C for 5–10 minutes.
5. Rinse the cells three times in PBS(+) and mount (see Chapter 4).

Notes

- For FITC-WGA, the excitation wavelength maximum will be near 495 nm and the emission wavelength maximum will be ~519 nm based on H_2O at pH 9 as the solvent for the unconjugated fluorophore. Maxima may vary upon binding and especially in different solvents or pH values. A fluorescein filter set should be used for observation. For FRAP and confocal microscopy, the 488-nm line of an argon-ion laser may be used for excitation.
- WGA is a plant lectin that exists as a dimer (m.w. 36,000) and is normally cationic. It selectively binds to sialic acid and *N*-acetylglucosaminyl residues that may be present on glycolipids or glycoproteins of the plasma membrane. Thus, it is a useful probe for the plasma membrane, although not a highly specific one.
- Concanavalin A, a tetramer (m.w. 104,000), is another plant lectin that can be used. However, it selectively binds to α-mannopyranosyl and α-glucopyranosyl residues.

LABELING THE CYTOSKELETON

The cytoskeleton is composed of a series of filamentous structures, including intermediate filaments, actin filaments, and microtubules (for details, see Alberts et al. 1994). Immunofluorescent staining has been most frequently used to study cytoskeletal components. However, it is also possible to fluorescently label isolated cytoskeletal proteins and either microinject them back into the cell or add them to fixed, permeabilized cells. Alternatively, it is possible to use the mushroom-derived fluorescinated toxins, phalloidin or phallacidin, to label F-actin of the cytoskeleton as detailed below.

Phalloidin is available labeled with different fluorophores. The choice of the specific fluorophore should depend on whether phalloidin labeling for actin is part of a double-label experiment. In most cells, the abundance of actin filaments should provide a very strong signal. In double-label experiments, the fluorofluor should be chosen to take this into account. In general, rhodamine labels are more resistant to photobleaching and can be subjected to longer exposures required for finer structures.

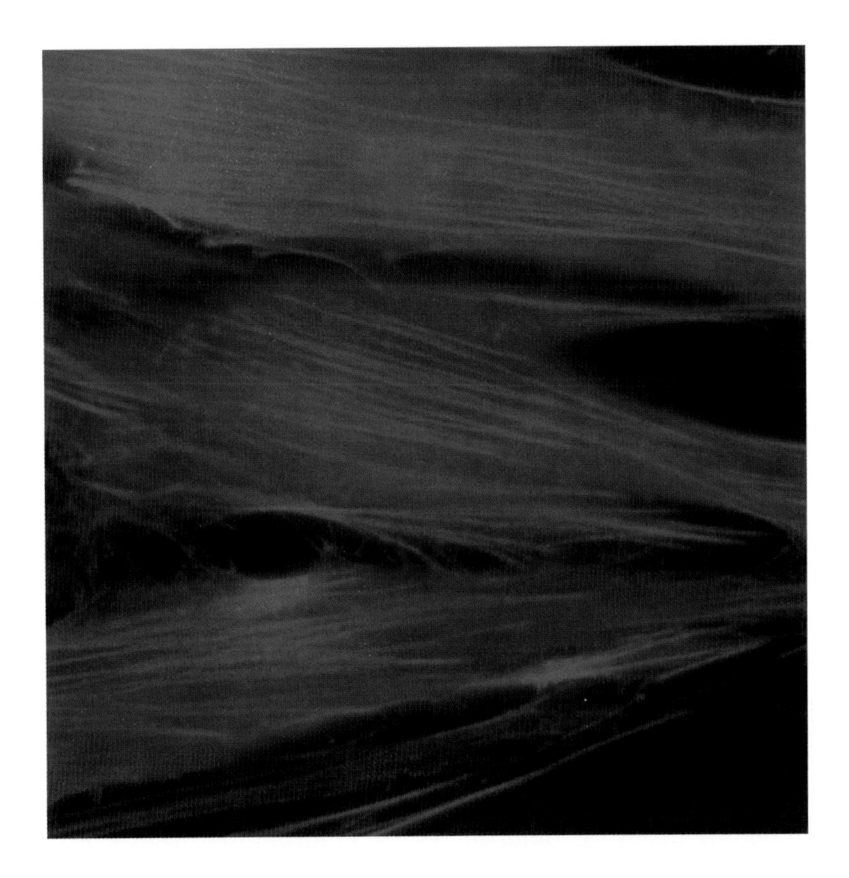

FIGURE 5.1

Detroit 551 cells (human fibroblasts) stained with rhodamine phalloidin to show the localization of F-actin. (Photo provided by J. McCann, Cold Spring Harbor Laboratory.)

Labeling F-actin of the Cytoskeleton with Rhodamine Phalloidin (Fig. 5.1) or Fluorescein Phalloidin

Labeling Fixed Cells

1 Prepare a 1:200 dilution of the stock solution (300 units/ml) of phalloidin using PBS.

2 Prepare a 3.7% **formaldehyde** solution in PBS/pH 7.4 for fixation. Also prepare a 0.1% Triton X-100 solution in PBS for permeabilization.

3 Aspirate the cell medium and rinse cells three times with PBS.

4 Fix cells for 10 minutes in the 3.7% formaldehyde solution.

5 Rinse fixed cells three times for 5 minutes each in PBS.

6 Permeabilize the cells for 5 minutes in the 0.2% Triton X-100 solution.

7 Rinse fixed cells three times with PBS.

8 Label with the fluorescent phalloidin at room temperature for 5–10 minutes.

9 Rinse cells three times in PBS for 5 minutes each time and mount (see Chapter 4).

CAUTION: Formaldehyde (see Appendix 2 for Caution)

Notes

- For FITC-phalloidin (RITC [rhodamine isothiocyanate]-phalloidin), the excitation wavelength maximum will be near 495 nm (550 nm) and the emission wavelength maximum will be approximately 520 nm based on H_2O at pH 8.0 as the solvent (575 nm with **methanol** as the solvent). The emission spectrum of fluorescein can depend on the environmental pH and/or ionic strength. The respective filter sets for fluorescein or rhodamine should be used for observation. For FRAP and confocal microscopy, the 488-nm line of an argon-ion laser may be used for fluorescein excitation, and the 514-nm line may be used for rhodamine excitation.

  CAUTION: Methanol (see Appendix 2 for Caution)

- Phalloidin staining will not work with methanol-fixed cells, probably because of the disruption of actin filament integrity, but it will work with cells fixed in 0.2% glutaraldehyde in PBS.

LABELING THE NUCLEUS

The nucleus contains almost all of the cell's DNA and is bounded by a double membrane. Inside and adjacent to the inner membrane of the nuclear envelope is the nuclear lamina. It is composed of a fibrous meshwork comprising one or more of three major intermediate filament-like polypeptides, lamins A, B, and C. The outer nuclear membrane is contiguous with the endoplasmic reticulum (for details, see Alberts et al. 1994). A number of fluorescent stains are available that label DNA and allow easy visualization of the nucleus in interphase cells and chromosomes in mitotic cells. These stains include Hoechst, DAPI (Fig. 5.2), ethidium bromide, propidium iodide, and acridine orange.

Labeling DNA Using DAPI

Labeling Fixed Cells

1 Prepare a stock solution at 10 mg/ml of **DAPI** (m.w. 350) in distilled H_2O, protect from light, and store at 4°C. Prepare a 5000-fold dilution in PBS to be used for labeling.

CAUTION: DAPI (see Appendix 2 for Caution)

2 Prepare a *fresh* 3.7% **formaldehyde** solution for fixation. Also prepare a 0.2% Triton X-100 solution for permeabilization.

CAUTION: Formaldehyde (see Appendix 2 for Caution)

3 Aspirate the cell medium.

4 Rinse cells three times with PBS(+).

5 Fix the cells for 10 minutes in 3.7% formaldehyde solution.

6 Aspirate and rinse the cells three times for 5 minutes each in PBS.

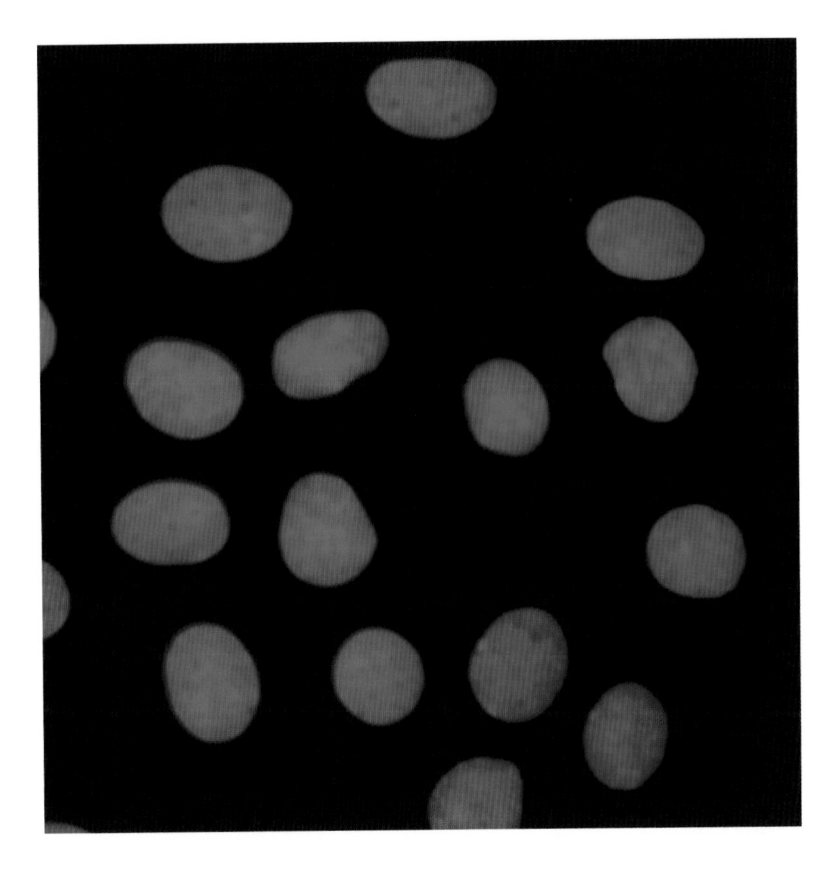

FIGURE 5.2

REF-52 cells stained with DAPI to show the localization of DNA. (Photo provided by J. McCann, Cold Spring Harbor Laboratory.)

7 Permeabilize the cells by immersion in 0.2% Triton X-100 for 5 minutes.

8 Aspirate and rinse three times for 5 minutes each in PBS.

9 Incubate the cells at room temperature for 1–5 minutes in the DAPI labeling solution.

10 Aspirate off the labeling medium, rinse three times in PBS, and mount (for mounting media, see Chapter 4).

■ *Notes*

- For DAPI, the excitation wavelength maximum will be near 359 nm and the emission wavelength maximum will be approximately 461 nm when bound to DNA. A UV filter set can be used for observation. For FRAP and confocal microscopy, a UV line of an argon-ion laser may be used for excitation.
- Although not as bright as the vital Hoechst stains for DNA, DAPI has greater photostability.
- The lactate salt of DAPI is more water-soluble than the chloride salt, but DAPI is not very soluble in PBS; therefore, use distilled H_2O to prepare the stock solution.
- It is believed that DAPI associates with the minor groove of double-stranded DNA with a preference for the adenine-thymine (AT) clusters. Fluorescence increases approximately 20-fold when DAPI is bound to double-stranded DNA.

- Cells that have been immunolabeled can be stained with DAPI by starting at step 9 above.

Labeling DNA with Hoechst 33342

Labeling Live Cells

1. Prepare a 10 µg/ml stock solution of Hoechst 33342 (m.w. 642) in dH_2O, protect from light, and store at 4°C. Prepare a 100-fold dilution in dH_2O to be used for labeling.
2. Aspirate the cell medium.
3. Rinse cells three times with PBS(+).
4. Incubate the cells at room temperature for 10–30 minutes in the Hoechst labeling solution.
5. Aspirate the labeling medium and rinse three times in PBS(+) and mount (for mounting media, see Chapter 4).

■ ***Notes***

- For Hoechst 33342, the excitation wavelength maximum will be near 343 nm and the emission wavelength maximum will be ~483 nm when bound to DNA. A UV filter set can be used for observation. For confocal microscopy, a UV line of an argon-ion laser may be used for excitation.
- One advantage of Hoechst 33342 over DAPI is that the former is membrane permeant and is therefore useful for living cells as it does not require cell fixation or permeabilization.
- Hoechst 33342 binds to AT-rich regions of DNA in the minor groove. Upon binding to DNA, the fluorescence greatly increases.
- Hoechst 33342 can also be used to stain fixed cells by following the DAPI protocol and substituting the Hoechst 33342 solution.

LABELING MITOCHONDRIA

The mitochondrion is a double-membraned organelle that functions in a variety of important metabolic processes, including oxidative phosphorylation and electron transport, Kreb's cycle, β-oxidation of fatty acids, and part of the urea cycle (for details, see Alberts et al. 1994). Several fluorochromes are available to label mitochondria. Most of the labeling protocols depend on dyes that are sensitive to the membrane potential of the respiring mitochondrion.

Labeling with Rhodamine 123 (Fig. 5.3)

Labeling Live Cells

1. Prepare a stock solution of 1 mg/ml rhodamine 123 (m.w. 381) in dH_2O, protect from light, and store at 4°C in a sealed vial.
2. Prepare a rhodamine-123-labeling solution at a concentration of 10 µg/ml (a 100-fold dilution) in PBS(+).

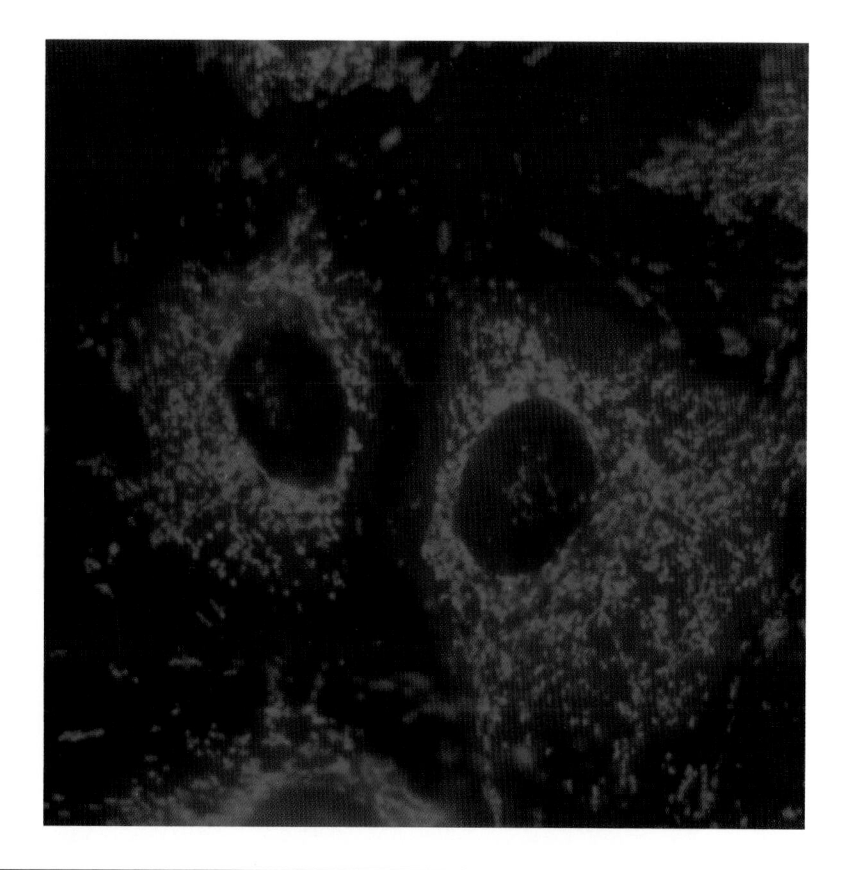

FIGURE 5.3

REF-52 cells stained with rhodamine 123 to show the localization of mitochondria. (Photo provided by J. McCann, Cold Spring Harbor Laboratory.)

3 Aspirate the culture medium and wash three times with PBS(+).

4 Incubate cells in the rhodamine-labeling solution at 37°C for 15 minutes.

5 Aspirate the labeling solution and rinse three times in PBS(+).

6 Mount as described in Chapter 4.

■ *Notes*

- For rhodamine 123, the excitation wavelength maximum will be near 504 nm and the emission wavelength maximum will be ~534 nm based on **methanol** as a solvent. Maxima may vary somewhat depending on the concentration of the dye in the mitochondrion and especially in different solvents. Isolated, energized mitochondria exhibit a redshift in the rhodamine 123 emission and also fluorescence quenching. A filter set for rhodamine can be used for observation. For FRAP and confocal microscopy, the 514-nm line of an argon-ion laser may be used for excitation.
- For more extensive details and references on the use of rhodamine 123, see Chen (1989).

CAUTION: Methanol (see Appendix 2 for Caution)

Labeling with TMRM or TMRE (Fig. 5.4)

Labeling Live Cells

1 Prepare a stock solution of TMRM (tetramethylrhodamine methy ester)(m.w. 501) or TMRE (tetramethylrhodamine ethyl ester) (m.w. 515) at a concentration of 600 μm in DMSO (dimethylsulfoxide), protect from light, and store sealed at –20°C.

2 Prepare a 1000-fold dilution (600 nM) of the stock TMRM solution in cell culture medium. Also prepare a 4000-fold dilution (150 nM) of the stock TMRM in cell culture medium.

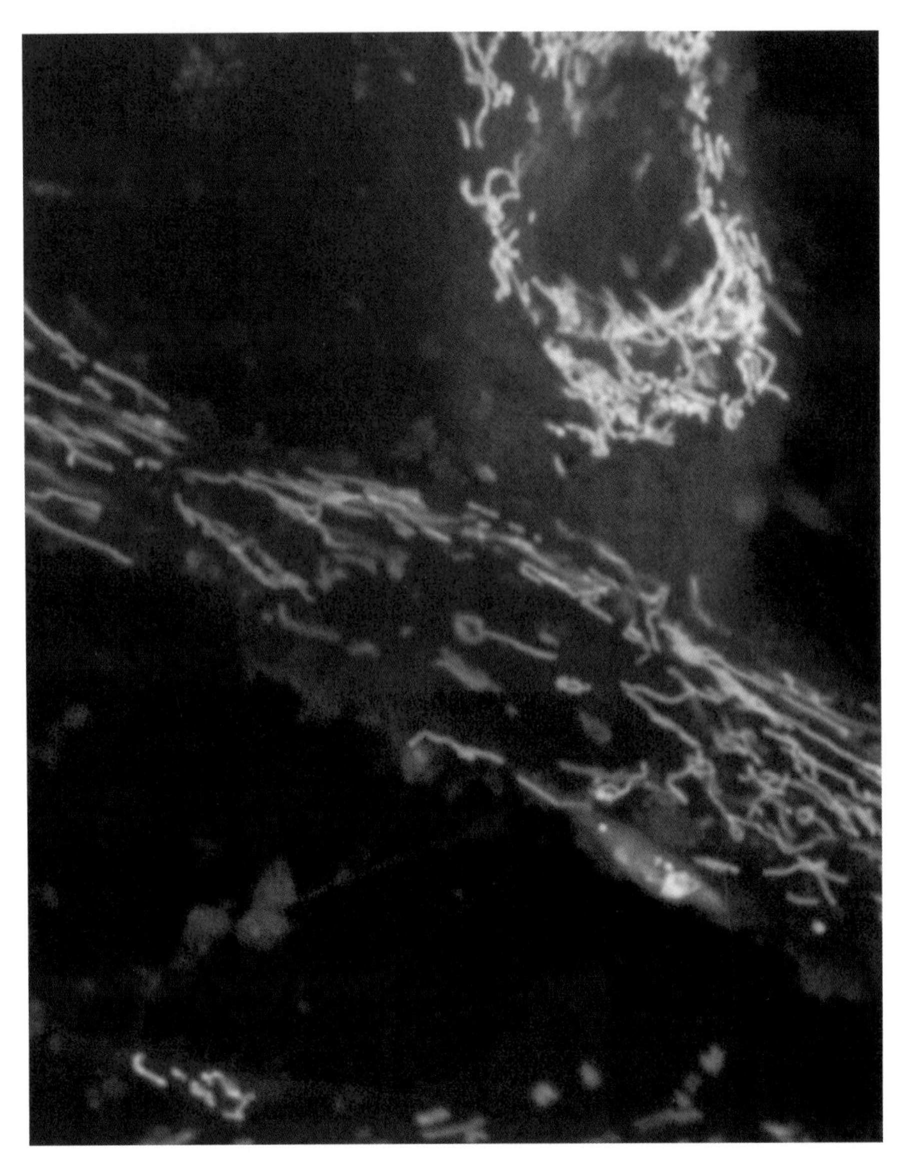

FIGURE 5.4

Confocal micrograph of living human BG-9 fibroblasts labeled with TMRM. The dye intensity is proportional to the membrane potential; that is, it follows a Nernst distribution. Therefore, mitochondria appear as the brightest objects having as much as a 10,000-fold higher TMRM concentration than outside the cells. (Photo provided by B. Chazotte, University of North Carolina.)

3 Aspirate off the culture medium from the cells.

4 Immerse the cells in the TMRM or TMRE (600 nM) containing culture medium at 37°C for 20 minutes to load the cells with the dye.

5 Aspirate the labeling medium and immerse the cells in culture medium containing 150 nM TMRM (or TMRE) to maintain the equilibrium distribution of the fluorophore. Mount as described in Chapter 4.

▪ *Note*

- For TMRM (or TMRE), the excitation wavelength maximum will be near 548 nm (549 nm) and the emission wavelength maximum will be approximately 573 nm (574 nm) based on **methanol** as a solvent. Maxima may vary somewhat depending on the concentration of the dye in the mitochondrion and especially in different solvents. A filter set for rhodamine should be used for observation. For FRAP and confocal microscopy, the 514-nm line of an argon-ion laser may be used for excitation.

CAUTION: Methanol (see Appendix 2 for Caution)

Labeling with Mito Tracker CMTMR (from Molecular Probes)

Labeling Live Cells

1 Prepare a 1 µM stock solution of the dye in **DMSO**. Protect from light and store at –20°C. *The reduced forms of the dye must additionally be stored under nitrogen or argon.*

2 Prepare a labeling solution at a working concentration between 25 nM and 500 nM in the desired cell culture medium and warm to the cell culture temperature.

3 Aspirate the culture medium.

4 Immerse the cells in the labeling medium at 37°C for 15–45 minutes.

5 Aspirate the labeling medium and rinse three times with culture medium.

6 Prepare for mounting as described in Chapter 4.

7 Check the cells using a fluorescence microscope. If the fluorescence staining is too low, either incubate for an additional 30 minutes in the normal culture medium to allow the thiol conjugation to proceed or try a higher labeling concentration initially.

CAUTION: DMSO (see Appendix 2 for Caution)

▪ *Note*

- For the tetramethylrhodamine "Mito Tracker dyes," the excitation wavelength maximum will be near 554 nm and the emission wavelength maximum will be ~576 nm based on **methanol** as a solvent. Maxima may vary somewhat depending on the concentration of the dye in the mitochondrion and especially in different solvents. A filter set for rhodamine can be used for observation. For FRAP and confocal microscopy, the 514-nm line of an argon-ion laser may be used for excitation.

CAUTION: Methanol (see Appendix 2 for Caution)

Labeling Fixed Cells

1. Prepare a labeling solution at a working concentration between 100 and 1000 nM in the desired cell culture medium and warm to the cell culture temperature.
2. Labeling is done on *live cells* and then the cells are fixed. Follow the same procedures as above but use the higher concentration.
3. Rinse cells three times with PBS(+).
4. Fix the cells for 10 minutes in 3.7% **formaldehyde** in PBS at pH 7.4.
5. Aspirate and rinse the cells three times for 5 minutes each in PBS.
6. Mount as described in Chapter 4.
7. Observe as in the above procedure for live cells.

   CAUTION: Formaldehyde (see Appendix 2 for Caution)

■ *Notes for Mitochondrial Rhodamine Dyes*

- Rhodamine 123, TMRE, and TMRM are membrane-potential-sensitive, cationic fluorophores. Therefore, the mitochondrion must be functioning and generating a membrane potential in order to *attract and maintain* the dyes in the mitochondrion. These dyes are also used as sensitive indicators of the mitochondrial membrane potential. Rhodamine 123 is specific for the mitochondrion. In contrast, TMRM, TMRE, and the carbocyanine dyes can also label the endoplasmic reticulum to some degree. Mito Tracker CMTMR has the added feature of chemical reactivity so that, once incorporated in the mitochondrion, it can chemically link to thiol groups and will not leave the mitochondrion when the membrane potential decreases as a result of fixation and/or cell death; hence, it can be used to locate mitochondria in specimens that will subsequently be fixed.
- The effect of submicromolar concentrations of mitochondrial inhibitors, e.g., KCN, rotenone, antimycin α, and uncouplers of oxidative phosphorylation, such as carbonylcyanide-*m*-chlorophenyl hydrazone (CCCP) and the related FCCP, can be examined on isolated mitochondria or mitochondria in intact cells.
- If the concentration of these membrane potential driven dyes is too high, mitochondrial function will become increasingly impaired. As a rule, the dye concentration should be kept below 1 mM.

LABELING THE GOLGI APPARATUS

The Golgi apparatus is composed of a series of flattened, disk-shaped cisternae typically located near the cell nucleus (for details, see Alberts et al. 1994). It is thought that the Golgi may be a principal organizer of macromolecular traffic in the cell since many molecules, such as secreted proteins, glycoproteins, glycolipids, and plasma membrane glycoproteins, pass through the Golgi during their maturation. The fluorescent probes of the Golgi make use of this function for labeling.

Labeling the Golgi Apparatus with BODIPY-FL-Ceramide (C_5-DMB-ceramide)

For more detailed information, see the review by Pagano (1989) that uses a related probe, C_6-NBD-ceramide. Both of these probes make use of the fact that cells incubated at low temperature with the probe and then subsequently washed and warmed to 37°C exhibit a strongly fluorescent Golgi apparatus, followed over time by a fluorescent plasma membrane. The latter result probably reflects the intracellular synthesis of fluorescent sphingomyelin and glucosylceramide analogs from the added fluorescent ceramide and their subsequent transport through the Golgi to the plasma membrane. Listed below are the advantages of BODIPY-FL-ceramide.

- At the high membrane concentrations the probe achieves in the *trans*-Golgi, the probe forms excimers, which allow a cleaner visualization of the Golgi when a long-pass red filter is used.
- It is brighter than NBD-ceramide.
- It is more fade resistant than NBD-ceramide.

Labeling Live Cells

1. Prepare a stock solution of 1 μM BODIPY-FL-ceramide (m.w. 576) (Molecular Probes) in cell culture medium. Protect from light in a sealed vial and store at –20°C.
2. Aspirate the cell culture medium and wash the coverslip containing the cells three times with PBS(+).
3. Aspirate the PBS(+) wash medium. Add the BODIPY-FL-ceramide solution (without dilution) to completely immerse cells attached to coverslips in a small petri dish. Incubate at 37°C for 10 minutes.
4. Aspirate the labeling solution and incubate 30 minutes at 37°C in DMEM with 10% FBS and penicillin/streptomyocin.
5. Mount the coverslip as described in Chapter 4.

■ *Note*

- For BODIPY-FL-ceramide, the excitation wavelength maximum will be near 464 nm and the emission wavelength maximum will be ~532 nm based on **methanol** as a solvent. Maxima may vary slightly for different carbon chain lengths and especially in different solvents. A filter set for fluorescein should be used for observation. For FRAP and confocal microscopy, the 488-nm line of an argon-ion laser may be used for excitation.

  CAUTION: Methanol (see Appendix 2 for Caution)

LABELING LYSOSOMES

Lysosomes, although diverse in shape and size, are membranous sacs containing more than 40 different acid hydrolases (for additional information, see Alberts et al. 1994). The enzymes operate optimally at acidic pH of the lysosome (~5) to break down various substances. It is thought that the highly glycolsylated nature of the proteins of the Golgi membrane help to protect them from degradation. A number of the fluorescent approaches to visualizing the lysosomes make use of their acidic pH. Commonly used probes include neutral red, *N*-(3-[2,4-dinitrophenyl amino] propyl)-*N*-(3-aminopropyl)methylamine (DAMP), and acridine orange (a DNA stain).

Labeling Live Cells

1 Prepare a 2 μM labeling solution in PBS(+).

2 Aspirate the culture medium and wash three times with PBS(+).

3 Immerse the cells in the labeling solution at room temperature for 10 minutes.

4 Aspirate the labeling solution and wash three times in PBS(+).

5 Mount as described in Chapter 4.

Notes

- For neutral red, the excitation wavelength will be near 541 nm and the emission wavelength will be ~640 nm based on ethanol as the solvent. A filter set for rhodamine should be used for observation. For confocal microscopy, the 514-nm line of an argon-ion laser may be used for excitation.
- The incubation time is dependent on the cell type. The longer the staining time, the greater the chance for nonspecific staining.

LABELING THE ENDOPLASMIC RETICULUM WITH DIO-C_6(3)

The endoplasmic reticulum is a highly convoluted, single membrane that is continuous with the outer nuclear membrane and is believed to form a contiguous closed sac within the cell cytosol (see Alberts et al. 1994). For more detailed information, see articles by Terasaki (1989, 1993). Although a probe specific for the endoplasmic reticulum currently does not exist, the DiO-C_6(3) probe is an effective label because of the distinctive intracellular morphology of the endoplasmic reticulum that distinguishes it from other intracellular organelles.

Labeling Live Cells

1 Prepare a stock solution of DiO-C_6(3) (m.w. 573), a cationic dye, at a concentration of 0.5 mg/ml in absolute ethanol. Protect from light in a sealed vial at room temperature.

2 For labeling, dilute the dye in cell culture medium. Depending on the particular cell type to be used, the specific conditions may vary and should be checked. For live cells, a general starting point would be a dye concentration of 0.5 μg/ml in the cell culture medium, i.e., a 1000-fold dilution of the stock medium.

3 Aspirate the cellular medium and follow with three rinses with PBS(+).

4 Stain at room temperature for 10 minutes. Shorter or longer times may be tried empirically to improve staining with a particular cell type.

5 Aspirate dye-containing cell culture medium. Rinse three times with PBS(+).

6 Mount as described in Chapter 4.

Note

- DiO-C_6(3) is viewed for microscopy using a fluorescein filter set. The excitation maximum wavelength is near 484 nm and the emission maximum is near 501 nm based on **methanol** as a solvent.

  CAUTION: Methanol (see Appendix 2 for Caution)

Labeling Fixed Cells

1. Prepare the labeling solution at a concentration of 2.5 μg/ml (a 2000-fold dilution of the stock).
2. Fix cells for 10 minutes with 0.5% **glutaraldehyde** (or 3.7% **formaldehyde**, but see second note below) in PBS.
3. Aspirate the fixation medium and rinse three times with PBS, allowing a 5-minute soaking for each rinse.
4. Stain at room temperature for 10 seconds.
5. Aspirate the labeling medium. Rinse three times with PBS.
6. Mount as described in Chapter 4.

   CAUTION: Formaldehyde; glutaraldehyde (see Appendix 2 for Caution)

Notes

- The DiO-C_5(3) probe has also been used to visualize the endoplasmic reticulum.
- Do not use methanol for fixation because it may extract membranes of the endoplasmic reticulum so that no pattern can be seen. Formaldehyde fixation can cause vesiculation of the endoplasmic reticulum.

LABELING RECEPTORS WITH TOXINS AND HORMONES

Integral protein receptors in the membrane of cells can be studied by using specifically fluoresceinated but still functional toxins or hormones that bind to specific receptors. A number of receptors may be studied using commercially available probes. The procedure for labeling the acetylcholine receptor using α-bungarotoxin (Fig. 5.5) (Axelrod et al. 1976) is given below. Some other receptors for which there are fluorescent probes include the benzodiazepine receptor, the Na^+-K^+-ATPase using anthroyl ouabain to the protein's α-subunit and the epidermal growth factor (EGF) receptor using fluorescent EGF, etc. (cf. Haugland 1992). For more detailed information, see Angelides (1989), Maxfield (1989), and Haugland (1992).

Labeling Acetylcholine Receptors Using Rhodamine α-Bungarotoxin

Labeling Live Cells

1. Prepare a labeling solution of 0.1 mM rhodamine α-bungarotoxin (m.w. ~8500) in PBS(+).
2. Aspirate the culture medium and wash three times with PBS(+).

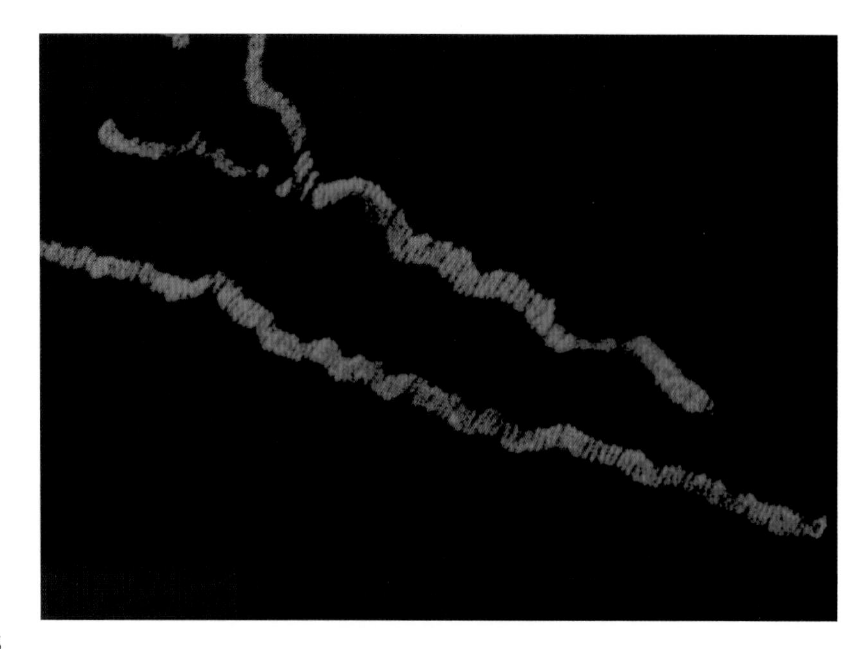

FIGURE 5.5

Confocal micrograph of acetylcholine receptors at frog neuromuscular junction labeled with Texas Red-X-α-bungarotoxin (Molecular Probes, Inc.). (Photo provided by D. Raciborska and M.P. Chariton, University of Toronto.)

3 Add the labeling solution and incubate at 37°C for 30 minutes.

4 Aspirate the labeling solution and rinse three times with PBS(+).

5 Incubate for 30 minutes in PBS(+) containing 2 mg/ml BSA.

6 Mount as described in Chapter 4.

Notes

- For rhodamine α-bungarotoxin, the excitation wavelength maximum will be near 550 nm and the emission wavelength maximum will be ~575 nm based on **methanol** as a solvent. A filter set for rhodamine can be used for observation. For FRAP and confocal microscopy, the 514-nm line of an argon-ion laser may be used for excitation.
- As a control, pretreatment of cells with 5×10^{-8} M α-bungarotoxin for 1 hour should prevent fluorescence labeling.
- The receptors should remain on the membrane surface for several hours after labeling.
- α-bungarotoxin binds with high affinity to the α-subunit of the nicotinic acetylcholine receptor of the neuromuscular junction.

CAUTION: Methanol (see Appendix 2 for Caution)

PINOCYTOTIC VESICLES AND CYTOPLASM STUDIED BY FLUORESCENT DEXTRANS OR FICOLLS

Dextran, a high-molecular-weight poly-D-glucose, and ficoll, a synthetic polymer of epichlorhydrin and sucrose, are electroneutral, hydrophilic polysaccharides whose size can be varied. This variability in size coupled with their membrane impermeability can make them useful for studying fluid-phase pinocytosis, the size of membrane pores (e.g., nuclear membrane pores), or the environmental conditions and size of a cell compartment. For more detailed information, see Luby-Phelps (1989).

FITC or TRITC Labeling of AECM-Ficoll or AECM-Dextran

1 This protocol is based on the Inman method as per Luby-Phelps (1989) and can be used if commercially labeled material is not available. The protocol requires 100 mg of AECM-ficoll or AECM-dextran, 3 ml of 10 mM carbonate–bicarbonate buffer (pH 9.2), 15 mg of either FITC (fluorescein isothiocyanate) or TRITC (tetramethylrhodamine isothiocyanate), 0.1 N NaOH for pH adjustment, and Sephadex G-25.

2 Prepare 100 mg of dextran or AECM-ficoll in 2 ml of carbonate-bicarbonate buffer at pH 9.2.

Carbonate-Bicarbonate Buffer

A. Make up a 0.2 M solution of anhydrous sodium carbonate (2.2 g/100 ml).
B. Make up a 0.2 M solution of sodium carbonate (1.68 g/100 ml).
Take 4 ml of A + 46 ml of B and bring up to 200 ml with DH_2O. The pH will be 9.2.

3 Prepare 15 mg of dye in 1 ml of carbonate-bicarbonate buffer.

a TRITC: Add 15 µg of TRITC to 400 µl of dimethyl formamide and then titrate the solution into 1 ml of buffer with constant stirring. Adjust the pH to 9.0 with 0.1 N NaOH.

b FITC: Add 15 µg of FITC to 1 ml of buffer with stirring. Continue stirring as the pH is adjusted to 9.0 with 0.1 N NaOH. The FITC will likely not completely dissolve until the proper pH has been reached.

4 Add the desired dye solution in a dropwise manner with constant stirring to either the ficoll or dextran solution.

5 Incubate the resultant solution for 30 minutes at 40°C.

6 Use a 1 x 30-cm column of Sephadex G-25 to remove unreacted dye by desalting.

7 Dialyze twice versus 1 liter of dH_2O.

8 Lyophilize for storage and protect from light at –20°C until used.

Pinocytotic Studies

Labeling Live or Fixed Cells

1 Prepare 5 µg/ml solution of labeled dextran or ficoll in complete cell medium, e.g., DMEM with 10% FBS.

2 Immerse cells attached to a glass coverslip with labeling medium in a small petri dish at 37°C for 30 minutes. Pinocytotic uptake is dependent on concentration, time, and cell cycle (greater uptake in mitotic over interphase).

3 For fixing cells, aspirate labeling medium and wash three times with PBS(+) and then fix immediately with 4% **formaldehyde** for 10 minutes.

4 For live cells, wash three times with DMEM (Dulbecco's modified Eagle's medium) and 10% FBS medium.

5 Mount fixed or live cells as described in Chapter 4.

CAUTION: Formaldehyde (see Appendix 2 for Caution)

Notes

- For FITC-labeled polysaccharides, the excitation wavelength maximum will be near 460 nm and the emission wavelength maximum will be ~534 nm based on ethanol as a solvent. It is important to note that the emission spectrum of fluorescein can depend on the environmental pH and/or ionic strength. A filter set for fluorescein should be used for observation. For FRAP and confocal microscopy, the 488-nm line of an argon-ion laser may be used for excitation.
- For additional information on this protocol, consult Berlin and Oliver (1980).

REFERENCES

Alberts B., Bray D., Lewis J., Raif M., Roberts K., and Watson J.D. 1994. *Molecular biology of the cell,* 2nd edition. Garland Publishing, New York.

Angelides, K.J. 1989. Fluorescent analogues of toxins. *Methods Cell Biol.* **29:** 29–58.

Axelrod D., Ravadin P., Koppel D.E., Schlessinger J., Webb W.W., Elson E.L., and Podleski T.R. 1976. Lateral motion of fluorescently labeled acetylcholine receptors in membranes of developing muscle fibers. *Proc. Natl. Acad. Sci.* **73:** 4594–4598.

Berlin R.D. and Oliver J.M. 1980. Surface functions during mitosis. II. Quantitation of pinocytosis and kinetic characterization of the mitotic cycle with a new fluorescence technique. *J. Cell Biol.* **85:** 660–670.

Chazotte B. 1994. Comparisons of the relative effects of polyhydroxyl compounds on local versus long-range motions in the mitochondrial inner membrane. Fluorescence recovery after photobleaching, fluorescence lifetime, and fluorescence anisotropy studies. *Biochim. Biophys. Acta* **1194:** 315–328.

Chen L.B. 1989. Fluorescent labeling of mitochondria. *Methods Cell Biol.* **29:** 103–124.

Hackenbrock C.R., Chazotte B., and Gupte S.S. 1986. The random collision model and a critical assessment of diffusion and collision in mitochondrial electron transport. *J. Bioenerg. Biomembr.* **18:** 331–368.

Haugland R. 1992. *Molecular Probes catalog.* Molecular Probes, Eugene, Oregon.

Herman B. and Jacobson K. 1990. *Optical microscopy for biology.* J. Wiley, New York.

Inoue S. 1986. *Video microscopy,* Plenum Press, New York.

Luby-Phelps K. 1989. Preparation of fluorescently labeled dextrans and ficolls. *Methods Cell Biol.* **29:** 59–74.

Mason W.T. 1993. *Fluorescence and luminescent probes for biological activity.* Academic Press, New York.

Maxwell, F.R. 1989. Fluorescent analogs of peptides and hormones. *Methods Cell Biol.* **29:** 103–124.

Pagano R.E. 1989. A fluorescent derivative ceramide: Physical properties and use in studying the golgi apparatus of animal cells. *Methods Cell Biol.* **29:** 75–87.

Periasamy A. and Herman B. 1994. Computerized fluorescence microscopic vision in the biomedical sciences. *J. Comput. Assisted Microsc.* **6:** 1–26.

Spink C.H., Yeager M.D., and Figenson G.W. 1990. *Biochem. Biophys. Acta* **1023:** 25–33.

Taylor D.L. and Wang Y.-L. 1989. Fluorescence microscopy of living cells in culture. Part B. Quantitative fluorescence microscopy—Imaging and spectroscopy. In *Methods in cell biology,* vol. 30. Academic Press, New York.

Terasaki M. 1989. Fluorescence labeling of endoplasmic reticulum. *Methods Cell Biol.* **29:** 125–136.

———. 1993. Probes for the endoplasmic reticulum. In *Fluorescence and luminescent probes for biological activity* (ed. W.T. Mason), pp. 120–123. Academic Press, New York.

Wang Y-L. and Taylor D.L. 1989. Fluorescence microscopy of living cells in culture. Part A. Fluorescent analogs, labeling cells and basic microscopy. In *Methods in cell biology,* vol. 29. Academic Press, New York.

CHAPTER 6

Introduction to Immunofluorescence Microscopy

George McNamara

The Saban Research Institute of Children's Hospital, Los Angeles, California

INTRODUCTION

Localization by immunofluorescence (IF) makes use of antibodies that are specific for biological molecules (antigens) of interest and allows the visualization and optional quantification of the amount of specific labeling in terms of the number of distinct fluorescence objects and/or their intensity. Fluorescence is due to the absorption of a photon by a "fluorochrome" molecule, which excites an electron to a higher energy level, followed by a loss of energy by emission of a longer-wavelength photon after a few nanoseconds. Optical filters can be used to separate the short-wavelength excitation photons from the longer-wavelength emission photons. Specific labeling of an antigen by a fluorochrome-conjugated antibody makes possible the localization of the antigen in the specimen, since only the specific antigen lights up. Fluorescence microscopy is a high-contrast technique because the bright fluorescence stands out in high contrast against the dark background of the specimen. Contrast is defined as the ratio of the absolute value of (specimen background) versus background intensity. To put this in context: It is easier to see stars in the sky at night than on a sunny day. The stars' intensities are the same, but the background is orders of magnitude brighter in the latter case. Fluorescence can easily be quantitated since fluorescence intensity is linearly proportional to the amount of excitation light and the amount of fluorochrome.

IF can be used to survey developing embryos, pathological tissues, cancer versus normal cellular phenotypes, plants, fungi, bacteria, viruses, and subcellular localization, and colocalization of antigens. The technique works particularly well on cytoskeletal antigens, which are often used to identify cell types and specific structures such as axons or dendrites. IF is complemented by fluorescence in situ hybridization (FISH) techniques, which aim to localize and/or count the number of RNAs or DNA signals in an organism, cell, or nucleus. IF and FISH can be used together with nonimmunologic fluorescent labeling of cell structures with fluorescent toxins or organelle-specific lipids and other reagents. The reagents and equipment can be adapted for use on live cells or extended for use at the molecular scale using atomic force or electron microscopy.

The limits to IF include specificity of the antibodies, specimen preparation, autofluorescence, and performance of the microscope and user. Specificity of the antibodies depends on the purity of the antigen used for immunization and can often be satisfied with either affinity-purified polyclonal antibodies or by monoclonal antibodies. Specimen preparation is discussed in Chapters 7–12. Autofluorescence is often a factor limiting detectability of fluorescent probes in cells and tissues. Autofluorescence can be minimized by spectral discrimination, which encompasses selecting probes and optical filters that maximize the probe fluorescence signal relative to the autofluorescence. Autofluorescence in mammalian cells is principally due to flavin coenzymes (FAD and FMN: absorption, 450 nm; emission, 515 nm) and reduced pyridine nucleotides (NADH: absorption, 340 nm; emission, 460 nm). In plants, autofluorescence is often due to lignins, which are a source of green autofluorescence, and porphyrins, such as chlorophylls, which result in the longer-wavelength red autofluorescence. Therefore IF can best be performed with a short-wavelength fluorochrome that emits in the blue. For fixed cells, autofluorescence can be significantly diminished by washing with 0.1% sodium borohydride in PBS (phosphate-buffered saline) for 30 minutes (Bacallao et al. 1995) prior to antibody incubation.

Optimum performance of the user depends on their skill level in using and maintaining the equipment as well as possible. This includes recognizing when something looks wrong with an image and how to fix it (or who to ask for help). If fluorescence microscopy is an important career skill, the technique can be learned in a few weeks at any of the microscopy courses taught for that purpose. Many excellent books on fluorescence and/or microscopy are also available, including Inoué (1986), Taylor and Wang (1989), Wang and Taylor (1989), Shotton (1993), Pawley (1995), Shapiro (1995), and a series of handbooks from the Royal Microscopy Society (see, e.g., Bradbury and Evennett 1996). Within the laboratory, an array of commercially available microscope standards can be used to help understand the performance (or lack thereof) of the fluorescence microscope. The Molecular Probes family of "Speck" standards (TetraSpeck, MultiSpeck, etc.) and multifluorescent probe slides (FluoCells, Fig. 6.1) are of great use in optimizing the instruments (and the user) for fluorescence microscopy (Haugland 1996). Small (100 nm) TetraSpeck fluorescent beads can be sprinkled on coverslips before mounting the specimen. These beads provide a quick optical quality test of the imaging system (and user), a marker for the coverslip, X, Y registration between fluorochromes and over time, and optional intensity calibration. The user's physiology also plays a part in image quality: Working in low light allows ones eyes to "dark-adapt," which allows imaging of low-light-emitting features. This can be readily appreciated by going from a bright area to a dark open field on a starry night: Within a few minutes, more stars become visible and they become colored. Unfortunately, many microscopes are housed in a lab environment that is not completely dark. To truly optimize the microscope setting, disable all overhead lights, screen the door with a black curtain, adjust all computer monitors to use a dark background, and keep all video monitors adjusted for dark-adapted eyes.

ANTIBODIES

Antigens can be detected by direct IF, where the primary antibody is directly coupled to a fluorophore, or by indirect IF, where the fluorophore is coupled to a secondary antibody that is used to recognize the primary antibody–antigen complex. Direct IF is rarely used because (1) the number of fluorochromes that can be attached to each antibody is limited and (2) the expense involved in labeling each primary antibody is high in terms of the researcher's time, amount of precious antibody required for each labeling reaction, and the quality control carried out for each batch of labeled antibody.

Most researchers use indirect IF, where the fluorescent label is on the secondary antibody. This permits the commercial preparation of large amounts of highly specific secondary antibody, which in turn permits the researcher to buy high-quality reagents at low cost. A further variation has been to use a signal amplification method, where an enzyme is conjugated to the secondary antibody, and the site of antigen–primary antibodies–secondary antibodies is detected by enzyme-mediated deposition of a fluorescent reaction product. Such amplification can increase the sensitivity of the assay, but it may compromise its spatial resolution and quantitativeness. Antibodies can also be conjugated to colloidal gold or fluorescently conjugated Nanogold (Powell et al. 1997) for imaging with both the light and electron microscopes. High-contrast imaging of such markers in the light microscope may require reflected light, video-enhanced contrast transmitted light, or fluorescent imaging modes.

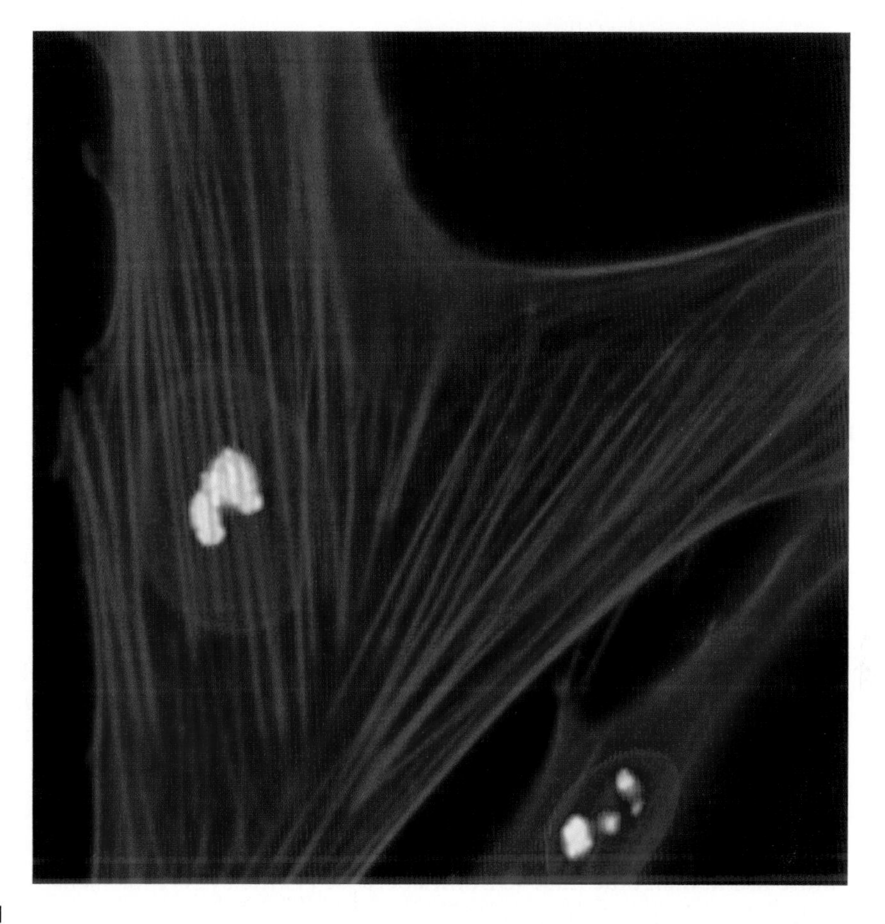

FIGURE 6.1

Triple-label fluorescence image showing the localization of F-actin (rhodamine phalloidin), DNA (DAPI), and a nucleolar protein (anti-fibrillarin, FITC [fluorescein isothiocyanate] goat-anti-human). Images were captured independently with a Photometrics SenSys cooled charge-coupled device (CCD) camera, pseudocolored, and overlaid using Oncor Image software. (Photo provided by J. McCann, Cold Spring Harbor Laboratory.)

A variety of correlative light microscopy–electron microscopy experiments are now (almost) routine. The goal is to test hypotheses over several magnitudes in scale, from the global (organism, tissue, cell interactions) to the molecular (nanometers, protein–protein interactions). The light microscope is used for obtaining an overall view of the cell or subcellular compartment, whereas the electron microscope is used for deep exploration into a smaller region of the cell or compartment. Deerinck et al. (1994) used eosin-labeled antibodies in an elegant correlative fluorescence and electron microscopic study. Eosin excites and emits between fluorescein and rhodamine, but it is not as bright as either. It is also used at much higher concentrations as an absorption dye, as in the standard H&E (hematoxylin and eosin) histological stain. Deerinck et al. (1994) conjugated eosin isothiocyanate to antibodies, carried out standard IF (in the absence of oxygen), and then took advantage of the ability of eosin's high quantum yield for oxygen radical formation (QY_{rad}) to photo-oxidize DAB (diaminobenzedine tetrahydrochloride), a common immunohistochemical reagent, which in the presence of light and oxygen forms a precipitate near the antigen. The DAB precipitate is then observed as a well-localized black label in transmitted light microscopy and most importantly at very high resolution in the transmission electron microscope (see Chapters 20 and 21).

Once specific binding to the antigen has been demonstrated by biochemical means (often by western blotting), the antibodies can be used to localize and quantify the antigen in whole mounts, tissue sections, and cells. Hundreds of tissue- and cell-type-specific antigens are known. Once the location of an antigen in a tissue or cell is known, questions of colocalization with other antigens are often raised. This can be efficiently addressed by a multiprobe experiment with two or more fluorochrome-antibody reactions. For cellular structures that do not have well-characterized specific antigens (yet), counterstains with organelle-specific fluorescent lipids or other reagents can be used for subcellular localization (see Chapter 5). Once colocalization has been observed at the light microscope level, it can be further characterized at the molecular (nanometer) scale by immunoelectron microscopy (see Chapter 21) and by high-resolution light microscope methods such as fluorescence resonance energy-transfer imaging.

FLUOROCHROMES

Fluorescein is the classic fluorochrome for IF. It can be excited by UV light (<400 nm wavelength) or blue light (usually 490 nm), and it emits green light (500–550 nm). This wavelength range is where the dark-adapted eye is most sensitive and where microscope optics are optimally corrected for. Chemically reactive FITC can be easily coupled to antibodies, although such secondary antibodies are most easily obtained by purchasing them from a commercial supplier. Fluorescein is an excellent fluorochrome because it has a high probability of absorbing a photon at the wavelengths mentioned, i.e., it has a high molar extinction coefficient (ϵ). Fluorescein also has a high probability of emitting a photon, i.e., it has a high fluorescence quantum yield (QY_f).

Rhodamine, eosin, tetramethylrhodamine, and Texas Red are related fluorochromes that excite from about 520 nm to 590 nm and emit from about 550 nm to 620 nm. They can also be coupled to antibodies through an isothiocyanate side group. In general, yellow and red fluorochromes are more useful than blue fluorochromes because (1) the eye is more sensitive to yellow-red light than to blue light and (2) blue-emitting fluorochromes often require UV excitation, which can damage many cellular structures and efficiently excites the major autofluorochromes of many cells.

The classic (and many new) fluorochromes have also been conjugated to many toxins, each of which binds specifically to one cellular target (see Chapter 5). For example, fluorescent phalloidin has frequently been used to specifically label filamentous actin (F-actin) in microinjected or fixed cells. This provides a nice way to "light up" cellular landmarks, such as the F-actin-based cytoskeleton of tissue culture cells. Other fluorochromes accumulate more or less specifically into specific organelles and can be trapped there by clever chemistry. This is the case for the MitoTracker Red labeling of mitochondria (see Chapter 5).

DAPI (4′,6-diamidino-2-phenylindole) has been the most common fluorochrome used for fluorescently labeling double-stranded DNA (Fig. 6.1). DAPI is about 20 times more fluorescent when bound to runs of A-T bases than when free in solution or bound to G-C bases (Kapuscinski 1995). DAPI has been used to visualize mycoplasma contamination and mitochondria, but it is most often used to visualize nuclei (see Chapter 5) and chromosomes (see Chapter 5). DAPI can be excited with 330–430 nm, and emits brightly from 400 nm to 500 nm, but does emit noticeably in the green wavelength range and weakly in the red. A problem with DAPI is that it can be so bright when excited at 360 nm that it overwhelms the other fluorescent signals in a multicolor imaging experiment. One simple approach to reducing the intensity of the DAPI signal is to excite DAPI with 405 nm light instead of 360 nm. The longer-wavelength light does not excite DAPI as well, results in less photodamage to the cells (an issue mostly for live experiments), and results in less photobleaching of the other fluorochromes.

Fluorescein, Texas Red, and DAPI are commonly used and are well matched for standard fluorescence microscopes. Development of new fluorochromes and their application to fluorescence microscopy are proceeding at a rapid pace. Some of the newer families of fluorochromes include the Alexa, BODIPY, and Cyanine dyes. Representatives from each family can be found that can substitute for fluorescein and rhodamine, but have better spectral characteristics for multiple probe experiments. The YOYO family (Molecular Probes) of DNA-binding dyes are also available in a number of colors, and are more sensitive than DAPI for microscopy or ethidium bromide for staining DNA gels. However, because DAPI is more than bright enough for common use, it remains the standard DNA fluorochrome.

MULTIPLE FLUOROCHROME EXPERIMENTS

The key to multiple probe IF is to use highly specific primary and secondary antibodies for each of the antigens of interest, and to minimize spectral cross talk between fluorochromes. A typical strategy is to use affinity-purified primary antibodies to each of the antigens. The primary antigens might be from different species or may be tagged with specific haptens, such as digoxigenin (dig), biotin, or DNP (dinitrophenol). The hassles of labeling primary antibodies with a hapten may (or may not) be worth the effort in terms of specificity. The secondary antibodies are labeled with different fluorochromes and are each specific for one primary antibody. Biotin can be detected with either avidin, streptavidin, or anti-biotin (each has advantages). Hapten conjugation kits and prelabeled fluorescent antibodies are available from many suppliers. Kits are also available for multiprobe FISH experiments.

An example of a multiple IF experiment is shown in Table 6.1. Substitute your antigens in the first column. An effort needs to be made to make sure that none of the secondary antibodies labels the specimen directly or interacts with the other primary antibodies.

An optimum choice for imaging such an experiment would be to use a fluorescence excitation filter set that would let one acquire separate images of the Cy5, Texas Red, fluorescein, and DAPI fluorescence, in that order, to minimize short-wavelength photodamage. Images can be acquired using a high-resolution digital cooled charge-coupled device (CCD) camera and an automated filter wheel to change the excitation wavelength. Depending on the equipment and fluorochromes available to you, you may be using a monochrome video camera, color video camera, confocal microscope, photographic film, or simply your eyes.

How much of a problem spectral cross talk will be depends on the fluorescence excitation and emission spectra of each fluorochrome (see Appendix 1), their relative amounts in the sample, the characteristics of the filter set(s), and the sensitivity of the detector(s) at each wavelength. For convenience, the peak excitation and emission wavelengths of fluorochromes are usually listed, i.e., fluorescein: 490/520 nm; however, you should recognize that these are simply two points on the fluorescence spectra. Another convenience is to refer to the difference (in nanometers) between the peaks as the Stokes shift, in honor of G.G. Stokes, who discovered fluorescence in the 1800s. For fluorescein, the Stokes shift is 30 nm. Some molecules have very small Stokes shifts, making it challenging to separate the fluorescence emission from scattered excitation light. Other molecules have Stokes

TABLE 6.1 Multiple fluorescence immunocytochemistry example

Antigen	Primary antibody	Fluorochrome-secondary antibody
Tubulin	mouse MAb IgG1 (or digoxigenin-MAb)	FITC-goat-anti-mouse IgG1 (or FITC anti-digoxigenin)
Actin	mouse MAb IgA (or biotinylated-MAb)	Texas Red-goat-anti-mouse-IgA (or Texas Red anti-biotin)
Vimentin	rat MAb IgM (or DNP-MAb)	Cy5-rabbit-anti-rat-IgM (or Cy5 anti-DNP)
DNA	—	DAPI

shifts of hundreds of nanometers; this makes separation of excitation and emission light trivial, but makes using multiple fluorochromes difficult. However, it is not necessary to excite or collect at the peak wavelengths. For tactical reasons, and with some types of spectral scanning emission wavelength filters, it is often worth using fluorochromes in the same family of dyes because they are likely to have similar fluorescence yields, Stokes shifts, and "shoulders" on their fluorescence spectra. (Caveat: Structurally related molecules may undergo dimer or "excimer" formation resulting in fluorescence quenching of the signal or a change in the emission spectrum. This could in turn lead to unpleasant surprises in the data or the need for yet more controls.)

Filters, Optics, and Light Budgets

The classic fluorescence microscope was on a very tight "light budget" because of nonoptimized optics and inefficient detectors. These microscopes needed to pump as many photons through the microscope as possible, to get a few photons to the eye or to film. Since only one fluorochrome was typically used, a short-pass excitation filter could be used to pass all wavelengths shorter than some target wavelength. Usually this wavelength was halfway between the excitation and emission maximum. A dichroic beam splitter was placed at a 45° angle to reflect the short-wavelength light from the mercury arc lamp to the objective lens, which then focused the light on the specimen. Fluorescence emission is then collected by the objective lens, is a longer wavelength, and so is transmitted through the beam splitter, through a long-pass emission filter, to the detector (typically someone's eye). The dichroic beam splitter could, in principle, be used alone. However, this does not work in practice because the excitation light reflects from the cover glass and is not completely reflected by the beam splitter. The excitation and emission filters provide most of the wavelength selection; the beam splitter directs the light to the specimen or to the detector (Fig. 6.2). For multiprobe experiments, you will almost certainly need to replace the short-pass excitation filter with narrower band-pass filter(s) and, likewise, replace the long-pass emission filter with band-pass emission filter(s). For example, instead of imaging DAPI with a <400-nm exciter, 410-nm beam splitter (reflecting <410 nm, transmitting >410 nm), and a >430-nm emission filter, you would use a 390–410-nm exciter, a 420-nm beam splitter, and a 440–470-nm emission filter (this is also an ideal filter set for the so-called blue fluorescent protein [BFP] spectral variant of green fluorescent protein [GFP]). The longer-wavelength exciter would reduce illumination intensity, which, in addition to moderating the amount of DAPI fluorescence, would also reduce fluorescein excitation and cellular photodamage. If you are looking by eye or using color film, a dual-, triple-, or quadpass filter cube could be used to capture multiple fluorochromes in a single look. However, balancing the intensities of multiple fluorochromes can be difficult, and photographic film has a low quantum efficiency (the percentage of photons that reach the detector that contribute to the recorded signal) compared to photomultiplier tubes (10–30%), the best video intensifiers (30%), or scientific-grade digital CCDs (30–90%). For monochrome electronic detectors, sequential acquisition of individual fluorescence channels will allow independent optimization of each fluorescent signal. The new multiple-wavelength confocal fluorescence microscopes allow the simultaneous recording of several fluorochromes more or less independently by using several photomultiplier tubes, each with a different wavelength filter (see Chapters 2 and 3; Brelje et al. 1993; Pawley 1995).

A low-numerical-aperture (NA) objective lens will collect less light than a high-NA lens. NA is also critical for spatial resolution: A high-NA lens will resolve finer details in a specimen than a low-NA lens. Magnification is also important: A high-magnification objective (100x) will collect less light than a 60x or 40x objective. The best and brightest images will therefore be obtained with a 40x or a 60x, 1.4 NA oil immersion objective, because these will collect the most light, resolve structures, and magnify them sufficiently for a detector to image them with high contrast (Inoué 1986).

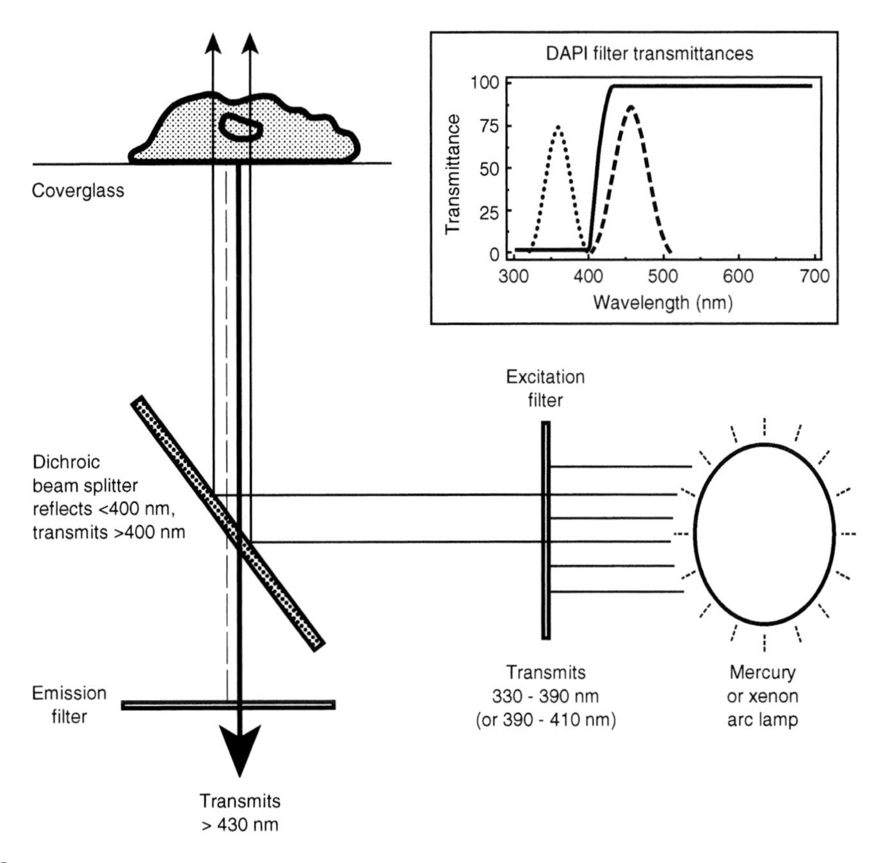

FIGURE 6.2

Epi-illumination light path of an inverted microscope. Light is emitted from a source (i.e., mercury or xenon lamp) and specific wavelengths are transmitted through an excitation filter. Specific excitation filters are used to allow the excitation of specific fluorochromes. Light then reaches the dichroic beam splitter, which reflects short-wavelength light to the specimen, which then emits longer-wavelength light that is transmitted by the beam splitter. The emission filter attenuates all of the light transmitted by the excitation filter and very efficiently transmits any fluorescence emitted by the specimen. The light is always of longer wavelength than the excitation color. *Inset:* Transmittance curves for a DAPI filter. *Dotted* line is excitation band pass; *solid* line is transmittance of the beam splitter; *dashed* line is emission band pass. (Photo provided by G. McNamara.)

For best results with such objectives, you need to use the correct thickness cover glass, look only at fluorescence directly under the cover glass, and/or image in a mounting medium whose refractive index (RI) matches that of the immersion medium (usually RI = 1.5). For imaging deep into aqueous samples, there are special water immersion objective lenses.

Choices for Multiple Fluorochromes

Optimization of fluorochromes in a multiprobe experiment really amounts to what fluorochromes, filters, and detectors you have on hand, how good they are, and how much it would cost to improve them. In many cases, your personnel budget, and current investment in antibodies, molecular cloning, and the "biological questions" that need to be answered will dictate what reagents and hardware you deal with. As a rule, what worked in the past is usually good enough, so fluorescein, Texas Red, and DAPI will usually work. These can be imaged very nicely using the Chroma Technology 84000 filter set.

Instrumentation

The choice of which fluorochromes and antibodies to use will be dictated by the questions that you are asking and the instrument(s) that are available to you. Most research laboratories have access to research light microscopes for fluorescence microscopy. This section briefly discusses questions to consider when upgrading an existing microscope or purchasing new equipment. First, are the illumination system, objective lenses, and detectors optimized for fluorescence? The single most important component is the objective lens, but the entire light path, including the immersion medium, cover glass, mounting medium, and specimen will contribute to image quality (or lack thereof). What additional imaging modes will be used in addition to fluorescence? Do these modes compromise the quality of the fluorescence images? Will the microscope also be used for densitometry (Feulgen stain analysis of DNA ploidy, immunoperoxidase-DAB staining), routine observation of cell cultures, or live cell BFP and GFP imaging? What is the optimal means of recording images (monochrome or color photography, color video camera, monochrome digital camera, photomultiplier tube)? Does the recording method permit sufficient intensity and spatial quantitation for immediate and future needs? What is the cost of consumables, such as film development or color prints? How much demand will there be for the instrument? Labs doing high-throughput screening or highly repetitive automated experiments should consider automating the rate-limiting (wavelength selectors) or precision-limiting (fast and accurate focus steps in Z-series) components of the microscope. Who is responsible for maintaining the instrument and training the users?

Automating specific components of the microscope can be extremely useful in increasing the throughput of the microscope, thus maximizing the investment of the equipment and making efficient use of the investigator's time (Table 6.2). Computer-controlled automation assists in the design, execution, and analysis of multidimensional experiments (Farkas et al. 1993; Thomas et al. 1996). These range from automating the fluorescence illumination filter sets for fixed-cell triple-label experiments to doing multiwavelength (CFP, YFP, and transmitted light) Z-series time-lapse experiments on live cells. Fast shutters, placed between each illumination source and the microscope body, can limit the exposure time to that needed to acquire an image. Optimized excitation filters for each fluorochromes can be placed in a computer-controlled wheel and used with a multipass beam splitter and emission filter.

CONCLUSION

Immunofluorescence is a powerful tool to analyze the location and associations of specific molecules at the organismal, cellular, and subcellular levels. It is complementary to in situ hybridization and live cell imaging. The results of immunofluorescence experiments can be verified and extended by transmission electron microscopy and other methods. Success depends on the specificity of the reagents, specimen preparation technique, autofluorescence, quality of the instrumentation, and skill in using the microscope and interpreting the results. Multiprobe immunofluorescence allows you to localize and quantify multiple antigens and possible interrelationships (e.g., are two antigens colocalized?). Counterstaining with DAPI or other organelle-specific reagents provides convenient landmarks for placing the specific immunofluorescence labels in their cellular context.

TABLE 6.2 Automation for light microscopes

Component	Description
Z-motor	A focus motor can be mounted on the fine focus knob and used to do fast Z-series acquisition. Precision is limited by the gears of the microscope focus. Some new automated microscopes use a DC motor to drive the microscope focus. For small step sizes ($<$100 nm) a microscope objective can be mounted directly on a piezoelectric focusing device (Physick Instruments). This is most useful for confocal microscopes or digital deconvolution stations (see Chapter 2) where high precision is needed and the microscope is intended for use with a single, special, objective lens such as a 60×, 1.2 NA water immersion lens for imaging aqueous specimens.
Excitation shutter	Fast (10-ms) blocking of epi-illumination source to minimize photobleaching of fluorochrome and photodamage to specimen
Transmission shutter	Block transmitted light source to allow independent acquisition of fluorescence and transmitted (phase-contrast, DIC [differential interference contrast], bright-field, dark-field) images
Excitation wavelength selection	Typically a filter wheel with interference filters to isolate optimal excitation wavelength for each fluorochrome. Some wheels can move between filter positions in $<$50 ms. Also useful for providing specific wavelengths for uncaging and interference reflection microscopy. A more expensive alternative is to use a monochromator. This has the advantage of allowing excitation spectral scanning.
Filter block turret or slider	Microscopes are now available that can automate up to eight filter cubes. Wavelength selection is much slower than excitation wheels but allows each beam splitter to be optimized for a single fluorochrome.
Emission wavelength selection	The multipass emission filter in the cube can be replaced with optimized emission filters in a wheel in front of an electronic detector. The filters are typically not accessible for viewing by eye (a multipass filter can still be slid into the path to the eyepiece for manual imaging). The filter wheel can also be loaded with a DIC analyzer and optimized filters for single or dual color densitometry and can be used to enable a high-resolution monochrome digital CCD to acquire "sequential color" images of histological and immuno-enzyme-labeled specimens. Such an instrument can also be used for imaging fluorescence resonance energy transfer.
Stage	High-throughput screening of multiwell (12, 24, 96, 384, 768) plates can benefit greatly from automation. High-resolution montage construction benefits from automatic alignment and intensity correction of digital images. Automation allows positions to be multiplexed over time (e.g., study of cell movement, division, or death in tissue culture over many hours) or in treatments (if destaining and restaining for sequential probes). Even tasks such as scanning an entire slide can benefit from a feature as simple as visual feedback of the areas of a slide that have already been visited.
Micromanipulation, microinjection, electrophysiology equipment	For live cell imaging, the addition of micromanipulator(s), pressure or iontophoretic microinjection accessories, and electrophysiology recording may be useful to both neurobiologists and cell biologists. The advent of GFPs is not going to eliminate the need to microinject cells with reagents. Usually these devices are controlled by the investigator with event triggering and data recording being handled by a computer.
Microscope internals	Microscope manufacturers offer completely automated microscopes, with the ability to change objective lenses, field, and NA diaphragms.
Cameras	Digital cooled CCD cameras acquire images rapidly, are more sensitive than film, are linear over three to five orders of magnitude of intra-scene brightness, and the results can be used for computerized enhancement and quantitation.
Computers	An inexpensive personal computer running commercial image acquisition and analysis software can be used to control the microscope, script each experiment, analyze the results, and prepare the manuscript for print or electronic publication. To maximize the investment in the microscope, one computer can be used to control the experiment; the image data can be stored on a lab, local area, or campus network and accessed from an analysis station.

REFERENCES

Bacallao R., Kiai K., and Jesaitis L. 1995. Guiding principles of specimen preservation for confocal fluorescence microscopy. In *Handbook of biological confocal microscopy* (ed. J.B. Pawley), pp. 311–325. Plenum Press, New York.

Bradbury S. and Evennett P.J. 1996. Contrast techniques in light microscopy. *Microsc. Handb.* **34:** 136.

Brelje T.C., Wessendorf M.W., and Sorenson R.L. 1993. Multicolor laser scanning confocal immunofluorescence microscopy. Practical applications and limitations. *Methods Cell Biol.* **38:** 97–181.

Deerinck T.J., Martone M.E., Lev-Ram V., Green D.P.L., Tsien R.Y., Spector D.L., Huang S., and Ellisman M.H. 1994. Fluorescence photooxidation with eosin: A method for high resolution immunolocalization and *in situ* hybridization for light and electron microscopy. *J. Cell Biol.* **126:** 901–910.

Farkas D.L., Baxter G., DeBiasio R.L., Gough A., Nederlof M.A., Pane D., Pane J., Patek D.R., Ryan K.W., and Taylor D.L. 1993. Multimode light microscopy and the dynamics of molecules, cells and tissues. *Annu. Rev. Physiol.* **55:** 785–817.

Haugland R.P. 1996. *Molecular probes: Handbook of fluorescent probes and research chemicals*, 6th edition. Molecular Probes, Eugene, Oregon.

Inoué S. 1986. *Video microscopy.* Plenum Press, New York.

Kapuscinski J. 1995. DAPI: A DNA-specific fluorescent probe. *Biotech. Histochem.* **70:** 220–226.

Niswender K.D., Blackman S.M., Rhode L., Magnuson M.A., and Piston D.W. 1995. Quantitative imaging of green fluorescent protein in cultured cells: Comparison of microscopic techniques, use in fusion proteins and detection limits. *J. Microsc.* **180:** 109–116.

Pawley J.B., ed. 1995. *Handbook of confocal microscopy*, 2nd edition. Plenum Press, New York.

Powell R.D., Halsey C.M.R., Spector D.L., Kaurin S.L., McCann J., and Hainfeld J.F. 1997. A covalent fluorescent-gold immunoprobe: Simultaneous detection of a pre-mRNA splicing factor by light and electron microscopy. *J. Histochem. Cytochem.* **45:** 947–956.

Shapiro H.M. 1995. *Practical flow cytometry*, 3rd edition. Wiley-Liss, New York.

Shotten D. Ed. 1993. *Electronic light microscopy: The principles and practice of video-enhanced contrast, digital intensified fluorescence, and confocal scanning light microscopy. Techniques in biomedical microscopy.* Wiley-Liss, New York.

Taylor D.L. and Wang Y.-L., eds. 1989. *Fluorescence microscopy of living cells in culture. B. Quantitative fluorescence microscopy—imaging and spectroscopy. Methods in Cell Biology,* vol. 30. Academic Press, San Diego.

Thomas C., DeVries P., Hardin J., and White J. 1996. Four-dimensional computer visualization of 3D movements in living specimens. *Science* **273:** 603–607.

Vershueren H. 1985. Interference reflection microscopy in cell biology: Methodology and applications. *J. Cell. Sci.* **75:** 279–301.

Wang Y.-L. and Taylor D.L., eds. 1989. *Fluorescence microscopy of living cells in culture. A. Fluorescent analogs, labeling cells, and basic microscopy. Methods in Cell Biology*, vol. 29. Academic Press, San Diego.

CHAPTER 7

Immunostaining of Microtubules, Microtubule-associated Proteins, and Intermediate Filaments

Satya Khoun[1] and Joanna Olmsted[2]

[1]*Northwestern University Feinberg School of Medicine, Chicago, Illinois*
[2]*University of Rochester, Rochester, New York*

INTRODUCTION

Microtubules are ubiquitous cytoskeletal organelles involved in structuring the cytoplasm and in forming the spindle apparatus necessary for mitosis. With the exception of the microtubules constituting the axonemes of cilia, flagella, and sperm tails, most microtubules are labile structures capable of undergoing dynamic changes during cellular events. The polymerization state of microtubules can also be altered by changes in the ionic strength of the medium and by temperature shifts. This inherent sensitivity of microtubule structure to various treatments delayed their fine-structural description until glutaraldehyde was introduced as a fixative (Sabatini et al. 1963) and persists as a consideration in generating images of well-preserved microtubule arrays at the light microscopic level.

Microtubule-associated proteins (MAPs) bind to the surface of microtubules (Olmsted 1986; Hyams and Lloyd 1994). In general, the same conditions that are satisfactory for visualizing tubulin subunits in microtubules are also suitable for viewing MAPs. However, since the affinity of the MAPs may vary, care must be taken such that MAPs are not extracted during processing; this usually involves testing several fixation protocols.

Intermediate filaments are major cytoskeletal constituents of vertebrate cells. They are composed of one or more of a large family of proteins encoded by more than 65 different genes. Within typical mammalian cells, intermediate filaments form extensive networks of 10-nm-diameter filaments that pervade the cytoplasm and appear to be closely associated with both the nuclear envelope and the plasma membrane (Chou et al. 1997). The various types of intermediate filament proteins are developmentally regulated with respect to their expression and, in adult organisms, they are differentially expressed in a tissue- and cell-type-specific fashion. For example, all early developing neurons form vimentin intermediate filaments and, during terminal differentiation, the vimentin intermediate filament network is converted into neuronal specific lintermediate filaments consisting of neurofilament triplet proteins, peripherin, etc. All intermediate filaments possess common structural features that enable them to form either 10-nm intermediate filaments or other types of oligomeric structures. However, they do differ with respect to their amino acid sequence, especially in their amino and carboxyl termini. These differences have permitted the production of specific non-cross-reactive antibodies directed against the various types of intermediate filament subunit proteins. Since intermediate filaments are major cellular proteins, antibodies directed against them have been used to distinguish different cell types both in embryos and in adult tissues such as skin and the central nervous system.

Fixation and Immunofluorescence Protocols for Microtubules

The following details four fixation protocols that have worked satisfactorily to view microtubules and MAPs in cultured cells. Choice of a protocol is largely dependent on the reactivity of the antibody following treatment with various fixatives and the importance in the experiments of preserving the most extractable or labile classes of microtubules or MAPs. It is useful to test antibodies with several fixation conditions.

General Considerations

Cell growth: Cells are grown on standard coverslips using conventional techniques of cell culture. It is often convenient to plate cells onto single coverslips within individual 35-mm dishes, although several coverslips can be arrayed in larger petri dishes.

Washing procedures: Pre- and postfixation rinses and those between antibody incubation steps can be carried out in 35-mm dishes. Liquid is added gently from the side and can be aspirated from the tipped dish with a pasteur pipette. If a large number of coverslips are to be processed, pipettes can be attached via a trap to a water aspirator or a vacuum line. Care should be taken to avoid having the sample dry at any time. Washes of four or more coverslips can be carried out efficiently by transferring coverslips into small Columbia staining jars (Arthur Thomas, Swedesboro, New Jersey). Liquid is changed by placing a finger over the opening, inverting the jar, and then replacing the liquid gently down the side. Alternatively, coverslips can be transferred into a coverslip staining rack (Arthur Thomas) that is suspended by the wire handle within a 150-ml beaker. The beaker contains the appropriate buffer, and the buffer is stirred gently with a magnetic stir bar at the bottom. To prevent drying and possible floating of the coverslips, the holder should be placed in the beaker and the coverslips loaded directly into it. When using either the staining jars or staining rack, it is important to mark the holder and orient the coverslips to keep track of the side of the coverslip with the cell layer.

Washing buffers: For most protocols, washes can be carried out in either PBS, pH 7.2, or Tris-buffered saline, pH 8.3 (TBS: 20 mM Tris-HCl, pH 8.3, 0.9% NaCl). Tween-20 at a concentration of 0.1% can be included in all washes except the last one prior to mounting the coverslips.

Incubations with antibodies: Antibodies diluted for use should be centrifuged briefly (5 min, ~2000g in an Eppendorf centrifuge) to sediment aggregates. Coverslips are incubated with small volumes (30–50 μl/22 x 22-mm coverslip) of primary or secondary antibodies within humidified chambers. Chambers are constructed by placing moistened filter paper within a petri dish or plastic box and placing a small glass plate covered with Parafilm on the paper. Coverslips can be positioned with cells facing down onto a pool of antibody, or antibody can be gently pipetted onto coverslips with the cells facing upward. If antibody is applied on top of the coverslip, wetting of the surface can be accomplished by using surface tension between the antibody and the edge of the disposable pipette tip to distribute the liquid. Care must be taken not to touch the surface of the coverslip, because cells will be scraped off. Antibody incubations can be carried out at room temperature (minimum of 1 hr for primary antibodies, 1/2 hr for secondary antibodies) or at 37°C (1/2-hr incubations). Incubations at 37°C are more likely to generate higher background and/or nonspecific aggregation of antibodies, but may be necessary if the affinity of the first antibody is low.

If double immunostaining is to be carried out for both tubulin and MAPs, the MAP primary antibody should be bound first, followed by the tubulin antibody. If the primary antibodies are derived from different animal species, then the secondary antibody incubations can be carried out simultaneously.

Antibody sources: Many antibodies for MAPs and some for tubulin are available only from individual investigators, and dilutions have to be determined depending on the source. Commercial mouse-anti-chicken tubulin antibodies that have worked on a number of different types of preparations are available from Amersham (anti-α-tubulin, N356; anti-β-tubulin, N357). Secondary antibodies generated in different species and conjugated with a variety of fluorophores are available from a number of companies. Those from Jackson ImmunoResearch Laboratories have been consistently satisfactory; dilutions are usually 1/100 to 1/250 depending on the conjugated fluorophore.

Mounting coverslips: Semipermanent mounts can be made using Gelvatol mounting medium (available from Monsanto, St. Louis, Missouri; see Chapter 4 for preparation). This medium is not autofluorescent, is viscous, and hardens slowly. An advantage of using this medium is that if it is desirable to perform additional staining on the same coverslip, a mounted coverslip can be released from the slide by immersion in phosphate-buffered saline (PBS) overnight.

GLUTARALDEHYDE FIXATION

This method is optimal for maximum preservation of microtubule structure. The disadvantages are that cytoplasmic background may be higher than with other methods, antibody penetration can be difficult because of the cross-linking, and washes generally have to be extensive. If the free pool of tubulin or MAPs is relatively high, the signal from the formed microtubule network will be less evident than with some other methods.

1 Rinse coverslip briefly in 0.1 M PIPES (piperazine-*N,N'*-bis(2-ethanesulfonic acid), pH 6.9.

2 Fix in 0.5% **glutaraldehyde** on 0.1 M PIPES, pH 6.9, for 10 minutes at room temperature.

CAUTION: Glutaraldehyde (see Appendix 2 for Caution)

Electron microscopy (EM)-grade glutaraldehyde obtained in sealed ampoules is preferred. Stock solutions more than 1 week old may give less satisfactory images.

3 Rinse the cells briefly three times with PBS.

4 Permeabilize the cells with 0.5% Triton-X 100 in PBS: three times, 5 minutes each.

The concentration of detergent and duration of permeabilization need to be determined empirically for the antibody used; the condition described is a good compromise for retaining MAPs and microtubule structure.

5 Rinse coverslip in PBS or TBS: three times, 5 minutes each.

6 Treat coverslip with 2.5 mg/ml **sodium borohydride** in 50% ethanol three times, 10 minutes each.

CAUTION: Sodium borohydride (see Appendix 2 for Caution)

Sodium borohydride reduces the free aldehyde groups, and this step is crucial in reducing nonspecific background. It is important that the $NaBH_4$ solution be made immediately before use. The solution will bubble vigorously, and should be changed and/or remade as soon as bubbles are no longer being generated. Although this treatment should not extend past 30 minutes, it is often preferable to change the solution more than three times during this period.

7 Rinse coverslip in PBS: three times, 5 minutes each.

8 Incubate coverslip in PBS + 10% NGS (normal goat serum) at room temperature for 10 minutes.

NGS is used to block nonspecific protein-binding sites. If a primary antibody is generated in goat, then serum from a non-cross-reacting species or 10% BSA (bovine serum albumin) can be used. 0.1% Tween-20 can also be substituted for the nonspecific protein blockers.

9 Incubate with primary antibody (e.g., mouse anti-tubulin) at room temperature for 1 hour or at 37°C for 30 minutes.

10 Rinse in PBS containing 0.1% Tween 20 at room temperature three times, 10 minutes each.

11 Incubate with appropriate secondary antibody (e.g., fluorescein-conjugated anti-mouse IgG) at room temperature for at least 30 minutes.

12 Rinse in PBS containing 0.1% Tween 20 at room temperature three times, 10 minutes each.

13 Rinse coverslip briefly in H_2O to remove salts.

14 Mount coverslip with cells down on a drop of Gelvatol (Monsanto; see Chapter 4 for preparation).

Gently wick off excess liquid and allow to dry overnight to obtain a hard mount. Drying can be hastened using a hair dryer with no heat; mounts that will remain stable under an oil immersion lens require ~30 minutes of drying by this method.

FORMALDEHYDE FIXATION

This method gives reasonable preservation of microtubules and MAP distributions and has the advantage of being a relatively quick procedure. The quality of the formaldehyde can significantly affect the apparent integrity of the microtubule network.

1 Rinse cells with PBS or 0.1 M PIPES, pH 6.9.

2 Fix in 3.7% **formaldehyde** in PBS for 10 minutes.

CAUTION: Formaldehyde (see Appendix 2 for Caution)

It is optimal to make formaldehyde freshly by dissolving the appropriate amount of EM-grade paraformaldehyde (Prill form, from Electron Microscopy Sciences) in PBS in a Pyrex bottle with a stir bar. The aldehyde goes into solution by heating on a hot plate in the hood at 60°C. Keep the bottle cap loosened so that pressure does not build up. Cool down to 20°C and pH to 7.4.

A freshly diluted stock (1:10) of reagent-grade formaldehyde (e.g., 37% formaldehyde, 10% formalin) can give satisfactory images, but lots vary widely.

3 Rinse three times, 10 minutes each in PBS.

4 Permeabilize in 0.5% Triton-X 100 for 10 minutes.

5 Rinse three times, 10 minutes each in PBS.

6 Proceed as in steps 8–14 listed in the glutaraldehyde fixation protocol.

A hybrid of the two protocols, in which fixation is carried out with 0.1% glutaraldehyde and 2% formaldehyde, is often advantageous in trying to achieve lower backgrounds and optimizing reactivity of antibodies. This procedure requires the $NaBH_4$ reduction described in step 6 of the glutaraldehyde protocol.

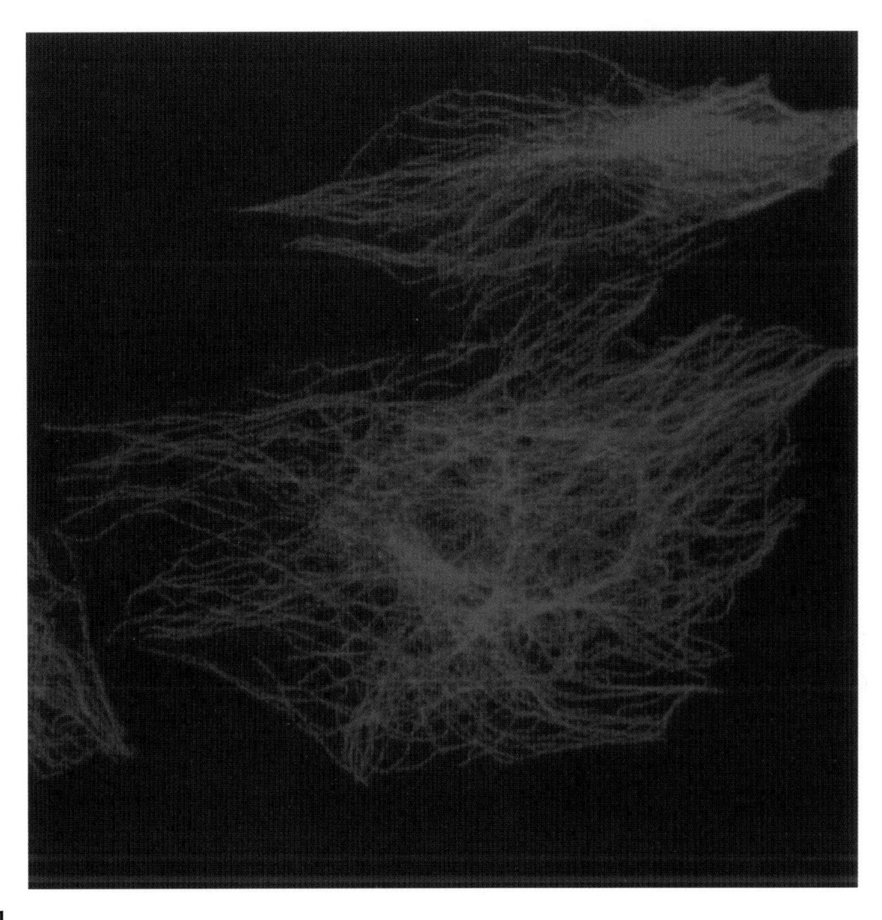

FIGURE 7.1

Confocal microscope image of a BHK cell fixed with methanol and processed for immunostaining with anti-tubulin primary antibody and Cy 3-conjugated secondary antibody. (Photo provided by K. Olson and J. Olmsted, University of Rochester.)

METHANOL FIXATION

This method is rapid and is useful if antibodies do not react well with aldehyde-fixed samples. Maintaining a subzero temperature is critical for reasonable preservation of structure. This method can often produce images of microtubules that appear as linear rows of dots rather than as continuous lines; this pattern may result from slight fragmentation during the fixation process (Fig. 7.1). The method is advantageous in preserving MAPs under conditions where the processing associated with aldehyde fixation and detergent extraction leads to loss of signal from injected or transfected MAPs.

1 Rinse coverslip very quickly in PBS.

Coverslips can be placed directly from the culture medium into ***methanol****, but unless the volume is large, precipitates can form on the surface of the cells.*

CAUTION: Methanol (see Appendix 2 for Caution)

2 Plunge coverslip into –20°C 100% methanol for 10 minutes: Keep coverslip in –20°C freezer.

It is critical that the volume of methanol is sufficiently large so that the temperature of the coverslip is reduced rapidly to –20°C. This can be accomplished by keeping Columbia staining jars and methanol in a –20°C freezer, and rapidly changing the methanol as each coverslip is introduced.

3 Rinse fixed coverslip in PBS and then incubate for 10 minutes with PBS + 10% NGS.

After this step, coverslips can be stored temporarily in 70% ethanol at 4°C in Columbia staining jars. Reasonable images can be obtained after up to a week of such storage.

4 Proceed as in steps 9–14 of the glutaraldehyde protocol above.

EXTRACTION FOLLOWED BY FIXATION

It is possible to extract cells with detergent and retain at least some of the normal distribution of microtubules and MAPs. This may be advantageous if antibody penetration is a problem, or for experiments in which semilysed models are needed. Steps 1–4 are best carried out with single coverslips in 35-mm dishes so the timing for the extraction can be tightly regulated.

1 Rinse coverslips with 0.1 M PIPES, pH 6.9.

2 Permeabilize cells for 1 minute by adding an extraction buffer of 2 M glycerol, 0.4% Triton X-100 in 0.1 M PIPES, pH 6.9. Add this mixture to the cells by tipping the petri dish at an angle, adding the extraction buffer, and then gently laying the dish flat.

The time of extraction is crucial: If it is too short, background will persist, and if too long, MAPs may dissociate and/or microtubules may depolymerize. Adjustment of the time of extraction allows determining whether subclasses of microtubules have different stabilities. Depending on experimental design, both the time and concentration of detergent can be altered. If extraction proceeds too rapidly, the detergent concentration can be reduced.

3 Add an equal volume of 1% **glutaraldehyde** in 0.1 M PIPES, pH 6.9. Perform this addition by tipping the petri dish, mixing the liquid with that already in the dish, and then laying the dish flat.

CAUTION: Glutaraldehyde (see Appendix 2 for Caution)

4 Fix for 10 minutes.

5 Proceed as in steps 3–14 of the glutaraldehyde protocol, *except* omit step 4 (extraction with Triton X-100).

▪ *Notes*

- All primary antibodies should be checked by immunoblot to be sure that they are specifically recognizing the antigen of interest. Immunofluorescence data obtained from an antibody that does not immunoblot or that gives even slight cross-reactivity with other proteins cannot be trusted.
- Primary antibody controls should be run with each experiment. The best control is to use preimmune serum in place of the primary antibody. If preimmune serum is not available, substitute normal immunoglobulin from the animal species in which the primary antibody was made. To control for autofluorescence, substitute both primary and secondary antibody incubations with BSA. Incubating cells only with secondary antibody controls for its nonspecific binding and for the filters used in your fluorescence microscope.
- If the antigen is available, it should be used for preabsorption with the primary antibody, as a control, prior to incubation with the cells.
- If you lose track of which side of the coverslip the cells are on, mark a corner with a felt-tip marker. Then examine the wet coverslip, in a petri dish containing buffer, with a tissue culture or dissecting microscope. Is the mark in focus with the cells?

Fixation and Immunofluorescence Protocols for Intermediate Filaments

In general, the procedures used for preparing antibodies, obtaining secondary antibodies, and growing and fixing cells for staining microtubules also apply to intermediate filaments.

METHANOL FIXATION OF INTERMEDIATE FILAMENTS

The following protocol for indirect immunofluorescence on cultured cells works well for the majority of different cytoskeletal intermediate filament types, as well as for the nuclear intermediate filament system, the nuclear lamins (Fig. 7.2).

1. Grow cells on #1 or #1.5 glass coverslips (22-mm square) in 35-mm-diameter plastic culture dishes to ~50% confluency in growth medium.

2. Prior to fixation, remove a coverslip with a jeweler's forceps. Rinse three times while carefully holding the coverslip and dipping it into a beaker containing PBS. Drain the PBS by touching an edge of the coverslip to a piece of Whatman #1 filter paper. It is important to remove most of the liquid, but do not let it air-dry.

   It is also important to keep track of the cell side of the coverslip throughout this step and each of the following steps.

3. Using the jeweler's forceps, rapidly plunge the moist coverslip into a Columbia coverslip staining jar (see above) that contains prechilled (stored at –20°C) absolute methanol. This step should be carried out in the freezer compartment of an explosion-proof refrigerator (because

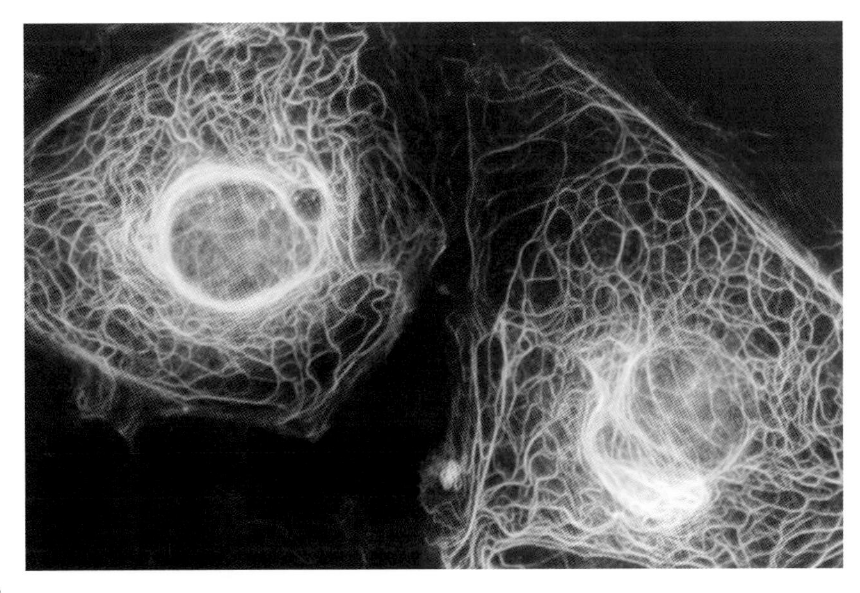

FIGURE 7.2

A typical pattern of keratin-rich tonofibrils in PTK-2 epithelial cells as seen by indirect immunofluorescence using anti-keratin antibody. (Photo provided by R.D. Goldman, Northwestern University Medical School.)

of the volatile nature of methanol). Close the freezer door quickly and let the cells fix for 5 minutes.

It is important to use dry absolute methanol, so it is best to store a stock solution of the methanol over Molecular Sieve in the freezer compartment. It is also important to change the methanol in the staining jars frequently to minimize buildup of water in the fixative.

4 Remove the coverslip and, while still wet, quickly transfer it to another staining jar containing PBS at room temperature for 3 minutes.

5 Remove the coverslip and drain most of the PBS by touching its edge to filter paper (see above). Place immediately on a moistened (but not soaking) piece of filter paper in a petri dish. Make certain that the cell side is up. Overlay 50–100 μl of 5% NGS (Sigma) in PBS to block nonspecific binding sites. Cover and incubate at 37°C for 20 minutes.

6 Remove the coverslip with jeweler's forceps and rinse for 3 minutes in a staining jar containing PBS with 0.05% Tween. Rinse twice more for 3 minutes each with PBS alone. Following the final rinse, drain excess liquid as described above.

7 Place the coverslip in the moist chamber and overlay the cell side with 50–100 μl of the primary antibody (e.g., monoclonal anti-vimentin, Sigma) that has been diluted with PBS. The antibody will be distributed over the usable surface of the coverslip if it is kept moist. Cover and incubate at 37°C for 20 minutes.

8 Rinse the coverslip three times in PBS as described above and drain excess liquid using filter paper.

9 Add 50–100 μl of the secondary antibody (e.g., goat anti-mouse IgG-FITC [fluorescein isothiocyanate]) as described above for the primary antibody. Incubate at 37°C as described above for primary antibody.

10 Rinse the coverslip as described above and drain excess liquid.

11 Mount the coverslips on a slide using DABCO/gelvatol (see Chapter 4 on mounting media) and observe in a fluorescence microscope.

DOUBLE-LABELING PROTOCOL

It is sometimes useful to carry out double-label observations in order to compare intermediate filament distribution with the distribution of microtubules, microfilaments, or other associated proteins or organelles within the same cell. In the case of microtubules, the methanol fix described above or the protocols described for microtubules can be used. Both primary antibodies (from different species) can be mixed for the first reaction with cells, and the same holds true for the secondary antibodies. The simplest procedure for observing microfilaments and intermediate filaments in the same cell is as follows.

1 Remove coverslips from petri dishes and rinse in PBS (complete) three times as described above.

2 Fix in 2% formaldehyde in PBS for 5 minutes at room temperature in a covered staining jar.

3 Extract in 0.05% NP-40 in PBS for 5 minutes at room temperature.

4 Wash three times in PBS and incubate in normal goat serum as described above.

5 Wash three times as above and add the primary intermediate filament antibody in PBS and incubate at 37°C as described above.

6 Add secondary antibody (an FITC conjugate) at the normal dilution, also containing 1:1000 dilution of rhodamine phalloidin (Molecular Probes). Incubate at 37°C for 30 minutes.

7 Rinse, mount, and observe as described above.

Note

- A large number of excellent intermediate filament antibodies are available from a number of suppliers. Examples include for vimentin, the monoclonal (V9) from Sigma and the chicken antibody from Covance; for neurofilament triplet intermediate filaments, the anti NF-L (68 kD) from Amersham; for desmin, the polyclonal from Zymed Laboratories; and for the keratins, antibodies from Miles-Yeda.

REFERENCES

Chou Y.H., Skalli O., and Goldman R.D. 1997. Intermediate filaments and cytoplasmic networking: New connections and more functions. *Curr. Opin. Cell Biol.* **9:** 49–53.

Hyams J.S. and Lloyd C.W., eds. 1994. *Microtubules.* Wiley-Liss, New York.

Olmsted J.B. 1986. Microtubule associated proteins. *Annu. Rev. Cell Biol.* **2:** 419–455.

Sabatini D., Bensch K., and Barrnett R. 1963. Cytochemistry and electron microscopy. The preservation of cellular ultrastructure and enzymatic activity by aldehyde fixation. *J. Cell Biol.* **17:** 19–36.

CHAPTER 8

Immunofluorescence Localization of Actin

Anna Spudich

National Center for Biological Sciences, Tata Institute of Fundamental Research, Bangalore, India

INTRODUCTION

Actin is a major component of all eukaryotic cells and is highly conserved across species. The different isoforms of actin show a very high degree of homology, and almost all actins bind cytochalasins, phallotoxins, and DNase I. In nonmuscle cells, actin is organized into different assembly states. Actin organization in cells is highly dynamic. The spatial and temporal control of assembly of actin from globular (G-) to filamentous (F-) actin in cells is regulated by several actin-binding proteins in the cytoplasm, and by signals that regulate them such as cell adhesion, receptor activation, and cell differentiation. Filamentous actin is found in the cell cortex under the plasma membrane, in dynamic structures such as microspikes, and in lamellopodia at the cell periphery and in stress fibers (Fig. 8.1). Approximately half of the cellular actin fails to sediment by high-speed centrifugation. This form of actin represents a soluble cytoplasmic pool. In some cells actin is also found organized into nonfilamentous structures at the cell cortex. Actin is important for maintaining cell shape and for myosin-based movements in cells. In addition, the actin cytoskeleton is involved in localization of other molecules in the cytoplasm and in cellular compartmentalization.

Several reagents are available for light microscopic localization of actin organized into the different assembly states. Polyclonal and monoclonal antibodies with different specificities are commercially available for labeling actin-containing structures in cells. Phallotoxins are cyclic peptides derived from the mushroom *Amanita phalloides* that bind with high affinity to filamentous actin (F-actin) from a variety of cells (see Chapter 5). The binding of phalloidin to actin stabilizes filamentous actin. Phalloidins do not bind to monomeric actin and are useful reagents for labeling purified F-actin and F-actin in cells (see Chapter 5). Phalloidins are available with different fluorescent markers and are valuable reagents in double-labeling experiments where an antibody is being used to localize other proteins in the same sample. DNase I has been shown to label G-actin pools in cells. DNase I forms a 1:1 complex with monomeric actin. The binding site for DNase I on the actin monomer does not overlap with the binding site for many other actin-binding proteins. This feature and the high affinity of the interaction make it a useful reagent for binding to G-actin in association with other actin-binding proteins in the cytosol. DNase I labeled with different fluorophores is available from Molecular Probes.

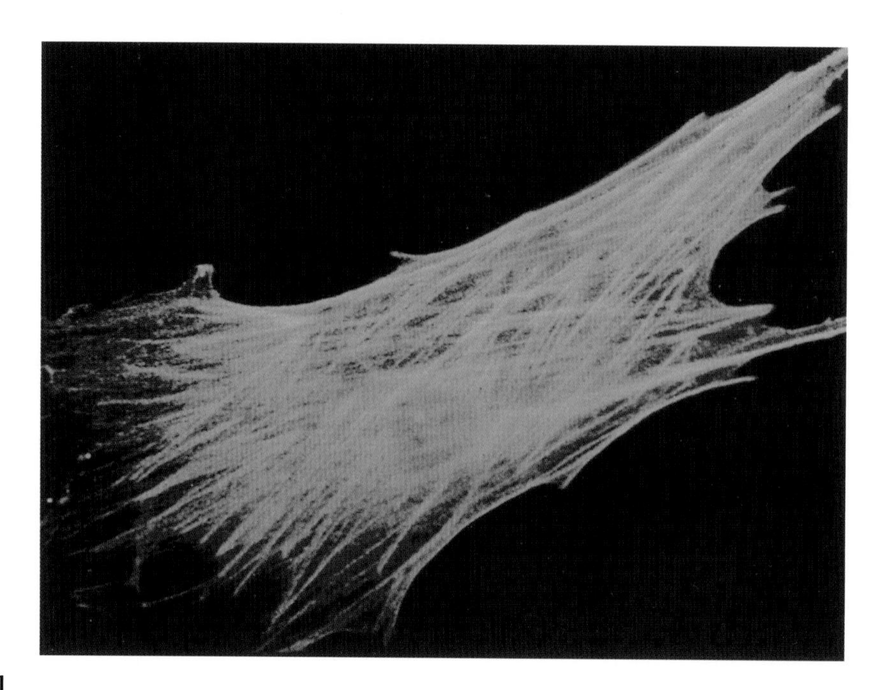

FIGURE 8.1

Human fibroblast cell stained with a monoclonal antibody to actin (mAbC4) and a fluorescein-conjugated goat anti-mouse immunoglobulin shows bundles of actin-containing stress fibers. Magnification, 800x. (Photo provided by J. Lessard, Division of Developmental Biology, Children's Hospital Medical Center, Cincinnati, Ohio).

Immunofluorescence Protocols

The protocols described below for immunolabeling actin work well for cells grown in tissue culture as monolayers and for cells grown in suspension cultures that can be attached to polylysine-coated coverslips. Conditions used for fixing cells should preserve actin filament integrity as well as preserve antigenicity. Many actin-containing structures are transient structures formed in response to specific signals. In general, buffers used for cell growth preserve cell morphology and actin filament organization better than fixation in PBS (phosphate-buffered saline). For example, electron microscopy showed that rat basophilic leukemia (RBL) cells fixed in a modified Ringer buffer containing 1 mM EGTA (ethyleneglycoltetra acetic acid) and 5 mM $MgCl_2$ had better cell morphology and actin filament integrity.

IMMUNOLABELING OF CELLS GROWN ATTACHED TO A SUBSTRATUM OR IN SUSPENSION WITH ACTIN ANTIBODY

In general, tissue culture cells that grow attached to substrates can be cultured on sterile glass coverslips in six-well plates. The fixation and washing steps below can be done either in six-well plates or by placing the coverslips in a porcelain staining rack for coverslips available from Thomas Scientific. Cells should be plated at low density to allow spreading so that cells can assume their normal morphology. (For details of how to sterilize coverslips and how to culture cells on coverslips, see Chapter 4.)

1. Remove the medium by three washes with PBS or the buffer of choice such as Ringer buffer. Cells grown in suspension cultures should be washed free of the culture medium with the buffer and attached to a glass coverslip previously coated with 0.5–1 mg/ml poly-L-lysine in H_2O at room temperature for 3–5 minutes as described (see Chapter 4). This works well for small tissue culture cells and for large cells, such as echinoderm eggs and embryos, provided close attention is paid to the time that cells are allowed to settle and attach to coverslips. Prolonged binding of cells to a polylysine-coated surface may cause distortion and disruption of cells in some cases.

   Other eukaryotic cells like *Dictyostelium* or yeast will also attach well to polylysine-coated glass coverslips. Centrifuge *Dictyostelium* cells grown in suspension at 2600 rpm for 3 minutes to pellet the cells. Remove the culture medium by two washes in 10 mM Tris, pH 7.5, and 0.1 M KCl. Resuspend cells at low density (10^5 cells/ml) and attach to polylysine-coated coverslips at room temperature for 15 minutes. Yeast cells are fixed in formaldehyde and treated with enzymes that remove the cell wall (see Chapter 10) prior to attachment on polylysine-coated glass coverslips.

2. Incubate the coverslips with cells in a 4% solution of ultrapure **formaldehyde** (16% solution, EM Grade, Electron Microscopy Sciences) in PBS or the buffer of choice at room temperature for 20 minutes. Alternatively, a fresh solution of formaldehyde can be prepared from **paraformaldehyde.** Dissolve the solid in 0.5 volume of distilled H_2O by heating to ~60°C in a

fume hood. Neutralize with 1 N **NaOH** to pH 7.0 and dilute with a 2X buffer solution to make the 4% solution.

CAUTION: Formaldehyde; paraformaldehyde; NaOH (see Appendix 2 for Caution)

Washing and fixation of the cells can also be done in a buffer solution with 1 mM EGTA. It is important to chelate calcium in the medium to inhibit the activity of calcium-activated actin-severing proteins present in most eukaryotic cells.

3 Remove the unreacted formaldehyde with one 5-minute wash in PBS and two washes for 5 minutes each with 0.5 M **glycine** in PBS.

CAUTION: Glycine (see Appendix 2 for Caution)

*Where autofluorescence of cells is a problem, treatment with 0.1% **sodium borohydride** in PBS for 10 minutes can be used to reduce excess aldehyde and to reduce autofluorescence.*

Caution: Sodium borohydride (see Appendix 2 for Caution)

4 Permeabilize cells with a 30-second extraction with cold **acetone** (–10°C). Acetone permeabilization is compatible with labeling F-actin with antibodies and with phalloidin.

CAUTION: Acetone (see Appendix 2 for Caution)

If double labeling for actin and another protein is contemplated, permeabilization with Triton X-100 may be more suitable because acetone treatment may extract or denature some antigens. Permeabilize fixed cells in blocking solution (see below) containing 0.2% Triton X-100 at room temperature for 5 minutes.

In some cases, it may be advisable to permeabilize and fix cells simultaneously to preserve very labile structures or transient associations in the cells (Kelley et al. 1996). Fix the cells in 0.5% **glutaraldehyde** and 0.2% Triton X-100 in a modified Ringer buffer containing 137 mM NaCl, 5 mM **KCl**, 2 mM **$MgCl_2$**, 2 mM EGTA, 5.5 mM glucose, and 5 mM HEPES ((*N*-2-hydroxyethyl)piperazine-*N′*-(2-ethanesulfonic acid)), pH 6.8, at room temperature for 5 minutes followed by a brief rinse in the buffer without glutaraldehyde and Triton. Fix the cells for another 10 minutes in the buffer containing 0.5% glutaraldehyde. Block the unreacted fixative by treatment with 0.1% sodium borohydride in PBS for 5 minutes.

CAUTION: Glutaraldehyde; KCl; $MgCl_2$ (see Appendix 2 for Caution)

5 Wash cells permeabilized with acetone three times with PBS for 3 minutes each to remove residual acetone. Block nonspecific binding of antibodies by incubating fixed cells with 1% BSA (bovine serum albumin) in PBS for 20 minutes (blocking solution). Blocking with a solution containing 1% BSA, 2% NGS (normal goat serum), and 0.2% saponin may be used for tissue sections prior to antibody staining.

6 Incubate cells with primary antibody diluted in blocking solution. Dilution of the antibody and time of incubation should be established for a given antibody. Place ~50 µl of the diluted primary antibody on the coverslip. With most anti-actin antibodies incubation at room temperature for 1–3 hours in a moist chamber is sufficient for good labeling. Samples can be incubated at 4°C overnight without significant increase in background staining, provided the blocking step was carried out earlier.

7 To remove unbound antibody, wash coverslips three times for 5 minutes each with PBS. If background staining is a problem, add 0.05% Triton X-100 to PBS used for washing the coverslips.

8 Incubate coverslips with fluorescently labeled secondary antibody, diluted in blocking buffer, at room temperature for 1 hour in the dark to reduce photobleaching. Concentration of the secondary antibody should be established using the manufacturer's suggestion as a starting point. In general, affinity-purified secondary antibodies should be used. To reduce nonspecific binding, secondary antibodies can be further purified by preabsorption against a preparation of formaldehyde-fixed cells as described by Fukui et al. (1987).

9 Remove unbound secondary antibody by three washes with PBS for 5 minutes each.

10 Mount the coverslips with 35 μl of a mounting medium containing antifade reagent on a clean glass slide (see Chapter 4 for mounting media).

PREPARATION OF SAMPLES EMBEDDED IN WATER-SOLUBLE RESINS FOR LABELING WITH ACTIN ANTIBODY

Thick (1-μm) sections of cell pellets and tissue samples fixed and embedded in water-soluble resins (Kreiner et al. 1986) can be labeled with actin antibodies.

1 Prepare cell pellets or tissue samples by washing in the buffer of choice. Fix pellets in 2% **formaldehyde** (refer to preparation of formaldehyde solution described above) in buffer of choice overnight in ice.

CAUTION: Formaldehyde (see Appendix 2 for Caution)

2 Remove fixative with two washes in 50 mM **ammonium chloride** in PBS.

CAUTION: Ammonium chloride (see Appendix 2 for Caution)

3 Dehydrate the sample by passing it through a graded ethanol series (50%, 60%, 70%, 80%, 95%, 100%) at –10°C for 15 minutes each.

4 Transfer the cell pellet to a 1:1 mixture of Lowicryl K4M resin (Pelco; prepared according to manufacturer's directions) and 100% ethanol for 30 minutes, followed by 100% Lowicryl for 30 minutes. Infiltrate overnight with 100% Lowicryl, all at –10°C.

5 Transfer the infiltrated cell/tissue samples to gelatin capsules (see Chapter 19) with fresh 100% Lowicryl. Polymerize with UV light in an atmosphere of N_2 for 24–72 hours (see Chapter 21).

6 Section the embedded sample to yield 1-μm-thick sections. Mount sections on gelatin-coated glass slides and dry, on a hot plate set at 37°C, for several hours. Drying the sections will bind them to the glass surface, and the sections can be stained using a glass staining jar used for holding glass slides.

7 Dilute the primary antibody in a mixture of 1% BSA, 0.2% saponin in PBS. The remaining steps for antibody labeling can be carried out as described above beginning with step 6 of the preceding protocol.

SUGGESTED READINGS

Numerous journal articles describe techniques for localization of actin in specific cell types. The citations below refer only to specific techniques cited and not fully described in the protocol above. For comprehensive reviews on actin and actin-binding proteins, see the following.

Pollard T.D., Almo S., Guirk S., Vinson V., and Lattman E.E. 1994. Structure of actin-binding proteins: Insights about function at atomic resolution. *Annu. Rev. Cell Biol.* **10:** 207–249.

Sheterline P. and Sparrow J.C. 1994. In *Actin in protein profile* (ed. P. Sheterline). Academic Press, London.

Stossel T.P., Chaponnier C., Ezzell R.M., Hartwig J.H., Janmey P.A., Kwiatkowski D.J., Lind S.E., Smith D.B., Southwick F.S., and Yin H.L. 1985. Nonmuscle actin-binding proteins. *Annu. Rev. Cell Biol.* **1:** 353–402.

REFERENCES

Fukui Y., Yumura S., and Yumura T.K. 1987. Agar overlay immunofluorescence: High-resolution studies of cytoskeletal components and their changes during chemotaxis. In Dictyostelium discoideum: *Molecular approaches to cell biology* (ed. J.A. Spudich), pp. 347–356. Academic Press, New York.

Kelley C.A., Sellers J.R., Gard D.L., Bui D., Adelstein R.A., and Baines I.C. 1996. *Xenopus* nonmuscle myosin heavy chain isoforms have different subcellular localizations and enzymatic activities. *J. Cell Biol.* **134:** 675–687.

Kreiner T., Sossin W., and Scheller R.H. 1986. Localization of *Aplysia* neurosecretory peptides to multiple populations of dense core vesicles. *J. Cell Biol.* **102:** 769–782.

Ng W.A., Doetschman T., Robbins J., and Lessard J.L. 1997. Muscle isoactin expression during in vitro differentiation of murine embryonic stem cells. *Pediatr. Res.* **41:** 285–292.

CHAPTER 9

Immunofluorescence Localization of Nuclear Proteins

David L. Spector

Cold Spring Harbor Laboratory, Cold Spring Harbor, New York

INTRODUCTION

A large number of nuclear proteins have been successfully localized using immunofluorescence microscopy (see Fig. 9.1). These proteins span all nuclear domains, including the nuclear envelope, nuclear lamina, nucleolus, chromatin-associated proteins, and proteins associated with RNA metabolism and nuclear bodies (for review, see Spector 1993). Among a variety of fixation protocols found in the literature, formaldehyde made fresh from paraformaldehyde has been widely used and is an excellent general fixative for the localization of most nuclear proteins. Paraformaldehyde in a "Prill" form (Electron Microscopy Sciences) is easier and safer to handle than the powder form of this fixative. Formaldehyde is a weak cross-linking fixative and therefore rarely masks the antigenic site of interest. In addition, it preserves cell structure extremely well at the light microscopic level. On occasion, methanol fixation has been used for those rare antigens that are not able to be localized after formaldehyde fixation. Methanol is a harsher fixative because it fixes by precipitation, and the cell is permeabilized and fixed simultaneously with this treatment. However, methanol fixation was particularly useful in an early study that examined the sites of DNA replication in mammalian cells (Bravo 1986; Bravo and MacDonald-Bravo 1987). Using this fixative, the authors were able to demonstrate that a population of the DNA replication factor, PCNA, was associated with replication foci and a more soluble population of this factor was removed or not recognized by the antibody after simultaneous fixation and permeabilization by methanol. This finding was not initially observed after formaldehyde fixation because the soluble pool of antigen masked the underlying replication foci. An alternative approach to examining the effects of a soluble pool of antigen masking an underlying pattern of localization is to permeabilize cells with detergent prior to formaldehyde fixation or to treat cells with cytoskeleton (CSK) buffer (Fey et al. 1986) prior to fixation. In summary, it can be useful to examine the localization of your antigen using different fixation methods.

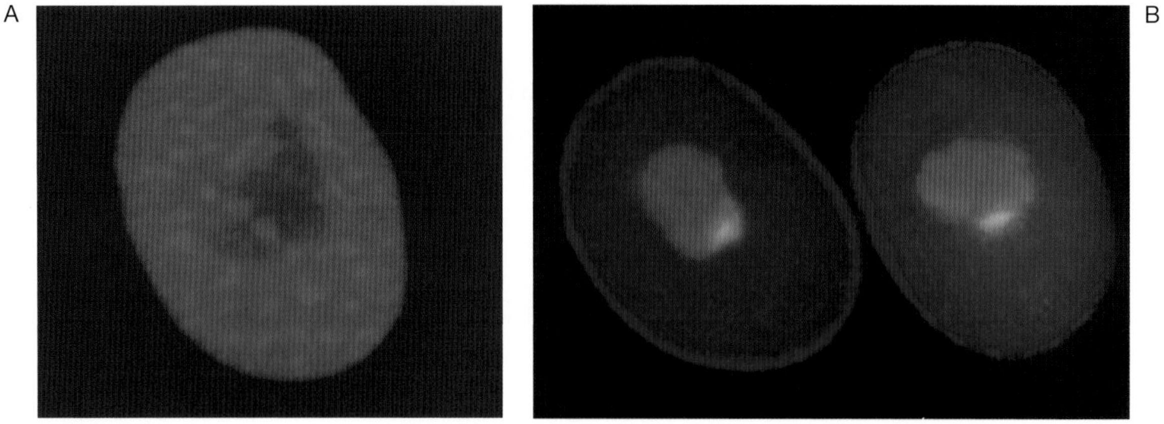

A B

FIGURE 9.1

(*A*) Immunofluorescence localization of a pre-mRNA splicing factor in the nucleus of a human fibroblast. The reporter is a Texas Red-conjugated secondary antibody. (Photo provided by D.L. Spector, Cold Spring Harbor Laboratory.) (*B*) Triple label immunofluorescence microscopy of two HeLa cells showing the localization of the nuclear lamina (*blue*), the nucleolus (*red*), and polypyrimidine tract binding protein, which is present in the perinucleolar compartment (*green/yellow*) and in a diffuse nucleoplasmic population (*green*). (Photo provided by S. Huang and D.L. Spector, Cold Spring Harbor Laboratory.)

Immunofluorescence Protocol

1 Grow cells on 22 x 22-mm #1.5 glass coverslips. One coverslip will fit into a 35-mm petri dish or in one well of a six-well dish. This coverslip thickness is optimal for most microscope lenses and is less fragile to handle than a #1 coverslip. Ideally, cells should be 50–70% confluent, but this may vary depending on the experimental study.

2 Fix cells in 2% **formaldehyde** in PBS (phosphate-buffered saline), pH 7.4, for 15 minutes at 20°C. Two percent formaldehyde is made up fresh prior to use by dissolving the appropriate amount of electron microscopy (EM)-grade **paraformaldehyde** (Prill form, Electron Microscopy Sciences) in PBS in a Pyrex bottle with a stir bar. The aldehyde goes into solution by heating on a hot plate in the hood at 60°C. Keep the bottle cap loosened so that pressure does not build up. Cool down to 20°C and pH to 7.4. Alternatively, cells may be fixed in 100% **methanol** at −20°C for 3 minutes. If methanol fixation is used, skip to step 5.

CAUTION: Formaldehyde; paraformaldehyde; methanol (see Appendix 2 for Caution)

If your primary antibody is an IgM to a nuclear protein, you may have to fix with methanol to allow penetration of this large molecule into the nucleus.

3 Wash in PBS, pH 7.4, three times, 10 minutes each.

Changes of buffer and coverslip transfers should be done quickly to avoid drying the cells. It is convenient to keep the coverslip in a 35-mm petri dish and aspirate off buffer washes and then add new buffer to the dish. When removing solutions, do not overaspirate; leave a slight layer of fluid on the cells. Do not allow the cells to dry out during any step; this is especially critical during the fluorescent probe incubation, because drying will deposit the probe nonspecifically on the specimen and significantly increase background fluorescence.

4 Permeabilize in 0.2% Triton X-100 plus 1% NGS (normal goat serum) in PBS for 5 minutes on ice.

The Triton X-100 concentration can be increased to 0.5% if you are experiencing problems with IgG antibody penetration into your particular cell type.

5 Wash in PBS + 1% NGS three times, 10 minutes each.

6 Incubate in the appropriate concentration of primary antibody for 1 hour at room temperature in a humidified chamber. If using 22 x 22-mm square coverslips, 30 μl of diluted antibody is placed on the coverslip and the coverslip is inverted onto a glass slide. The slide is then placed in the humidified chamber, which is incubated at room temperature.

Remove the coverslip from the petri dish containing buffer, as removing it from a bufferless dish may result in breaking the coverslip because of surface tension effects. Wick excess buffer from the coverslip by removing it from the petri dish and carefully touching the edges of the coverslip with a piece of Whatman #1 filter paper; gingerly dry the top of the coverslip (i.e., the side with no cells). Place the 30 μl of antibody on the coverslip and invert it onto a glass slide.

A humidified chamber can be easily made using a 140-mm petri dish. Place a moist paper towel at the bottom of the dish, put two wooden dowels on top of the towel, and place the slides across the dowels so that they are not in direct contact with the towel. A 140-mm dish will hold four slides.

7 Wash in PBS + 1% NGS three times, 10 minutes each.

To remove the coverslip from the slide, flood the slide with buffer and, using forceps, gently pull the coverslip straight toward you without drastically breaking the surface tension, as this would remove cells from the coverslip.

8 Incubate in secondary antibody (e.g., fluorescein isothiocyanate [FITC]- or Texas Red-conjugated) at a dilution of 4 μg/ml at room temperature for 1 hour in a humidified chamber.

Once the fluorescent conjugated antibody has been applied to the cells, keep them in the dark (i.e., cover with foil) to minimize photobleaching of the fluorophore.

9 Wash in PBS four times, 10 minutes each.

10 Mount coverslip with a drop of mounting media (see Chapter 4) and seal coverslip with clear nail polish to prevent drying and movement under the microscope.

After the last rinsing step, leave the buffer in the petri dish and remove the coverslip from the dish. Wick excess buffer from the coverslip by carefully touching the edges of the coverslip with a piece of Whatman #1 filter paper and gingerly dry the top of the coverslip (i.e., the side with no cells). Place a drop of mounting medium on a slide and gently place the front edge of the coverslip at an angle over the drop so that as you lower the coverslip, any air will be pushed out leaving no air bubbles behind.

Notes

- All primary antibodies should be checked by immunoblot to be sure that they are specifically recognizing the antigen of interest. Immunofluorescence data obtained from an antibody that does not immunoblot or that gives even slight cross-reactivity with other proteins cannot be trusted.
- Primary antibody controls should be run with each experiment. The best control is to use preimmune serum in place of the primary antibody. If preimmune serum is not available, substitute normal immunoglobulin from the animal species in which the primary antibody was made. To control for autofluorescence, substitute both primary and secondary antibody incubations with BSA (bovine serum albumin). Incubating cells only with secondary antibody controls for its nonspecific binding and for the filters used in your fluorescence microscope.
- If the antigen is available, it should be used for preabsorption with the primary antibody, as a control, prior to incubation with the cells.
- If you lose track of which side of the coverslip the cells are on, mark a corner with a felt-tip marker. Then examine the wet coverslip, in a petri dish containing buffer, with a tissue culture or dissecting microscope. Is the mark in focus with the cells?
- If you need to attend a seminar during this protocol, cells can be left in any of the wash steps for several hours.

REFERENCES

Bravo R. 1986. Synthesis of the nuclear protein cyclin (PCNA) and its relationship with DNA replication. *Exp. Cell Res.* **163:** 287–293.

Bravo R. and MacDonald-Bravo H. 1987. Existence of two populations of cyclin/proliferating cell nuclear antigen during the cell cycle: Association with DNA replication sites. *J. Cell Biol.* **105:** 1529–1554.

Fey E.G., Krochmalnic G. and Penman S. 1986. The nonchromatin substructures of the nucleus: The ribonucleoprotein (RNP)-containing and RNP-depleted matrices analyzed by sequential fractionation and resinless section electron microscopy. *J. Cell Biol.* **102:** 1654–1665.

Spector D.L. 1993. Macromolecular domains within the cell nucleus. *Annu. Rev. Cell Biol.* **9:** 265–315.

CHAPTER 10

Immunofluorescence Methods for *Saccharomyces cerevisiae*

Pamela Silver

Harvard University Medical School, Boston, Massachusetts

INTRODUCTION

There is currently no better organism than the budding yeast *Saccharomyces cerevisiae* in which to carry out the initial characterization of a novel gene and its corresponding gene product. The ease of genetic analysis, combined with the sophisticated cell biological methods that are available, makes this an ideal model system. Furthermore, only in a genetically tractable organism is it possible to use epitope-tagged proteins to the best advantage. In *S. cerevisiae* it is a trivial matter to determine whether a modified version of a protein is functional. This is accomplished by testing whether the tagged protein can complement a null phenotype or rescue a conditional phenotype. These possibilities allow the investigator to correlate the intracellular localization of the protein with its function in vivo.

A few factors must be considered when carrying out indirect immunofluorescence studies in budding yeast. The first is simply its small size relative to many other eukaryotic cells. This may be somewhat limiting for intracellular localization, but as shown in the accompanying figures, it is possible to localize proteins to specific organelles. A second consideration is the cell wall of the organism. The cell wall must be removed in order for antibodies to diffuse into the cell. This is easily accomplished by incubating the fixed cells with enzymes that digest away this feature of the fungi (detailed in the protocol below).

Green Fluorescent Protein in Yeast

The use of the *Aequorea victoria* green fluorescent protein (GFP) (Prasher et al. 1992) to form fusion proteins that can be localized in living cells has gained great popularity in recent years (Kahana and Silver 1996; for a more detailed discussion of GFP, see Goldman and Spector 2005). Fusions between the protein of interest and the 28-kD GFP protein can be viewed in living cells directly through a fluorescein isothiocyanate (FITC) filter or with a filter set that has been modified for use with GFP. As mentioned above, one advantage of carrying out such studies in a genetically tractable organism is that it is possible to determine whether the fusion protein generated is functional in vivo. This approach can be used to localize a protein with indirect immunofluorescence, but it also has the advantage that it can be carried out in living cells.

When functional fusion proteins are generated, it is then possible to integrate the gene encoding the tagged protein into the genome and to replace the endogenous copy of the gene. This creates an informational yeast strain where an organelle of interest is constitutively labeled. This approach can be useful for screening for conditional mutants in various cellular pathways of interest to investigators.

Protein Localization Studies

Proteins can be localized either by indirect immunofluorescence or through the use of GFP. Each method has its own strengths and weaknesses. Indirect immunofluorescence can be used to localize the endogenous protein, whereas GFP methods require the construction of a functional fusion protein with the 28-kD GFP fused to the protein of interest. However, once the appropriate GFP fusion protein is generated, it can be viewed directly in living cells and, as such, results in superior images of the intact living cell as compared to fixed cells. Finally, GFP fusion proteins can be used to carry out temporal studies in which a particular cell or group of cells can be observed not only throughout the cell cycle, but also over several generations (Kahana et al. 1995). Hence, the use of in vivo labeling with GFP is rapidly becoming a method that complements the use of traditional indirect immunofluorescence.

Indirect Immunofluorescence

The following is a detailed protocol for performing indirect immunofluorescence with *S. cerevisiae* (adapted from Adams and Pringle 1984 and Kilmartin and Adams 1984). The basic protocol requires that cells be grown exponentially and then fixed. (This protocol is designed for 5 ml of exponentially growing cells.)

The standard fixation method employs formaldehyde, but other variations may be more appropriate depending on the protein of interest. This is a point to consider before beginning the protocol, as some steps differ depending on the method chosen (see Table 10.1 for details). Following fixation, cells are digested to remove the cell wall and then are adhered to slides. Subsequent steps involve the application of primary and secondary antibody and the final processing of the slide. Examples of results obtained using the protocol presented here are shown in Figure 10.1. More in-depth discussion of the theory of indirect immunofluorescence, as well as the use of *S. cerevisiae* for cell biological experiments, can be found in Pringle et al. (1991) or Rose et al. (1990).

1 Grow yeast cells in rich or defined media to a concentration of 1×10^7 to 5×10^7 cells/ml.

If the experiment requires a temperature shift, it should be performed after the cells reach this density (for greater detail, see Lee et al. 1996).

2 Fix the cells in 2% **formaldehyde** by adding 0.6 ml of 37% formaldehyde directly to the culture (5 ml) and incubate with gentle shaking.

CAUTION: Formaldehyde (see Appendix 2 for Caution)

Fixation conditions can be varied according to the requirements of the experiment. Short fixation times may work well for some proteins but not others. For most purposes, 30 minutes is more than sufficient. Sometimes only a few minutes of fixation are required to obtain a similar signal as with longer fixation periods. If it is not known how the desired protein responds to formaldehyde fixation, try a time course. Use the highest purity of formaldehyde available and always use the same brand for consistent results. It is best to buy small bottles of 37% formaldehyde and then prepare small aliquots to prevent deterioration of the formaldehyde.

Some antigens will not respond to formaldehyde fixation. In this case, a solution of 30% formaldehyde (w/v) plus 10% ***glutaraldehyde*** *(w/v) should be prepared fresh before the experiment. This solution is added to the cell culture to a final concentration of 3% formaldehyde and 1% glutaraldehyde (Adams and Pringle 1984; Kilmartin and Adams 1984). As with formaldehyde, time of fixation may be varied to suit the need of the experiment.*

CAUTION: Glutaraldehyde (see Appendix 2 for Caution)

TABLE 10.1 Methods of fixation

Fixation condition	Permeabilization condition	Types of antigens	References
Formaldehyde	methanol/acetone	most proteins	Adams and Pringle (1984); Kilmartin and Adams (1984)
Formaldehyde	NP-40	nuclear pore proteins	Davis and Fink (1990)
Paraformaldehyde	methanol/acetone	tubulin	Kilmartin and Adams (1984)
None	methanol/acetone	spindle proteins	Rout and Kilmartin (1990); Osborne et al. (1994)

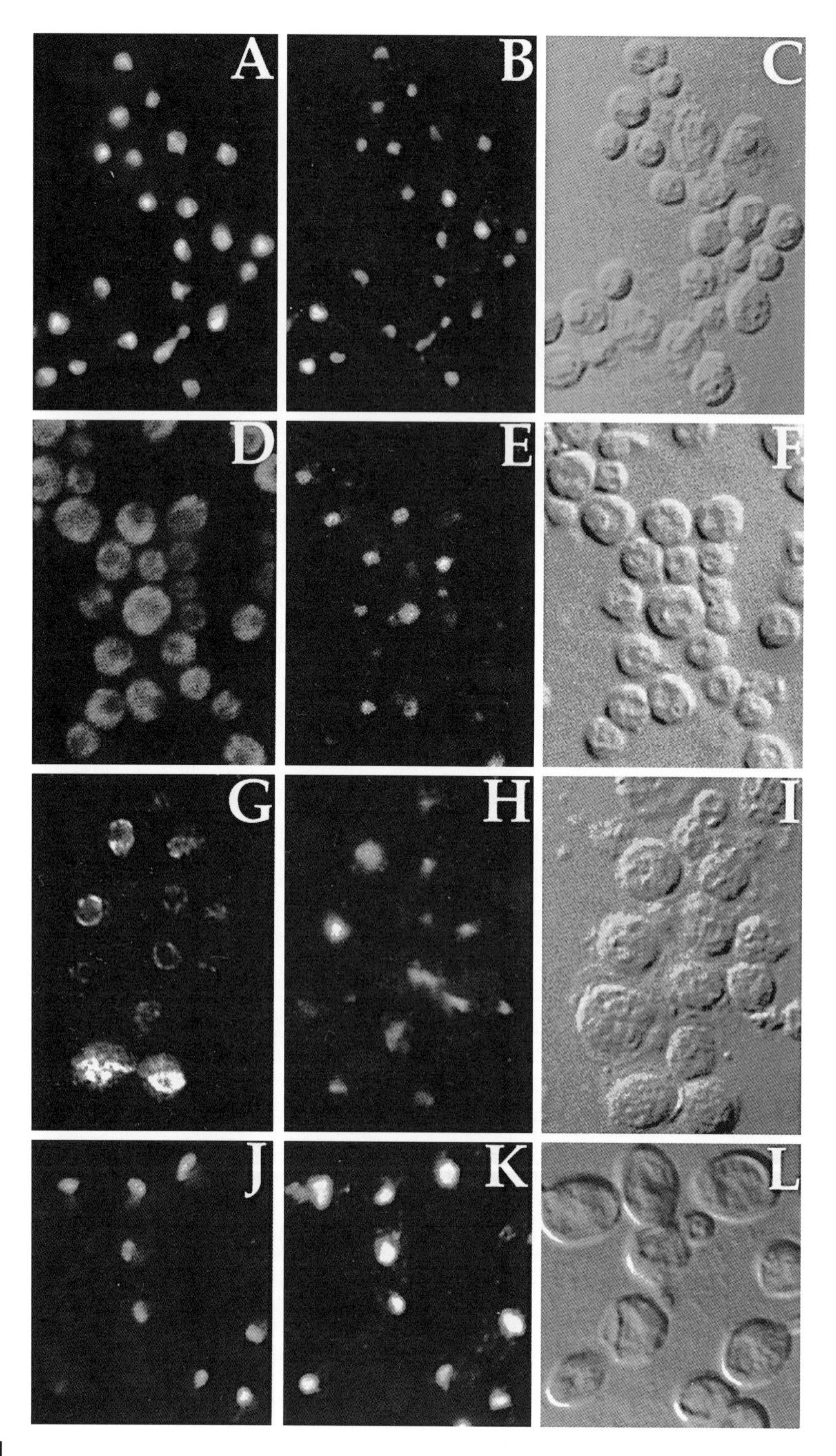

FIGURE 10.1

Examples using indirect immunofluorescence of various types to identify cellular compartments in *S. cerevisiae.* In the first column (*A, D, G, J*), cells are stained for a particular antigen. In the second column (*B, E, H, K*), cells are stained with DAPI to localize DNA. In the third column (*C, F, I, L*), images of the cells using Nomarski optics are shown. Cells in *A* are stained with an antibody to the nuclear protein Npl3p (Bossie et al. 1992) and cells in *D* show the localization of the cytosolic protein Rna1p (Hopper et al. 1990; Koepp et al. 1996). The staining in *G* corresponds to the nuclear envelope localization of importin-β (Koepp et al. 1996) and the cells in *J* were stained with the monoclonal antibody 9C4 that recognizes a nucleolar antigen (Lee et al. 1996). (Photos provided by P. Silver, Dana Farber Cancer Institute, Boston, MA.)

A few antigens will not stain if fixed with aldehydes. In this case, some success has been obtained when the fixation step is omitted entirely (Rout and Kilmartin 1990; Osborne et al. 1994). The protocol should be carried out as described, except step 2 (fixation) is omitted and at step 11, the methanol/acetone method of permeabilization must be used.

3 Transfer cells to a 15-ml conical tube and pellet by centrifugation at room temperature for 3 minutes at 800*g*.

4 Aspirate supernatant and wash cells with 1 ml of 0.1 M potassium phosphate, pH 6.5.

5 Repellet cells, aspirate, and wash with 1 ml of P solution.

P Solution

0.1 M potassium phosphate, pH 6.5, in 1.2 M sorbitol
This solution is easily contaminated. It is best to prepare a 50-ml aliquot from a 1-liter stock.

6 Pellet cells, aspirate, and resuspend in 1 ml of P solution. Transfer to a microfuge tube.

7 To digest the cell wall, add 15 μl of 10 mg/ml zymolyase and 5 μl of **β-mercaptoethanol**. Incubate the cells at 30°C with shaking until digestion is sufficient.

CAUTION: β-mercaptoethanol (see Appendix 2 for Caution)

10 mg/ml Zymolyase in P Solution

Vortex 10 mg of dry zymolyase in 1 ml of P solution and then spin at 13,000 rpm in a microfuge. Allow precipitate to settle and use supernatant. This solution may be stored at –20°C and used more than once.

Digestion can be monitored by observing the cells under phase contrast microscopy (see Chapter 1). Undigested cells glow, whereas cells that have had the cell walls removed appear dark.

8 While the cells are digesting, coat the wells of a multiwell slide with 0.3% polylysine. Deliver 50 μl of 0.3% polylysine onto each well and incubate for several minutes. Wash once with H_2O and allow to air-dry.

0.3% Poly-L-lysine (Sigma P 1524)

Prepare 1-ml aliquots and store at –20°C.
Slides may be prepared several hours in advance if desired.

9 Spin cells at room temperature in a microcentrifuge at 800*g* for 3 minutes, aspirate and discard the supernatant, and resuspend pellet in ~1 ml of P solution. The volume can be varied depending on the desired density of cells on the slide.

10 Place 25 μl of cell suspension onto each well of a prepared slide. Aspirate excess cells after several minutes.

11 Immediately immerse slide in ice-cold **methanol** for 6 minutes. Remove and immerse in ice-cold **acetone** for 30 seconds. Allow slides to air-dry.

CAUTION: Methanol; acetone (see Appendix 2 for Caution)

An alternative method of permeabilizing cells uses detergent instead of methanol (Davis and Fink 1990). This method is less stringent and causes less damage to the cells, but the cells do not adhere as tightly to the slide. To use this method, after aspirating the excess cells from the slide, add 50 μl of 0.5% NP-40 to each well. Incubate for 5 minutes and then aspirate. Wash the wells once with P solution and then proceed to step 12.

12 Beginning with this step, never let the wells dry out. Wash the slides once with PBS-BSA (phosphate-buffered saline/bovine serum albumin). Aspirate and add primary antibody (diluted in PBS-BSA). Incubate in a moist chamber for at least 2 hours.

PBS-BSA
5 mg/ml BSA in PBS

This incubation may be performed overnight if desired.

13 Aspirate the primary antibody and wash the cells four times with PBS-BSA. Add secondary antibody (diluted in PBS-BSA) to wells. Incubate in a dark, moist chamber for at least 2 hours.
Exposure of the slide to light should be kept to a minimum to prevent bleaching of the secondary antibody.

14 Aspirate and wash the wells three times with PBS-BSA. Wash once with PBS. To stain DNA in addition to the antigen you are localizing, dilute the DAPI (4′,6-diamidino-2-phenylindole) solution to 1 μg/ml in PBS and add to wells. Incubate for at least 30 seconds. Aspirate and wash once with PBS.

DAPI Stock Solution: 1 mg/ml
Store at –20°C.

15 Aspirate PBS and place a drop of antifade solution in each well. Place coverslip on slide and wipe up excess antifade solution (see Chapter 4). Seal the edges with clear nail polish.
Slides should be stored in the dark at –20°C or –70°C. They may be kept for several weeks or months without significant loss of signal.

REFERENCES

Adams A.E.M. and Pringle J. 1984. Relationship of actin and tubulin distribution to bud growth in wildtype and morphogenetic-mutant *Saccharomyces cerevisiae. J. Cell Biol.* **98:** 934–945.

Bossie M.A., DeHoratius C., Barcelo G., and Silver P. 1992. A mutant nuclear protein with similarity to RNA binding proteins interferes with nuclear import in yeast. *Mol. Biol. Cell* **3:** 875–893.

Davis L.I. and Fink G.R. 1990. The *NUP1* gene encodes an essential component of the yeast nuclear pore complex. *Cell* **61:** 965–978.

Goldman R.D. and Spector D.L. 2005. *Live cell imaging: A laboratory manual.* Cold Spring Harbor Laboratory Press, Cold Spring Harbor, New York.

Hopper A.K., Traglia H.M., and Dunst R.W. 1990. The yeast *RNA1* gene product necessary for RNA processing is located in the cytosol and apparently excluded from the nucleus. *J. Cell Biol.* **111:** 309–321.

Kahana J.A. and Silver P.A. 1996. Use of *A. victoria* green fluorescent protein to study protein dynamics *in vivo.* In *Current protocols in molecular biology* (ed. F.M. Ausubel et al.), pp. 9.6.13–9.6.19. Wiley, New York.

———. 1997. The uses of green fluorescent proteins in yeasts. In *GFP: Green fluorescent protein strategies and applications* (ed. M. Chalfie and S. Kain). John Wiley, New York.

Kahana J.A., Schnapp B.J., and Silver P.A. 1995. Kinetics of spindle pole body separation in budding yeast. *Proc. Natl. Acad. Sci.* **92:** 9707–9711.

Kilmartin J. and Adams A.E.M. 1984. Structural rearrangements of tubulin and actin during the cell cycle of the yeast *Saccharomyces cerevisiae. J. Cell Biol.* **98:** 922–933.

Koepp D.M., Wong D.H., Corbett A.H., and Silver P.A.

1996. Dynamic localization of the nuclear import receptor and its interactions with transport factors. *J. Cell Biol.* **133:** 1163–1176.

Lee M.S., Henry M., and Silver P.A. 1996. A protein that shuttles between the nucleus and the cytoplasm is an important mediator of RNA export. *Genes Dev.* **10:** 1233–1246.

Osborne M.A., Schlenstedt G., Jinks T., and Silver P.A. 1994. Nuf2, a spindle pole body-associated protein required for nuclear division in yeast. *J. Cell Biol.* **125:** 853–866.

Prasher D.C., Eckenrode V.K., Ward W.W., Prendergast F.G., and Cormier M.J. 1992. Primary structure of the *Aequorea victoria* green-fluorescent protein. *Gene* **111:** 229–233.

Pringle J.R., Adams A.E.M., Drubin D.G., and Haarer B.K.. 1991. Immunofluorescence methods for yeast. In *Guide to yeast genetics and molecular biology* (ed. C. Guthrie and G.R.S. Fink), pp. 565–602. Academic Press, San Diego.

Rose M.D., Winston F., and Hieter P., Eds. 1990. *Methods in yeast genetics: A laboratory course manual.* Cold Spring Harbor Laboratory Press, Cold Spring Harbor, New York.

Rout M.P. and Kilmartin J.V.. 1990. Components of the yeast spindle and spindle pole body. *J. Cell Biol.* **111:** 1913–1927.

CHAPTER 11

Immunofluorescence Methods for *Drosophila* Tissues

Jason Swedlow

University of Dundee, Dundee, United Kingdom

INTRODUCTION

The fruit fly *Drosophila melanogaster* has long been used to study the development of an organism using genetics. More recently, a number of laboratories have developed the use of *Drosophila* for biochemical and cell biological approaches. The ability to localize molecules within a organism that allows genetic manipulation can be quite useful.

In this section, protocols and issues are presented for the analysis of various *Drosophila* tissues by immunofluorescence microscopy. Special treatment is given to the handling of embryos, because these are frequently used and require special consideration. References to many specific fixation protocols are included.

ISSUES IN IMMUNOFLUORESCENCE IN *DROSOPHILA*

Fixation

There is no single fixation protocol that answers the question, "How should I fix this?" As with other types of samples, with *Drosophila* tissues the proper fixation protocol depends on which structure needs to be visualized, the degree of preservation required, the preservation of antigenicity of the molecules of interest, and the level of resolution of the subsequent imaging. In addition, the fixation of thick tissues requires a protocol that effectively fixes the interior of the sample while not cross-linking the matrix so heavily that antibodies or other probes cannot penetrate efficiently and be washed out of the tissue. The optimal protocol is usually a compromise between these many competing issues. Table 11.1 shows fixatives for a number of different *Drosophila* tissues.

It is important to note that when examining the distribution of a novel molecule, a number of different protocols should be tried. Note also that these protocols are specific to immunofluorescence. Other visualization techniques such as in situ hybridization (see Chapters 15–18) or electron microscopy (see Chapters 19–21) require quite different protocols.

Fixation of tissues from larvae or adults usually requires dissection or some other method to remove the organ of interest from the whole organism. Methods for dissection are listed in Ashburner (1989b). Some tissues can be removed using much cruder (and easier) techniques. For example, egg chambers can be isolated in good yield and reasonable purity by simply grinding adult flies in a blender (Theurkauf 1994). In general, however, tissues are dissected into modified Robb's saline and then fixed (see Table 11.1).

TABLE 11.1 Fixation conditions for *Drosophila* tissues

Structure	Tissue	Fixative	Buffer	References
Cytoskeleton	embryo	methanol	–	Kellogg et al. (1988)
Cytoskeleton[a]	embryo	37% CH_2O	–	Theurkauf (1994)
Nuclei and chromosomes	embryo	3.7% CH_2O	PIPES/KCl/polyamine	Hiraoka et al. (1990)
Chromosomes and microtubules[b]	embryo	methanol/0.1% CH2O	–	Hiraoka et al. (1990)
Cytoskeleton/membrane proteins	egg chambers	8% CH_2O	modified Robb's[c]	Theurkauf et al. (1992); Ruohola et al. (1991)
Signaling proteins	imaginal disks	4% CH_2O/lysine	phosphate[d]	Gaul et al. (1992)
Polytene nuclei—whole mount	salivary gland	4% CH_2O	PIPES/KCl/polyamine	Urata et al. (1995)
Polytene chromosomes—squashes[e]	salivary gland	formaldehyde/acetic acid	PBS	Ashburner (1989b)

The conditions listed in this table are a sample of those reported in the literature and are by no means an exhaustive account. For any given tissue, a number of conditions have been reported. Those listed are established methods, but refinement may be necessary.

[a]The use of high formaldehyde concentrations may be generally useful for many embryo antigens. Nuclei and chromosomes are also fixed well by this method (J. Swedlow, unpubl.)

[b] Fixing chromosomes and microtubules simultaneously and preserving their structure is notoriously difficult. This method appears to produce a good compromise for embryos.

[c]Modified Robb's is 100 mM HEPES, pH 7.2, 55 mM potassium acetate, 40 mM sodium acetate, 100 mM sucrose, 10 mM glucose, 1.2 mM magnesium chloride, 1 mM calcium chloride. Tilney et al. (1996) have used a glutamate saline buffer that mimics the osmolality of *Drosophila* hemolymph for dissection of egg chambers. Samples were fixed with glutaraldehyde for subsequent examination by electron microscopy. This method gave superior preservation of the actin cytoskeleton, so it may be useful for examination of cytoskeletal components by immunofluorescence in other tissues.

[d]This method uses a solution of 37 mM $NaPO_4$, pH 7.2, 2% CH_2O, 0.01 M $NaIO_4$, 75 mM lysine for a fixative. This method uses periodate to oxidize carbohydrates, and the combination of formaldehyde and lysine to cross-link effectively and efficiently. It has found increasing use as a fixative for tissues.

[e] Fixation depends on antigen.

PIPES, piperazine-*N,N'*-bis(2-ethanesulfonic acid); PBS, phosphate-buffered saline.

Embryo Fixation

Two aspects of the *Drosophila* embryo require special attention for proper fixation. First, the embryo is surrounded by two specialized structures: the chorion or egg shell, and the vitelline membrane, a thick waxy layer that allows O_2 and CO_2 but little else to enter or exit the embryo (for a complete description of the structure of the embryo, see Ashburner [1989a]). Embryos incubated in an aqueous buffer containing fixative will not be fixed because the vitelline membrane prevents entry of the fixative into the membrane. Entry of fixative is facilitated by incubating embryos in an immiscible mixture of heptane and fixative solution. After fixation, the vitelline membrane is removed either by hand or by treatment with methanol.

Second, the embryo requires a continuous supply of O_2 for respiration to support the rapid events of early development. When fixing large amounts of embryos, embryos can pile up next to or on top of one another, causing local depletions of O_2. Anoxic embryos arrest their development (Foe and Alberts 1985) and condense their chromatin. Serious anoxia is easy to recognize—embryos show anaphase bridges and heavily condensed chromatin during interphase. However, less severe cases can be recognized as thin wispy fibers during interphase and the absence of embryos showing gradients of progression through the mitotic cycle running from the pole toward the center of the embryo (Foe and Alberts 1983). Anoxia can be avoided by minimizing the number of embryos processed at any one time, fixing them in vessels that allow a single layer of embryos to form during treatment with fixative, and constantly agitating the embryos throughout the fixation. A typical result is shown in Figure 11.1.

Overfixation

It is quite easy to overfix samples, that is, to fix so hard that antibodies cannot freely diffuse into or out of the sample. It is therefore important to attempt a number of fixation trials and to minimize

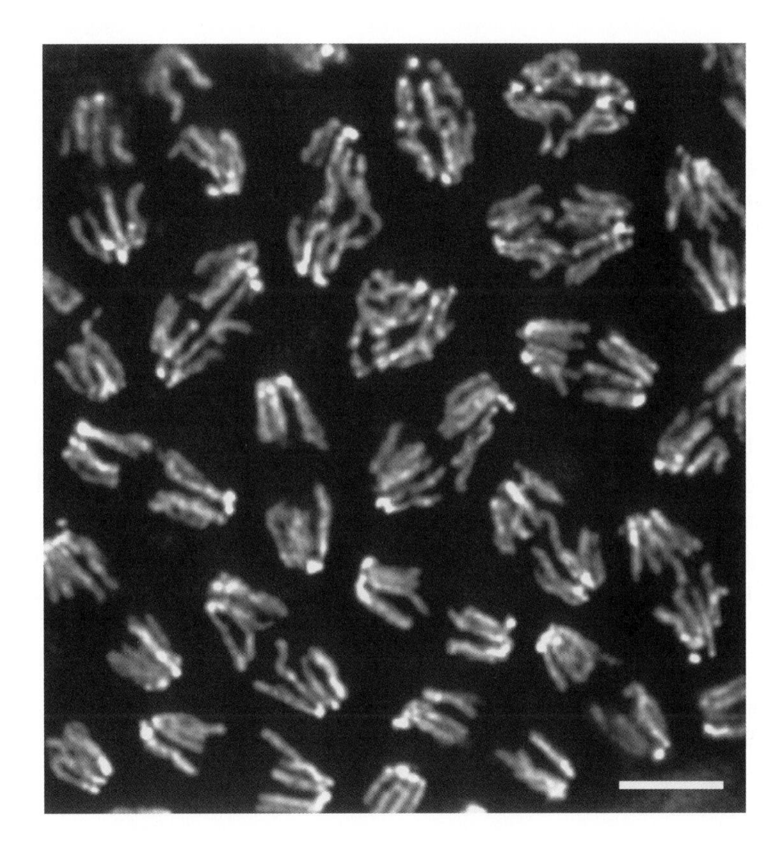

FIGURE 11.1

Proper fixation of *Drosophila* embryos preserves mitotic gradients. The events of mitosis occur as a wave that progresses over the surface of the embryo (Foe and Alberts 1983). When embryos are made anoxic during handling prior to fixation, the cell cycle arrests and gradients are not observed in the fixed material. Embryos that are fixed slowly because of low concentrations of fixative or slow processing time often show collapsed anaphase figures. The figure shows a nuclear cycle-12 embryo that was fixed in 10 ml heptane/10 ml 37% CH_2O for 5 minutes, dechorionated using methanol/EGTA (ethyleneglycoltetra acetid acid), and stained with DAPI. A gradient in the progression through mitosis is evident from the top right to the bottom left of the figure. Moreover, the anaphase figures in the bottom of the figure have not collapsed during fixation and are well separated. Bar, 5 μm. (Photo provided by J. Swedlow, University of Dundee, Dundee, United Kingdom.)

the length of time the sample is exposed to fixative. In general, if poor labeling of interior structures is observed, *decrease* the fixation time.

Antibodies for Immunofluorescence

As always, reliable results are obtained only when high-specificity antibodies are used for detection of an antigen. It is always advisable to use affinity-purified antibodies when possible (Harlow and Lane 1988). This is especially true when performing immunofluorescence on tissue samples, because the increased volume means nonspecific binding (or, more accurately, localization of antibody to antigens other than the one of interest) and can result in significant levels of background signal that can hide the real signal. Moreover, the embryonic cytoplasm contains significant amounts of most proteins that the embryo requires during early development. Observing the localization of antigen to a specific structure in a sea of cytoplasmic signal can sometimes be difficult (see below).

Fluorophore-conjugated secondary antibodies should be of high quality. Previously, it was necessary to adsorb secondary antibodies to fixed embryos to reduce background signal to acceptable levels (Karr and Alberts 1986). Affinity-purified secondary antibodies are now commercially available from a number of vendors. We have successfully used affinity-purified fluorophore-conjugated antibodies from Jackson Immunoresearch, Inc., without embryo adsorption. The cleanest antibodies for work in *Drosophila* appear to be those raised in the donkey.

Microscopy

The exact choice of microscope and objective lens depends heavily on the nature of the experiment. Simply closing the field diaphragm of the microscope can give enough contrast for most purposes. However, given the high level of cytoplasmic signal that can occur in tissues (this problem is especially severe in embryos that contain large pools of cytoplasmic proteins), the use of a microscopy method that reduces or eliminates out-of-focus light can help generate images that emphasize the localization of the antigen. In general, either a confocal or a wide-field optical sectioning microscope can be used for this type of imaging (see Chapters 2 and 3).

Preparation and Immunolabeling of Embryos

Fixation Reagents

For methanol fixation and embryo devitellinization, reagent-grade methanol can be used. Formaldehyde is commercially available, but upon storage at room temperature, formaldehyde forms formic acid and methanol. The best success is achieved with freshly made formaldehyde using this protocol based on Hayat (1989). Some labs make this and dilute to 4% in buffer and then store at –20°C in aliquots. Typically, however, fresh formaldehyde works best. The method below for the preparation of 37% formaldehyde is based on the hydrolysis of paraformaldehyde catalyzed by base. To a large glass tube, add

1.85 g of **paraformaldehyde** (Polysciences)
3.5 ml of H_2O
80 μ of 1 N **NaOH**

CAUTION: Paraformaldehyde; NaOH (see Appendix 2 for Caution)

Heat a beaker full of H_2O to boiling in a microwave oven. Remove from microwave and heat the above mixture with occasional mixing. The **formaldehyde** will dissolve within 5 minutes. Filter the solution using a 0.22-μm syringe filter. Use immediately after the solution comes to room temperature.

CAUTION: Formaldehyde (see Appendix 2 for Caution)

Reagents and Buffers

The choice of buffer for fixation depends on the structure that is to be preserved during fixation. A representative list of fixation reagents and buffers for a variety of structures and tissue is given in Table 11.1.

Processing of samples after fixation for immunofluorescence is done in PBS. Mounting media are available commercially from a number of sources. For fluorescence microscopy, the medium should not affect the sample nor quench the fluorophores, but have as high a refractive index as possible. A number of recipes for mounting media can be found in Chapter 4. Good results are obtained with 1 mg/ml ***p*-phenylenediamine** (Sigma) dissolved in PBS in 90% glycerol/10% H_2O. The *p*-phenylenediamine is dissolved by bubbling N_2 through the glycerol solution. The medium is stored in syringes at –20°C wrapped in aluminum foil.

CAUTION: *p*-phenylenediamine (see Appendix 2 for Caution)

EMBRYO FIXATION

All steps are performed at room temperature except where indicated.

1 Collect embryos on juice agar plates (Ashburner 1989b).

2 Wash embryos from plates with 0.1% Triton X-100, 0.7% NaCl (embryo wash) and collect in a filter chamber made by placing a #40 mesh filter cup inside a #220 mesh filter cup (Bellco Glass). The coarse mesh collects dead flies and detritus; the fine mesh collects the embryos.

3 Dechorionate the embryos by rinsing them into a 400-ml beaker filled with a solution of 50% bleach, 50% embryo wash. Incubate for 90 seconds, keeping the embryos suspended by occasional stirring with a spatula.

4 Pour the embryos back into the #220 mesh filter and wash with a steady stream of embryo wash until the smell of bleach subsides, usually after 45–60 seconds of washing. Rinse the embryos briefly with deionized H_2O.

5 Rinse the embryos with 10 ml of heptane (reagent grade) in a glass vessel. The choice of vessel depends on the number of embryos being fixed. The important issue is to use a vessel that is wide enough so that the embryos do not form large masses during fixation.

For fixing 50–500 embryos, we have used 50-μl Kimax tubes (Fisher Scientific). These have a screw top that seals the tube, allowing agitation during fixation. Erlenmeyer flasks sealed with Parafilm can also be used.

6 Add an equal volume of fixative to the heptane/embryo mixture. See Table 11.1 for choice of fixative and length of fixation. The embryos will sit at the interface between the lower fixative layer and the upper heptane layer. Immediately begin to agitate the embryos. Incubate at room temperature with gentle but thorough agitation. We have had good success with securing the glass tube to a laboratory shaker and shaking at 100 rpm.

7 If the embryos were fixed in methanol, the fixed and devitellinized embryos must be rehydrated (see en masse method 1, step 2 below). If the embryos were fixed in formaldehyde, the vitelline membrane must now be removed. This can either be accomplished en masse or by hand. The en masse method uses methanol, which helps clear the embryos, but affects some structures. For example, filamentous actin cannot be efficiently localized with phalloidin after methanol treatment. The en masse method requires a reasonable number of embryos (<50); otherwise, do this by hand.

EN MASSE DEVITELLINIZATION

These methods are based on those originally described by Mitchison and Sedat (1983), with some modification. Method 1 is easier, but typically gives somewhat lower yields of devitellinized embryos. The yield is a function of fixation method used.

Method 1

1 Add 10 ml of **methanol** (reagent grade) to the fixed embryos. Shake rapidly for 30 seconds to 1 minute. This ruptures the vitelline membrane. Some embryos will retain their membranes and remain at the **heptane**/methanol interface. Devitellinized embryos fall to the bottom of the flask.

CAUTION: Methanol; heptane (see Appendix 2 for Caution)

2 Remove the heptane, methanol, and all embryos at the interface. Wash the embryos at the bottom of the flask twice with methanol, then rehydrate the embryos by washing once each with 90% methanol/10% H_2O, 70% methanol/30% PBS, 50% methanol/50% PBS, 30% methanol/70% PBS, 10% methanol/90% PBS. Wash two times with PBS. For each wash, add the solution, mix, and allow the embryos to settle to the bottom of the tube. Remove the supernatant and add the next solution in the series. The embryos are now ready for processing for immunofluorescence.

Method 2

1 Add a mixture of 9 ml of methanol and 1 ml of 10 mM EGTA, pH 7.4, to the fixed embryos. Pack the tube in dry ice and agitate on a rotating shaker for 10 minutes. Remove the vessel from the dry ice and warm the mixture in a stream of warm tap water while shaking vigorously.

The rapid temperature change helps break the vitelline membranes.

Devitellinized embryos sink to the bottom of the tube.

2 Remove the **heptane, methanol,** and all embryos at the interface. Wash the embryos at the bottom of the flask twice with methanol, then rehydrate the embryos by washing once each with 90% methanol/10% H_2O, 70% methanol/30% PBS, 50% methanol/50% PBS, 30% methanol/70% PBS, 10% methanol/90% PBS. Wash two times with PBS. For each wash, add the solution, mix, and allow the embryos to settle to the bottom of the tube. Remove the supernatant and add the next solution in the series. The embryos are now ready for processing for immunofluorescence.

CAUTION: Heptane; methanol (see Appendix 2 for Caution)

MANUAL DEVITELLINIZATION

When small numbers of embryos are available, manual devitellinization is required. F-actin is disrupted by methanol, so if phalloidin is to be used to detect F-actin, the vitelline membrane must be removed manually. There are two methods.

Method 1

1 Place double-sided sticky tape (3M) in the bottom of a small glass petri dish. Cover the tape with PBS.

2 Transfer the fixed embryos to a filter and blot excess fixative away. Rinse embryos with PBS. Transfer embryos to the petri dish and allow them to stick on the tape. Puncture the vitelline membrane with a tungsten needle and tease the embryo free. Transfer to a fresh tube with PBS. The embryos are now ready for immunolabeling.

Method 2

This method is from Theurkauf (1994). It takes some practice, but works well. Practice on wild-type embryos before using precious mutant embryos.

1 Transfer the fixed embryos to the frosted part of a slide; remove excess heptane/fixative.

2 Pull the edge of a 22 x 40-mm coverslip across the embryos. Then slide the flat surface of the coverslip over the embryos in a circular motion. Check the removal of the vitelline membranes with a dissecting scope. Repeat the procedure if necessary.

3 Wash the devitellinized embryos into a fresh tube with 0.1% Triton X-100/PBS (PBST). The embryos are now ready for immunolabeling.

IMMUNOLABELING OF *DROSOPHILA* TISSUES AND EMBRYOS

This is a standard method for labeling tissues for immunofluorescence. Samples can be labeled in 0.5-ml tubes using solution volumes of 0.2 ml. For all steps, continuous gentle agitation is *absolutely required* to fully expose tissues to antibodies and wash solutions. For delicate or easily aspirated samples, solutions can be manipulated with plastic gel loading tips.

Once fixed, the samples are fairly stable, but we have noticed some degradation in structure in samples that have been stored at room temperature overnight. If the staining protocol is rapid, samples can be handled at room temperature. Any overnight incubations should be done at 4°C. We generally block for 1–2 hours, then incubate with primary antibody overnight.

To determine the specificity of fluorophore labeling, a control sample processed without primary antibody should be processed in parallel with other samples. The specificity of primary antibody labeling can be tested by performing the primary antibody incubation in the presence of excess antigen.

1 Block with 5% NGS (normal goat serum) (Jackson Immunoresearch) in PBST. Incubate 1 hour to overnight. The volume used in this step and all subsequent ones is enough to keep the embryos well dispersed during continuous agitation.

2 Remove blocking solution. Add an appropriate amount of a primary antibody in 5% NGS/PBST. The amount of antibody used must be determined empirically for each antibody. A good starting concentration is 1 µg/ml, but many antibodies work well at significantly lower concentrations. Incubate 1 hour to overnight. If labeling with more than one antibody, all primary antibodies can be added simultaneously.

3 Remove the primary antibody solution and wash the sample four times with PBST for 15 minutes each.

4 Remove the last wash solution and add an empirically determined dilution of fluorophore-conjugated antibody (secondary antibody) in PBST. Doing this incubation in 5% NGS/PBST helps reduce nonspecific binding with some secondary antibodies.

5 Wash the sample four times with PBST for 15 minutes each. Wash twice with PBS. If staining of nuclei and chromosomes is required, add 0.5 µg/ml DAPI (Sigma) and incubate for 10 minutes. Wash twice more with PBS.

6 Mount the samples on slides with a few drops of mounting medium and seal with a coverslip and nail polish.

Samples can be stored at –20°C.

REFERENCES

Ashburner M. 1989a. Drosophila: *A laboratory handbook.* Cold Spring Harbor Laboratory Press, Cold Spring Harbor, New York.

———. 1989b. Drosophila: *A laboratory manual.* Cold Spring Harbor Laboratory Press, Cold Spring Harbor, New York.

Foe V.E. and Alberts B.M. 1983. Studies of nuclear and cytoplasmic behavior during the five mitotic cycles that precede gastrulation in *Drosophilia* embryogenesis. *J. Cell Sci.* **61:** 31–70.

———. 1985. Reversible chromosome condensation induced in *Drosophila* embryos by anoxia: Visualization of interphase nuclear organization. *J. Cell Biol.* **100:** 1623–1636.

Gaul U., Mardon G., and Rubin G.M. 1992. A putative Ras GTPase activating protein acts as a negative regulator of signaling by the Sevenless receptor tyrosine kinase. *Cell* **68:** 1007–1019.

Harlow E. and Lane D.P. 1988. *Antibodies: A laboratory manual.* Cold Spring Harbor Laboratory, Cold Spring Harbor, New York.

Hayat M.A. 1989. *Principles and techniques of electron microscopy.* CRC Press, Boca Raton, Florida.

Hiraoka Y., Agard D.A., and Sedat J.W. 1990. Temporal and spatial coordination of chromosome movement, spindle formation, and nuclear envelope breakdown during prometaphase in *Drosophila melanogaster* embryos. *J. Cell Biol.* **111:** 2815–2828.

Karr T.L. and Alberts B.M. 1986. Organization of the cytoskeleton in early *Drosophila* embryos. *J. Cell Biol.* **102:** 1494–1509.

Kellogg D.R., Mitchison T.J., and Alberts B.M. 1988. Behaviour of microtubules and actin filaments in living *Drosophila* embryos. *Development* **103:** 675–686.

Mitchison T.J. and Sedat J.W. 1983. Localization of antigenic determinants in whole *Drosophila* embryos. *Dev. Biol.* **99:** 261–264.

Ruohola H., Bremer K.A., Baker D., Swedlow J.R., Jan L.Y., and Jan Y.N. 1991. Role of neurogenic genes in establishment of follicle cell fate and oocyte polarity during oogenesis in *Drosophila. Cell* **66:** 433–449.

Theurkauf W.E. 1994. Immunofluorescence analysis of the cytoskeleton during oogenesis and early embryogenesis. *Methods Cell Biol.* **44:** 489–505.

Theurkauf W.E., Smiley S., Wong M.L., and Alberts B.M. 1992. Reorganization of the cytoskeleton during *Drosophila* oogenesis: Implications for axis specification and intracellular transport. *Development* **115:** 923–936.

Tilney L.G., Tilney M.S., and Guild G.M.. 1996. Formation of actin filament bundles in the ring canals of developing *Drosophila* follicles. *J. Cell Biol.* **133:** 61–74.

Urata Y., Parmelee S.J., Agard D.A., and Sedat J.W. 1995. A three-dimensional structural dissection of *Drosophila* polytene chromosomes. *J. Cell Biol.* **131:** 279–295.

CHAPTER 12

Immunofluorescence Methods for *Caenorhabditis elegans*

Sarah Crittenden[1,2] and Judith Kimble[2]

[1]*Howard Hughes Medical Institute and* [2]*University of Wisconsin, Madison, Wisconsin*

INTRODUCTION

The use of antibodies to visualize the distribution and subcellular localization of gene products powerfully complements genetic and molecular analysis of gene function in *Caenorhabditis elegans*. The challenge to immunolabeling *C. elegans* is finding the fixation and permeabilization methods that effectively make antigens accessible without destroying the tissue morphology or the antigen. Embryos are surrounded by a chitinous eggshell and larvae and adults are surrounded by a collagenous cuticle, each of which must be permeabilized to allow penetration of antibodies. In addition, antigens and antibodies are sensitive to different fixing and permeabilizing conditions. For example, some antibodies do not work well on paraformaldehyde-fixed samples, and others are sensitive to incubation in acetone. There are many protocols used in the *C. elegans* field; additional protocols are summarized in Miller and Shakes (1994) and on the *C. elegans* World Wide Web page (http://elegans.swmed.edu/).

Preparation and Immunolabeling of *C. elegans*

The following protocols describe the fixation of embryos, larvae, or adults for immunolabeling with antibodies.

FIXATION OF ANIMALS AND TISSUE

The organism may be fixed using either a whole-mount freeze-cracking method or tissue extrusion.

Whole-Mount Freeze-cracking Method

This method (Strome and Wood 1982; Albertson 1984; Bowerman et al. 1993) is a good starting point; it is easy and it works well with most antibodies and with embryos, larvae, and adults.

1 Assemble the following items before starting the procedure:

25 gauge syringe needle
M9 buffer

M9 Buffer (1 liter)

KH_2PO_4	3.0 g
Na_2HPO_4	6.0 g
NaCl	0.5 g
NH_4Cl	1.0 g

Bring to 1 liter with dH_2O

CAUTION: Na_2HPO_4; NH_4Cl (see Appendix 2 for Caution)

Formaldehyde, if used
TBSB

TBSB (enough for 1 or 2 staining experiments)

Add 9 ml of 1x TBS (Tris-bufferred saline) to 1 ml of 5% BSA (bovine serum albumin).

Coplin jars containing methanol and acetone precooled on dry ice (~10 min)
Metal plate precooled on dry ice (~10 min)
Subbed slides
Coverslips

2 Prepare subbed slides:

a Prepare subbing solution:

Bring 200 ml of dH_2O to 60°C.
Add 0.4 g of gelatin and cool to 40°C.
Add 0.04 g of Chrome Alum.

Add **sodium azide** to 1 mM.
Add polylysine (molecular weight [m.w.] >300,000; Sigma F 1524) to 1 mg/ml.

CAUTION: Sodium azide (see Appendix 2 for Caution)

b Put the subbing solution in a Coplin jar and store it at 4°C.

c Soak clean slides in subbing solution for 5 minutes to 1 hour, air-dry, and store at room temperature.

Subbed slides can be used for weeks. Several batches of slides can be subbed in the same subbing solution. When slides become less sticky, it is time to make a new solution.

3 Place animals into 6 μl of M9 on a subbed slide. Cut the animals open with a 25-gauge syringe needle if early embryos or extruded germ lines or intestines are to be stained.

Use more than ten adults or more than 40 larvae; some animals or tissues will be lost from the slide during the staining procedure.

4 Add 2 μl of 5% formaldehyde.

5% Formaldehyde (5 ml)

Add 0.25 g of **paraformaldehyde** to 4.2 ml dH_2O.
Add 2 μl of 4 N **NaOH.**
Heat at 65°C until dissolved.
Add 0.5 ml of 10x PBS, bring volume to 5 ml with dH_2O.
Filter and store at 4°C for no more than one week.

CAUTION: Paraformaldehyde; NaOH (see Appendix 2 for Caution)

Concentrations of between 1% and 5% formaldehyde are commonly used (e.g., Bowerman et al. 1993; Evans et al. 1994; Lin et al. 1995); adjust concentration for ideal staining.

Formaldehyde can be omitted if it interferes with antibody binding.

Formaldehyde fixation improves morphology.

5 Set an 18 x 18-mm #1 coverslip carefully on top of the animals. Use a needle to apply gentle pressure several times over each animal or region of the slide. The animals will flatten; usually a few burst. This procedure aids in opening the eggshell or cuticle. Alternatively, a Kimwipe can be used to wick excess liquid from under the coverslip until the worms flatten.

6 If formaldehyde was added, let the sample incubate 30 seconds to 30 minutes at room temperature in a humidified chamber (e.g., Bowerman et al. 1993; Evans et al. 1994; Lin et al. 1995); adjust time for ideal staining.

A humidified chamber can be made from a plastic petri dish (or other container) with a wet paper towel taped to the lid.

7 Put the slide on a metal plate on top of dry ice for at least 10 minutes.

We use plates 1/4″ thick. The plates are used to make a cold, flat surface for the animals to sit on.

8 Pop the coverslip off with a razor blade and immerse the slide immediately in 100% cold **methanol** for 5 minutes, followed by 100% cold **acetone** for 5 minutes.

CAUTION: Methanol; acetone (see Appendix 2 for Caution)

For some antigens, it is better to omit the acetone incubation.

For some antigens, incubation in cold DMF (N,N-dimethylformamide) works better than methanol or acetone (Lin et al. 1998).

9 Air-dry the slide for 5 minutes. This enhances the adhesion of the animals to the slide. Rehydrate the sample through a series of increasingly aqueous solutions. Either methanol (90%, 70%, 50%, followed by 1x TBS; Lin et al. 1995), ethanol (2 minutes each in 95%, 70%, 50%, and 30%, followed by 1x TBS; Miller and Shakes 1994), or acetone (Goldstein and Hird 1996) series have been used.

10 Gently drop 200 µl of TBS containing 0.5% BSA (TBSB) onto the animals and incubate at room temperature for 30 minutes.

11 Follow the antibody incubation procedure (see below).

■ Notes

- Larvae and adults can either be picked from plates or washed off with M9. Young embryos (1–50 cells) are easily obtained by cutting open gravid hermaphrodites. Older embryos can be obtained by adding M9 to a plate, washing off the adults and larvae, then scraping the remaining embryos off with a pasteur pipette into additional M9. The worms or embryos are then pelleted by spinning for 1–2 minutes at 1000 rpm in a microfuge. To remove *Escherichia coli*, more M9 is added, and the worms are pelleted again. Then 6–8 µl of concentrated worms can be dropped onto a slide.
- When staining larvae, it helps to stage the animals so that they are similar in size. This way the amount of pressure can be adjusted for the size of the worms being fixed. For example, if there are lots of large adults (or larvae), it is difficult to permeabilize L1s without completely squashing the adults. It gets increasingly more difficult to effectively permeabilize the worms as they get older.
- For tissues that can be extruded from the cuticle, such as germ lines and intestines, the morphology is generally better using a non-freeze-crack method (see extrusion method).
- Another commonly used procedure for whole mounts is the reduction/oxidation method of Finney and Ruvkun (1990). This method is described in detail in Miller and Shakes (1995).

Tissue Extrusion Method

Using this method (Crittenden et al. 1994), gonads and intestines, which are extruded from the carcass, are well fixed and permeabilized (Fig. 12.1A,B). Tissues remaining in the carcass are not usually stained well.

1 Assemble the following items before starting the procedure:

0.25 mM levamisole (Sigma)

> ***0.25 M Levamisole (5 ml)***
>
> Dissolve 0.3 g of levamisole in 5 ml of M9. Aliquot and store at –20°C.
> For 0.25 M levamisole, add 1 µl of 0.25 M levamisole to 1 ml of M9. Store at room temperature.

25 gauge syringe needle
1.0% formaldehyde
TBSB

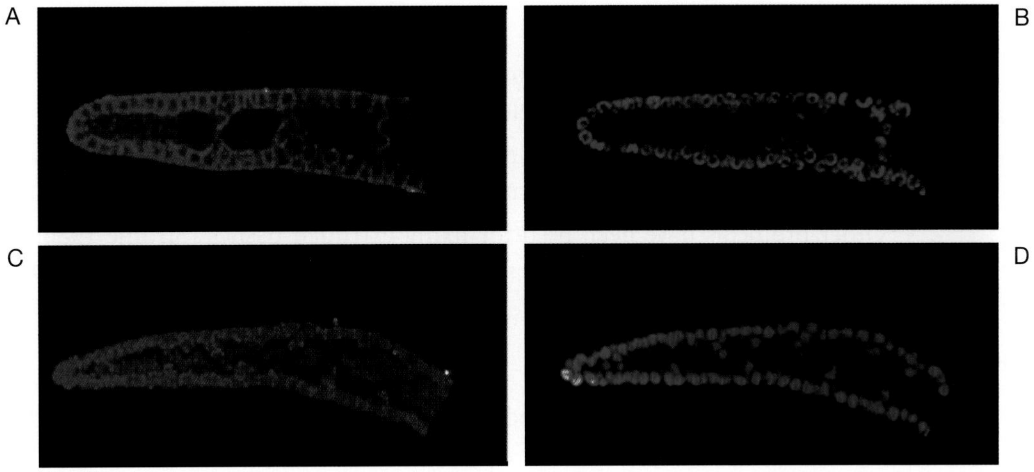

A B C D

FIGURE 12.1

Gonads from wild-type *C. elegans* hermaphrodites were fixed according to the tissue extrusion method. Primary antibodies were rat anti-GLP-1 (Crittenden et al. 1994) and mouse anti-DNA (Chemicon, MAB030). Secondary antibodies were Cy3-conjugated donkey anti-rat and FITC-conjugated donkey anti-mouse (Jackson Immunoresearch). Images were obtained on a Bio-Rad MRC 1024 laser scanning confocal microscope. (*A,B*) Well-fixed gonad. The transmembrane receptor, GLP-1, is membrane-associated in a crisp honeycomb pattern. DNA staining is crisp, and discrete structures within the nucleus are visible. (*C,D*) Poorly fixed gonad. GLP-1 is punctate and diffuse. In addition, there is a faint red haze in the nuclei. DNA staining is bright, but nuclear morphology is poor; the nuclei are fuzzy and discrete structures are not discernible. The poor morphology was due to a contaminant in the primary antibody solution. (Photo provided by S. Crittenden, University of Wisconsin.)

TBSB (enough for 1 or 2 staining experiments)

Add 9 ml of 1x TBS to 1 ml of 5% BSA.

TBSBTx (TBSB with 0.1% Triton X-100)
Subbed slides

2 Put five to ten adult hermaphrodites into 5 μl of M9 containing 0.25 mM levamisole on a subbed slide. Using a 25-gauge syringe needle, cut off the heads or tails of the animals, allowing the gonad and intestine to extrude from the animal.

Levamisole causes the animals to contract, which results in their germ lines and intestines being extruded efficiently.

3 Gently drop 100 μl of 1.0% **formaldehyde** onto the cut animals. Incubate at room temperature for 10 minutes in a humidified chamber.

CAUTION: Formaldehyde (see Appendix 2 for Caution)

4 Remove the formaldehyde and add 50 μl of TBSB containing 0.1% Triton X-100 (TBSBTx) at room temperature for 5 minutes.

5 Remove the TBSBTx and wash two times with 200 μl of TBSB.

6 Incubate samples in 200 μl of TBSB at room temperature for ~30 minutes.

7 Follow the antibody incubation procedure.

Notes

- This fix works well for at least some membrane proteins (Crittenden et al. 1994; Henderson et al. 1994), but not for the cytoskeletal proteins actin and tubulin.
- For cytoskeleton, try fixing first in 100% methanol at room temperature for 5 minutes followed by 1% formaldehyde at room temperature for 25 minutes (Crittenden et al. 1994).
- Other protocols have been used for fixing extruded tissues (Strome 1986; Francis et al. 1995) and/or the cytoskeleton (Strome 1986; Waddle et al. 1994; Francis et al. 1995).

ANTIBODY INCUBATION PROCEDURE

Worms fixed according to the preceding protocols are incubated overnight with primary antibody, subsequently exposed to secondary antibody, and mounted for viewing.

1. Incubate fixed worms with primary antibodies overnight at 4°C or for several hours at room temperature in a humidified chamber. Use 30–50 μl of antibody solution per slide.

2. Wash by gently covering the worms with 200 μl of TBSB. Wash three times for 15 minutes each at room temperature. Alternatively, the slides can be immersed in a Coplin jar if the worms are well attached to the slide.

3. Dilute secondary antibodies to the recommended concentration in TBSB. Use 100 μl per slide and incubate at room temperature for 1–2 hours.

   We use Jackson Immunoresearch's purified IgG coupled to Cy3, Cy5, FITC (fluorescein isothiocyanate), or rhodamine.

   To reduce nonspecific background, add ~1 mg of worm acetone powder/200 μl of secondary antibody solution.

   Acetone Powder (modified from Harlow and Lane 1988)

   Homogenize worms in Dounce homogenizer; use about 1 g of worms/ml of M9.
   Set on ice for 5 minutes.
   Add 4 ml of –20°C **acetone**/ml of worm suspension. Mix vigorously.
   Set on ice for 30 minutes with occasional vigorous mixing.
   Spin at 10,000*g* for 10 minutes.
   Resuspend pellet with fresh –20°C acetone.
   Mix vigorously on ice for 10 minutes.
   Spin at 10,000*g* for 10 minutes.
   Spread pellet on clean filter paper and allow to dry at room temperature. When dry, break up chunks in a mortar and pestle, then transfer powder to a microfuge tube and store at 4°C.

   CAUTION: Acetone (see Appendix 2 for Caution)

   Incubate the secondary antibody/worm acetone powder mix at 4°C for 15 minutes to 1 hour. Then centrifuge at 10,000 rpm for 5 minutes in a microfuge to pellet the acetone powder. Use the supernatant as the secondary antibody solution.

4. Wash worms as in step 2. If desired, add the DNA stain, DAPI (4′,6-diamidino-2-phenylindole), to 0.5 μg/ml in the final wash.

5. After removing the last wash, wick off excess moisture from the slide by touching the edges with a Kimwipe. Add 8 μl of mounting medium (see Chapter 4), put an 18 x 18-mm #1 coverslip over the worms, and seal with nail polish.

Notes

- Worms generally do not stick well to slides, so start with plenty of animals and be gentle when doing washes.
- If background is high, determine whether it is due to the primary or secondary antibody. Try diluting the antibodies further, affinity-purify the primary antibody, and preabsorb the primary antibody with worm or bacterial acetone powder or with fixed worms; null mutants should be used so that the specific antibody will not be depleted.
- Do not let the worms dry after they have been fixed and rehydrated; this tends to give a nonspecific haze to the nuclei and cytoplasm.
- If the morphology looks poor, try to fix the worms more quickly. Alternatively, contaminated solutions can cause poor morphology. DAPI-stained DNA should look well defined and crisp; if it does not, be suspicious (e.g., Fig. 12.1C,D).
- Common background problems include intestine autofluorescence on the DAPI and fluorescein channels, causing a dim nuclear stain.
- Using different fixatives or making small changes in concentration of fixative or time of fixation can make a big difference in the quality of staining.

CONTROLS

- Use an antibody that is known to work to test for morphological preservation, permeability, and fixation. Some useful control antibodies are anti-DNA monoclonal mAb 030 (Chemicon, MAB030), anti-actin clone C4 (ICN Biochemicals, 69-100-1) (Strome 1986; Evans et al. 1994), anti-β-tubulin (Amersham, N357) (Crittenden et al. 1994; Evans et al. 1994; Waddle et al. 1994).
- Stain a null mutant and look for loss of staining.
- Compete for staining with proteins that contain the antigen used to raise the antibodies.
- If a null mutant is not available, it is possible to abolish the antigen in embryos from animals that have been injected with antisense RNA (Guo and Kemphues 1995).

REFERENCES

Albertson D.G. 1984. Formation of the first cleavage spindle in nematode embryos. *Dev. Biol.* **101:** 61–72.

Bowerman B., Draper B.W., Mello C.C., and Priess J.F. 1993. The maternal gene *skn-1* encodes a protein that is distributed unequally in early *C. elegans* embryos. *Cell* **74:** 443–452.

Crittenden S.L., Troemel E.R., Evans T.C., and Kimble J. 1994. GLP-1 is localized to the mitotic region of the *C. elegans* germ line. *Development* **120:** 2901–2911.

Evans T.C., Crittenden S.L., Kodoyianni V., and Kimble J. 1994. Translational control of maternal *glp-1* mRNA establishes an asymmetry in the *C. elegans* embryo. *Cell* **77:** 183–194.

Finney M. and Ruvkun G. 1990. The *unc-86* gene product couples cell lineage and cell identity in *C. elegans. Cell* **63:** 895–905.

Francis R., Barton M.K., Kimble J., and Schedl T. 1995. *gld-1*, a tumor suppressor gene required for oocyte development in *Caenorhabditis elegans. Genetics* **139:** 579–606.

Goldstein B. and Hird S.N. 1996. Specification of the anteroposterior axis in *Caenorhabditis elegans. Development* **122:** 1467–1474.

Guo S. and Kemphues K.J. 1995. *par-1*, a gene required for establishing polarity in *C. elegans* embryos, encodes a putative Ser/Thr kinase that is asymmetrically distributed. *Cell* **81:** 611–620.

Harlow E. and Lane D. 1988. *Antibodies: A laboratory manual.* Cold Spring Harbor Laboratory, Cold Spring Harbor, New York.

Henderson S.T., Gao D., Lambie E.J., and Kimble J. 1994. *lag-2* may encode a signaling ligand for the GLP-1 and LIN-12 receptors of *C. elegans. Development* **120:**

2913-2924.

Lin R., Hill, R.J., and Priess J.R. 1998. POP-1 and anterior–posterior fate decisions in *C. elegans* embryos. *Cell* **92:** 229–239.

Lin R., Thompson S., and Priess J.R. 1995. *pop-1* encodes an HMG box protein required for the specification of a mesoderm precursor in early *C. elegans* embryos. *Cell* **83:** 599–609.

Miller D.M. and Shakes D.C. 1995. Immunofluorescence microscopy. In Caenorhabditis elegans. *Modern biological analysis of an organism* (ed. H.F. Epstein and D.C. Shakes), pp. 365–394. Academic Press, San Diego.

Strome S. 1986. Fluorescence visualization of the distribution of microfilaments in gonads and early embryos of the nematode *Caenorhabditis elegans. J. Cell Biol.* **103:** 2241–2252.

Strome S. and Wood W.B. 1982. Immunofluorescence visualization of germ-like-specific cytoplasmic granules in embryos, larvae, and adults of *Caenorhabditis elegans. Proc. Natl. Acad. Sci.* **79:** 1558–1562.

Waddle J.A., Cooper J.A., and Waterston R.H. 1994. Transient localized accumulation of actin in *Caenorhabditis elegans* blastomeres with oriented asymmetric divisions. *Development* **120:** 2317–2328.

CHAPTER 13

Analyzing DNA Replication: Nonisotopic Labeling

Dean Jackson[1] and Peter R. Cook[2]

[1]*University of Manchester, Manchester, United Kingdom*
[2]*University of Oxford, Oxford, United Kingdom*

INTRODUCTION

The number of cells traversing the cell cycle and the rate of progression through it provide important indices of cell growth and tumorigenicity. The number of cells in S-phase has traditionally been determined by autoradiography after incorporation of [^{3}H]thymidine. S-phase cells can also be identified by their high content of DNA polymerase and PCNA (proliferating cell nuclear antigen), a component of the leading-strand polymerase, but both of these markers, which can be detected rapidly and conveniently using the appropriate antibodies, are not found exclusively in S-phase cells. Immunolabeling after incorporation of modified DNA precursors (e.g., BrdU [5-bromodeoxyuridine]; Gratzner 1982) allows more rapid and precise detection of cells in S-phase of the cell cycle. S-phase lasts for at least 6 hours in mammalian cells, so incubation with a precursor for a few minutes gives a labeling index that reflects the proportion of cells in S-phase; longer incubations also label cells that have entered and left S-phase. In living cells, BrdU is incorporated into replication sites that can then be detected using fluorochrome or enzyme-coupled antibodies. Less than 1% of cells in a population that are synthesizing DNA can be reliably detected. Using two fluorochromes or enzymes that generate products with different colors allows double labeling (e.g., using FITC [fluorescein isothiocyanate] to detect incorporated BrUMP and Texas Red to detect PCNA). Because the enzymatic products are diffusible, cells detected in this way are usually scored as positive or negative without reference to subnuclear detail.

A better way of labeling sites of DNA synthesis at high resolution is to incubate cells with analogs of the natural precursors of DNA and then label the incorporation sites with fluorochrome-tagged antibodies. Cells labeled in this way either in vivo (Nakamura et al. 1986) or in vitro (Nakayasu and Berezney 1989) display a few hundred discrete nuclear sites early in S-phase, with distinct patterns of DNA replication that are characteristic of different stages of S-phase (Fig. 13.1) (Humbert and Usson 1992; O'Keefe et al. 1992). "Foci" labeled after very short incubations correspond with sites where many replicons are duplicated simultaneously within massive protein complexes (Hozák et al. 1993).

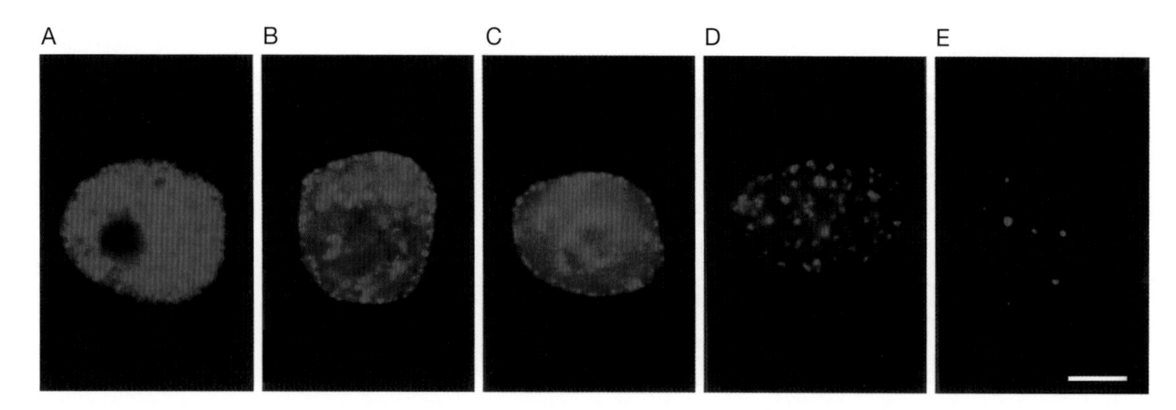

FIGURE 13.1

HeLa cells display five distinct patterns of DNA replication. Early in S-phase, numerous foci of replication are observed scattered throughout the nucleoplasm (*A*). Later in S-phase, larger clusters of DNA appear to replicate (*B–E*) with chromosome-specific α-satellite DNA sequences replicating in mid S-phase (*C*). Bar, 10 μm. (Reprinted, with permission, from O'Keefe et al. 1992; © Rockefeller University Press.)

Labeling Procedure

ADMINISTRATION AND LABELING WITH BROMODEOXYURIDINE

BrdU is phosphorylated by cells to give BrdUTP, and this precursor is incorporated into DNA instead of dTTP (deoxythimidine triphosphate).

Whole Animals (Mouse/Rat)

1 Dissolve 20–50 mg of **BrdU** (Sigma) in PBS and filter sterilize.
BrdU (see Appendix 2 for Caution)

2 Inject 5–50 mg/kg of body weight intraperitoneally.

3 After 1–5 hours (usually 1–2 hr), sacrifice the animal and recover appropriate tissue.

4 Rinse tissue in PBS (phosphate-buffered saline).

5 Fix samples and immunolabel as described below (see pp. 207–209).

Tissue Samples

Labeling can be performed after sacrifice.

1 Sacrifice the animal and remove appropriate tissue.

2 Mince or otherwise disperse the tissue.

a Cut into convenient pieces, commonly 1-mm slices or 2-mm cubes, and return to medium.

b Disperse tissue mechanically; treat with collagenase if necessary (conditions depend on tissue). Collect single-cell suspension, resuspend in medium, and seed on coverslips coated with gelatin or poly-L-lysine.
Nonadherent cells can be purified and transferred to appropriate medium or encapsulated in agarose microbeads (see p. 214).

3 Incubate cells in medium containing 20 μM BrdU for 15 minutes at 37°C.
Adjust concentration and time of BrdU incubation as required.

4 Rinse the cells in ice-cold PBS.

5 Fix samples and immunolabel as described below (see pp. 207–209).

Cell Cultures

Cells should be growing logarithmically.

Adherent Cells

1 Place a clean, 22-mm square glass coverslip in a 35-mm diameter petri dish.

2 Seed approximately 2×10^5 cells in 2 ml of fresh medium.

3 Grow cells for 24–48 hours until cells reach ~50% confluency.

4 Replace growth medium with fresh medium containing 20 μM BrdU and incubate for 15 minutes at 37°C.

5 Rinse cells in ice-cold PBS.

6 Fix samples and immunolabel as described below (see pp. 207–209).

Nonadherent Cells

1 Encapsulate cells in agarose microbeads at a density of approximately 2×10^6/ml (see p. 214).

2 Regrow cells in fresh medium for 1 hour.

3 Add 20 μM BrdU and regrow for 15 minutes.

4 Rinse cells in ice-cold PBS.

5 Fix samples and immunolabel as described below (see pp. 207–209).

▪ *Notes*

- For continual labeling of cells in situ (over 2–7 days), low doses of BrdU are usually administered with an osmotic minipump.
- An intravenous dose of 100–200 mg/m^2 BrdU is sufficient for unambiguous labeling of S-phase cells in human tumors after 1–2 hours (Hoshino et al. 1986; Yanik et al. 1992).
- Cell cycle parameters are best assessed using two different pulse labels (e.g., BrdU with [^{3}H]thymidine; Repka and Adler 1992). Because antibodies are available that have different affinities for different incorporated halogens, cells can also be double-labeled using iododeoxyuridine (IdU) and chlorodeoxyuridine (Manders et al. 1992) or IdU and BrdU (Yanik et al. 1992).
- It is sometimes informative to use synchronized S-phase cells.
- Adherent cells can be labeled after encapsulation in agarose microbeads; this is particularly useful for morphological analysis or if many subsequent manipulations are required (e.g., Hozák et al. 1993).

LABELING IN VITRO WITH PHOSPHORYLATED PRECURSORS

For many applications, labeling in vitro provides an appealing versatility. Labeling in vitro with phosphorylated precursors has the following advantages:

- After permeabilization, precursor pools can be depleted by washing.
- Elongation rates can be modified at will by adjusting precursor concentrations.

- A range of different precursors is available (see Notes below). Some label sites of DNA synthesis directly and so can be used without fixation (Hassan and Cook 1993), and others do not require DNA denaturation during subsequent detection (Nakayasu and Berezney 1989); this allows better preservation of nuclear morphology. (Note that incorporated BrdU can only be detected after the DNA is denatured unless one immunolabels thin sections for electron microscopy [O'Keefe et al. 1992].)

1 Prepare cells on coverslips (as described above) or encapsulated in agarose microbeads (see p. 214).

2 Prepare physiological buffer (PB) and 10× initiation mix (IM).

PB

100 mM KCH_3COOH
30 mM **KCl**
10 mM Na_2PO_4
1 mM **$MgCl_2$**
1 mM Na_2ATP
1 mM **DTT** (dithiothreitol)
Adjust pH to 7.4 with 100 mM KH_2PO_4 if necessary, and add protease and nuclease inhibitors as required.

CAUTION: KCl; $MgCl_2$; DTT (see Appendix 2 for Caution)

10x IM

10x IM is PB with the following concentrations of phosphorylated precursors and $MgCl_2$:
250 μM dATP (deoxyadenine triphosphate)
250 μM dCTP (deoxycytosine triphosphate)
250 μM dGTP (deoxyguanine triphosphate)
100 μM CTP (cytosine triphosphate)
100 μM GTP (guanine triphosphate)
100 μM UTP (uridine triphosphate)
10–100 μM dTTP analog
$MgCl_2$ in a molarity equal to that of the triphosphates

3 Permeabilize cells in PB containing 0.1–0.2% Triton X-100 for 5 minutes.

4 Incubate coverslips or microbeads in PB at 33°C for 5 minutes.

5 Add 0.1 volume of 10x IM, mix, and incubate at 33°C for 2–60 minutes.

6 Wash three times in more than 10 volumes of ice-cold PB.

7 Fix samples and immunolabel as described below (see pp. 207–209).

NOTES

- The following modified precursors support replication by endogenous eukaryotic DNA polymerases when added instead of the equivalent unmodified precursor, but at a reduced rate (elongation is typically 5–20% of the normal level):

biotin-16-dUTP (deoxyuridine triphosphate) ((Boehringer Mannheim; Sigma)
biotin-11-dUTP (Sigma)
biotin-7-dATP (Sigma)
biotin-11-dCTP (Sigma)
digoxigenin-11-dUTP (Boehringer Mannheim)
fluorescein-12-dUTP (Boehringer Mannheim; DuPont)
fluorescein-11-dUTP (Amersham International)
rhodamine-4-dUTP (Amersham International)
Texas Red-5-dUTP (DuPont)
coumarin-4-dUTP (Amersham International)
coumarin-5-dUTP (DuPont)

Closely related compounds not listed above must be tested but can be assumed to be acceptable substrates.

- The concentration of modified precursor and duration of labeling can be adjusted to suit individual requirements. The following provides some guidelines. (i) With 20 μM biotin- or digoxigenin-coupled precursors, 15-minute incubations give good indirect immunofluorescence signals and longer incubations give correspondingly stronger signals. (ii) With 100 μM biotin-16-dUTP, 5- and 2-minute incubations allow detection by light and electron microscopy, respectively, using standard detection protocols. (iii) With 20-μM fluorescent precursors, incorporated label can be detected after 30–60-minute incubations.
- If inhibitors are to be used, incubate cells in them at 0°C for 15 minutes and at 33°C for 5 minutes prior to addition of 10x IM.
- Modified precursors incorporated into DNA may be removed by the repair pathway (see, e.g., Huijzer and Smerdon 1992), but such repair is inefficient in permeabilized HeLa cells (see, e.g., Jackson et al. 1994).

Fixation and Processing

Formaldehyde- or alcohol-based (e.g., Carnoy's fluid) fixation and paraffin embedding are used frequently and are described below. Formaldehyde, a mild protein cross-linking reagent, preserves cell structure well, and washing with a nonionic detergent or organic solvent permeabilizes membranes and allows antibodies access to the interior of the cell.

TISSUES

1 Prepare fixative as follows:

For 4% formaldehyde: Dissolve 4 g of **paraformaldehyde** (prill form obtained from Electron Microscopy Sciences, Inc.) in 50 ml of distilled H_2O in a 50-ml tube. Heat at 60°C until dissolved. Add a few drops of 2 M NaOH to help dissolve, if necessary. Cool to room temperature, add an equal volume of 2x concentrated PBS, and filter (0.22-μM pore disposable filter unit).

CAUTION: Paraformaldehyde (see Appendix 2 for Caution)

Carnoy's fixative

glacial acetic acid	10 ml
chloroform	30 ml
ethanol	60 ml

CAUTION: Glacial acetic acid, acids and bases that are concentrated; chloroform (see Appendix 2 for Caution)

2 After labeling, place tissue sample in freshly prepared 4% formaldehyde in PBS or Carnoy's fixative, replace fixative once, and incubate at 20°C for 2–16 hours to allow the fixative to permeate the sample.

3 Embed the sample in paraffin.

4 Collect 3–5-μm sections on clean glass slides.

5 Place slides at 60°C for 30 minutes.

6 Dewax sections in chloroform (two changes, 3 min each).

7 Rehydrate sections through an ethanol series of 100%, 95%, and 70%; two changes for 3 minutes each.

8 Rinse sections in distilled H_2O and perform antibody labeling as described below (see p. 209).

Notes

- Double-labeling can be performed on cells grown in BrdU. This is useful to assess the cell cycle dependency of different antigens (e.g., PCNA). Some modification of standard techniques (such as microwave treatment) may be required to optimize antigenicity (Connolly and Bogdanffy 1993).
- Frozen sections can be useful if sites of replication are to be labeled in conjunction with fixation-sensitive antigens.

- Samples of soft or dispersed tissues, needle aspirates, or blood can be spread on slides using standard techniques, air-dried, fixed with organic fixatives, rinsed in PBS, and stained using the standard procedure described below. Such preparation does not preserve tissue architecture well but can be useful in preliminary screens.
- If single-cell suspensions can be prepared, cells can be fixed directly and labeling indices can be assessed by flow cytometry (see, e.g., Dolbeare et al. 1983; Landberg and Roos 1991). This approach allows the labeling intensity of large populations to be determined but neglects morphological detail.

CELLS GROWN ON GLASS

1 Prepare 4% **formaldehyde** in PBS as described in step 1 above.

CAUTION: Formaldehyde (see Appendix 2 for Caution)

2 Rinse samples in PBS and then place coverslips or slides in 4% formaldehyde at room temperature for 10 minutes.

3 Wash coverslips or slides twice with PBS.

4 Permeabilize samples by incubating them in 0.2% Triton X-100 in PBS at room temperature for 5 minutes.

5 Gently wash the samples in PBS (three changes over 5 min) and perform antibody labeling as described below (see p. 209).

Notes

- Cells labeled in vitro can be fixed as described above, but they should first be washed to remove unincorporated precursors. If background due to unincorporated precursors is high, perform steps 4 and 5 before steps 2 and 3.
- If washing detaches the cells from the glass, try an organic fixative such as 90% ethanol in distilled H_2O for 10 minutes at room temperature, followed by steps 2–4 above, which gives reasonable morphological preservation. Alternatively, fixation in methanol or acetone may be acceptable.

ENCAPSULATED CELLS

1 Prepare 4% **formaldehyde** in PB as described above (see pp. 204–205 and 207).

CAUTION: Formaldehyde (see Appendix 2 for Caution)

2 After labeling samples in vitro, remove unincorporated precursors by washing beads three times for 5 minutes each in 10 volumes of ice-cold PB.

3 Permeabilize the samples by incubating them in 0.2% Triton X-100 in PB for 10 minutes at 0°C.

4 Wash the samples three times for 5 minutes each in 10 volumes of ice-cold PB.

5 Fix the samples in ice-cold 4% formaldehyde in PB for 10 minutes.

6 Wash the samples three times in 10 volumes of ice-cold PBS and perform antibody labeling as described below (see p. 209).

Antibody Labeling

After incorporating BrdU into DNA, the duplex must be denatured to allow antibodies access to the analog. (There is no need to denature DNA after incorporating the nonhalogenated precursors listed on pp. 204–205, e.g., biotin-16-dUTP, digoxygenin-11-dUTP.) Commercially available anti-BrdU antibodies (see Notes below) react very poorly with BrdUMP in native DNA, but denaturation increases antigenicity 10–20-fold. The two widely used denaturation procedures are described.

ACID DENATURATION

1 Rinse slides or coverslips in distilled H_2O.

2 Denature DNA in 2 N **HCl** at room temperature for 1 hour.

Caution: Acids and bases that are concentrated (see Appendix 2 for Caution)

3 Neutralize DNA in 0.1 M $Na_2B_4O_7$ in distilled H_2O.

4 Rinse samples two times in PBS.

5 Add primary antibody (e.g., mouse anti-BrdU, mouse anti-digoxigenin, goat anti-biotin) diluted 1/50–1/500 (determined empirically) in PBS containing 0.5% BSA (Fraction V; Sigma) and 0.1% Tween 20. Incubate in a humidified chamber at room temperature for 1–2 hours.

6 Wash the samples three times at room temperature for 10 minutes each in PBS/BSA/Tween.

7 Incubate cells in fluorochrome or enzyme-labeled secondary antibody (e.g., FITC goat anti-mouse, alkaline phosphatase goat anti-mouse) diluted 1/50–1/1000 in PBS containing 0.5% BSA and 0.1% Tween 20. (Alkaline phosphatase-labeled antibodies should be diluted in Tris-buffered saline.) Incubate in a humidified chamber at room temperature for 1–2 hours.

8 Wash the samples three times at room temperature for 15 minutes each in PBS (or Tris-saline).

9 Develop enzyme-coupled reagents using standard protocols.

10 Prepare samples for microscopy using an appropriate mounting medium (see Chapter 4).

Specimens are now ready for microscopic examination.

Notes

- DNA can also be denatured by incubating samples with 50% deionized formamide in 2x SSC at 65°C for 10 minutes.
- If problems are experienced with nonspecific antibody binding, samples should be incubated prior to primary antibody labeling, with 5% normal serum (diluted in PBS) from the animal species in which the secondary antibody was generated.

- Commercial anti-BrdU antibodies are usually mouse monoclonals. Sera-Lab supplies a rat monoclonal that is useful for double labeling.
- Enzyme-dependent detection systems are generally sensitive but give poor resolution (e.g., the different and characteristic S-phase replication patterns are usually obscured). Similar sensitivity, but with higher resolution, can be achieved using 1-nm gold-conjugated secondary antibody and silver enhancement (Meyer et al. 1989) or postembedding immunogold labeling, which requires no DNA denaturation step (O'Keefe et al. 1992).
- Fluorochrome-coupled secondary antibodies provide high-resolution detection and can be analyzed using confocal laser scanning microscopes or sensitive, cooled charge-coupled-device cameras. Fluorescence-based detection systems allow convenient multiple labeling (see, e.g., Manders et al. 1992; O'Keefe et al. 1992; Hassan et al. 1994).

NUCLEASE-DEPENDENT DENATURATION

Acid denaturation is accompanied by some loss of morphology. If morphological considerations are critical, nuclease-dependent detection systems are preferred.

1 Rinse slides or coverslips in PBS.

2 Add primary antibody (e.g., mouse anti-BrdU, mouse anti-digoxigenin, goat anti-biotin) diluted 1/50-1/500 (determine empirically) in PBS containing 0.5% BSA (Fraction V; Sigma), 0.1% Tween 20, 50 μg/ml DNase I (Sigma), and 2.5 mM $MgCl_2$. Incubate in a humidified chamber at 37°C for 1–2 hours.

3 Perform steps 6–10 of the acid denaturation procedure above.

▪ *Note*

Nuclease-dependent detection systems improve morphology at the expense of sensitivity. Sensitivity can be increased using a biotinylated secondary antibody followed by streptavidin coupled to alkaline phosphatase or a fluorochrome.

ENCAPSULATED CELLS

1 Wash beads containing fixed cells two times in ice-cold PBS containing 0.5% BSA and 0.1% Tween 20.

2 Add primary antibody (e.g., mouse anti-digoxigenin, goat anti-biotin), mix 100 μl of beads with 400 μl of PBS/BSA/Tween containing 1/400 dilution of appropriate primary antibody, and incubate at 0°C for 2 hours with periodic mixing.

3 Wash three times for 10 minutes each in 10 volumes of ice-cold PBS/BSA/Tween.

4 Add secondary antibody, mix the bead pellet with 400 μl of PBS/BSA/Tween containing 1/400 dilution of appropriate fluorochrome-coupled secondary antibody (e.g., FITC-donkey anti-goat IgG, Texas Red-donkey anti-mouse IgG), and incubate at 0°C for 2 hours with periodic mixing.

5 Wash three times for 10 minutes each in 10 volumes of ice-cold PBS/BSA/Tween.

6 Wash three times at room temperature for 5 minutes each in 10 volumes of PBS. Add 0.02 μg/ml **DAPI** (4′,6-diamidino-2-phenylindole) to the second wash.

CAUTION: DAPI (see Appendix 2 for Caution)

7 Mount by mixing 5 μl of beads with an equal volume of mounting medium (Vectashield, Vector Laboratories), apply coverslip with gentle pressure to eliminate excess fluid, and seal with nail polish.

Specimens are now ready for microscopic examination.

REFERENCES

Connolly K.M. and Bogdanffy M.S. 1993. Evaluation of proliferating cell nuclear antigen (PCNA) as an endogenous marker of cell proliferation in rat liver: A dual-stain comparison with 5-bromo-2′-deoxyuridine. *J. Histochem. Cytochem.* **41:** 1–6.

Dolbeare F., Gratzner H., Pallavicini M.G., and Gray J.W. 1983. Flow cytometric measurement of total DNA content and incorporated bromodeoxyuridine. *Proc. Natl. Acad. Sci.* **80:** 5573–5577.

Gratzner H.G. 1982. Monoclonal antibody to 5-bromo and 5-iododeoxyuridine: A new reagent for detection of DNA replication. *Science* **218:** 474–475.

Hassan A.B. and Cook P.R. 1993. Visualization of replication sites in unfixed human cells. *J. Cell Sci.* **105:** 541–550.

Hassan A.B., Errington R.J., White N.S., Jackson D.A., and Cook P.R. 1994. Replication and transcription sites are colocalized in human cells. *J. Cell Sci.* **107:** 425–434.

Hoshino T., T. Nagashima, K.G. Cho, J.A. Murovic, J.E. Hodes, C.B. Wilson, M.S.B. Edwards, and L.H. Pitts. 1986. S-phase fraction of human brain tumors in situ measured by uptake of bromodeoxyuridine. *Int. J. Cancer* **38:** 369–374.

Hozák P., Hassan A.B., Jackson D.A., and Cook P.R. 1993. Visualization of replication factories attached to a nucleoskeleton. *Cell* **73:** 361–373.

Huijzer J.C. and Smerdon M.J. 1992. Characterization of biotinylated repair regions in reversibly permeabilized human fibroblasts. *Biochemistry* **3:** 5077–5084.

Humbert C. and Usson Y. 1992. Eukaryotic DNA replication is a topographically ordered process. *Cytometry* **13:** 603–614.

Jackson D.A., Balajee A.S., Mullenders L., and Cook P.R. 1994. Sites in human nuclei where DNA damaged by ultraviolet light is repaired: Visualization and localization relative to the nucleoskeleton. *J. Cell Sci.* **107:** 1745–1752.

Landberg G. and Roos G. 1991. Antibodies to proliferating cell nuclear antigen (PCNA) as S-phase specific probes in flow cytometric cell cycle analysis. *Cancer Res.* **51:** 4570–4575.

Manders E.M.M., Stap J., Brakenhoff G.J., van Dreil R., and Aten J.A. 1992. Dynamics of three-dimensional replication patterns during the S-phase, analysed by double labeling of DNA and confocal microscopy. *J. Cell Sci.* **103:** 857–862.

Meyer J.S., Nauert J., Koehm S., and Hughes J. 1989. Cell kinetics of human tumors by in vitro bromodeoxyuridine labeling. *J. Histochem. Cytochem.* **37:** 1449–1454.

Nakamura H., Morita T., and Sato C. 1986. Structural organisation of replicon domains during DNA synthesis phase in the mammalian nucleus. *Exp. Cell Res.* **165:** 291–297.

Nakayasu H. and Berezney R. 1989. Mapping replication sites in the eukaryotic cell nucleus. *J. Cell Biol.* **108:** 1–11.

O'Keefe R.T., Henderson S.C., and Spector D.L. 1992. Dynamic organization of DNA replication in mammalian cell nuclei: Spatially and temporally defined replication of chromosome-specific α-satellite sequences. *J. Cell Biol.* **116:** 1095–1110.

Repka A.M. and Adler R. 1992. Accurate determination of the time of cell birth using a sequential labeling technique with [^{3}H]-thymidine and bromodeoxyuridine ("window labeling"). *J. Histochem. Cytochem.* **40:** 947–953.

Yanik G., Yousuf N., Miller M.A., Swerdlow S.H., Lampkin B., and Raza A. 1992. In vivo determination of cell cycle kinetics of non-Hodgkin's lymphomas using iododeoxyuridine and bromodeoxyuridine. *J. Histochem. Cytochem.* **40:** 723–728.

CHAPTER 14

Analyzing RNA Synthesis: Nonisotopic Labeling

Dean Jackson[1] and Peter R. Cook[2]

[1]*University of Manchester, Manchester, United Kingdom*
[2]*University of Oxford, Oxford, United Kingdom*

INTRODUCTION

Although autoradiography can be used to detect sites of transcription after incubating cells with radiolabeled RNA precursors (e.g., [^{3}H] uridine), this approach has some disadvantages.

- It is rather specialized and technically demanding.
- Detection of transcription sites requires very short pulses because the transcription rate is so rapid (transcripts are extended by 1000 nucleotides/minute in vivo). Because pools of unlabeled triphosphates are relatively high (e.g., many cells contain ~0.5 μM UTP [uridine triphosphate]), little radiolabel is incorporated during short pulses, necessitating lengthy autoradiographic exposures.
- The path length of the particles emitted by ^{3}H is so long that autoradiographic grains can lie hundreds of nanometers away from the incorporation site.
- Labeling additional markers is technically difficult.
- The resultant grains sit on top of the cell and provide only a two-dimensional localization of the sites of incorporation.

Therefore, the ability to label sites of transcription in permeabilized cells using nonisotopic RNA precursors simplifies the analysis of transcription sites and allows simultaneous three-dimensional immunodetection of proteins or sites where DNA is replicated or repaired. Sites of nascent transcription in permeabilized HeLa cells are shown in Figure 14.1.

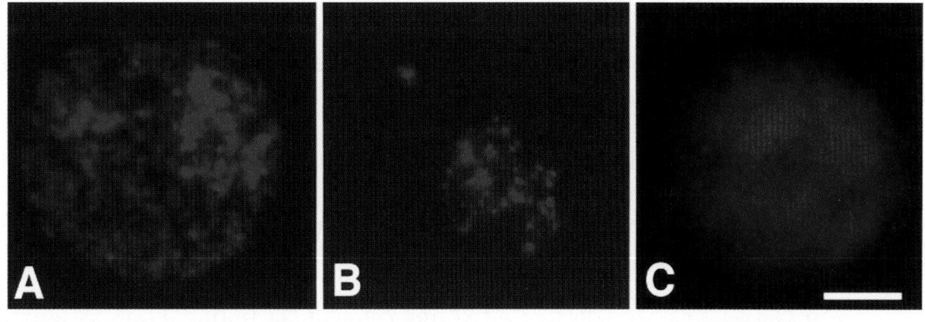

FIGURE 14.1

Sites of nascent transcription observed by fluorescence microscopy. Encapsulated HeLa cells were permeabilized with saponin and incubated with BrUTP for 10 min. Cells were fixed and stained using mouse anti-BrdU (5-bromodeoxyuridine) (Boehringer Mannheim) and a Texas Red-conjugated secondary antibody. (*A*) Transcription pattern showing strong nucleolar and weaker nucleoplasmic incorporation. (*B*) Inhibition of RNA polymerase II transcription with α-amanitin eliminates much of the nucleoplasmic signal. (*C*) DAPI (4′,6-diamidino-2-phenylindole) staining of cell in *B*. Bar, 2.5 μm. (Photo provided by P.R. Cook, University of Oxford.)

Preparation of Cells

IMMOBILIZING CELLS

Procedures for labeling transcription sites require various manipulations and are applied with difficulty to cells free in suspension. These approaches can be more successfully applied to immobilized cells; this section describes the preparation of cells attached to coverslips or trapped in agarose microbeads.

Preparing Cells Grown on Coverslips

1 Prepare acid-washed coverslips as described in Chapter 4.
For most lenses, coverslips having a thickness of 0.16–0.19 mm (#1–1/2) are best.

2 Remove an acid-washed coverslip from the ethanol storage solution, flame it to burn off the ethanol, and place it in a 35-mm diameter petri dish.

3 Seed cells (~20% confluency) in the dish and allow to adhere 1–2 days.

Notes

- Best results are obtained with well-spread cells, covering 30–50% of the coverslip during labeling.
- Manipulation detaches some permeabilized cells; attachment can be improved by coating coverslips with gelatin or poly-L-lysine (0.5 mg/ml) prior to plating.

Preparing Cells Encapsulated in Agarose Microbeads

Encapsulation in agarose microbeads (Jackson and Cook 1985) provides a convenient way of protecting cells from damage during manipulation. Encapsulated cells can be permeabilized using a wide range of treatments, including buffers that cause nonencapsulated cells to aggregate into an unworkable mess. Both adherent and nonadherent cells can be permeabilized and labeled after encapsulation in agarose microbeads.

1 Warm 10 ml of liquid paraffin (BDH 29436; Merck 7162) to 37°C.

2 Heat 0.25 g of low-gelling agarose (e.g., Sigma type VII) in 10 ml of PBS at 95°C until dissolved; then cool to 37°C.

3 Resuspend 1×10^7 cells in 4 ml of PBS (phosphate-buffered saline) at 37°C in a 100-ml round-bottomed flask.

4 Add 1 ml of agarose solution at 37°C to 4 ml of cell suspension at 37°C and mix thoroughly.

5 Add 10 ml of paraffin at 37°C, seal flask with plastic film, and immediately shake (by hand or at 800 cycles/min using a flask shaker) until a creamy emulsion forms (about 15 sec).

6 Cool flask by periodic rotation in ice-cold H_2O for 10 minutes; this allows spherical droplets of molten agarose suspended in the paraffin to gel.

7 Add 35 ml of ice-cold PBS, mix, and transfer to a 50-ml plastic centrifuge tube.

8 Pellet microbeads by spinning at 1000 rpm on a benchtop centrifuge at 20°C for 5 minutes.

9 Aspirate the supernatant and wash pelleted microbeads once in PBS. If some beads remain at the H_2O/paraffin interface, remove most paraffin, mix thoroughly, and respin.

10 Encapsulated cells can now be regrown in medium or permeabilized directly.

Notes

- A cell density of 2×10^6/ml is convenient for most labeling procedures, but densities up to 10^8/ml can be used.
- Small volumes of microbeads can be prepared by homogenizing 50–500 μl of cell/agarose mixture with 1 ml of paraffin in a 50-ml round-bottomed tube.
- Adequate emulsions can be prepared by shaking manually (shake as fast as possible for 10–15 sec). Microbead quality can be assessed microscopically; acceptable microbeads should be spherical, relatively uniform in size (r = 25–75 μm) with evenly dispersed cells.
- Microbead preparations should pass freely through tips used with automatic pipettes (10–200 μl). If not, large beads can be removed by filtration through monofilament nylon filters (R. Cadisch and Sons) using a Swinex filter (Millipore).
- To prevent losses during aspiration, samples should be transferred to 10-ml plastic centrifuge tubes at the earliest convenient point.
- Cells can be grown following encapsulation; simply resuspend beads in medium and incubate at 37°C. Encapsulated HeLa cells grow with normal cell-cycle kinetics for at least one generation.

CELL PERMEABILIZATION

After permeabilization, precursor pools of NTPs can be depleted by washing, and nucleotide levels can be adjusted to give the required rate of elongation. In the absence of the natural triphosphate, any modified triphosphate that is recognized as a substrate by the endogenous RNA polymerase is then incorporated relatively efficiently.

Permeabilization with Detergents

Permeabilization can be achieved by treatment with any of a variety of detergents. It is critical to control carefully both the concentration of detergent used and the timing of the treatment to permeabilize about 95% of the cells.

1 Prepare H_2O treated with **DEPC** (diethyl pyrocarbonate). Add 0.5 ml of DEPC to 500 ml of distilled H_2O, mix, let stand at 37°C overnight, and autoclave.

CAUTION: DEPC (see Appendix 2 for Caution)

Alternatively, molecular-biology-grade H_2O can be obtained commercially (e.g., 6211-440, VWR).

2 Prepare fresh PB from stock solutions prepared in sterile purified H_2O and cool on ice.

Chromatin structure is especially sensitive to changes in its ionic environment, and choice of the buffer to use with permeabilized cells is particularly important. Many buffers are in common use, but the following "physiological" buffer preserves considerable nuclear structure and function.

Physiological Buffer (PB)

100 mM KCH_3COOH
30 mM **KCl**
10 mM Na_2PO_4
1 mM **$MgCl_2$**
1 mM Na_2ATP (adenosine triphosphate)
1 mM **DTT** (dithiothreitol)
100 mM KH_2PO_4 is added, if required, to give pH 7.4.
Protease and nuclease inhibitors are added as required.

CAUTION: KCl; $MgCl_2$; DTT (See Appendix 2 for Caution)

3 Wash cells on coverslips, or cells encapsulated in agarose, twice with PBS.

Coverslips: Place in convenient container (e.g., 24-well flat-bottomed tissue culture plate; Falcon 3047), add 2 ml of PBS to each well, wait for 2 minutes, and then aspirate; repeat once.

Microbeads: Transfer 1 ml of cells in agarose beads and 9 ml of PBS into a 10-ml plastic centrifuge tube, centrifuge at 1000 rpm in a benchtop centrifuge for 30 seconds to pellet, aspirate supernatant, replace with appropriate buffer, and invert tube to mix.

4 Wash cells twice with ice-cold PB.

5 Transfer to ice-cold PB containing 0.01–0.1% detergent for 1–2 minutes (coverslips) or 2–3 minutes (microbeads). To define conditions, use a twofold dilution series of detergent in PB.

The following (and related) detergents can be used; guideline concentrations are indicated:

0.02–0.05% Triton X-100 (Pierce 28314)
0.01–0.02% digitonin (Sigma D 1407)
0.02–0.05% lysolecithin (Sigma L 4129)
0.01–0.02% saponin (Sigma S 7900)

6 Assess the level of permeabilization using trypan blue exclusion (the extent of lysis is critical).

a Add 50 μl of 1% trypan blue in PB to cells on a coverslip or to 50 μl of packed microbeads; wait 2 minutes.

b Inspect by light microscopy; score the percentage of permeabilized (dark-blue) cells.

c Choose the detergent concentration that permeabilizes >95% cells. If cells detach from coverslips during washing, use a lower concentration of detergent.

7 Wash detergent-treated cells three times with ice-cold PB. Cells are now ready for use in the transcription reaction.

Notes

- It is sometimes useful to mark the backs of coverslips to ensure the correct orientation.
- If the buffer composition is to be varied, the following factors should be considered:
  (1) Various combinations of monovalent anion can be used. Different chloride/acetate/glutamate (or polyglutamate) combinations support almost identical rates of transcription or replication. However, acetate/glutamate is preferred to the smaller Cl^-, which is more damaging to tertiary protein structure.
  (2) The concentration of divalent cation must be carefully controlled. As little as 0.5 mM free Mg^{++} causes the visible (by electron microscopy [EM]) collapse or aggregation of chromatin. The equimolar Mg/ATP combination used here preserves chromatin structure and supports the action of Mg-dependent enzymes.
  (3) DTT, protease inhibitors, and ribonuclease inhibitors protect the sample and preserve cell morphology.

Permeabilization with Proteins

Streptolysin O and α-toxin assemble multimeric pore complexes in cell membranes, thereby permeabilizing them. Streptolysin generates 15-nm pores in cholesterol-containing membranes, sufficient to allow passage of large proteins like antibodies. Because excess protein can be removed prior to lysis, internal membranes should remain intact. α-toxin pores are only 2 nm, allowing passage of small molecules, such as nucleic acid precursors, but few proteins.

1 Prepare PB, as described in steps 1–2 above.

2 Wash cells on coverslips or cells encapsulated in agarose with ice-cold PBS three times.

3 Lyse with streptolysin O or α-toxin:

Lyse with streptolysin O

a Prepare streptolysin O (Gibco BRL 3493SA; Sigma S 5265) in PBS as instructed by supplier. Cool on ice.

b Bind streptolysin O to cells on coverslips or in agarose beads.

Coverslips: Immerse in streptolysin O/PBS for 30 minutes on ice.

Microbeads: Mix 1 ml of beads with 9 ml of streptolysin O/PBS and incubate for 30 minutes on ice. Mix periodically.

c Wash cells once in ice-cold PBS to remove unbound reagent.

d Wash cells once in ice-cold PB.

e Transfer cells to PB at 33°C for 3 minutes to lyse membranes.

f Wash cells once in ice-cold PB.

Lyse with α-toxin

a Dissolve α-toxin (Gibco BRL 3463SA) as instructed by supplier.

b Add 5–10 volumes of PB and warm to 33°C.

c Treat cells with α-toxin.

Coverslips: Immerse cells on coverslips in α-toxin/PB and incubate at 33°C for 10–30 minutes.

Microbeads: Mix 1 ml of cells in agarose beads with 9 ml of α-toxin/PB and incubate at 33°C for 10–30 minutes. Mix periodically.

d Wash cells twice in ice-cold PB.

Cells are now ready for use in the transcription reaction.

Notes

- Because cholesterol inhibits permeabilization by streptolysin O, it is important that samples contain no serum (which usually contains cholesterol) during binding.
- Concentrations of streptolysin O and α-toxin should be titrated to optimize permeabilization (monitored by trypan blue uptake, as described in preceding protocol).
- Permeabilization by α-toxin is relatively slow. Shorter lysis times are recommended and incubation longer than 30 minutes should be avoided.

Labeling and Processing

TRANSCRIPTION IN VITRO

Cells permeabilized in PB and incubated at 33°C resume RNA synthesis once the necessary precursors are added. Unfortunately, the rate of RNA synthesis in vivo is not known, so the relative efficiency in vitro cannot be established. In the presence of optimal concentrations of triphosphates, however, permeabilized cells do synthesize DNA at the in vivo rate, and, because (under optimal conditions) they synthesize tenfold more RNA than DNA, it is probable that most RNA polymerase complexes engaged prior to lysis survive permeabilization. Under these conditions, there is little or no initiation.

1 Prepare fresh PB supplemented with 0.5 mM **PMSF** (phenylmethylsulfonyl fluoride).

CAUTION: PMSF (see Appendix 2 for Caution)

2 Prepare 10x IM (initiation mix).

10x IM

PB/PMSF
500 µM CTP (cytosine triphosphate)
500 µM GTP (guanine triphosphate)
10–500 µM modified UTP
$MgCl_2$
(molarity equal to molarity of added triphosphates)
Warm to 33°C.

3 Incubate permeabilized cells at 33°C for 5 minutes.

Coverslips: Place in wells of 24-well tissue culture plate with 0.45 ml of PB.

Microbeads: Mix 200 µl of beads with 250 µl of PB in a 1.5-ml microfuge tube.

4 Add 50 µl of prewarmed 10x IM and mix.

5 Incubate cells at 33°C for 1–30 minutes, as required.

6 Terminate transcription by washing the reaction mixture five times with ice-cold PB over a 30-minute period.

Samples can now be fixed and stained.

Notes

- A range of precursors is available, but of those tested, only one, BrUTP (Sigma B 7166), is used efficiently by the endogenous RNA polymerase complex (Jackson et al. 1993; Wansink et al. 1993). Biotin-14-CTP is also incorporated into RNA, but less efficiently (Iborra et al. 1996).

Biotin-11-UTP, digoxigenin-11-UTP, fluorescein-12-UTP, coumarin-5-UTP, and lissamine-5-UTP are not incorporated (or incorporated very poorly) by permeabilized HeLa cells.

- Great care must be taken to ensure that only transcription sites are labeled. The concentration of modified precursor can be adjusted to control the rate of elongation. At 33°C in PB, 50 μM BrUTP supports an elongation rate of 50 nucleotides/minute, and then a 10-minute incubation probably ensures that >90% of the incorporated label remains at the site of polymerization (Jackson et al. 1993). If higher concentrations or longer times are used, a greater proportion of labeled RNA might move away from transcription sites.
- The elongation rate can also be reduced using a lower temperature.
- If inhibitors are to be used, incubate them for 15 minutes at 0°C and 5 minutes at 33°C prior to addition of 10x IM.

FIXATION AND ANTIBODY BINDING

Paraformaldehyde-based fixatives preserve nuclear structure well and are preferred for light microscopy, whereas glutaraldehyde-based fixatives provide more stable cross-linking and are therefore used for electron microscopy.

1 Prepare PB+ buffer by adding 0.5 mM **PMSF** and 2.5 units/ml HPRI (human placental ribonuclease inhibitor) (Amersham International E 2310Y) to PB (see pp. 215–216).

   CAUTION: PMSF (see Appendix 2 for Caution)

2 Prepare 4% **formaldehyde.**

   CAUTION: Formaldehyde (see Appendix 2 for Caution)

   a Dissolve 4 g of paraformaldehyde in 50 ml of distilled H_2O in a 50-ml tube.

   b Heat at 60°C until dissolved; add a few drops of 2 M NaOH to help dissolve, if necessary.

   c Cool to room temperature and add an equal volume of 2x concentrated PB+.

   d Filter through nitrocellulose (0.22-μm pore size) and cool on ice.

3 To washed samples of cells from the preceding transcription protocol, add 1 ml of ice-cold 4% formaldehyde in PB+ to coverslips in 24-well tissue-culture plates or 200 μl microbeads in a 1.5-ml microfuge tube. Incubate on ice for 15 minutes.

4 Wash fixed cells twice in ice-cold PB+.

5 Replace PB+ with ice-cold PB+ containing 0.25% Triton X-100 and incubate on ice for 10 minutes.

6 Repeat step 3.

7 Add primary antibody to fixed cells.

   Coverslips: Cover samples with a few drops of PB+ containing 0.5% acetylated BSA (bovine serum albumin) (Sigma B 2518), 0.05% Tween 20, and 1/50 to 1/250 dilution anti-BrdU antibody, and incubate 1–2 hours on ice.

Microbeads: Mix 50–100 µl of beads with an equal volume of PB+/acetylated BSA/Tween containing anti-BrdU antibody and incubate 1–2 hours on ice; mix periodically.

Many antibodies to BrdU are commercially available. Cross-reactivities with BrUMP incorporated into RNA are indicated below (reactivity: poor [–] to good [+++]):

Boehringer Mannheim (BMC 9318; 1170 376)+++
Sera-Lab (clone BU1/75 MAS 250p)++
Becton Dickinson (clone B44; No. 7580)+/++
Sigma (clone BU-33; B 2531)+
Caltag (clone BR-3)+/–
Dako (clone Bu20a; M744)–

These antibodies are all mouse monoclonals (IgG1) except for the product from Sera-Lab, which is a rat monoclonal. Products from other sources should be tested for binding.

8 Wash cells three times in ice-cold PB+/BSA/Tween over a 15-minute period.

9 Add secondary antibody.

Coverslips: Cover samples with a few drops of PB+/BSA/Tween and 1/500 dilution of fluorochrome-labeled second antibody (e.g., Texas Red-donkey anti-mouse; Jackson Laboratories 715-075-137). Incubate 1–2 hours on ice.

Microbeads: Mix 50–100 µl cells in agarose beads with 400 µl of PB+/BSA/Tween and 1/500 dilution of second antibody. Incubate 1–2 hours on ice.

10 Wash samples three times in ice-cold PB+/BSA/Tween, over a 15–30 minute period.

11 Wash samples three times in ice-cold PB+, for 5 minutes each time. To the second wash add 0.02 µg/ml **DAPI** (Boehringer Mannheim 236 276).

CAUTION: DAPI (see Appendix 2 for Caution)

12 Mount samples and seal with nail polish.

For recipe for mounting media, see Chapter 4.

Specimens are now ready for observation.

Notes

- It is important to protect the labeled RNA from nuclease digestion. Always use DEPC-treated or molecular biology grade H_2O to prepare buffers. A ribonuclease inhibitor should also be added following transcription; adding HPRI (or equivalent) at 2.5 units/ml should be sufficient during most steps, but this can be increased to 25 units/ml during antibody binding.
- BSA *must* be essentially nuclease-free. The use of acetylated BSA in buffers is recommended; acetylation inactivates nucleases.
- Commercial reagents with cross-reactivity designated +++ or ++ should be used for routine staining. Antibodies with weaker binding will give signals close to the detection limits of unenhanced light microscopy. Weak cross-reactivities are sensitive to DTT (required by the ribonuclease inhibitor). Staining under nonreducing conditions can be beneficial (the antibody from Sera-Lab, for example, can give much improved staining in the absence of DTT).
- If antibodies with moderate cross-reactivity are used, it may be necessary to increase sensitivity using a biotin-conjugated secondary antibody followed by streptavidin coupled to a fluo-

rochrome (Wansink et al. 1993). Texas Red-streptavidin usually gives less background staining than FITC (fluorescein isothiocyanate)-streptavidin.

- Standard techniques can be used to visualize transcription sites by EM (Hozak et al. 1994; Iborra et al. 1996).

LABELING IN VIVO

The distribution of transcription sites labeled in vitro can be confirmed if cells are first labeled in vivo. This can be achieved simply by growing cells in medium supplemented with bromouridine: 50 μM for 10 minutes for EM analysis (Hozak et al. 1994) or 100 μM for 15 minutes for flow cytometry (Jensen et al. 1993). Because bromouridine added directly to cells will be incorporated into DNA as well as RNA, short labeling periods, preferably outside S-phase, should be used. However, sites of transcription can be labeled during S-phase if the DNA is not denatured.

REFERENCES

Hozak P., Cook P.R., Schofer C., Mosgoller W., and Wachtler F. 1994. Site of transcription of ribosomal RNA and intranucleolar structure in HeLa cells. *J. Cell Sci.* **107:** 639–648.

Iborra F.J., Jackson D.A., and Cook P.R. 1996. Active RNA polymerases are localized within discrete transcription "factories" in human nuclei. *J. Cell Sci.* **109:** 1427–1436.

Jackson D.A. and Cook P.R. 1985. A general method for preparing chromatin containing intact DNA. *EMBO J.* **4:** 913–918.

Jackson D.A., Hassan A.B., Errington R.J., and Cook P.R. 1993. Visualization of focal sites of transcription within human nuclei. *EMBO J.* **12:** 1059–1065.

Jensen P.O., Larsen J., Christiansen J., and Larsen J.K. 1993. Flow cytometric measurement of RNA synthesis using bromouridine labelling and bromodeoxyuridine antibodies. *Cytometry* **14:** 455–458.

Wansink D.G., Schul W., van der Kraan I., van Steensel B., van Driel R., and de Jong L. 1993. Fluorescent labeling of nascent RNA reveals transcription by RNA polymerase II in domains scattered throughout the nucleus. *J. Cell Biol.* **122:** 283–293.

CHAPTER 15

Fluorescence In Situ Hybridization to DNA

Peter Lichter,[1] Stefan Lampen,[2] and Johanna Bridger[2]

[1]*DKFZ-Deutsches Krebsforschungszentrum Organisation Komplexer Genome, Heidelberg, Germany*
[2]*German Cancer Research Center, Heidelberg, Germany*

INTRODUCTION

In situ hybridization (ISH) is a technique used to visualize defined nucleic acid sequences in cellular preparations by hybridization of complementary probe sequences. Thus, ISH provides a unique link among cell biology, cytogenetics, and molecular genetics, taking advantage of the increasing availability of DNA probes that can be used to visualize target regions of interest at the single cell level. It is presently used by many laboratories for diagnosis of infectious diseases and cytogenetic analyses, as well as for basic cell biology research to study structural and dynamic properties of tissues, cells, and subcellular entities (Lichter et al. 1991; Trask 1991; Joos et al. 1994).

The most popular ISH protocol utilizes fluorescence detection, and this chapter presents procedures currently used in fluorescence in situ hybridization (FISH). A flowchart of the sequence of these techniques is presented in Figure 15.1. Nonisotopic labels include enzymes, haptens such as biotin or digoxigenin, and fluorochromes. Commonly, FISH is performed with biotinylated or dioxigenin-labeled probes detected via a fluorochrome-conjugated detection reagent, such as an antibody, or other hapten-specific reactive compounds, such as avidin or streptavidin when biotin is incorporated. Nucleic acid probes directly labeled with fluorochromes are utilized increasingly for the detection of large target sequences. In these cases, the ISH technique requires less time, and background staining is reduced considerably, but at the cost of signal intensity. When performing FISH with certain reporter-labeled probes, a number of signal amplification steps can be employed whereby layers of detection reagents are built up, resulting in an exponential signal amplification. Using these procedures, it is possible to obtain high sensitivity, with detection of single-copy sequences on chromosomes reported with probes <0.8 kb (see, e.g., Korenberg et al. 1992).

ISH has many applications, from basic gene mapping and diagnosis of chromosomal aberrations (Lichter and Ward 1990; Bentz et al. 1993b) to detailed studies on cellular structure and function, such as the painting of chromosomes in three-dimensionally preserved nuclei (Zirbel et al. 1993; Kurz et al. 1996). Table 15.1 provides an overview of ISH applications.

Probes for In Situ Hybridization

A large variety of nucleic acid probes are used for FISH. The length of individual probes can vary from ~12 base pairs (bp) for synthetic oligonucleotides to over 1 Mbp for yeast artificial chromosomes (YACs). In general, probes can be defined by the complexity of their target sequences (see Table 15.2 and Fig. 15.2).

Probes containing repeat elements are often used to detect clustered repetitive DNA in heterochromatin blocks and centromeric regions of individual chromosomes that consist of alphoid and/or satellite sequences. They are particularly useful in the determination of

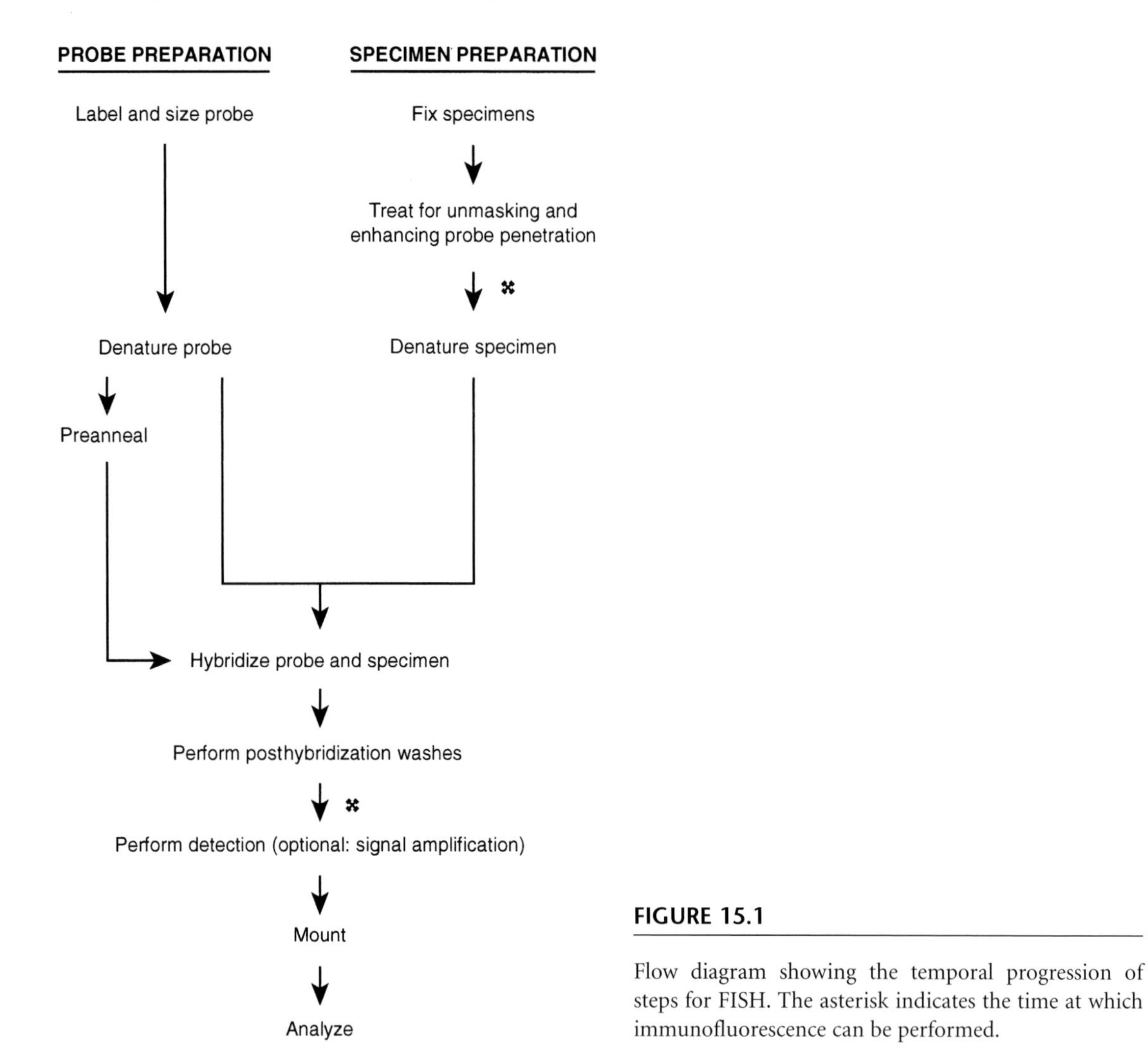

FIGURE 15.1

Flow diagram showing the temporal progression of steps for FISH. The asterisk indicates the time at which immunofluorescence can be performed.

TABLE 15.1 Different applications of nonisotopic in situ hybridization

	Nonisotopic in situ hybridization applications
Clinical	prenatal diagnosis, for inherited chromosome abberations postnatal diagnosis of carriers of genetic disease infectious disease diagnosis, viral and bacterial tumor cytogenetic diagnosis biological dosimetry detection of aberrant gene expression
Research	mapping of chromosomal genes (Human Genome Project) evolution of genomes (Zoo FISH) nuclear organization, visualization chromosomal territories, and chromatin in interphase cell analysis of dynamic nuclear processes somatic hybrid cell analysis developmental biology, e.g., temporal expression of genes during differentiation and development temporal analysis of replication fluorescence cell and chromosome sorting microbiological studies tumor biology

TABLE 15.2 Different types of probes for in situ hybridization

Probe detecting	Probe type
Clusters of repetitive DNA	PCR amplification of repeat sequences of total DNA oligonucleotides specific for the repeat elements cloned repeat element
Single loci	cDNA cloned in plasmid or phage vectors genomic DNA cloned in plasmid, phage, cosmid, P1, BAC, YAC vectors IRS-PCR-amplified sequences from a YAC-containing yeast clone universal PCR from purified YAC in vitro transcription of cloned DNA
Chromosomes or chromosomal regions	chromosome-specific DNA libraries: + from sorted chromosomes + cloned from microdissected chromosomes universal PCR from: + sorted chromosomes + microdissected chromosomes whole genomic DNA from somatic hybrid cell lines IRS-PCR products from somatic cell hybrid DNA

PCR, polymerase chain reaction; BAC, bacterial artificial chromosome; YAC, yeast artificial chromomosome; IRS, interspersed repetitive sequence.

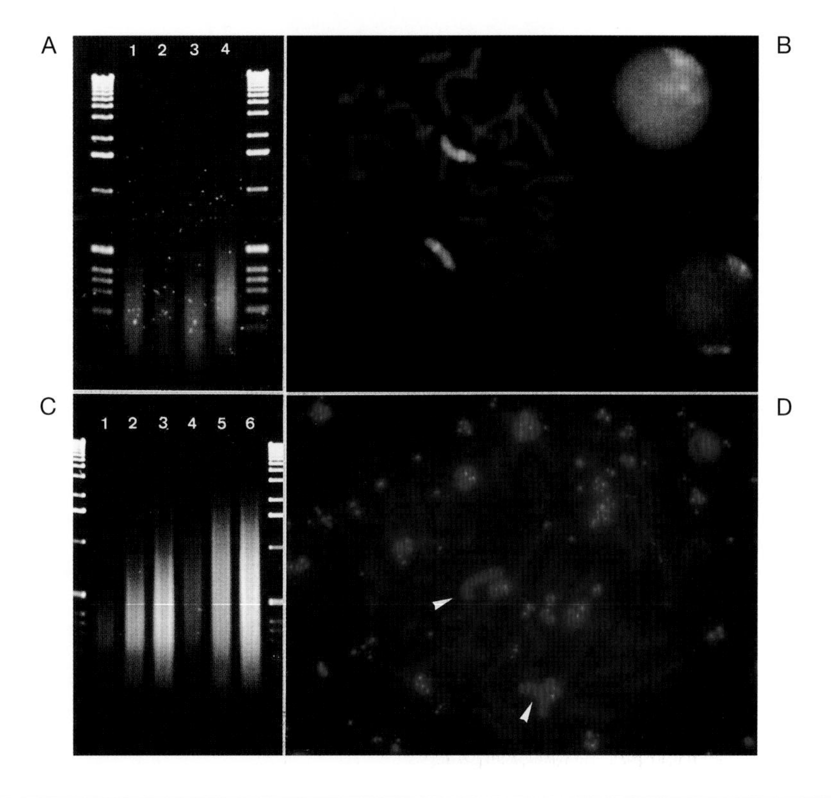

FIGURE 15.2

Optimal DNA probe sizes for chromosome painting. Panels *A* and *C* display the size range of probe molecules after labeling and digestion by nick translation. (*A*) Comparison of smears of probe DNA fragments with size markers shows probe fragments to be 200–500 nucleotides (nt) in length, ideally suited for FISH. Using such a size of labeled probe (pBS library for chromosome 4) results in good quality chromosome painting (*B*). Note the delineation of individual chromosomes in metaphase and interphase cells. (*C*) An agarose gel of a pBS library for chromosome 4, wherein the probe DNA has been digested with DNase I for varying lengths of time (the longer the incubation the shorter the fragments). When FISH was performed with a probe (*C*, lane 6) that was too long, painting is weak (*D, arrowheads*), but most of all, there are a large number of bright spots that are star-like in appearance (see troubleshooting; Table 15.5). (Photos provided by A. Rätsch and EMBO FISH course participants from 1994.)

numerical chromosome aberrations, for example, in clinical samples (see Fig. 15.3). The large size of the target yields a high detection efficiency and so allows rapid evaluation of nuclei. Chromosome-specific repetitive probes can be obtained by standard cloning strategies or by using selective oligonucleotides. Alternatively, primers flanking the alphoid sequences of individual human chromosomes can be used for amplifying repetitive sequences from whole human genomic DNA (Dunham et al. 1992).

Single-copy DNA probes can be cloned into different vector systems (plasmids, phages, cosmids, P1, BACs, YACs) that allow a range of insert sizes (100 bp–1 Mbp). Large probes often contain interspersed repetitive sequences (IRSs) such as SINE (short interspersed nucleotide element) and LINE (long interspersed nucleotide element) sequences that can cause background staining because of the wide distribution of these sequences throughout the genome. This background staining can be avoided by a saturation of these sequences by hybridization with an excess of unlabeled repetitive DNA (Landegent et al. 1985; Lichter et al. 1988; Pinkel et al.

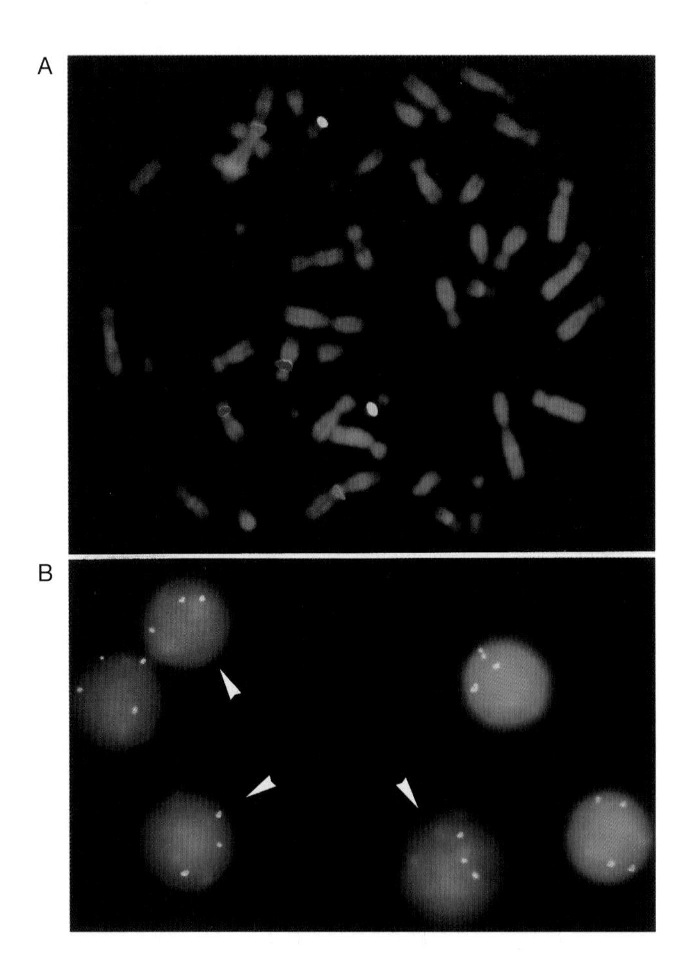

FIGURE 15.3

Multicolor FISH with methanol/acetic acid fixed metaphase chromosomes (*A*) and on interphase nuclei (*B*). (*A*) FISH with DNA probes targeting a chromosome 1q12, centromere of chromosome 12, and centromere of chromosome 17 detected with fluorescein (FITC [fluorescein isothiocyanate], *green*), rhodamine (TRITC [tetramethylrhodamine isothiocyanate], *red*), or a combination of both (*yellow*), respectively. (*B*) A clinical diagnostic application of dual-color FISH, with lymphocyte interphase nuclei from a patient with chronic lymphocytic leukemia. A control probe was visualized via TRITC and a probe binding to the retinoblastoma tumor suppressor gene 1 (RB-1) was visualized via FITC. In the nuclei with arrows, only one RB-1 signal is observed, indicating a deletion of RB-1 oncogene in these cells (see Stilgenbauer et al. 1993). Chromosomes and nuclei were counterstained with DAPI (4′,6-diamidino-2-phenylindole). (Photos provided by T. Fink and S. Stilgenbauer.)

1988). This procedure has also been termed chromosomal in situ suppression (CISS) (Lichter et al. 1988). Because the efficiency of detecting a hybridized probe correlates with the size of the target sequence, vector systems containing larger inserts generally yield stronger signals, thus facilitating the detection of single loci. Plasmid probes for hybridization to single gene loci were the first single-copy probes available for ISH. Plasmids containing inserts of >2–3 kbp are detected by standard FISH detection methods; smaller probes often require a series of amplification steps. Phages and cosmid probes with insert sizes of up to 20 kbp and 40 kbp, respectively, usually result in the detection of 60–95% of the target sequences.

When DNA probes from yeast containing a YAC are used, high background staining often results from the large amount of labeled, extraneous yeast DNA. Purification of these probes can be performed by isolating the YAC DNA from a pulsed-field gel, followed by amplification via universal polymerase chain reaction (PCR), resulting in lower background staining. Alternatively, species-specific amplification of insert sequences by IRS-PCR (Nelson et al. 1989; Ledbetter et al. 1990) results in a probe DNA that is highly enriched in a subset of insert sequences.

The first painting probes for delineation of individual chromosomes were available from the American Type Culture Collection (ATCC) (Lichter et al. 1988; Pinkel et al. 1988). These probes were generated from fragments of flow-sorted chromosomes cloned in phage vectors. These early painting probes had a low ratio of probe to vector sequences, which also resulted in background problems. Recloning these sequences into plasmids (Fuscoe et al. 1989) resulted in an improved ratio of probe DNA to vector sequences, and hence lower background. Painting probes, cloned in pBluescribe vectors, have been available for all human chromosomes since 1991 (Collins et al. 1991). Alternatively, painting probes can be generated by applying various PCR amplification procedures for a small number of flow-sorted chromosomes; for example, by PCR using degenerate oligonucleotide primers (DOP-PCR) (Carter et al. 1992; Telenius et al. 1992) or amplification of interspersed repetitive sequences (IRS-PCR) (Suijkerbuijk et al. 1992) or linker adapter PCR (Chang et al. 1992). Somatic hybrid cell lines containing only one human chromosome (Boyle et al. 1990; Kievits et al. 1990) can also serve as a source for painting probes. Similar to YAC probes, these hybrid cell DNAs also contain a high proportion of labeled sequences not contributing to the hybridization signal. By applying IRS-PCR, the species-specific DNA can be amplified (Lichter et al. 1990b; Lengauer et al. 1991a).

Painting probes for chromosomal subregions can be generated from microdissected chromosomes. Fragments obtained in this manner are either cloned directly or after linker adapter PCR (Lengauer et al. 1991a,b). Alternatively, an amplification by universal PCR can be performed (Bohlander et al. 1992; Melzer et al. 1992).

Modification of probe DNA by detectable components like haptens or fluorochromes using enzymatic incorporation of labeled nucleotides is generally more efficient than labeling with chemically reactive compounds such as photobiotin. Probes produced by PCR procedures can be labeled by incorporation of modified nucleotides during amplification of template DNA.

Oligonucleotide probes incorporating nucleotides conjugated to haptens or fluorochromes can be synthesized directly. Haptens widely used for indirect labeling are biotin, digoxigenin, and DNP (dinitrophenol), all of which yield comparable sensitivities. The sensitivity of a probe directly labeled with a fluorochrome is much lower but is sufficient for the detection of repetitive probes and probes with large targets. For the simultaneous detection of two or more probes, each probe must be labeled. Combinations of haptens or fluorochromes for a single probe increase the number of probes that can be visualized simultaneously (Dauwerse et al. 1992; Ried et al. 1992). Multicolor labeling can be expanded further by combining reporter molecules at different ratios (Wiegant et al. 1993).

Labeling of Probes

The following procedures describe the labeling of probes for ISH using either nick translation or various PCR techniques. A method for amplification using standard conditions with labeled nucleotides is followed by the "universal" PCR labeling techniques of sequence-independent amplification (SIA) of DNA (Bohlander et al. 1992) and DOP-PCR (Telenius et al. 1992).

NICK TRANSLATION

For most applications of FISH, DNA probes are labeled by nick translation (Langer et al. 1981). This procedure includes two different enzymatic reactions: (1) DNase I introduces nicks in the double-stranded DNA, which serves as a starting point for the second enzyme *Escherichia coli* polymerase I. (2) Using the second DNA strand as a template, this enzyme replaces all nucleotides with modified nucleotides, using its endonuclease and polymerase activities. Because the labeled probe can be stored for a long time at ≤ 20°C without affecting probe quality, a large amount of probe can be labeled in one reaction with a concentration of template DNA ≤ 20 ng/μl. A standard reaction mix contains 2 μg of probe DNA in 100 μl volume.

1 Prepare a reaction according to the following pipetting scheme:

DNA solution containing 2 μg of probe DNA	
10x NT buffer	10 μl
0.1 M **β-mercaptoethanol**	10 μl
10x nucleotide stock	10 μl
E. coli DNA polymerase I	20 units
DNase stock solution (see below for dilution)	
ddH_2O to a final volume of 100 μl	

CAUTION: β-mercaptoethanol (see Appendix 2 for Caution)

10x NT Buffer

0.5 M Tris-HCl, pH 8.0
50 mM **$MgCl_2$**
0.5 mg/ml BSA (bovine serum albumin)

CAUTION: $MgCl_2$ (see Appendix 2 for Caution)

10x Nucleotide Stock

0.5 mM dATP (deoxyadenosine triphosphate)
0.5 mM dGTP (deoxyguanine triphosphate)
0.5 mM dCTP (deoxycytosine triphosphate)
0.5 mM nucleotides containing dUTP (deoxyuridine triphosphate) conjugate of choice:
- 0.4 mM **biotin**-11-dUTP, 0.1 mM dTTP (deoxythymidine triphosphate)
- 0.125 mM **digoxigenin**-11-dUTP, 0.375 mM dTTP
- 0.4 mM **DNP**-11-UTP, 0.1 mM dTTP

CAUTION: Biotin; digoxigenin; DNP (see Appendix 2 for Caution)

DNase Stock Solution

0.15 M NaCl
1 mg/ml DNase I
50% glycerol
Store at –20°C. As a rule of thumb, the stock is diluted 1:1000 in ice-cold ddH$_2$O immediately before use, from which aliquots of 2–10 µl are added to the standard reaction.

The amount to be used from this DNase stock must be tested for each new batch by performing a series of nick-translation reactions without polymerase. Test the size of resulting fragments in this series by gel electrophoresis. For nick translations, the volume of DNase resulting in 100–500 nt fragment length is used for further labeling (see Fig. 15.2A,C).

2 Incubate the reaction for 2 hours at 15°C; place on ice.

The length of probe molecules is critical for probe diffusion and hybridization to the specific target sequence. The optimal size of probe molecules is between 100 and 500 nt. Fragments larger than 500 nt exhibit suboptimal penetration and decrease hybridization efficiency. Fragments <100 nt tend to bind nonspecifically, resulting in an increased background. Therefore, the size of denatured fragments should be checked after the labeling reaction by agarose gel electrophoresis (see Fig. 15.2).

3 Remove an aliquot of 5–10 µl of probe, add agarose gel loading buffer, denature for 3 minutes at 95°C, and check fragment size by running on a 1% agarose gel (15 V/cm) together with an appropriate size standard.

Agarose Gel Loading Buffer

10 mM EDTA (ethylenediaminetetraacetic acid)
80% glycerol
0.1% **bromophenol blue** powder
CAUTION: Bromophenol blue (see Appendix 2 for Caution)

The probe DNA will be visible as a smear. The majority of the fragments should be between 100 and 500 nt. When this range is observed, stop the reaction by enzyme inhibition. If the fragments are larger, add another aliquot of DNase and incubate for a further 15–30 minutes at 15°C. If the DNA is digested slightly or not at all, purify the probe again. If all fragments are <100 nt, purify the DNA template and repeat the labeling reaction.

4 Add 3 µl of 0.5 M EDTA and 1 µl of 10% **SDS** and incubate for 15 minutes at 68°C to inactivate enzymes.

CAUTION: SDS (see Appendix 2 for Caution)

After the labeling reactions, the enzymes involved in the labeling procedures should be inactivated before the probes can be used for ISH experiments.

5 Separate probe from unincorporated nucleotides by ethanol precipitation (see p. 247) and proceed to step 6.

For most applications, ethanol precipitation will satisfactorily separate probe from unincorporated nucleotides. However, for some applications, for example, when using highly preserved cellular preparations, additional purification using a Sephadex column is required.

a Fill a 1-ml syringe, plugged with siliconized glass wool, with Sephadex G-50 and spin at 1600*g* in a clinical centrifuge for 5 minutes.

Column Buffer

10 mM Tris-HCl (pH 8.0)
1 mM EDTA
0.1% SDS (sodium dodecyl sulfate)

Sephadex G-50

30 g of Sephadex G-50
500 ml of column buffer
Incubate at 95°C for 1 hour

b Refill the volume of the syringe twice and spin as before.

c Rinse the column with column buffer and spin at 1600*g* for 5 minutes. Repeat two times.

d Place the probe in the center of the upper surface of the Sephadex, spin again, and collect the sample eluted from the column.

6 Assess the quality of labeling by a dot-blot assay.

The quality of labeling is one of the most important factors in hybridization and therefore must be tested. The number of incorporated reporter molecules is critical for the sensitivity of probe detection and can be checked with a dot-blot assay. After enzyme-based detection, the efficiency of biotin or digoxigenin incorporation in a labeled probe is estimated by colorimetric evaluation of intensities of colored precipitates on a nitrocellulose membrane. Commercial kits for dot-blot assays are available.

a Prepare dilutions of a labeled standard DNA and the probe DNA containing 0, 1, 2, 5, 10, and 20 pg/μl in DNA dilution buffer. As a standard, use an adequately labeled probe (commercially available) labeled with the same hapten.

DNA Dilution Buffer

6x standard saline citrate (SSC)
0.1 mg/ml sheared salmon sperm DNA

20x Standard Saline Citrate (SSC)

3 M NaCl
300 mM $NaC_6H_5Na_3O_7$

b Spot 1-μl aliquots from each solution of test and standard DNA in parallel on a piece of nitrocellulose filter. Bake the filter at 80°C for 60 minutes.

c Wash the filter in AP 7.5 buffer at room temperature for 1 minute.

AP 7.5 Buffer

0.1 M Tris-HCl, pH 7.5
0.1 M NaCl
2 mM **$MgCl_2$**

CAUTION: $MgCl_2$ (see Appendix 2 for Caution)

d Block the membrane in AP 7.5 buffer containing 3% BSA (w/v) at room temperature for 20 minutes.

e Incubate the membrane with 1 μg/ml streptavidin alkaline phosphatase for biotinylated probes, or 1 μg/ml anti-digoxigenin alkaline phosphatase for digoxigenin-labeled probes, at room temperature for 30 minutes.

f Wash the filter three times for 2 minutes with AP 7.5 buffer and equilibrate in AP 9.5 buffer, changing buffer twice.

AP 9.5 Buffer

0.1 M Tris-HCl, pH 9.5
0.1 M NaCl
50 mM **$MgCl_2$**

CAUTION: $MgCl_2$ (see Appendix 2 for Caution)

g Prepare AP/NBT/BCIP by adding the following to AP 9.5 buffer in the order below and mixing gently:

nitrobluetetrazolium-chloride (**NBT**) (stock 75 mg/ml in 70% **DMF**)	33 μl
5-bromo-4-chloro-2-indolyl-phosphate (**BCIP**) (stock: 50 mg/ml in DMF)	25 μl

CAUTION: NBT; DMF; BCIP (see Appendix 2 for Caution)

h Incubate the filter in AP/NBT/BCIP buffer and stop the reaction when the color development is suitable by rinsing the filter in H_2O and air-drying. Visualize and compare the DNA spots.
Usually 15–30 minutes of incubation is sufficient; longer incubation often results in high background staining. The color signals of the test DNA should be similar to the signals obtained from the standard DNA of a corresponding concentration. The 1-pg spot of the test DNA should be visible.

i Proceed with hybridization on p. 244.

PROBE LABELING WITH PCR

DNA probes can also be labeled by PCR techniques. These labeling procedures are suitable when probe sequences are small and when only small amounts of probe DNA are available. No single PCR protocol will be appropriate for all experimental regimes. Therefore, a standard protocol is given here followed by variations known as universal PCR techniques.

Standard Procedure

1 Prepare a reaction that combines the components according to the following pipetting scheme:

10x PCR buffer	2.5 μl
each dNTP (ATP, CTP, GTP)	200 μM
dTTP	150 μM
Biotin-16-dUTP	50 μM
Both primers	0.4 μM
Taq polymerase	5 units
ddH_2O to a final volume of 24 μl	
template DNA (~100 ng)	1 μl

10× PCR Buffer

20 mM **$MgCl_2$**
500 mM **KCl**
100 mM Tris-HCl, pH 8.4
0.1 mg/ml gelatin

CAUTION: $MgCl_2$; KCl (see Appendix 2 for Caution)

Positive and negative controls should always be included during PCR reactions. The negative control is performed without template DNA. A positive control is performed with, for example, 20 pg of placental DNA.

2 Spin the reaction in a microfuge, overlay with 100 μl of mineral oil, and carry out PCR in a thermal cycler using the following cycles:

first cycle
95°C for 5 minutes

30 cycles
94°C for 30 seconds
60°C for 30 seconds
72°C for 30 seconds

After the last cycle, maintain the reaction for 7 minutes at 72°C and then cool down to 4°C.

3 Check the size of the resulting fragments by agarose gel electrophoresis as in step 3 of the nick translation procedure (see p. 229; probe denaturation is not required).

4 Follow steps 5 and 6 of the nick translation procedure (see pp. 229–230) to separate unincorporated nucleotides and to assess quality of labeling.

5 Proceed with hybridization on p. 244.

Universal PCR

Routinely used universal PCR techniques for labeling of DNA are SIAs of DNA (Bohlander et al. 1992) and DOP-PCR (Telenius et al. 1992). Both techniques consist of three steps. The first step is a linear amplification of probe DNA under low-stringency conditions, which allows annealing of the degenerate primers and introduction of specific flanking sequences for a second reaction. In the second step, the amplification reaction is performed under high-stringency conditions, which results in an exponential DNA amplification using the previously synthesized sequences. During the third step, labeled nucleotides are included in the DNA synthesis reaction, which generates labeled DNA strands. Typically, DNA from purified YACs or microdissected chromosomes is amplified and labeled by universal PCR.

SEQUENCE-INDEPENDENT AMPLIFICATION (SIA)

Linear DNA Amplification by T7 Polymerase under Low-Stringency Conditions

1 Sterilize microfuge tubes and pipetter tips by UV irradiation.

2 Prepare 6.5 μl of buffer A for each labeling reaction as follows.

Buffer A

10× MAP buffer	0.6 μl
BSA (1 mg/ml)	0.3 μl
3 mM each dNTP in	0.6 μl
ddH_2O	5 μl

10× MAP Buffer

400 mM Tris-HCl, pH 7.5
100 mM **$MgCl_2$**
500 mM NaCl
50 mM **DTT**

CAUTION: $MgCl_2$; DTT (see Appendix 2 for Caution)

3 Eliminate DNA contamination of materials and solutions by irradiating them with UV light for ~10 minutes.

4 Add 0.375 μl of primer A to buffer A to a final concentration of 1.25 μM.

Primer A
sequence: 5′-TGGTAGCTCTTGATCANNNNN-3′
stock concentration: 20 μM

5 Pipette 4 μl of buffer A containing primer A into a sterile tube and add the template DNA. (For plasmid and cosmid probes, use 0.1–1 ng of template DNA in 5 μl of TE [Tris-EDTA buffer]; for YACs, use 50–100 pg of purified probe.)

6 Overlay this reaction mix with 100 μl of mineral oil, denature in a thermal cycler for 2 minutes at 97°C, cool down to 4°C, and place on ice.

7 Add 0.14 μl of T7 polymerase (13 units/μl; e.g., Sequenase USB) to the remaining buffer A and place on ice.

8 Add 0.5 μl of the T7 polymerase/buffer A (step 7) to the reaction mix (step 6), spin down, and place in a thermal cycler.

9 Carry out the following time/temperature regime:

13°C for 1 minute
37°C for 2 minutes
97°C for 50 seconds

Cool down to 4°C and place on ice.

10 Add another 0.5-μl aliquot of T7 polymerase/buffer A and repeat step 9.

11 Repeat step 10 two more times to achieve a total number of four cycles with T7 polymerase.

Exponential Amplification by Taq *Polymerase under High-Stringency Conditions*

12 Combine 3.75 µl of primer B (20 µM) to a final concentration of 1.25 mM, 0.4 µl of *Taq* polymerase (5 units/µl), and 55 µl of buffer B. Add it to the first PCR with T7 polymerase after four cycles are ended (step 11).

Primer B
sequence: 5′-AGAGTTGGTAGCTCTTGATC-3′
stock concentration: 20 µM

Buffer B

10x BP buffer	5.4 µl
3 mM each dNTP	2.4 µl
ddH_2O	47.2 µl

Total volume 55 µl
Irradiate to eliminate contaminants.

10x BP Buffer

66 mM Tris-HCl, pH 9.0
550 mM **KCl**

CAUTION: KCl (see Appendix 2 for Caution)

13 Microfuge to mix reaction components, overlay with 100 µl of mineral oil, and carry out PCR in a thermal cycler using the following conditions:

first cycle
94°C for 2 minutes

5 cycles
94°C for 45 seconds
44°C for 3 minutes
within 2 minutes to 72°C for 3 minutes

30 cycles
94°C for 45 seconds
56°C for 45 seconds
72°C for 4 minutes

After the last cycle, maintain the reaction for 7 minutes at 72°C and then cool down to 4°C.

Labeling Reaction with Taq *Polymerase under High-Stringency Conditions*

14 Prepare 29 µl of Bio Master Mix as follows:

Bio Master Mix

10x PCR buffer	3.0 µl
dATP, dCTP, dGTP each	150 µM
dTTP	110 µM
Bio-11-dUTP	40 µM
Taq polymerase (5 units/µl)	0.2 µl
primer B (see step 12)	1.25 µM

H_2O to a final volume of 29 µl

10x PCR Buffer

20 mM **$MgCl_2$**
500 mM KCl
100 mM Tris-HCl, pH 8.4
0.1 mg/ml gelatin

15 Add 1 µl of primary PCR product, overlay with 100 µl of mineral oil, and amplify using the following cycles:

first cycle
94°C for 2 minutes

22 cycles
94°C for 45 seconds
56°C for 45 seconds
72°C for 3 minutes

After the last cycle, maintain the reaction for 7 minutes at 72°C and then cool down to 4°C.

16 Check the size of the resulting fragments by agarose gel electrophoresis as in step 3 of the nick translation procedure (see pp. 229–230; probe denaturation is not required).

17 Follow steps 5 and 6 of the nick translation procedure (see pp. 229–230) to separate unincorporated nucleotides and to assess quality of labeling.

18 Proceed with hybridization on p. 244.

DEGENERATE OLIGONUCLEOTIDE PRIMER (DOP)-PCR

In principle, the DOP-PCR method follows a strategy similar to SIA-PCR amplification, but only primer is included in the reaction mix with no template.

Amplification of Probe DNA

1 Set up the reaction according to the following pipetting scheme:

10x DOP-PCR buffer	5.0 µl
each dNTP (2 mM stock)	5.0 µl
DOP	5.0 µl
ddH_2O	34 µl
Taq polymerase	2.5 units

10x DOP-PCR Buffer

20 mM **$MgCl_2$**
500 mM **KCl**
100 mM Tris-HCl, pH 8.4
0.1 mg/ml gelatin

DOP
sequence: 5′-CCGACTCGAGNNNNNNATGTGG-3′
stock concentration: 20 μM

2 Add 1 μl of probe DNA (~100–900 pg), overlay with 100 μl of mineral oil, and carry out amplification using the following parameters:

first cycle
93°C for 10 minutes

5 cycles
94°C for 1 minute
30°C for 1 minute 30 seconds
72°C for 3 minutes

35 cycles
94°C for 1 minute
62°C for 1 minute
72°C for 3 minutes, add 1 second per cycle

After the last cycle, maintain the reaction for 10 minutes at 72°C and then cool down to 4°C.

Biotinylation of Primary PCR Products

3 Set up the reaction according to the following pipetting scheme:

primary PCR product	5.0 μl
10x DOP-PCR buffer	5.0 μl
dCTP, dGTP, dATP each (2 mM stock) to a final concentration of 200 μM each	5.0 μl
biotin-16-dUTP (2 mM stock) to a final concentration of 200 μM	5.0 μl
dTTP (2 mM stock) to a final concentration of μM	2.0 μl
DOP	5.0 μl
ddH_2O	22 μl
Taq polymerase	2.5 units
Final volume 50 μl	

4 Carry out the following amplification protocol.

first cycle
94°C for 5 minutes

30 cycles
94°C for 1 minute
62°C for 1 minute
72°C for 1 minute 30 seconds

After the last cycle, maintain the reaction for 10 minutes at 72°C and then cool down to 4°C.

5 Check the size of the resulting fragments by agarose gel electrophoresis as in step 3 of the nick translation procedure (see p. 229; probe denaturation is not required).

6 Follow steps 5 and 6 of the nick translation procedure (see pp. 229–230) to separate unincorporated nucleotides and to assess quality of labeling.

7 Proceed with hybridization on p. 244.

Specimen Preparation

FISH using tissue sections requires specialized pretreatment and fixation procedures. The critical steps are attachment of the sections to slides and permeabilization of the specimen to allow a sufficient penetration of probe to the target. The sections are pretreated on the slide after attachment or, alternatively, the pretreatments are performed in a test tube before the sections are mounted. A number of different protocols have been developed for this purpose, all of which were successful in the author's laboratory. Specimens that are subjected to special pretreatment are described in the following series of procedures. Important features of these protocols as well as references are listed in Table 15.3.

PREPARATION OF METAPHASE CHROMOSOME SPREADS

Metaphase chromosome spreads may be prepared either from mitotic cells grown as short-term blood cultures, other suspension cell cultures, or from mitotic cells from adherent cultures.

1 Prepare mitotic cells from short-term blood cultures and other suspension cultures or from adherently growing cells.

From suspension cultures

a Incubate 1 ml of heparinized whole blood with 10 ml of RPMI 1640 full medium containing 1.5% phytohemagglutinin for 71 hours at 37°C in a cell culture incubator, or subculture established cell cultures 1 day before preparation of chromosomes in order to increase the number of mitotic cells.

TABLE 15.3 Different preparations of tissue sections prior to ISH

	Material		
	Fresh biopsy	**Paraffin-embedded tissue sections**	**Frozen tissue sections**
Prefixation treatment		dewaxed by 3× 100% xylene for 10 min 2× methanol for 5 min	
Fixation	4% PFA/picric acid fixation for 4–18 hr, followed by sectioning	air-dry section bake onto poly-L-lysine-coated slides overnight at 56°C	air-dry onto poly-L-lysine-coated slides overnight fixation in methanol acetone, 1:1 for 20 min at –20°C
Permeabilization	equilibration in 20% glycerol/PBS 4–6× freeze and thaw	incubation in 1 M Na isothiocyanate at 80°C for 10 min incubation 5–60 min at 37°C 4 mg/ml pepsin in 0.2 HCl	incubation with 50–400 mg/ml pepsin for 10 min in 0.01 M HCl, postfix in 1% formaldehyde in PBS
Citation	Manuelidis and Borden (1988)	Hopman et al. (1988); Scherthan and Cremer (1994)	Hopman et al. (1992)

PFA, paraformaldehyde; PBS, phosphate-buffered saline.

RPMI 1640 Full Medium

RPMI 1640 medium
10% FCS (fetal calf serum)
1% penicillin (5000 units/ml)
streptomycin (5000 μg/ml)
1% glutamine (0.2 M)

White blood cells in peripheral blood must be stimulated with a mitogen inducing cell division as a prerequisite for preparation of cells in metaphase. In preparations of peripheral human blood cells, T-lymphocytes are stimulated with phytohemagglutinin. Established cell cultures growing in suspension should be subcultured the day before metaphase preparations are harvested in order to obtain actively proliferating cells.

b Add **ethidium bromide** to a final concentration of 10 μg/ml and incubate for 30 minutes at 37°C.

CAUTION: Ethidium bromide (see Appendix 2 for Caution)

The ethidium bromide intercalates into DNA, inhibiting condensation of metaphase chromosomes to some degree.

c Add colcemid to the culture to a final concentration of 0.1 μg/ml. Incubate for 30 minutes at 37°C.

Colcemid is added to arrest cells in mitosis.

d Proceed to step 2 to continue with the hypotonic treatment for chromosome preparation.

From adherently growing cells

a Subculture adherent cells for 1 or 2 days before cell preparation in order to achieve ~70% confluency of actively growing cells. The presence of mitotic cells should be checked before proceeding.

b Add colcemid to the culture to a final concentration of 0.1 μg/ml. Incubate for 60–90 minutes at 37°C.

c Perform a modest trypsinization combined with a sharp tap on the culture flask to remove the rounded, mitotic cells.

d Proceed to step 2 to continue with the hypotonic treatment for chromosome preparation.

2 Centrifuge the cells at 800*g* for 5–10 minutes. Remove medium completely except for 0.2–0.6 ml of supernatant remaining above the cell pellet.

3 Resuspend cells in the remaining medium and add carefully approximately half the culturing volume of prewarmed (37°C) 0.075 M **KCl** while agitating gently.

CAUTION: KCl (see Appendix 2 for Caution)

4 Incubate for 12–20 minutes at 37°C.

This hypotonic treatment causes a swelling of the cells; the optimal time of treatment varies for different cell types and must be determined empirically.

Adding 10 drops of freshly prepared fixative on top of hypo helps eliminate clumping.

5 Centrifuge the cells as in step 2 and resuspend.

It is important to remove as much hypo as possible before resuspending.

6 Fix the cells by the dropwise addition of 10 ml of freshly prepared ice-cold fixative with gentle agitation.

Fixative
methanol/glacial acetic acid (3:1)

CAUTION: Methanol; glacial acetic acid (see Appendix 2 for Caution)

7 Incubate cells on ice for 5 minutes.

8 Centrifuge the cells as in step 5 and repeat the fixation (step 6). Incubate the cells on ice for 30 minutes.

9 Repeat the fixation procedure (steps 5 and 6) more than five times. An incubation on ice between fixations is not required.

10 Drop the suspension onto a prewashed (with ethanol [±1% ether] and then H_2O) microscope slide. This should be carried out in a humid atmosphere, for example, over a hot water bath.

11 Dehydrate the slides through an ethanol series of 70%, 90%, and 100%, 5 minutes each, and then air-dry.

The slides can be stored for a few days to several weeks at room temperature. Alternatively, slides can be stored for over a year in a dry container with drierite at –20°C or –80°C. Once thawed, the slides should not be refrozen. Metaphase chromosome preparations are shown in Figures 15.2B,D and 15.3A.

PREPARATION OF SAMPLES FROM STAINED BLOOD SMEARS

Blood smear samples can be used for FISH experiments after routine Wrights staining (Bentz et al. 1993a). However, the blood smears must be destained. A subsequent enzymatic permeabilization may facilitate the penetration of the probe.

1 Destain blood smear preparations in 100% methanol at room temperature for 10 minutes.

2 Incubate the slides in 0.02% pepsin in 0.01 M HCl at 37°C for 5 minutes.

Permeabilization of the cells with pepsin may enhance penetration of the probe.

3 Wash slides in PBS three times at room temperature for 5 minutes.

Phosphate Buffered Saline (PBS)

1.4 M NaCl
27 mM **KCl**
210 mM **Na_2HPO_4**
15 mM KH_2PO_4

CAUTION: KCl; Na_2HPO_4 (see Appendix 2 for Caution)

4 Fix samples in 6% **paraformaldehyde**/PBS (w/v) at room temperature for 5 minutes and wash again three times for 5 minutes in PBS.

CAUTION: Paraformaldehyde (see Appendix 2 for Caution)

5 Dehydrate slides through an ethanol series of 70%, 90%, and 100% and air-dry.

ISOLATION OF NUCLEI FROM FROZEN TISSUE SECTIONS

Although morphologic information will be lost, an easy way to obtain a high efficiency FISH with tissue section material is to isolate nuclei from frozen material.

1 Place ~50-μm sections of frozen tissue on a microscope slide.

2 Add 200–300 μl of digestion solution and incubate at 37°C for 20 minutes.

Digestion Solution

0.01 N **HCl**
100 μg/ml **pepsin**

CAUTION: HCl; acids and bases that are concentrated; pepsin (see Appendix 2 for Caution)

3 Triturate the solution with a pipette to suspend the disintegrated material including nuclei and transfer to an Eppendorf tube.

4 Dilute nuclei to a concentration appropriate for later analyses.

5 Centrifuge ~200 μl of suspended nuclei onto a slide in a cytospin at 300–500*g* at room temperature for 5 minutes.

6 Air-dry the slides, dehydrate through an ethanol series of 70%, 90%, and 100%, and air-dry again.

7 Fix the nuclei in 1% paraformaldehyde/PBS at room temperature for 5–10 minutes and dehydrate again through an ethanol series.
The number and quality of nuclei can be checked by DAPI staining before FISH (see Chapter 5).

PREPARATION OF INTERPHASE CELLS PRESERVING THE THREE-DIMENSIONAL MORPHOLOGY OF NUCLEI

To study the positioning and organization of whole chromosomes, subchromosomal regions, or specific sequences such as genes or RNA within nuclei by ISH, it is imperative that, after fixation, cells retain as much as possible their in vivo morphology (Manuelidis 1985; Lichter et al. 1988; Zirbel et al. 1993). To this end, cells are prepared using a specialized protocol that includes an efficient permeabilization procedure. This is necessary to enable labeled probes to enter nuclei. For preservation of the three-dimensional morphology of cells, fixation is performed with 4% paraformaldehyde and permeabilization is performed with Triton-X 100/saponin, a freeze and thaw procedure, and, in some cases, with hydrochloric acid (modified from Manuelidis 1985). See Figures 15.4 and 15.5 for a display of nuclei fixed and permeabilized in this way.

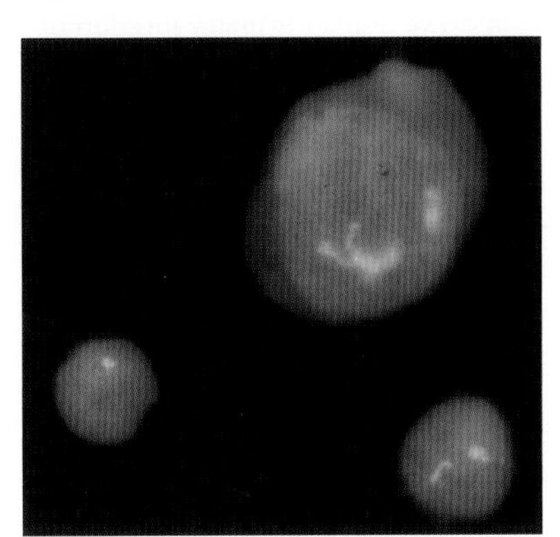

FIGURE 15.4

RNA localization from Epstein-Barr virus (EBV) genome integrated into Namalwa cells. FISH was performed with a cosmid containing EBV sequences detected via FITC. Only RNA is visualized since no denaturation was performed. Nuclei were counterstained with PI (propidium iodide). (Photo provided by U. Mathieu and R. Zirbel, German Cancer Research Center, Heidelberg.)

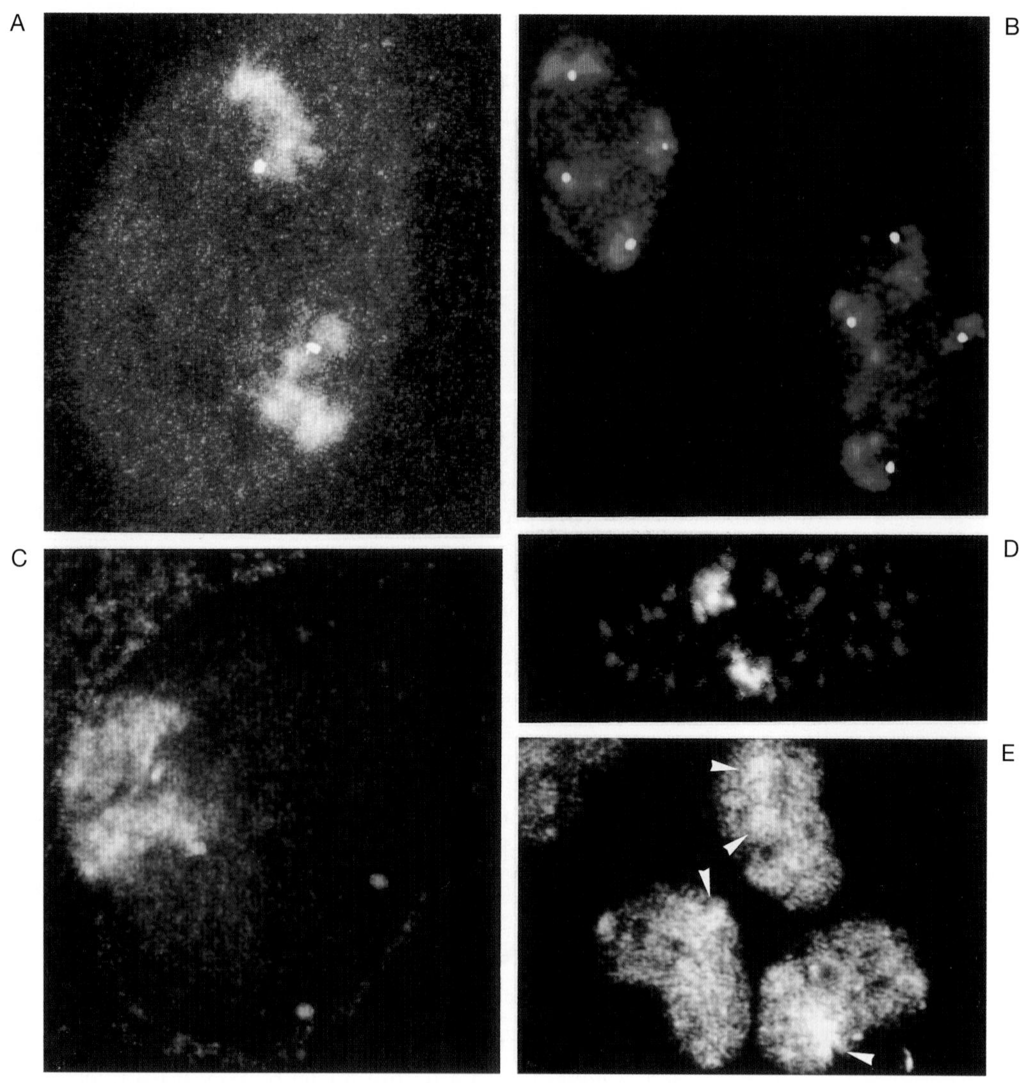

FIGURE 15.5

Chromosomal territories visualized in parallel with other nuclear entities in three-dimensionally preserved human nuclei. Cells were fixed with 4% paraformaldehyde and subjected to permeabilization techniques (see pp. 242–243). (*A,B*) Dual-color FISH, delineating chromosomal territories (TRITC), in conjunction with specific subsequences. Duchenne's muscular dystrophin gene in *A* and DIZ2 on chromosome 1p in *B*, both visualized via fluorescein. Note that *B* shows tetraploid nuclei. (*C,D*) Combination of FISH (revealing chromosomal territories, FITC) and indirect immunolocalization of "nuclear bodies" (***C***) and accumulations of Sm antigen (***D***), both via TRITC. (*E*) Chromosome 4 painting with a high background staining. Since the probe was not purified from unincorporated nucleotides, chromosomal territories are barely visible; see arrowheads. *A–C* and *E* display HeLa cells and *D* a human primary foreskin fibroblast. All panels present optical sections acquired by confocal laser scanning microscopy using a Zeiss LSM. (Photos provided by A. Kurz, J. Tajbakhsh, and R. Zirbel, German Cancer Research Center, Heidelberg.)

1 Grow or attach cells to glass microscope slides that can survive high and low temperatures, such as "Superfrost" glass microscope slides.

For suspension cells

Apply suspension cells in medium onto adequately coated (e.g., with 0.04 g/ml poly-L-lysine or other reagents) glass microscope slides at a density of ~5×10^4/cm^2. Incubate for 1–3 hours at 37°C.

For adherent cells

Place adherently growing cells on glass microscope slides and culture at 37°C to achieve a final density of ~5 x 10^4/cm^2.

2 Place glass microscope slides with attached cells in glass Coplin jars and rinse in PBS three times.

3 Fix the cells by incubating the slides in 4% **paraformaldehyde** in 1x PBS at room temperature for 10 minutes.

CAUTION: Paraformaldehyde (see Appendix 2 for Caution)

The above fixation procedure was chosen because it has been shown to preserve the three-dimensional morphology of the nucleus. This has been tested by comparing the distribution of kinetochores after a light fixation (2% paraformaldehyde) using CREST antiserum with the distribution of the centromeric alphoid DNA sequences detected by FISH after the whole fixation and permeabilization procedure. The complete colocalization of both signals indicates that the three-dimensional morphology is maintained (Kurz et al. 1996).

4 Wash the cells three times in PBS.

5 Incubate the slides in a mixture of 0.5% Triton-X 100 (v/v) and 0.5% saponin (w/v) for 20 minutes. This treatment will permeabilize the cells.

Triton-X 100/Saponin Buffer

saponin	0.5 g
Triton-X 100	0.5 ml

1x PBS to a final volume of 100 ml
Make fresh for each experiment.

6 Wash three times in PBS.

7 Place the slides in 0.1 N **HCl** for 10 minutes.

This last step may be omitted if one intends to perform immunohistochemistry on the material and acid precludes detection of an antigen of interest

CAUTION: HCl; acids and bases that are concentrated (see Appendix 2 for Caution)

8 Place the slides in a solution of 20% glycerol in 1x PBS (v/v) for at least 30 minutes and immerse in liquid nitrogen for ~30 seconds to allow the material to freeze.

20% Glycerol

glycerol	20 ml
10x PBS	10 ml

H_2O to a final volume of 100 ml

9 Let the cells thaw slowly at room temperature.

10 Repeat steps 8 and 9 one to ten times. When repeating more than five times, preequilibrate again (step 8).

Slides can be stored at –80°C after one of the freezing procedures.

Fluorescence In Situ Hybridization

This procedure describes fluorescence in situ hybridization (FISH) of biotin- or digoxygenin-labeled probes (prepared in the procedures on pp. 228–237) to denatured metaphase chromosomes and interphase nuclei (prepared in the procedures on pp. 238–243). The hybridized probes are detected and visualized using fluorochrome-conjugated reagents. Variations in the preparation of probe and in stringencies of hybridization and washing conditions are provided in order to optimize the procedure for particular applications.

Stringencies of denaturation and renaturation: The stringency of denaturation/renaturation is a relationship between temperature, salt, and formamide concentration. Stringency can be increased by raising the temperature, increasing the concentration of formamide, or decreasing the number of monovalent cations (i.e., lowering the concentration of SSC). For a FISH procedure, stringency is important for the dissociation of bound nucleic acids with sequences of incomplete homology. High stringency can be applied in any of the FISH protocol steps (i.e., during denaturation, during hybridization, or in posthybridization washes). In performing the FISH procedure, there are different preferences with respect to stringency. The stringency of the hybridization step can be high, with posthybridization washing steps less stringent, or the stringency of the hybridization step can be low, compensated by high-stringency conditions in the posthybridization washes.

Denaturation of specimen: Although alkaline treatment and exonuclease activities have been used to produce single-stranded chromosomal target DNA, heat denaturation is preferred because it produces homogeneously distributed single-stranded areas throughout the genome without a major loss of target sequences. Conditions for heat denaturation of nucleic acids within cellular specimens are a compromise between achieving efficient formation of single-stranded target sequences and damaging the architecture of cells and cellular components. The stringency of the denaturation procedure to be applied is dependent on the fixation protocol; for example, a high degree of cross-linking requires high stringencies during denaturation.

Amplification of signal: For small target sequences or for small probes, it is sometimes necessary to perform amplification steps to increase the intensity of the signal. This entails building layers of detection reagents; for example, for a biotinylated probe, biotin-conjugated anti-avidin antibody is followed by another layer of fluorochrome-conjugated avidin/streptavidin (Pinkel et al. 1986). See Table 15.4 for details of signal detection and amplification; see Table 15.5 on p. 246 for troubleshooting of FISH.

PREPARATION OF THE HYBRIDIZATION COCKTAIL

The following recipes for the composition of probe are for a 10-μl volume, which is sufficient for one hybridization reaction covering an area of 18 x 18 mm^2. For larger volumes, adjust accordingly (e.g., for 22 x 40 mm^2 use 25–30 μl).

Two recipes are presented: (1) for hybridization of probes targeting clustered repetitive DNA and (2) for single-locus probes containing IRS and chromosome painting probes.

Preparation of Probes Targeting Clustered Repetitive DNA

For chromosome-specific repeat clusters such as alphoid DNA, many regions with partial homology exist. Higher stringencies, such as increasing the percentage of formamide to 60–70%, eliminate the cross-reaction of sequences (step 3).

TABLE 15.4 Signal detection and amplification schemes using a variety of commonly used detection reagents

Reporter molecules for labeling probes	First layer	Second layer	Third layer
Biotin	avidin/streptavidin-fluoro* anti-biotin antibody-fluoro* anti-biotin antibody-strep/avidin anti-biotin antibody	species-specific antibody-fluoro* anti-strep/avidin antibody-fluoro* anti-strep/avidin antibody-biotin species-specific antibody	species-specific antibody-fluoro* anti-biotin antibody-fluoro*
Digoxigenin	anti-digoxigenin-fluoro* anti-digoxigenin antibody	species-specific antibody-fluoro* species-specific antibody-digoxigenin species-specific antibody	anti-digoxigenin antibody-fluoro* species-specific antibody-fluoro*
Dinitrophenol	anti-dinitrophenol antibody-fluoro* anti-dinitrophenol antibody	species-specific antibody-fluoro* species-specific antibody	species-specific antibody-fluoro*
Halogens BrdU, FdU	anti-halogen antibody-fluoro* anti-halogen antibody	species-specific antibody-fluoro* species-specific antibody	species-specific antibody-fluoro*

Possible fluorochromes (Fluoro*) include rhodamine, Texas red, cyanine 3, cyanine 3.5, cyanine 5 (all red); fluorescein, eosin (both green); AMCA (blue).

Note: When amplifying a signal in dual-color FISH, do not use a hapten or species of antibody that has already been used in the experiment. This will lead to gross cross-reactions between the two signals.

It is possible to use other reporter molecules in an amplification other than the original one used; i.e., with a biotinylated probe, the first layer could be anti-biotin antibody conjugated to digoxigenin with a further amplification step of anti-digoxigenin antibody conjugated to a fluorochrome. One must be very careful when having to amplify two separate signals as to which haptens and detection reagents one uses.

BrdU, 5-bromodeoxyuridine; FdU, 5-fluoro-2′-deoxyuridine.

1 Prepare the probe cocktail by combining in a microfuge tube:

probe DNA for a chromosome-specific repeat cluster	10–20 ng
salmon sperm DNA	1–3 μg

2 Lyophilize the DNA in a Savant SpeedVac or on a heating block.

3 Resuspend the DNA in 6–7 μl of deionized formamide and mix well (for at least 30 min).

Deionized Formamide

Incubate 250 ml of **formamide** with ion-exchange resin (e.g., 10–15 g of Analytical Grade Mixed Bed Resin AGR 501-XB [D] [BioRad], 20–50 mesh) for 2–3 hours with constant stirring. Filter the mixture through 3MM paper and store in 50-ml aliquots at –20°C. The conductivity of the solution should be <100 μs. Deionized formamide is also commercially available from Ambion.

CAUTION: Formamide (see Appendix 2 for Caution)

4 Add to the probe cocktail for a final volume of 10 μl:

1 μl 20x SSC to give a final concentration of 2x SSC
dextran sulfate to give a final concentration of 10% (*optional*)
25 mM sodium phosphate

Mix well for a further 30 minutes.

Preparation of Single-Locus Probe Containing IRSs and Chromosome Painting Probes

When the probe DNA contains IRSs, CISS hybridization is performed (Lichter et al. 1988; Pinkel et al. 1988). This prevents binding of labeled IRS within the probe to sequences other than the target-

TABLE 15.5 Troubleshooting FISH

Problem	Probable cause	Solution
High background staining on chromosomes and chromatin	probe too short	check probe size on 1% agarose repeat labeling
Starry background	probe too long	check probe size on 1% agarose digest further with DNase I or repeat labeling
	dirt on slides or in buffers	ensure buffers are clean—filter clean slides and coverslips with ethanol and deionized H_2O
	aggregation of detection reagent	spin in microfuge and use only supernatant
General high background	unincorporated nucleotides present	pass through Sephadex G-50 spin column (note: Fig. 15.5E displays a FISH experiment with a painting probe for chromosome 1, but the background is so high because of unincorporated nucleotides that the chromosomal territories are hard to define)
	nonspecific binding of detection reagent	increase wash time add Tween-20 to wash buffer use new batch of detergent
	blocking step insufficient	try other blocking reagents (e.g., dried milk powder, serum, or BSA) perform blocking step for a longer time
	poor chromosome or specimen preparation	incubate with pepsin to remove cell debris
Problems in detecting signal	insufficient incorporation of reporter into probe nucleic acid	amplify signal repeat probe labeling
	target or probe size too small for a one-step detection procedure	amplify signal
	incorrect size of labeled probe molecules	check probe size
	efficiency of detection reagents decreased, such as deconjugated antibodies	check by gel filtration obtain new batches of detection reagents
	denaturation procedure inadequate	try different stringency conditions, i.e., change denaturation temperature, formamide concentration, or concentration of SSC in wash buffer B
	instrumentation problem	survey optical equipment
	poor permeabilization of probe (in particular, in 3D conserved preparations)	increase number of freeze and thaw steps place slides in HCl longer use a higher molarity of HCl increase hybridization time make sure all buffers, reagents, slides, and coverslips are RNase- or DNase-free

ed locus. Competitor DNA is added to the probe mix and a preannealing step is performed. The competitor DNA can be total genomic DNA (e.g., human placenta DNA) sized to an average of 500 nt in length. Alternatively, the C_0t1 fraction of DNA (commercially available for human and mouse) makes an excellent competitor, since it is enriched for IRS (Nisson et al. 1991).

1 Prepare probe cocktail as follows:

a Select the appropriate probe from the following table and pipette the corresponding amount indicated (10 µl final volume) into a microfuge tube.

Probe DNA	Amount of DNA
Single cosmids/plasmids/phage	20–60 ng
Band-specific probe DNA	100–200 ng
Purified YAC DNA	50 ng
Total yeast DNA plus YAC	1 μg
PCR amplified YAC DNA	100–150 ng
Total DNA derived from single sorted chromosomes with a vector-to-insert ratio of 2:1 (pBS libraries)	150–500 ng
Total DNA derived from single sorted chromosomes with a vector-to-insert ratio of 10:1 (λl libraries from ATCC)	0.5–2 μg

b Add 3–7 μg of competitor DNA (C_0t1 DNA).
When IRS-PCR amplified probe DNA is used, add >30 μg of C_0t1 DNA.

c Add 0–3 μg of carrier DNA (salmon sperm DNA) (*optional*).
There is no requirement for the carrier salmon sperm DNA when adding large amounts of C_0t1 DNA.
For different chromosome painting probe DNAs, varying amounts need to be used, depending on the size of the target chromosome.

2 Add 1/20th volume of 3 M sodium acetate, pH 5.0, and 2 volumes of 100% ethanol; precipitate DNA for at least 30 minutes at –80°C.

3 Microfuge at 10,000*g* for 10 minutes and discard the supernatant by inversion.

4 Wash the pellet in 400 μl of 70% ethanol by pipetting the liquid up and down in an Eppendorf tube.

5 Repeat the spinning and washing steps once and centrifuge as in step 3.

6 Lyophilize the pellet by placing on a heating block for 40 minutes at 37°C or in a Savant SpeedVac for a shorter time.

7 Resuspend the dry pellet in 5 μl of deionized formamide and mix well for at least 30 minutes.

8 Add 5 μl of hybridization buffer and mix again for at least 30 minutes.

Hybridization Buffer

4x SSC
20% dextran sulfate
50 mM sodium phosphate

DENATURATION OF PROBE AND SPECIMEN

Denaturation of the probe as well as of the specimen are critical steps in hybridization. Chromosomes and nuclei can be examined with a phase-contrast microscope after denaturation; overdenatured nuclei and chromosomes will have a "hollow" appearance, whereas underdenatured chromosomes will appear as untreated specimens. The quality of formamide and the temperature, pH, and time course are all important parameters of denaturation. Denaturing the probe will separate double-stranded sequences and/or reduce secondary structure. If hybridizing to tissue sections, blood smears, or isolated nuclei from frozen tissues, use the procedure on p. 250 to simultaneously denature probe and specimen.

Denaturation of the Probe Mixture

For hybridization of a double-stranded DNA probe to the target sequences, the two strands must be dissociated. This is achieved by denaturation with formamide and high temperature. Probes containing single-stranded DNA and RNA are heated to 60°C to release intramolecular interactions.

1 After complete resuspension of lyophilized nucleic acids, denature the double-stranded probe mixture for 5 minutes at 75°C either in a water bath or on a heating block.

2 *For repetitive probes where no C_0t1 DNA has been added*

Place the probe mixture directly on ice, and then go to the following procedures for denaturation of specimens.

For probe DNA that contains IRS sequences and has C_0t1 DNA in the mixture

Incubate the probe mixture at 37°C as a preannealing step.

C_0t1 DNA associates with IRS sequences within the probe DNA. The length of time for preannealing can be altered depending on the probe type. This incubation can be performed for 1 minute to several hours. In general, 10–20 minutes is used.

Denaturation of Acetic Acid-fixed Nuclei and Chromosomes

Fixed nuclei and chromosomes are denatured by treatment with formamide at high temperature followed by dehydration before the probe is delivered onto the specimen. Properly denatured chromosomes stained with DAPI are shown in Figure 15.3A; Figure 15.6 shows chromosomes that have been overdenatured.

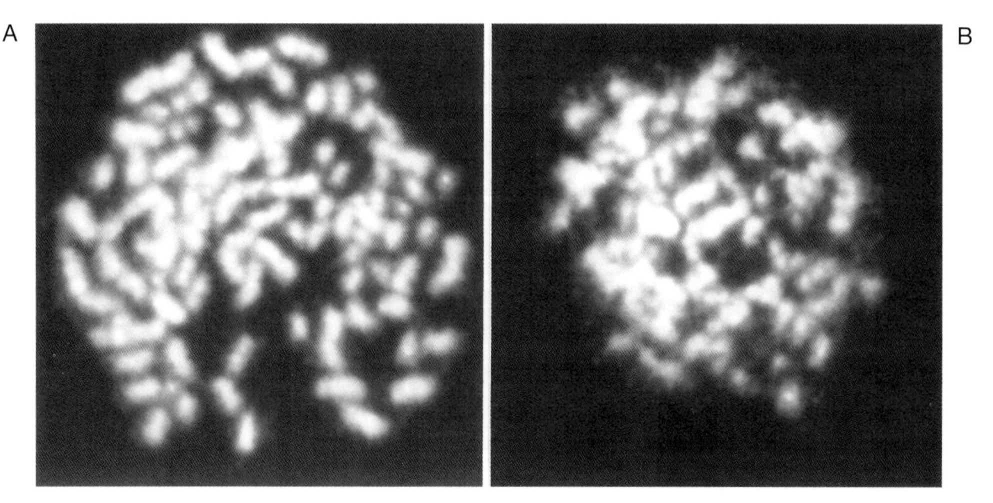

FIGURE 15.6

Overdenatured chromosomes visualized by DAPI. (*A*) The chromosome outlines are fuzzy because of spooling out of DNA. (*B*) Overdenaturation was so strong that chromosomes can barely be defined. (Photos provided by P. Lichter, German Cancer Research Center, Heidelberg.)

1 Denature fixed chromosomes by immersing slides in denaturation buffer A for 2 minutes at 70°C (Coplin jar in 70°C water bath).

Denaturation Buffer A

70% **formamide**
2x SSC
Adjust pH to 7.0 with 1 N **HCl.** Add 50 mM sodium phosphate to maintain pH (*optional*).

CAUTION: Formamide; HCl; acids and bases that are concentrated (see Appendix 2 for Caution)

Since the DNA is to be denatured at 70°C, the decline in temperature when the microscope slides are added to the denaturation solution should be considered, and hence the temperature of the buffer should be 1–2°C above 70°C for each slide to be denatured. Do not place more than four slides in a Coplin jar at any one time.

2 Dehydrate slides through an ice-cold ethanol series of 70%, 90%, and 100% for 3 minutes each and air-dry.

3 Apply 10 µl of the denatured probe onto a prewarned (37°C) 18 x 18-mm^2 glass coverslip and lower the slide onto the probe mix in such a way as to avoid trapping air bubbles between coverslip and slide.

Denaturation of Three-dimensionally Conserved Tissue Culture Cells

Three-dimensionally conserved cultured cells are denatured in formamide before the probe is delivered onto the specimen.

1 Incubate the slides in denaturation buffer A for exactly 3 minutes at 73°C. Although this time period might vary for different preparations, the optimal denaturation time is within a narrow window and must be determined.

Denaturation Buffer A

70% **formamide**
2x SSC
Adjust pH to 7.0 with 1 N **HCl.** Add 50 mM sodium phosphate to maintain pH (*optional*).

CAUTION: Formamide; HCl; acids and bases that are concentrated (see Appendix 2 for Caution)

Since the DNA is to be denatured at 73°C, the decline in temperature when the microscope slides are added to the denaturation solution should be considered, and hence the temperature of the buffer should be 1–2°C above 73°C for each slide to be denatured. Do not place more than four slides in a Coplin jar at any one time.

2 Place the slides in a second Coplin jar containing denaturation buffer B for ~1 minute at 73°C. During this minute, pipette 10 µl of probe onto a prewarmed (37°C) 18 x 18-mm^2 glass coverslip.

Denaturation Buffer B

50% formamide
2x SSC
Adjust pH to 7.0 with 1 N HCl. Add 50 mM sodium phosphate to maintain pH (*optional*).

3 Remove the slides from denaturation buffer B and wipe clean all areas of the slide except where the probe is to be presented. Remove the remainder of buffer by flicking it off with a quick wrist movement. Lower the slide gently onto the center of the drop containing the probe, which then spreads across the coverslip.

It is imperative to work as fast as possible in this last stage.

4 Seal coverslip with rubber cement.

Denaturation of Probes with Tissue Sections, Blood Smears, and Isolated Nuclei from Frozen Tissue

The denaturation process for these three specimen types is similar and can be performed simultaneously with denaturation of probe. The denaturation temperature is usually higher for these types of specimens because of the fixation and embedding procedures that they have undergone.

1 Place the resuspended probe (in formamide and hybridization mix, 10 μl) onto a specific area on the pretreated slide, which has normally been air-dried after dehydration through an ice-cold ethanol series of 70%, 90%, and 100%.

2 Cover the area of hybridization with an 18 x 18-mm^2 glass coverslip.

3 Apply rubber cement to seal around the coverslip. Mark the area hybridized underneath by scoring the glass with a diamond pen.

4 Denature the probe and specimen.

The denaturation temperature and length of denaturation stated in different protocols varies from 75°C to 95°C and from 4 to 15 minutes. Therefore, the optimum temperature and time for each specimen will have to be determined. Bear in mind that tissue sections require the higher denaturation temperatures and nuclei from frozen tissue the lowest.

HYBRIDIZATION

The hybridization of probe with endogenous nucleic acids is usually performed at a temperature of ≥37°C and with a fairly long incubation time. In some cases, 2–3 days are required. There are mathematical equations to characterize hybridization reactions. However, these equations are based on the behavior of DNA in solution where there are no problems of probe penetration, conglomerations of protein, or reannealing of the template strands of DNA in close proximity. Hence, stringency and duration of hybridization should be optimized by the individual. For example, the length of hybridization for abundant sequences such as chromosome-specific repetitive probes could be <1 hour compared with 48 hours for low-abundancy sequences of low complexity.

Posthybridization washes aid in the removal of nonspecifically bound probe, greatly reducing background staining. Furthermore, the stringency of FISH experiments can be adjusted by posthybridization washes if they are performed under more severe conditions than during hybridization.

1 Place slides with samples to be hybridized in a humidified chamber in a hot-air oven for 1–18 hours at 37°C or 42°C, according to the specimen and probe.

2 Prepare materials and reagents for posthybridization washes: Preheat Coplin jars to 42°C and 60°C, and wash buffer A to 42°C and wash buffer B to 60°C.

Wash Buffer A

50% **formamide**
2x SSC
Adjust pH to 7.0 with 1 N **HCl.**

CAUTION: Formamide; HCl; acids and bases that are concentrated (see Appendix 2 for Caution)

Wash Buffer B

0.1–1x SSC (depending on stringency required)
Adjust pH to 7.0 with 1 N HCl.

3 Remove rubber cement from slides with forceps and carefully remove coverslips with washing.

4 Place slides in preheated wash buffer A in a shaking water bath at 42°C. Replace the buffer every 5 minutes for 15 minutes.

5 Place slides in wash buffer B (preheated to 60°C). Replace the buffer every 5 minutes for 15 minutes.

6 Exchange wash buffer B for 4x SSC at room temperature.

DETECTION

The detection of labeled probe DNA is subject to the labeling hapten used and to the fluorochrome of choice (several common detection regimes are presented in Table 15.4). Depending on which hapten is used or whether amplification of signal is needed, select the required detection system.

1 Place 50–200 µl of blocking solution on the cells, cover the area with a coverslip, and incubate for 20 minutes at room temperature. The blocking step is performed before detection of bound probe DNA.

Blocking Solution

4% BSA or dried milk powder
4x SSC

2 Make up a solution of the detecting compound of 1–20 µg/ml in blocking buffer containing 1% BSA or other blocking compound. Consult Table 15.4 for selection of appropriate reagents.

3 Place 50 µl of this solution on a 24 x 40-mm^2 glass coverslip.

4 Lower the slide over the coverslip and present the hybridized area to the solution. Incubate in a humidified chamber for 1 hour at room temperature or for 30 minutes at 37°C.

5 Wash slides in wash buffer C three times for 5 minutes at 42°C in a shaking water bath.

Wash Buffer C

2x SSC
0.1% Tween 20
Counterstaining dyes such as PI, DAPI, and Hoechst can be included in wash buffer C (see Chapter 5).

MOUNTING OF SPECIMEN

Mounting of slides is usually performed in 90% glycerol. For FISH, such mounting media should contain reagents such as DABCO (1,4-diazabicyclo-[2.2.2]-octane), *p*-phenylenediamine-dihydrochloride, or isopropylgallate. Mowiol provides a self-sealing property to a mounting medium and hence no sealant is required. If chromosome banding is desired (see pp. 253–255), it can be performed either before or during mounting.

1 Apply 10–50 µl of appropriate mounting medium to a glass coverslip.

Mounting Medium A

2.5% **DABCO** or isopropylgallate or 1% **phenylenediamine-dihydrochloride**
200 mM Tris-HCl, pH 8.6
90% glycerol

CAUTION: DABCO; *p*-phenylenediamine (see Appendix 2 for Caution)

Mounting Medium B

2.5% DABCO or isopropylgallate
30% glycerol
12% Mowiol
200 mM Tris-HCl, pH 8.6

Mounting Medium C

Vectashield (available from Vector Laboratories)
Counterstains such as DAPI and PI can be included in these mountants at a final concentration of 0.01–1 mg/ml to visualize DNA in samples.

2 Lower slide over mounting medium and gently press excess mounting medium out from under the coverslip (cover with a paper tissue first).

3 If Mowiol is not included in the mounting medium, seal the slide with clear nail polish or rubber cement.

Slides can be stored for a few days to many months in the dark at –20°C or –70°C.

Chromosome Banding in Combination with FISH

With certain dyes and staining procedures, it is possible to visualize the characteristic banding patterns present on each chromosome. Many of these procedures are compatible with FISH. Posthybridization banding is often preferred because it allows simultaneous visualization of hybridization signals and banding patterns. Several such protocols are provided. Alternate procedures in this section describe quinacrine mustard staining for Q bands, DAPI/Hoechst staining, and BrdU incorporation for chromosome banding and labeling of replication sites. (See Fig. 15.3 for examples of counterstaining chromosomes and nuclei with DAPI.)

QUINACRINE MUSTARD STAINING FOR Q-BANDS

Quinacrine mustard was one of the first fluorescence banding stains used and, hence, the resulting banding pattern was given the first initial of quinacrine mustard. The bright Q-bands correspond to the Giemsa-dark (G-) bands. Quinacrine displays a fluorescent green coloration when excited. The procedure uses fixed metaphase chromosomes or fixed metaphase chromosomes that have been processed through the complete FISH procedure but have not yet been mounted.

1 Place slides in 0.05% quinacrine staining solution for 15–30 minutes in the dark at room temperature.

Quinacrine Staining Solution

4 mM $H_3C_6H_7$
12 mM Na_2HPO_4
0.05% **quinacrine orange**

CAUTION: Na_2HPO_4; quinacrine (see Appendix 2 for Caution)

2 Rinse the slides in PBS three times for 5 minutes.

3 Mount stained specimen (see p. 252) and view chromosomes with excitation light of 450–500 nm.

DAPI/HOECHST STAINING

The stains generally known as DAPI and Hoechst 33258 are the most common DNA intercalating dyes and are used extensively to visualize DNA of both metaphase and interphase chromosomes. They also preferentially stain the Q-bands heavily. Although DAPI banding alone is often performed with addition of other reagents such as distamycin to enhance the banding, these reagents are not required when applying the staining procedure to denatured chromosomes prepared for FISH.

1 Rinse the slides in PBS.

2 Incubate the slides in PBS containing 0.01–1 µg/ml **DAPI** for 30 minutes at room temperature or in PBS containing 0.5 µg/ml Hoechst 33258 dye for 10–15 minutes at room temperature.

CAUTION: DAPI (see Appendix 2 for Caution)

3 Rinse the slides in PBS and then mount stained specimen (see p. 252).

BrdU INCORPORATION FOR CHROMOSOME BANDING AND LABELING OF REPLICATION SITES

To detect nascent DNA synthesis in S-phase cells or for an R-banding distribution on metaphase chromosomes, a halogenated nucleotide base analog, such as BrdU, can be incubated in the medium of actively proliferating tissue culture cells. To obtain higher yields, including a protocol for cell synchronization is recommended (see Chapter 13). To visualize replication sites, incubation is performed for a short time immediately prior to fixation. The incorporated BrdU is detected by a fluorochrome-conjugated anti-BrdU antibody. To reveal R-banding in metaphase chromosomes, BrdU is incubated with cells for 6 hours before harvesting and detected via antibodies (step 4d below) or via dyes such as Hoechst or PI (Takahashi et al. 1990) that differentially stain BrdU-containing DNA.

1 Mix the following components together to make analog mix:

4.2 mg/ml BrdU stock solution	0.7 ml
100 μg/ml FdU stock solution	0.1 ml
1 mg/ml uridine stock solution	0.2 ml

BrdU Stock Solution

BrdU	30 mg
ddH_2O	7 ml

Store frozen at −20°C.

FdU Stock Solution

FdU	1 mg
ddH_2O	10 ml

Store frozen at –20°C.

Uridine Stock Solution

uridine	10 mg
ddH_2O	10 ml

Store frozen at –20°C.

2 Add 100 μl of analog mix to 10 ml of cell culture and incubate for 2–5 minutes at 37°C to locate replication sites and 5–6 hours to reveal R-bands.

When performing R-banding, the duration of BrdU incubation must be optimized for different cultures.

3 Perform a fixation protocol suitable for your experimental design (see pp. 239–240).

4 Detect replication sites or fluorescent banding.

For replication sites:

a Denature the fixed and permeabilized specimen by incubation with 1 N HCl for 20 minutes at room temperature.

b Rinse the slides in PBS three times for 5 minutes at room temperature.

c Proceed with the FISH protocol (see p. 244).

d After hybridization and concomitant with the detection step for FISH (see pp. 244–252), incubate the cells with a fluorochrome-conjugated anti-BrdU antibody.

For fluorescence banding to BrdU-containing metaphase preparations:

a Rinse the slides in 2x SSC for 5 minutes.

b Stain the slides in 1 μg/ml bisbenzimide in 2x SSC for 20 minutes. (Dilute a 1 mg/ml bisbenzimide stock solution 1:1000.)

c Rinse the slides in 2x SSC three times for 5 minutes each at room temperature.

d Irradiate the slides under UV light (254 nm) for 20 minutes.

e Proceed with the FISH procedure for metaphase chromosomes.

f Rinse the slides in PBS three times for 5 minutes at room temperature or counterstain with 0.1–1 μg/ml PI in PBS.

Alu-banding can also be performed and Alu sequences can be detected with a repetitive probe specific for the Alu repeats (Lichter et al. 1990b).

Immunolocalization in Combination with FISH

Immunohistochemistry is a technique that allows the visualization and localization of specific cellular or tissue components by the binding of an antibody to an antigen within the cell or tissue. The binding of antibodies to an endogenous cellular component is conducted through highly specific antigen/antibody interactions. The localization of the antigen/antibody complex is normally performed by a secondary antibody recognizing the primary antibody conjugated to a reporter molecule. This is commonly a fluorochrome or an enzyme. The secondary antibody reacts with the species-specific part of the primary antibody. Alternatively, the primary antibody can be directly labeled, precluding the need for a secondary antibody.

To study cytoplasmic or nuclear structure or specific cellular processes, such as transcription, DNA replication, or signal transduction, it is sometimes necessary to combine visualization of nucleic acids by FISH and visualization of other cellular components by indirect immunofluorescence (IF) (Fig. 15.5C,D; Fig. 15.7). This is not always easy because the denaturation procedure for FISH can destroy or distort the antigen in such a way that it cannot be recognized by the antibody after denaturation. In addition, the acid treatment used for permeabilization can alter the epitope recognized by the antibody. Hence, in these cases, it is possible to omit the acid treatment. Some antigens are particularly sensitive to the fixation protocol used, and therefore they may still be recognized in organic solvent-fixed cells but not in inorganically fixed cells, and vice versa. Furthermore, distributions of some antigens can vary in cells fixed with different methods. Hence, one should test various fixation protocols.

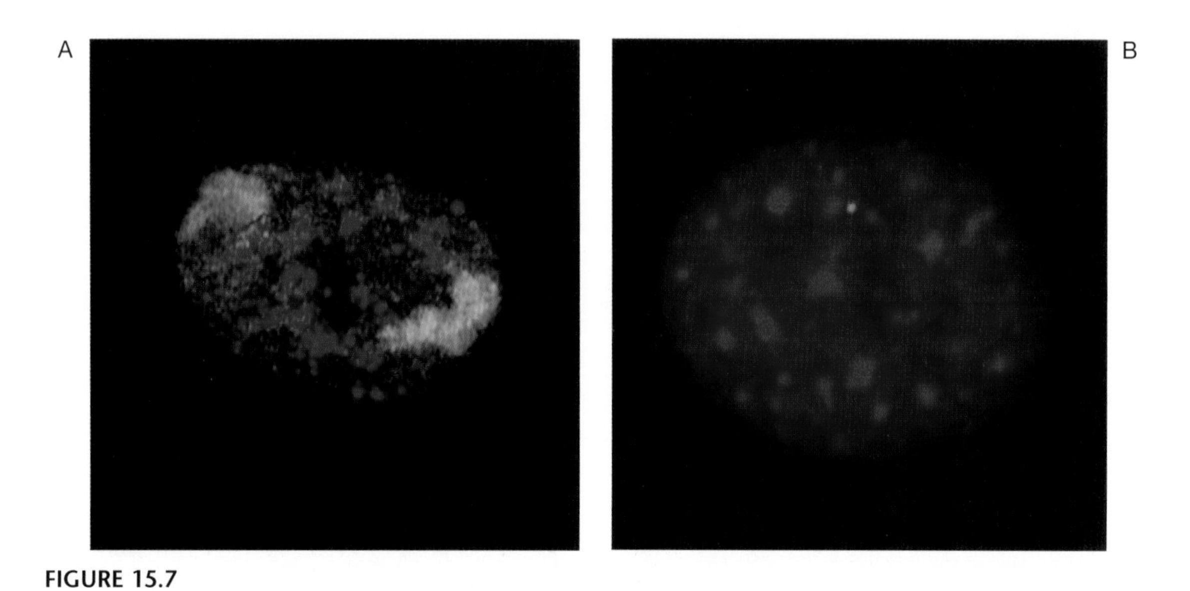

FIGURE 15.7

(*A*) Visualization of the two X chromosomes (*green*) in a female fibroblast by chromosome painting and a pre-mRNA splicing factor (*red*) by indirect immunofluorescence microscopy. Only the transcriptionally active X chromosome (*right*) is associated with splicing factors. (*B*) Visualization of the stably integrated cytomegalovirus immediate-early genes (*green*) by FISH and a pre-mRNA splicing factor (*red*) by indirect immunofluorescence microscopy. (Photos provided by T. Misteli and D.L. Spector, Cold Spring Harbor Laboratory.)

To combine FISH and IF can be quite difficult, but for some antigens there are no problems. There are four different possibilities in performing combined FISH and IF.

- The complete IF procedure can be performed after the hybridization step.
- The complete IF procedure can be performed prior to denaturation.
- The IF can be split; in other words, the primary antibody incubations can be performed before specimen denaturation for FISH, with the secondary antibody incubation during the FISH detection step.
- Primary or primary and secondary antibodies conjugated to a reporter incubation can be performed before specimen denaturation.

IMMUNOLOCALIZATION FOLLOWING SPECIMEN DENATURATION

1. After the posthybridization washes and blocking step of FISH, place on the slide 50–200 μl of 1–20 μg/ml detection reagent reacting against the labeled probe; for example, rhodamine-conjugated streptavidin or fluorescein-conjugated anti-digoxigenin, plus 1–20 μg/ml primary antibody in 4x SSC containing 1% BSA or other blocking reagent. Cover with a glass coverslip and place in a humidified chamber.

   Alternatively, the entire IF can be performed after the detection steps for ISH.

   Do not use the same species origin of antibody for ISH detection procedure and the primary antibody because the secondary antibody will react with both.

2. Incubate the slides for 30 minutes at 37°C or 1 hour at room temperature.

3. Wash the slides in wash buffer C (see p. 251) three times for 5 minutes each at 42°C.

4. Place on the slides 50 μl of 10–20 μg/ml secondary antibody solution made up in either 4x SSC or PBS containing 1% BSA (w/v) or other blocking reagent.

5. Wash the slides three times for 5 minutes in PBS (+1% Tween 20, *optional*).

6. Mount specimens as described on p. 252 and view.

IMMUNOLOCALIZATION PRIOR TO SPECIMEN DENATURATION

With some antigens, it is necessary to perform the indirect IF technique entirely prior to the denaturation of DNA and to completely immobilize the primary and secondary antibodies with an additional fixation step. Unfortunately, this method often reduces the intensity of the immunostaining.

1. After fixation and pretreatment of slides (see p. 238), rinse slides in PBS a few times.

2. Place 50–200 μl of the primary antibody solution on the slide and cover with a glass coverslip. Incubate for 30 minutes at 37°C, or 1 hour at room temperature, or overnight at 4°C.

3. Wash the slides in PBS three times for 5 minutes. Place 50 μl of secondary antibody solution on the slide. Incubate for 30 minutes at 37°C, or 1 hour at room temperature, or 4 hours at 4°C.

4. Wash the slides in PBS three times for 5 minutes.

5 Place the slides in 4% **paraformaldehyde** (w/v) for 10–20 minutes at room temperature.

CAUTION: Paraformaldehyde (see Appendix 2 for Caution)

6 Rinse the slides in PBS and then continue with denaturation of cells for ISH.

Fluorochromes are now present on the slides, so slides should be kept in the dark as much as possible.

PRIMARY ANTIBODY INCUBATION PREDENATURATION; SECONDARY ANTIBODY INCUBATION POSTDENATURATION

This procedure is preferable to the entire IF process being performed before denaturation because of the reduction in fluorescence intensity from the secondary antibody caused by the denaturation process. This is the method of choice if an overnight incubation of primary antibody is required, because sometimes antigens are only revealed with long primary antibody incubations (Bridger et al. 1993).

1 After fixation and pretreatment of the slides (see Chapter 4), place them in PBS.

2 Perform primary antibody incubation as described in step 2 (bottom of p. 237).

3 Fix primary antibody with 4% **paraformaldehyde** for 10–20 minutes.

CAUTION: Paraformaldehyde (see Appendix 2 for Caution)

4 Rinse slides in PBS and proceed with FISH denaturation step (see pp. 248–250).

5 Perform secondary antibody incubation either simultaneously with, or subsequent to, the detection of the labeled probe (see p. 251).

6 Mount specimen as described in Chapter 4 and view.

REPORTER-MODIFIED ANTIBODY INCUBATION PRE-FISH

In this modification of the protocol, the primary antibody is directly conjugated to a reporter molecule that can survive the denaturation procedure. Such haptens include biotin, digoxigenin, or DNP.

The protocol for this procedure is similar to the procedures presented on pp. 257–258. A secondary detection reagent, incubated postdenaturation, is required when using an antibody labeled with biotin, digoxigenin, or DNP.

REFERENCES

Bentz, M., Döhner H., Schröder M., Pohl S., and Lichter P. 1993a. Detection of chromosomal abnormalities on previously stained peripheral blood and bone marrow smears. In *Acute leukemias IV: Prognostic factors and treatment strategies*, pp. 38–42. Springer Press, Heidelberg.

Bentz M., Schröder M., Herz M., Stilgenbauer S., Lichter P., and Döhner H. 1993b. Detection of trisomy 8 on blood smears using fluorescence in situ hybridization. *Leukemia* **7:** 752–757.

Bohlander S.K., Espinosa R., Le Beau M., Rowley J.D., and Diaz M.O. 1992. A method for the rapid sequence-independent amplification of microdissected chromosomal material. *Genomics* **13:** 1322–1324.

Boyle A.L., Lichter P., and Ward D.C. 1990. Rapid analysis of mouse-hamster hybrid cell lines by in situ hybridization. *Genomics* **7:** 127–130.

Bridger J.M., Kill I.R., O'Farrell M., and Hutchison C.J. 1993. Internal lamin structures within G1 nuclei of human dermal fibroblasts. *J. Cell Sci.* **104:** 297–306.

Carter N.P., Ferguson-Smith M., Perryman M.T., Telenius H., Pelmear A.H., Leversha M.A., Glancy M.T., Wood S.L., Cook K., and Dyson H.M. 1992. Reverse chromosome painting: A method for the rapid analysis of aberrant chromosomes in clinical cytogenetics. *J. Med. Genet.* **29:** 299–307.

Chang K.S., Vyas R.C., Deaven L.L., Trujillo J.M., Stass S.A., and Hittelman W.N. 1992. PCR amplification of chromosome-specific DNA isolated from flow cytometry-sorted chromosomes. *Genomics* **12:** 307–312.

Collins C., Kuo W.L., Segraves R., Fuscoe J., Pinkel D., and Gray J. 1991. Construction and characterization of plasmid libraries enriched in sequences from single human chromosomes. *Genomics* **11:** 997–1006.

Dauwerse J.G., Wiegant J., Raap A.K., Breuning M.H., and van Ommen G.J.B. 1992. Multiple colors by fluorescence in situ hybridization using radiolabelled DNA probes create a molecular karyotype. *Hum. Mol. Genet.* **1:** 593–598.

Dunham I., Lengauer C., Cremer T., and Featherstone C. 1992. Rapid generation of chromosome specific alphoid DNA probes using the polymerase chain reaction. *Hum. Genet.* **88:** 457–462.

Fuscoe J.C., Collins C.C., Pinkel D., and Gray J.W. 1989. An efficient method for selecting unique-sequence clones from DNA libraries and its application to fluorescent staining of human chromosome 21 using in situ hybridization. *Genomics* **5:** 100–109.

Hopman A.H.N., Poddighe P., Moesker O., and Ramaekers F.C.S. 1992. Interphase cytogenetics: An approach to the detection of genetic aberrations in tumours. In *Diagnostic molecular pathology: A practical approach* (ed. H. McGee), pp. 142–167. IRL Press Inc., Oxford.

Hopman A.H.N., Ramaekers F.C.S., Raap A.K., Beck J.L.M., Devilee P., van der Ploeg M., and Vooijs G.P. 1988. In situ hybridization as a tool to study numerical chromosome aberrations in solid bladder tumors. *Histochemistry* **89:** 307–316.

Joos J., Fink T.M., Rätsch A., and Lichter P. 1994. Mapping and chromosome analysis: The potential of fluorescence in situ hybridization. *J. Biotech.* **35:** 135–153.

Kievits T., Devilee P., Wiegant J., Wapenaar M.C., Cornelisse C.J., van Ommen G.J.B., and Pearson P.L. 1990. Direct nonradioactive in situ hybridization of somatic cell hybrid DNA to human lymphocyte chromosomes. *Cytometry* **11:** 105–109.

Korenberg J.R., Yang-Feng T., Schreck R., and Chen X.N. 1992. Using fluorescence in situ hybridization (FISH) in genome mapping. *Trends Biotechnol.* **10:** 27–32.

Kurz A., Lampel S., Nickolenko J.E., Bradl J., Benner A., Zirbel R.M., Cremer T., and Lichter P. 1996. Active and inactive genes localize preferentially in the periphery of chromosome territories. *J. Cell Biol.* **135:** 1195–1205.

Landegent J.E., Jansen in de Wal N., van Ommen G.-J., Baas F., de Vijlder J.J.M., van Duijn P., and van der Ploeg M. 1985. Chromosomal localization of a unique gene by non-autoradiographic in situ hybridization. *Nature* **317:** 175–177.

Langer, P.R., Waldrop A.A., and Ward D.C. 1981. Enzymatic synthesis of biotin-labeled polynucleotides: Novel nucleic acid affinity probes. *Proc. Natl. Acad. Sci.* **78:** 6633–6637.

Lawrence J.B., Singer R.H., and Marselle L.M. 1989. Highly localized tracks of specific transcripts within interphase nuclei visualized by in situ hybridization. *Cell* **57:** 493–502.

Ledbetter S.A., Garcia-Heras J., and Ledbetter D.H. 1990. "PCR-karyotype" of human chromosomes in somatic cell hybrids. *Genomics* **8:** 614–622.

Lengauer C., Lüdecke H.-J., Wienberg J., Cremer T., and Horsthemke B. 1991a. Comparative chromosome band mapping in primates by in situ suppression hybridization of band specific DNA microlibraries. *Hum. Evol.* **6:** 67–71.

Lengauer C., Eckelt A., Weith A., Endlich N., Ponelies N., Lichter P., Greulich K.O., and Cremer T. 1991b. Painting of defined chromosomal regions by in situ suppression hybridization of libraries from laser-microdissected chromosomes. *Cytogenet. Cell Genet.* **56:** 27–30.

Lichter P. and Ward D.C. 1990. Is non-isotopic in situ hybridization finally coming of age? *Nature* **345:** 93–95.

Lichter P., Boyle A.L., Cremer T., and Ward D.C. 1991. Analysis of genes and chromosomes by nonisotopic in situ hybridization. *Genet. Anal. Techn. Appl.* **8:** 24–35.

Lichter P., Jauch A., Cremer T., and Ward D.C.. 1990a. Detection of Down syndrome by in situ hybridization with chromosome 21 specific DNA probes. In *Molecular genetics of chromosome 21 and Down syndrome: Proceedings of the Sixth Annual National Down*

Syndrome Society Symposium (eds. D. Paterson and C.J. Epstein), pp. 69–78. Wiley, New York.

Lichter P., Ledbetter S.A., Ledbetter D.H., and Ward D.C. 1990b. Fluorescence in situ hybridization with Alu and L1 polymerase chain reaction probes for rapid characterization of human chromosomes in hybrid cell lines. *Proc. Natl. Acad. Sci.* **87:** 6634–6638.

Lichter P., Cremer T., Borden J., Manuelidis L., and Ward D.C. 1988. Delineation of individual human chromosomes in metaphase and interphase cells by in situ suppression hybridization using recombinant DNA libraries. *Hum. Genet.* **80:** 224–234.

Manuelidis L. 1985. Individual interphase chromosome domains revealed by in situ hybridization. *Hum. Genet.* **71:** 288–293.

Manuelidis L. and Borden J. 1988. Reproducible compartmentalization of individual chromosome domains in human CNS cells revealed by in situ hybridization and three-dimensional reconstruction. *Chromosoma* **96:** 397–410.

Melzer P.S., Guan X.Y., Burgess A., and Trent J.M. 1992. Rapid generation of region specific probes by chromosome microdissection and their application. *Nature Gen.* **1:** 24–28.

Nelson D.L., Ledbetter S.A., Corbo L., Victoria M.F., Ramirez-Solis R., Webster T.D., Ledbetter D.H., and Caskey C.T. 1989. Alu polymerase chain reaction: A method for rapid isolation of human-specific sequences from complex DNA sources. *Proc. Natl. Acad. Sci.* **86:** 6686–6690.

Nisson P.E., Rashtchian A., and Watkins P.C. 1991. Rapid and efficient cloning of Alu-PCR using uracil DNA glyosylase. *PCR Methods Appl.* **1:** 120–123.

Pinkel D., Straume T., and Gray J. 1986. Cytogenetic analysis using quantative, high sensitivity, fluorescence hybridization. *Proc. Natl. Acad. Sci.* **83:** 2934–2938.

Pinkel D., Landegent J., Collins C., Fuscoe J., Segraves R., Lucas J., and Gray J.W. 1988. Fluorescence in situ hybridization with human chromosome-specific libraries: Detection of trisomy 21 and translocations of chromosome 4. *Proc. Natl. Acad. Sci.* **85:** 9138–9142.

Raap A.K., van de Rijke F.M., Dirks R.W., Sol C.J., Boom R., and van der Ploeg M. 1991. Bicolor fluorescence in situ hybridization to intron and exon mRNA sequences. *Exp. Cell Res.* **197:** 319–322.

Ried T., Baldini A., Rand T.C., and Ward D.C. 1992. Simultaneous visualization of seven different DNA probes by in situ hybridization using combinatorial fluorescence and digital imaging microscopy. *Proc. Natl. Acad. Sci.* **89:** 1388–1392.

Scherthan H. and Cremer T. 1994. Methology of non-isotopic in situ hybridization in paraffin-embedded tissue sections. In *Methods in molecular genetics*, vol. 2: *Chromosome and gene analysis* (ed. K.W. Adolph). Academic Press, New York.

Stilgenbauer S., Döhner H., Bulgay-Mörschel M., Weitz S., Bentz M., and Lichter P. 1993. High frequency of monoallelic retinoblastoma gene deletion in B-cell chronic lymphoid leukemia shown by interphase cytogenetics. *Blood* **81:** 2118–2124.

Suijkerbuijk R.F., Matthopoulos D., Kearney L., Monard S., Dhut S., Cotter F., Herbergs J., van Kessel A.G., and Young B.D. 1992. Fluorescent in situ identification of human marker chromosomes using flow sorting and Alu element-mediated PCR. *Genomics* **13:** 355–362.

Takahashi E.I., Hori T.A., O'Connell P., Leppert M., and White R. 1990. R-microbanding and non-isotopic in situ hybridization: Precise localization of the human type II collagen gene (COL2A1). *Hum. Genet.* **86:** 14–16.

Telenius H., Carter N.P., Bebb C.E., Nordenskjöld M., Ponder B.A.J., and Tunnacliffe A. 1992. Degenerate oligonucleotide-primed PCR: General amplification of target DNA by a single degenerate primer. *Genomics* **13:** 718–725.

Trask B. 1991. Fluorescence in situ hybridization: Applications in cytogenetics and gene mapping. *Trends Genet.* **7:** 149–154.

Wiegant J., Wiesmeijer C.C., Hoovers J.M.N., Schuuring E., d'Azzo A., Vrolijk J., Tanke H.J., and Raap A.K. 1993. Multiple and sensitive fluorescence in situ hybridization with rhodamine-, fluorescein-, and coumarin-labeled DNAs. *Cytogenet. Cell. Genet.* **63:** 73–76.

Zirbel R.M., Mathieu U.R., Kurz A., Cremer T., and Lichter P. 1993. Evidence for a nuclear compartment of transcription and splicing located at chromosome domain boundaries. *Chromosome Res.* **1:** 93–106.

CHAPTER 16

Whole-Mount Fluorescence In Situ Hybridization to *Drosophila* Chromosomal DNA

Abby Dernburg

Lawrence Berkeley National Laboratory, Berkeley, California

INTRODUCTION

Diverse questions in cell biology require information about the subnuclear localization of specific chromosomal regions. Such questions can be addressed by probing specific sequences using fluorescence in situ hybridization (FISH) of labeled probes to whole cells and tissues. Here, this general experimental technique is referred to as 3D FISH. The optimization of reliable methods for sample preparation and hybridization has made this technique nearly as straightforward as immunofluorescence. Fluorescence-based detection is advantageous because light-emitting probes can be detected with high sensitivity against a black background, because several different probes may be discriminated in the same sample, and also because fluorescence microscopy can provide excellent spatial resolution.

3D FISH allows investigation of chromosome organization in a wide variety of tissues and developmental stages (Fig. 16.1). Although individual tissues may require specialized preparation methods, the same general principles guide synthesis of probes, hybridization procedures, and detection strategies for all 3D FISH experiments. A major advantage of the techniques presented here is that they allow simultaneous localization of proteins by immunofluorescence and chromosomal sequences by FISH.

These methods have been subdivided into three sections: The first section discusses probe design considerations and synthesis procedures. In the second section, a general approach to tissue fixation is presented, and the third section focuses on the hybridization and fluorescent detection of labeled probes.

Probes for 3D FISH

The choice of sequence targets depends on individual experimental goals, but some principles apply to probes used in any whole-mount FISH experiment. First, regardless of the length of the target sequence, the probe must be fragmented so that individual molecules can easily diffuse through fixed tissue. Empirically, probe fragments of less than 150 bases are optimal. If fragments are too long, they will tend to generate a punctate fluorescent background, particularly at nuclear and cell boundaries. Second, an efficient labeling method should be used; one that has proven reliable is presented here. Third, if multiple probes are to be detected in the same specimen, consideration must be given to the labeling scheme so that all signals can be distinguished.

Both single-copy regions and repetitive sequences can be targeted in 3D FISH experiments. In general, probes to abundant repetitive sequences work more reliably and give brighter signals, even relative to the total chromosomal representation of the sequence, than single-copy probes.

The lower limit of target size that can be detected will be determined by several factors, including the tissue and its preparation, the probe labeling and detection scheme, and the

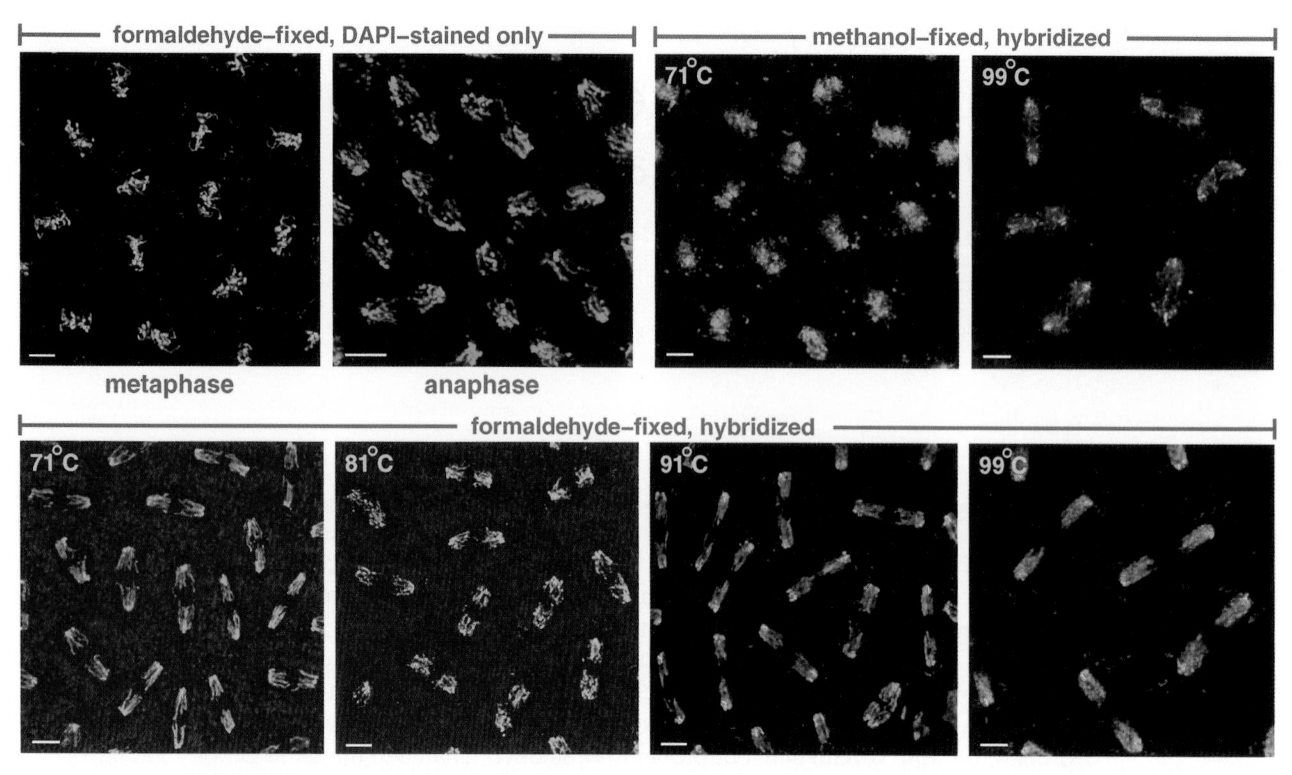

FIGURE 16.1

Effects of fixation method and denaturation temperature on 3D FISH in *Drosophilia* embryos. In situ hybridization requires the denaturation of chromosomal DNA, here achieved by brief heat treatment. To accomplish this denaturation with minimal perturbation of the chromosome morphology, the tissue must be fixed appropriately, and the denaturation temperature chosen to optimize signal/background. Chromosome preservation can be assayed visually by comparison to samples that have been fixed but not subjected to hybridization. The first two panels show metaphase and anaphase chromosomes in syncytial *Drosophila* embryos fixed with buffered formaldehyde (as described in this chapter) and stained with the DNA-binding dye DAPI. Well-preserved mitotic nuclei should reveal distinct, individual chromosome arms. Anaphase figures often appear somewhat "collapsed" when fixed by this procedure, probably because of depolymerization of the mitotic spindle during the fixation. All other panels show embryos that have been subjected to the in situ hybridization procedure described in this chapter, varying only the type of fixation and temperature of denaturation. The probe was a double-stranded polymerase chain reaction product resulting from amplification of the 359-bp repeat (Hsieh and Brutlag 1979), which makes up ~10 Mb of the heterochromatin of the X chromosome (embryos showing only one signal in every nucleus are thus male). Because this probe recognizes a highly abundant sequence, it can be detected even under suboptimal experimental conditions. The probe was digested with restriction enzymes and terminally labeled with biotin, as described in this chapter. Detection was performed using FITC-conjugated UltraAvidin (Leinco Technologies). Fixation in methanol-EGTA is compatible with in situ hybridization, but the chromosome preservation is poor (top, right two panels). Nevertheless, such fixation is adequate for many experimental goals in which the presence or absence of a signal is being scored, as in karyotyping or sexing embryos. For optimal chromosome preservation following FISH, fixation in formaldehyde (*bottom panels*) is preferable. With formaldehyde fixation, the denaturation temperature has a marked effect on signal/background: Denaturation at 71°C results in poor signal intensity and strong cytoplasmic fluorescence, but increasing the temperature over the range of 71–91°C both increases the signal and reduces the background staining, without adversely affecting morphology. With higher temperatures or longer denaturation times (*right panel*), the morphology begins to deteriorate, and little if any further increase in signal is seen. Each of these images is a projection through ~2 µm of a three-dimensional data stack, following restoration of out-of-focus information by mathematical deconvolution. Scale bars, 5 µm.

microscope used to record the fluorescent signals. Typically, a minimum of 5–50 kb of chromosomal target is required for reproducible detection. In some cases, the limit is substantially higher, because of poor permeability of the tissue, condensed and thus less-accessible chromatin, or high autofluorescence in the sample.

Although RNA probes may be useful in particular cases, their disadvantages generally outweigh any advantage they might provide. The probe synthesis method given here is designed to label DNA, which can be synthesized and purified by a wide variety of techniques.

Preparation and Labeling of Probes

A variety of methods exist to label DNA, most involving incorporation of modified nucleotides by enzymes (these include random priming, nick translation, and 5′- and 3′-end labeling; see Chapter 82 of Spector et al. 1998). However, they are not all equal in their ability to generate probes of high quality reproducibly. For some applications this is not critical, but, particularly when making probes to low-abundance sequences, care should be taken to ensure that probes are optimally labeled. A particularly reliable scheme employs the enzyme terminal deoxynucleotidyl transferase (TdT) to incorporate labeled nucleotides onto the 3′ end of DNA fragments. With this method, DNA must be fragmented prior to labeling so that the resulting probe molecules are sufficiently small.

If the sequence target comprises a short, tandemly repeated sequence, it is most convenient to generate probes starting from either a polymerase chain reaction (PCR) product (if the repeat is moderately long) or a synthetic oligonucleotide of 20–50 bases in length, corresponding to one or more copies of the repeated sequence. If an oligonucleotide or a PCR product of less than 200 bp is used this fragmentation step is unnecessary.

DNA FRAGMENTATION

DNA to be labeled for use as probe may be fragmented using either restriction endonucleases or DNase I.

DNA Fragmentation with a Mixture of Restriction Endonucleases

Either site-specific or nonspecific endonucleases may be used to fragment probe DNA prior to labeling. Using restriction endonucleases is technically simpler but more costly.

1 To a 1.5-ml tube on ice add:

   20–30 μg of plasmid DNA
   50 μl of 5x 4-base-cutter buffer
   (40 mM Tris-HCl, pH 7.5, 250 mM NaCl, 40 mM **$MgCl_2$**)
   2.5 μl 5 mg/ml BSA (bovine serum albumin) (50 μg/ml final)
   2.5 μl 100 mM **DTT** (dithiothreitol) (1 mM final)
   H_2O to about 225 μl (depending on volume of enzymes needed; total reaction volume should be 250 μl)

   CAUTION: $MgCl_2$; DTT (see Appendix 2 for Caution)

2 Add 50 units each of the following 4-base cutters (all available from New England Biolabs):

   AluI
   HaeIII
   MseI
   MspI
   RsaI
   Sau3AI if DNA is Dam methylated or DpnI (which is cheaper) if it is not

3 Mix the reaction components and incubate at 37°C for at least 2 hours.

All of these enzymes are fairly stable in a reaction, so longer digestion with lower concentrations of enzyme is an option.

4 Check the size distribution by electrophoresis through a 2% agarose gel; compare to pBR322-*Msp*I markers (New England Biolabs) or other appropriate size standards. Average fragment size should be ~100–150 bp.

Optional: Extract the reaction mixture with aqueous-saturated ***phenol****.*

CAUTION: Phenol (see Appendix 2 for Caution)

5 Ethanol-precipitate the DNA using sodium acetate and resuspend to 1 µg/µl DNA in TE′ (10 mM Tris-HCl, 0.1 mM EDTA).

The EDTA concentration is low to avoid inhibiting the subsequent terminal transferase reaction.

CHOICE OF PROBE LABELS FOR 3D FISH

There are currently a wide variety of commercially available dyes that are useful for FISH applications. The development of DNA microarrays has motivated a great deal of creative and commercial investment into fluorescent DNA-labeling technologies, so the range of options is constantly broadening. In choosing a labeling and detection scheme, several factors should be considered.

First, probes can be directly labeled with fluorescently conjugated nucleotides or they can be hapten-labeled and then detected following hybridization using fluorescent antibodies or hapten-binding proteins such as avidin. Both scenarios have advantages and disadvantages. Directly labeled probes are simpler to use, since they require no secondary staining step. Directly labeled probes tend to give optimal signal-to-background fluorescence in FISH experiments. Conversely, the use of secondary detection reagents tends to increase fluorescent background, but on the other hand the absolute intensity of the signals is also brighter. This may be particularly advantageous if laser illumination will be used to image the specimens, because shorter exposure times can be used and less photobleaching will occur. As a rule, "sandwiching" techniques to amplify signals (using multiple layers of avidin/biotin, secondary antibodies, etc.) are problematic in whole-mount tissues. The fluorescence background often increases at least in proportion to the signal, which can make probes more difficult to detect, rather than easier. Similarly, amplification schemes employing alkaline phosphatase or horseradish peroxidase to precipitate fluorescent products cause vastly increased background and irreproducible staining. Generally it is best to use the simplest and most direct detection: either the probe is labeled fluorescently or it is detected with one layer of fluorescently conjugated antibody or avidin.

An additional consideration is that samples are most often counterstained with fluorescent DNA dyes, and the probe detection scheme must be compatible with the dye that is chosen. With wide-field fluorescence microscopy employing mercury arc lamp illumination the most useful counterstain is DAPI. Probes labeled with dyes requiring ultraviolet illumination (such as AMCA [Amersham]) have yielded poor results, and so we reserve this channel for the DAPI counterstain and use other wavelengths to detect the probes. However, laser-scanning confocal microscopes are often restricted to longer wavelength illumination, so counterstains such as propidium iodide (PI) or OliGreen (Molecular Probes) are more suitable and probes must be labeled so as to fluoresce in other portions of the spectrum.

With the development of dye-conjugation labeling methods, a wide variety of fluorescent labeling reagents are available for conjugating to amine-modified DNA. The Alexa dyes from Molecular Probes, as well as the CyDyes from GE Healthcare (formerly Amersham Biosciences), have proven very reliable, photostable, and bright. DNA that has been chemically aminated or enzymatically synthesized with the incorporation of aminoallyl-dU can be conjugated with the amine-reactive forms of these dyes, either N-hydroxysuccinimidyl esters (GE) or Alexa-488 TFP ester (Molecular Probes/Invitrogen).

To label probes with haptens, both biotin-dNTPs (various suppliers) and digoxigenin-dUTP (Boehringer Mannheim) have proven to be reliable.

Biotinylated probes are most successfully detected with fluorescently conjugated, chemically modified avidin or streptavidin. Native avidin is both highly charged at neutral pH and also glycosylated, and each of these factors can contribute to nonspecific binding of the protein. Streptavidin usually generates somewhat lower fluorescent background in whole-mount specimens, as do modified avidins such as NeutrAvidin (Molecular Probes). These reagents are used at dilutions of approximately 1:5,000 from the supplied stock solutions, stored at 4°C. Fluorescent anti-biotin antibodies (Jackson Immunoresearch) also perform very well for FISH applications. To detect digoxigenin-labeled probes, the reagents of choice are fluorescently conjugated anti-digoxin antibodies from Jackson Immunoresearch.

The labeling strategy is particularly critical for experiments involving multiple probes. The number of probes that can be discriminated in any given sample will depend on the excitation and emission filters available. However, the fluorophores must also be selected so as to minimize overlap between both their excitation and emission spectra.

LABELING

To label DNA fragments with the TdT enzyme, a mixture of unlabeled and labeled nucleotides is used, typically 2:1 unlabeled to labeled. This is because TdT, like most polymerases, works more processively if unmodified nucleotides are included, and also because incorporation of a label at too high a density will actually reduce the resultant fluorescence, as a result of quenching. Here I provide a procedure for incorporating an amine-modified nucleotide, aminoallyl-dUTP, which can subsequently be reacted with a wide variety of amine-reactive fluorescent dyes. This procedure can also be used to incorporate fluorescently conjugated nucleotides, but these reagents are extremely costly and are poorly incorporated by enzymes. Probes labeled by amino-allyl dUTP incorporation followed by dye conjugation are thus more economical to generate and also tend to produce better results in FISH experiments.

Note that amine-reactive fluorophores will react with TRIS or ammonium ions as well as with the amine-modified DNA, so the DNA should contain minimal concentrations of such contaminants prior to dye conjugation.

1. Dilute 10 μg of digested DNA or oligonucleotide with water to a total volume of 68 μl.
   This is to 10 μg scale—scale up or down as desired.

2. If DNA is double-stranded, cap the tube and place the tube in a 95°C water bath for 2 minutes, then chill immediately on ice; spin briefly to collect condensed water from the sides of the tube (single-stranded DNA is a preferred substrate for TdT).

3. Add (at room temperature)

   20 μl Roche 5x TdT reaction buffer
   3.3 μl 1 mM aa-dUTP (Sigma A0410; 10 mM stock should be prepared from the lyophilized powder in molecular biology grade water, aliquoted, and stored at –80°C for long-term stability; a working aliquot can be stored at –20°C and diluted tenfold shortly before use)

6.6 μl 1 mM unlabeled dTTP (freshly diluted from 100 mM stock solution)
2 μl (800 units) TdT (Roche 3333566)

Allow reaction to proceed for 1 hour at 37°C. Add EDTA to 5 mM final concentration.

4 Add 1 μl of 20 mg/ml glycogen (Roche, molecular biology grade).
This acts as a carrier that helps to precipitate small probe fragments.

5 Add NaOAc (do not add NH_4OAc because it will react with the dye) to 0.3 M, then add 2.5 volumes of absolute ethanol to precipitate the DNA. Chill briefly at –80°C.

6 Spin 15 minutes at maximum speed in a microcentrifuge.

7 Wash with ice-cold 75% ethanol.

8 Dry and resuspend pellet in TE (10 mM Tris-HCl, pH 7, 1 mM EDTA).
It is convenient to store probes at appropriate concentrations so that 1 μl is used per sample in a hybridization experiment. However, this amount varies: For simple sequence probes, a concentration of about 50 ng/μl is suitable; for complex single-copy probes it can be as high as 200–500 ng/μl.

Sample Preparation for 3D FISH

The key to success in applying 3D FISH to any tissue is to identify isolation and fixation conditions that preserve the sample morphology while permitting probe diffusion, DNA denaturation, and hybridization to occur. In general, the best fixative is **formaldehyde. Glutaraldehyde** fixation is generally not compatible with hybridization, probably because it cross-links the chromosomal DNA and prevents denaturation. In cases where light formaldehyde fixation is not suitable, e.g., if detection of an antigen that is not preserved by such fixation is required, other approaches may be taken, but these usually give inferior morphological preservation.

CAUTION: Formaldehyde; glutaraldehyde (see Appendix 2 for Caution)

Formaldehyde can be purchased as a stabilized solution, but more consistent results may be obtained if it is prepared fresh from **paraformaldehyde** (EM grade, Polysciences). To make 5 ml of 37% formaldehyde, add 1.85 g of paraformaldehyde and 3.5 ml of H_2O to a large glass test tube. Loosely cap and place in a boiling water bath in a fume hood. Add 90 µl of 1 N NaOH, and shake the tube for ~1 minute, until the solution has nearly cleared. Remove from the water bath, allow to cool, and immediately filter through a nonsterile 0.45-µm syringe filter into a glass vial with an air-tight lid. Use the same day, and discard if any precipitate forms.

CAUTION: Paraformaldehyde (see Appendix 2 for Caution)

Formaldehyde is both an irritant and carcinogenic. The vial containing formaldehyde solution should be kept capped at all times.

The tissue may be carried through the hybridization procedure either in suspension in liquid in small tubes or affixed to microscope slides. When possible, it is simpler to perform the procedure in tubes since the volume of solutions required is much lower, and the temperature and duration of the denaturation step can be easily and precisely controlled using thermal cycler technology. However, fixation on a microscope slide is a good solution to the tissue-handling difficulties presented by single cells or tissues that are too small or delicate to handle in tubes.

As in immunostaining procedures, sometimes physical barriers must be removed in order for probes to penetrate a particular tissue. The vitelline membrane of a *Drosophila* embryo is one such obstacle, but efficient methods have been devised to remove it. The cell walls of plants and fungi and the cuticle of *Caenorhabditis elegans* larvae and adults are other examples of barriers that must be removed by either physical or enzymatic methods to attain successful hybridization.

DROSOPHILA EMBRYOS

For *Drosophila* embryos, the tissue for which these procedures were originally developed, standard formaldehyde fixation identical to that used for immunocytochemical staining is ideal.

1. Dechorionate embryos en masse using sodium hypochlorite (commercial bleach) (see Chapter 11), rinse free of bleach, and fix by shaking for 10 minutes in a 1:1 mixture of **heptane** and 3.7% **formaldehyde** in Buffer A (15 mM PIPES buffer, pH 7, containing 80 mM **KCl**, 20 mM NaCl, 2 mM EDTA, and 0.5 mM EGTA, 1 mM **DTT**, plus the polyamines spermine and spermidine at 0.15 and 0.5 mM, respectively). PBS (phosphate-buffered saline) or other buffers may be substituted.

2 Transfer the fixed embryos to a prechilled (–20°C) mixture of 7.5 ml of heptane, 7.1 ml of **methanol**, and 0.4 ml of 100 mM EGTA in a 25 x 200-mm test tube for 5–10 minutes.

CAUTION: Methanol (see Appendix 2 for Caution)

3 Heat-shock the embryos by running the bottom of the tube under warm water (~50°C for 15 sec) while shaking.

This results in the removal of the vitelline membranes, and devitellinized embryos sink to the bottom of the methanol (lower) layer.

4 Rehydrate the embryos through a series of 9:1, 7:3, 5:5, 3:7, 1:9 parts methanol:Buffer A, followed by two washes with 100% Buffer A (10 min per wash).

5 Perform a final wash with Buffer A containing 0.1% Tween-20, which prevents the embryos from sticking to glass and plastic (and helps to permeabilize the tissue). Aliquot the rehydrated embryos into 0.5-ml polypropylene tubes (PCR tubes are ideal), ~30–40 µl bed volume of embryos per sample.

Notes

- *For a small number of embryos.* An alternative procedure may be employed if the number of embryos is very limited (heptane-formaldehyde fixation is difficult to carry out with high yield on small numbers of embryos) or if tubulin immunofluorescence is to be performed. This method fixes the embryos and devitellinizes them in one step: Following bleach dechorionation, the embryos are transferred to a glass tube containing 10 ml of heptane, and then, while vortexing the tube, 10 ml of methanol-EGTA is added (3 ml of 500 mM EGTA, pH 6, added to 47 ml of methanol). Devitellinized embryos sink to the bottom. These should be washed with two further changes of methanol-EGTA and then rehydrated in one step by aspirating the methanol and replacing with buffer solution.
- *Other whole-mount samples.* For all other samples, the general approach that has worked best is to dissect or otherwise isolate the tissue in an isotonic buffer (PBS will usually work, as will various insect "Ringer's" solutions, etc.). Formaldehyde is then added to 3.7% (w/v) concentration in the same buffer. Fixation for short times (5–10 min) at room temperature is generally adequate and optimal, although this time may be extended if it proves insufficient to preserve the tissue. The fixative is then removed by several washes in buffer containing 0.1% Tween-20, which may increase the permeability of the sample.
- *Fixing samples on microscope slides.* If the sample is to be prepared on microscope slides, the dissected tissue is transferred to the center of a clean slide, the fixative (3.7% buffered formaldehyde) is added, and a siliconized (SurfaSil, Pierce) coverslip, typically 18 x 18 mm, is gently placed over the tissue, preferably avoiding air bubbles. Excess liquid is wicked out using absorbent paper. After appropriate fixation time, the slide is immersed in liquid nitrogen for 15–30 seconds to freeze the solution, the coverslip is quickly cracked off with a fresh razor blade, and the slide is transferred to prechilled (–20°C) methanol in a Coplin jar or slide staining dish. The samples are then brought to room temperature in the methanol bath and transferred to aqueous buffer solution.

  It may be helpful to pretreat the slides with agents that improve tissue adhesion. This will depend on the sample, but polylysine (see Chapter 4), TESPA (aminoalkylsilane), and other "subbing" procedures can be tested for their efficacy.

- *Tissue culture cells.* For hybridization to cultured cells, it is simplest to grow the cells directly on microscope coverslips. To do this, sterile TESPA-treated 22 x 40-mm coverslips are placed in small petri dishes, and cells in fresh medium are added and then grown to the desired density. The cells can then be fixed by gently aspirating the medium and replacing with 3.7% formaldehyde in PBS. Incubate in the fixative for 5 minutes. Gently aspirate the fixative and add 100% methanol. Incubate 2 minutes. Remove methanol and replace with PBS. Remove the coverslip from the plate with forceps. This will damage cells near the edge, but cells elsewhere on the coverslip should be fine. Transfer to a coverslip staining jar (Thomas Scientific) filled with PBS.

  To treat coverslips with TESPA: Immerse clean coverslips in a 1–2% (v/v) solution of TESPA (3-aminopropyltriethoxysilane; Sigma) in toluene for 30–60 seconds, rinse twice in toluene, and air-dry.

  Small variations in these methods have improved hybridization results for particular samples. For example, for hybridization to Drosophila *imaginal disks, a combination of 0.1% Triton X-100 and 0.1% DOC (deoxycholate) added to the dissection buffer improves tissue permeability dramatically. For* Drosophila *egg chambers, prewarming the fixative to 30–37°C gives better results. Variations in the fixative concentration (from ~1% to 10% formaldehyde) may also be worth testing on other tissues.*

Hybridization Procedures

Performing in situ hybridization on whole-mount samples is in principle no different from any other type of DNA–DNA hybridization procedure. To denature the target DNA, any hybridization protocol will include a step involving either chemical (usually acid or base) or heat treatment. For tissues fixed as described above, an optimized combination of heat and formamide efficiently denatures the DNA without significant damage to the cellular morphology at the level of the light microscope. To accomplish this, the tissue is stepped from aqueous buffer (2x saline sodium citrate [SSC]) into increasingly high concentrations of **formamide**, and finally equilibrated in 50% formamide in 2x SSC. The probe is added in the same solution (plus dextran sulfate, which improves hybridization), and the chromosomal DNA and probe are denatured simultaneously by a brief heat treatment. Over several hours, the probe is then allowed to anneal at an appropriate temperature, usually 37°C. Washing steps are performed to remove the unbound probe. If the probe is labeled for use with an indirect detection method (e.g., **biotin**/avidin or **digoxigenin**/antidigoxigenin), the sample is then blocked and stained with appropriate fluorescent detection reagents, and usually counterstained with a fluorescent DNA dye such as **DAPI** or **PI**. Cellular proteins may also be localized at this stage by preparation for immunofluorescence.

CAUTION: Formamide; digoxigenin; biotin; DAPI; PI (see Appendix 2 for Caution)

These reagents and solutions will be needed for the following protocols:

20x SSC, pH 7.0. 3 M NaCl, 0.3 M sodium citrate—the pH does not require adjustment. Autoclave and store at room temperature.

Formamide (Fluka, 47670; substitute other brands at your own risk)

Formamide may be used straight from the bottle, which should be stored at 4°C. Formamide is considered toxic and possibly mutagenic.

Tween-20 (Pierce, "Surfact-Amps" 10% aqueous Tween-20 solution)

Dextran sulfate (Pharmacia)

Labeled probe(s)

See section on probe labeling (pp. 266–267).

Blocking reagents

Samples should be "blocked" after hybridization but before staining with antibodies or other detection reagents; this is accomplished by incubation with a 0.5% (w/v) protein solution. BSA, normal serum protein (e.g., NGS [normal goat serum], Jackson Immunoresearch), and Boehringer Mannheim's Blocking Reagent for Nucleic Acid Hybridization (1 096 176) have been used interchangeably. BSA and NGS are typically stored in frozen aliquots at –20°C, and Boehringer Mannheim blocking reagent is made up as a 10% (w/v) stock in maleate buffer, according to the manufacturer's directions, and stored at room temperature.

Fluorescence antifade mounting solution

See Chapter 4.

Fluorescent detection reagents (for use with biotin- or digoxigenin-labeled probes)

Streptavidin, conjugated to

Cy2 (Jackson ImmunoResearch 016-220-084)

or Alexa Fluor 488 (Invitrogen S-11223)

Cy3 (Jackson ImmunoResearch 016-160-084)
Cy5 (Jackson ImmunoResearch 016-170-084)
Monoclonal mouse antidigoxin antibodies, conjugated to
Cy2 (Jackson ImmunoResearch 200-222-156)
Cy3 (Jackson ImmunoResearch 200-162-156)
Cy5 (Jackson ImmunoResearch 200-172-156)

Note: These fluorophores are provided as examples; you should pick reagents that will best match your available fluorescence filters.

Other equipment
Aspirator with a fine tip, equipped with a liquid trap
Microscope slides and coverslips

HYBRIDIZATION TO WHOLE-MOUNT TISSUES IN TUBES

This protocol has been used successfully for *Drosophila* embryos and egg chambers, and also for maize meiocyte columns.

You will need a thermal cycler or other suitable device to heat tubes to 91°C.

Washing is carried out by allowing the samples to settle to the bottom of the tube, aspirating most of the solution, and replacing with fresh solution. All washes are ~500 µl unless otherwise stated.

1 Wash tissue in 2x SSCT (2x SSC, 0.1% Tween-20).

2 Aliquot the samples into 0.5-ml tubes, one tube per hybridization.

3 Wash with two further changes of 2x SSCT.

When probing highly transcribed sequences such as ribosomal RNA genes, it may be beneficial to treat the samples with RNase A to ensure that any signal detected results from hybridization to chromosomes. This step may be performed at this point by adding 100 µg/ml boiled RNase in 2x SSCT to the samples and incubating for 30 minutes at 37°C. Following RNase treatment, wash with two further changes of 2x SSCT.

4 Exchange gradually into 2x SSCT containing 50% **formamide**.

CAUTION: Formamide (see Appendix 2 for Caution)

The salt concentration should remain constant at 0.3 M NaCl, 0.03 M sodium citrate.

Wash once in 2x SSCT/20% formamide for 10 minutes.
Wash once in 2x SSCT/40% formamide for 10 minutes.
Wash once in 2x SSCT/50% formamide for 10 minutes.
Wash once more in 2x SSCT/50% formamide.

It is normal for the tissue to appear translucent and to swell somewhat while immersed in high concentrations of formamide.

5 Incubate at 37°C for at least 15 minutes but up to several hours, until you are ready to add probe.

The slow exchange into and preincubation in 50% formamide/2x SSCT are to ensure that the samples are completely equilibrated in this solution.

6 For each sample, make up 40 µl of a solution of the probe(s) in hybridization solution (3x SSC/50% formamide/10% [w/v] dextran sulfate).

The amount of each probe required depends on its complexity and is determined empirically. Highly repeated, simple-sequence probes work well at lower concentrations: typically 50 ng probe/40 µl (1.25 ng/µl). Complex single-copy probes may need to be applied at tenfold higher concentration. Note that nonspecific competitor DNA is not included, since it has not proven to be beneficial.

It is convenient to prepare a stock hybridization solution containing

formamide	5 ml
20x SSC	1.5 ml
dextran sulfate	1 g

Bring to a volume of 9.0 ml with dH_2O.

The dextran sulfate will take a while to dissolve even with constant mixing. This solution can be stored at 4°C and used for at least several months.

Up to 4 μl of probe(s) can be added to 36 μl of this solution. The volume is made up to 40 μl with H_2O, and all the components are then at the correct concentration.

7 Carefully aspirate solution away from tissue (remove as much as possible without aspirating the samples) and add 40 μl of hybridization solution containing probe(s). Mix gently by tapping the tube.

8 Place tube(s) in a thermal cycler. Run a program that will denature the sample(s) by heating to 91°C (controlled with a reference tube) for 2 minutes, then drop to an appropriate hybridization temperature and remain at that temperature overnight. An annealing temperature of 37°C is suitable for most probes, but very AT-rich DNA may require lower temperatures (with a *Drosophila* satellite probe comprising the sequence $(AATAT)_n$ a temperature of 30°C has worked well). Incubate the tube(s) overnight at the annealing temperature, shielded from light. Alternatively, perform the denaturation by manually transferring the tubes, first to a heat block at 91°C for 2 minutes, then to one set at the annealing temperature.

Denaturation for a period of 2 minutes has empirically been optimal; shorter or longer times result in reduced signal, and longer times may also adversely affect sample morphology.

The annealing step is probably complete within a few hours, particularly with repetitive, simple-sequence probes. Overnight incubation is generally convenient, but shorter times have also given good results, particularly with abundant, repetitive sequences.

Once fluorescent probes or detection reagents are introduced into the procedure, it is advisable to protect the samples from excessive exposure to room light, at least during the longer incubations. During the incubation steps the tubes can simply be placed inside a light-tight container such as a film canister.

9 Wash samples.

Warm several milliliters of 2x SSCT/50% formamide to the annealing temperature. Add 500 μl of 2x SSCT/50% formamide to each sample and mix by inversion.

The tissue will float in the dense hybridization solution; it cannot therefore be aspirated until it is diluted with the first wash.

Allow samples to settle; aspirate liquid and add fresh 2x SSCT/50% formamide. Incubate for a total of 1 hour at the annealing temperature, with at least one further change of 2x SSCT/50% formamide.
Wash once with 2x SSCT/25% formamide at room temperature.
Wash three times with 2x SSCT.

If the probes are directly labeled with fluorescent nucleotides, steps 10–12 are not performed; skip to step 13.

10 Add blocking reagent (0.5% [w/v] BSA or other protein) in 2x SSCT and incubate for 30 minutes to 1 hour at room temperature with mixing, e.g., on a nutator or rotating mixer.

11 Stain samples.

Dilute fluorescent avidin/streptavidin (1 mg/ml stock) 1:3000 and/or antidigoxigenin F(ab) fragments 1:1000 into 2x SSCT. Aspirate the blocking solution and add staining solution containing fluorescent avidin and/or antibody. Incubate at least two hours at room temperature with mixing, protected from light.

12 Wash three times with 2x SSCT: twice for 5–10 minutes and then once for 1 hour.

13 Stain with 0.5 µg/ml **DAPI** or appropriate concentration of a different DNA dye in 2x SSCT for 15 minutes. Destain by washing in two changes of 2x SSCT without DAPI for at least 30 minutes.

Note that if PI will be used, an RNase treatment step should first be performed if it was not done prior to hybridization.

CAUTION: DAPI; PI (see Appendix 2 for Caution)

14 Mount in fluorescence mounting medium on microscope slides (see Chapter 4).

Hybridized samples do not keep very well once mounted in glycerol, though this may depend on the mounting medium and fixation method. However, they can be stored in buffer at 4°C, protected from light, and examined days later (addition of a small amount of EDTA may help to prevent deterioration).

HYBRIDIZATION TO WHOLE-MOUNT TISSUES ON SLIDES

As described in the section on sample preparation above, with many kinds of tissues it is easier to handle samples fixed onto microscope slides rather than in suspension. In such cases, the above procedure is modified as follows. Washes are performed in Coplin jars or slide-staining dishes. Heat denaturation is performed by sealing the probe solution under a coverslip and placing the slide on a prewarmed heat block. Specialized equipment such as thermal cyclers that accommodate slides (such as those used for PRINS [primed in situ labeling] or in situ PCR) may be used but are not necessary. Probe detection and immunostaining are carried out by pipetting small volumes of solution onto the slide and covering with small Parafilm squares.

The following equipment will be needed:

Flat-tipped filter forceps (Millipore)

Slide/coverslip washing containers

For slides, it is convenient to perform washes in either Wheaton slide-staining jars with removable slide holders (Thomas Scientific) or Coplin jars (Thomas Scientific). For Coplin jars, each wash is 50–70 ml; staining jars require about 200 ml to cover the slides.

For samples on coverslips, use glass coverslip-staining jars (Thomas Scientific); each wash is 8–10 ml.

Heat block set at 95°C

Rubber cement

Humid chamber, typically a plastic box with watertight lid (e.g., Tupperware containing a few layers of damp paper towels).

1 Rehydrate the fixed samples (if still in a methanol bath) by transferring to a container filled with 2x SSCT (0.3 M NaCl, 0.03 M sodium citrate, 0.1% Tween-20).

2 Exchange the slide(s) into 2x SSCT/50% **formamide**:

Wash once in 2x SSCT/20% formamide for 10 minutes.
Wash once in 2x SSCT/40% formamide for 10 minutes.
Wash once in 2x SSCT/50% formamide for 10 minutes.
Wash once more in 2x SSCT/50% formamide.

CAUTION: Formamide (see Appendix 2 for Caution)

3 Incubate for at least 15 minutes in 2x SSCT/50% formamide at 37°C. This is easily accomplished by placing the entire container in a water bath and allowing it to reach 37°C slowly.

4 Prepare the probe(s). Make up probes in hybridization solution 3x SSCT/50% formamide/10% dextran sulfate) as in the above protocol, step 6. For each sample prepare 15 µl of hybridization solution containing 20–200 ng of each probe.

5 Deliver the probe.

If the sample is on a slide: Pipette the probe(s) in 11–15 µl of hybridization solution onto a clean 18 x 18-mm or 22 x 22-mm coverslip (depending on how much area the specimen covers). Remove a slide from the prehybridization solution (wear gloves, as formamide is toxic). Remove as much of the liquid as possible by wiping the edges and aspirating carefully around the sample. Touch the specimen to the drop of hybridization solution to pick up the coverslip. Quickly invert the slide; the probe solution should spread out under the coverslip.

If the sample is on a coverslip, do the reverse: Pipette the probe solution onto a clean microscope slide. Remove the coverslip from the prehybridization solution and blot the back side carefully with Kimwipes. Touch the drop of probe solution to the sample to pick up the coverslip and invert.

6 Seal the coverslip to the slide with rubber cement. One way to do this is to fill a syringe with rubber cement and to use a large gauge needle. Be generous; make a wide gasket to seal the edges of the coverslip. Avoid getting cement on the bottom of the slide, because it will reduce thermal contact with the heat block (step 8).

7 Prewarm a heat block to 95°C. A simple version with a removable aluminum block works well. When a thermometer placed in one of the wells stabilizes at 95°C, invert the block so that its flat side is facing up.

8 Denature the samples by placing each slide on the heat block for 3 minutes, slide side down. Air bubbles will tend to force their way out from under the coverslip during the heating, but this is normal.

9 Transfer the denatured slides to a humid chamber prewarmed to the annealing temperature (usually 37°C, but see notes in step 8 of the preceding protocol). The chamber is floated in a large water bath set to the annealing temperature.

10 Allow to anneal overnight.

11 Immerse each slide in a beaker containing 2x SSC. Using forceps, carefully pull off the rubber cement. The coverslip may come off too; if not, it will probably drop off soon thereafter.

12 Transfer slides or coverslips to a staining jar filled with 2x SSCT/50% formamide at the annealing temperature. Wash for 1 hour with a total of three changes of this solution at the annealing temperature.

13 Wash once in 2x SSCT/25% formamide at room temperature.

14 Wash in 2–3 changes of 2x SSCT.

15 To stain slides or coverslips, use the same blocking and staining solutions and incubation times as for samples in suspension, above. Place the slides in a humid chamber. Pipette

50–100 µl of solution onto the specimen and cover with a small square of Parafilm. Carry out washes and DNA counterstaining in staining jars. Just before mounting, rinse each sample in 50 mM Tris-HC1, pH 7.5, to remove most of the salt; otherwise it will precipitate onto the glass as it dries.

16 To mount slides for microscopy, remove as much liquid as possible without allowing the sample to dry out. Pipette a small drop of mounting medium onto a 22 x 22-mm coverslip (or slide if the sample is on a coverslip). Touch the specimen to the drop of mounting medium, invert the slide, and seal with nail polish (see Chapter 4 for various mounting media).

Notes: Troubleshooting

- The two major problems that face the investigator with FISH experiments are (1) undetectable or weak fluorescent signals and (2) prohibitively high background fluorescence. The most common source of such problems in 3D FISH experiments is overfixation of the sample. If only a weak signal or no signal can be detected, this is the first aspect of the procedure that should be altered. Whole-mount hybridization necessarily involves a trade-off between the preservation of the sample and the intensity of signal. Lighter fixation or fixatives other than formaldehyde (e.g., alcohol or alcohol/acid or other non-cross-linking fixatives such as Streck Laboratories' M.B.F. reagent) may provide adequate preservation for the requirements of a particular experiment, while increasing signal intensity relative to formaldehyde.
- With formaldehyde fixation, the optimal denaturation temperature has empirically been found to be 90–91°C. Lower temperatures will result in reduced signal and increased background, and higher temperatures will noticeably affect the nuclear morphology, resulting in anomalous detection of multiple signals for single-copy probes. When fixing with agents other than formaldehyde, lower denaturation temperatures in the range of 70–90°C should be tested.
- When attempting 3D FISH for the first time in a new tissue, it is valuable to have as a positive control a probe comprising a very abundant genomic sequence, since this will offer the best possible chance of detecting hybridization. All eukaryotic organisms possess genomic sequences of high copy number. Simple-sequence "satellite" repeats are particularly suitable, especially since probes to such elements are easy to synthesize. The ribosomal DNA (rDNA) may also be a good choice (particularly for *Saccharomyces cerevisiae* and *C. elegans*, both of which possess few other high-copy-number sequences). Once a signal can be reliably detected, it becomes possible to evaluate the consequences of variations in the procedure. Incorporation of biotinylated or digoxigenin-labeled nucleotides may be assayed by Southern or dot-blot procedures, using enzyme-linked detection reagents.

REFERENCES

Hsieh T. and Brutlag D. 1979. Sequence and sequence variation within the 1.688 g/cm^3 satellite DNA of *Drosophila melanogaster*. *J. Mol. Biol.* **135:** 465–481.

Spector D.L., Goldman R.D., and Leinwand L.A. 1998. *Cells: A laboratory manual*. Cold Spring Harbor Laboratory Press, Cold Spring Harbor, New York.

CHAPTER 17

In Situ Hybridization to RNA

Sui Huang[1] and Robert Singer[2]

[1]*Northwestern University Medical School, Chicago, Illinois*
[2]*Albert Einstein College of Medicine, Bronx, New York*

INTRODUCTION

Nonisotopic in situ hybridization is a powerful tool for evaluating the relative levels of expression and the distribution of specific RNA transcripts within cells and tissues (Figs. 17.1 and 17.2) (Singer et al. 1986; Lawrence and Singer 1991), as well as for mapping specific nucleic acid sequences on chromosomes (see Chapter 15). The methodologies described in this chapter have been used to study the localization of cytoplasmic RNAs (Sundell and Singer 1991; Latham et al. 1994), the sites where transcription and RNA processing occur (Jiménez-García and Spector 1993; Xing et al. 1993; Dirks et al. 1995; Huang and Spector 1996), and the localization of transcription patterns during development (see Chapter 18; Tautz and Pfeifle 1989; Haramis et al. 1995). The use of nonradioactive probes has several advantages over radiolabeled probes for in situ hybridization. Primary among these is the high resolution obtained by the use of probes detected at their site of hybridization. In addition, the probes are more convenient, the result of the entire hybridization experiment can be obtained more quickly, and the probes are ecologically sound. In specific cases, where the spatial concentration of the target is high, they are also more sensitive.

In situ hybridization involves a series of steps, each of which is critical to the final success of RNA detection. These steps include:

- Preparation of the probe
- Preparation of the tissue or cells
- Hybridization of the probe
- Detection of the probe

All of these steps are described in detail in the protocols presented in this chapter. In addition, alternatives are presented for preparing different types of probes, biological samples, and methods of detection. Finally, at the end of the chapter, a troubleshooting guide is presented to help in dealing with problems that may occur during localization of specific RNA transcripts.

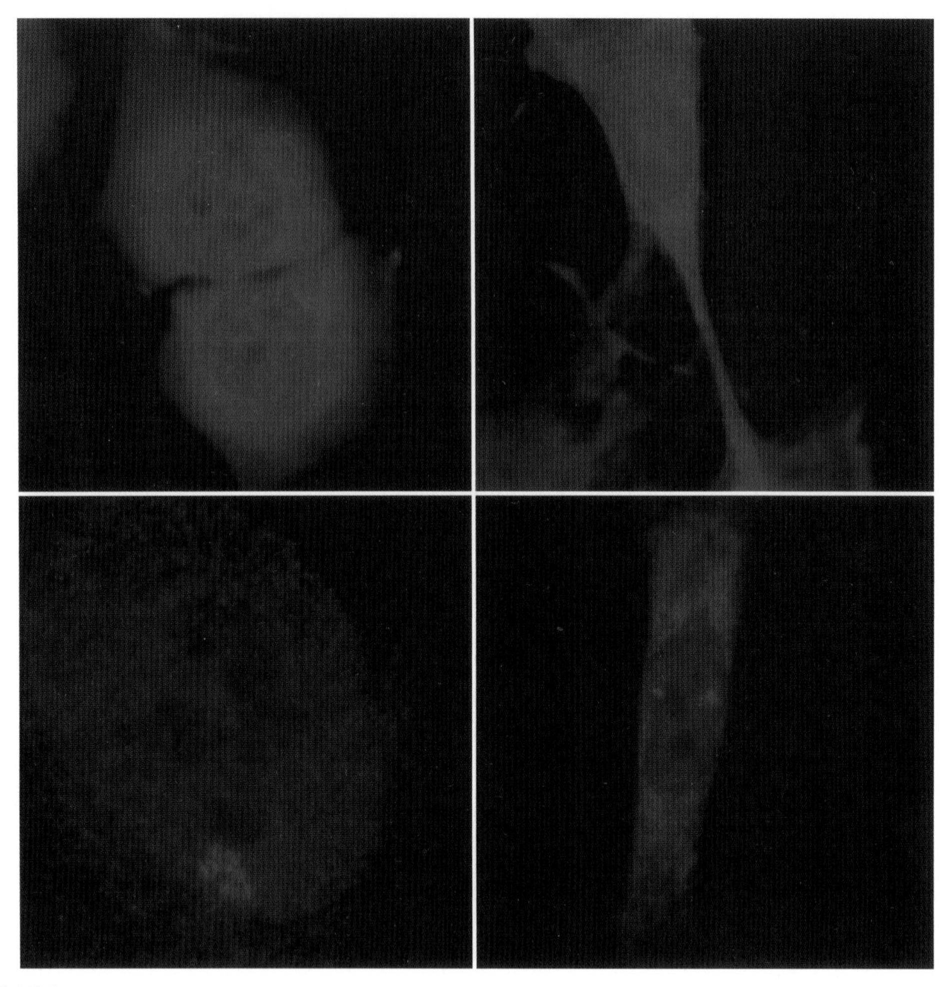

FIGURE 17.1

(*Top left*) Localization of poly(A)$^+$ RNA using a biotinylated oligo dT(50)mer followed by Texas Red avidin. These RNAs are present in a nuclear speckled and diffuse pattern, and they are diffusely distributed throughout the cytoplasm. The nucleoli in each of the nuclei do not contain these RNAs. (*Top right*) Localization of rRNA using a nick-translated probe followed by Texas Red avidin. (*Bottom left*) Localization of XIST RNA coincident with the inactive X chromosome territory in a female fibroblast. (*Bottom right*) Localization of fibronectin RNA in a human fibroblast. The transcription sites are seen as two nuclear dots. The RNA is also diffusely distributed throughout the cytoplasm. (Photo provided by J. McCann and D.L. Spector, Cold Spring Harbor Laboratory.)

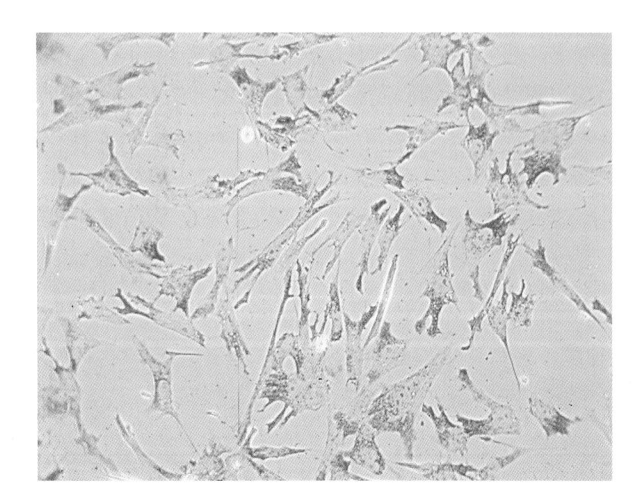

FIGURE 17.2

Localization of β-actin mRNA in fibroblasts using alkaline phosphatase detection. The RNA is concentrated at the leading edge of each cell. (Photo provided by R. Singer, Albert Einstein College of Medicine.)

Preparation of Probes

Probe Types

A number of different kinds of probes can be used for in situ hybridization:

- dsDNA probe (genomic or cDNA clones)
- ssRNA probe (enzymatic synthesis)
- Oligonucleotide (DNA or RNA) probe (chemical synthesis)

These probes may be used for detection, localization, and quantitation of DNA, nuclear pre-mRNA, or cytoplasmic mRNA in single cells, tissue sections, or chromosome preparations. The advantages and disadvantages of each of these types of probes are discussed below. This chapter focuses on hybridization to RNA targets.

Double-stranded DNA probes generated by nick translation from genomic or cDNA clones are commonly used to localize RNAs, because they may be in hand or are easy to obtain. Their advantage is that they will be complementary to a large region of the target RNA, and, therefore, there is a high degree of probability that part of the target sequence will be accessible to the probe. However, their disadvantage is that their specific activity may be lower than a highly labeled oligonucleotide probe, and their larger-size fragments may present penetration problems into cells or tissue under certain fixation regimes.

Single-stranded antisense RNA probes have been used extensively on tissue sections and whole-mount preparations. They form stronger hybrids with the RNA target than do DNA probes. In addition, one can also easily generate sense strand probes that can be used as excellent controls. However, in some cases, they may result in a high degree of nonspecific labeling to nucleoli because of the high RNA content within this nuclear region. The production and use of ssRNA probes is described in Chapter 18.

Oligonucleotide probes have an advantage over cDNA probes because they can be designed to distinguish specific sequences (Carmo-Fonseca et al. 1991; Carter et. al. 1991; Huang et al. 1994) such as two closely related, homologous mRNAs (e.g., mRNA isoforms). Their small size provides superior penetration into the nucleus and tissue sections. They are readily available through automated synthesis. The fact that they are single stranded excludes the possibility of renaturation. However, an oligonucleotide has a limited target size and, therefore, such probes result in fewer detection groups per molecule of target, and thus less signal is generated as compared to cDNA or genomic probes. However, the target abundance is sometimes sufficient so that, even considering the lower stoichiometry of detection groups using oligonucleotide probes, the target can be detected (e.g., some snRNAs or histones). An easy rule to follow is that if target size times abundance equals 200 kb, detection is possible. The signal obtained with oligonucleotide probes can be increased by pooling oligonucleotide probes made from different *cis*-related sequences. Oligonucleotide probes can also be used to detect two or more sequences simultaneously within the same cell by using different fluorochromes (see below).

Probe Size

The size of the probe is a critical factor in the hybridization efficiency (signal) and background. This is because the probe must be small enough to penetrate the cell or tissue, and it must not be so large that it sticks randomly and gives high nonspecific signal, an especially important consideration for detection of low-copy-number sequences. For nick-translated probes, the median probe size should be no larger than 200 nt, with a range to no larger than 600 nt. The background increases significantly when this size is exceeded. For oligonucleotides, 15–20-mers are routinely used.

LABELING PROBES

The probe can be labeled with a variety of detection groups including:

- Biotin
- Digoxigenin
- Fluorochromes
- Enzymes

Biotin and digoxigenin are routinely used to label probes for RNA detection. Biotin (vitamin H) can be detected by avidin, a glycoprotein from egg white that has a binding constant of 10^{-15}/mole at 25°C. Digoxigenin is a steroid isolated from digitalis plants. Since the flowers and leaves of this plant are the only known source of digoxigenin, anti-digoxigenin antibodies will not cross-react when used in most cell or tissue systems. Both biotin and digoxigenin can be detected with fluorochromes or enzymes (e.g., alkaline phosphatase or horseradish peroxidase) conjugated to avidin or anti-digoxigenin antibodies, respectively. Enzymes such as alkaline phosphatase have the advantage of continuing to precipitate substrate with time, and the reaction can be terminated when the signal-to-noise ratio is optimal. Fluorescent detection is less sensitive than enzymatic detection, but has advantages of direct (single-step) detection, superior resolution, and simultaneous detection of multiple probes chemically labeled with spectrally discriminable fluorophores. In addition, since no secondary reporter is used, these probes result in less background than probes labeled using an indirect approach. In situations where target sequence is highly concentrated, the sensitivity of nonradioactive hybridization far exceeds radioactivity because, in microscopic terms, the spread of the nuclide decay is considerable. For instance, the tritium β-particle is not detected until it contacts the emulsion, some distance away from the hybridization site. This increases the area of detection and, hence, decreases the concentration of signal. In addition, sensitivity of fluorochrome detection can be increased by the use of photon-sensitive devices such as cooled charge-coupled device (CCD) cameras, where the photon captured on a chip generates an electric charge coupled to a visual signaling device. The integration of the signal over time and the ability to subtract noise provides digital enhancement of the concentrated signal. For additional methods of probe preparation, see Chapters 15, 16, and 18.

Nick Translation of Cloned Probes

Nick translation of genomic or cDNA probes with either bio-16-dUTP (Boehringer Mannheim) or digoxigenin (Boehringer Mannheim) can be easily performed according to the protocol provided in Chapter 15 or by using a commercially available nick translation kit (Boehringer Mannheim).

Labeling Oligonucleotide Probes with Biotin or Digoxigenin

1 Design one or more oligonucleotides (15–20-mers) that are complementary to regions of the target RNA that are thought to be accessible. This may be assessed by evaluating the secondary structure and/or potential protein-binding sites on the respective target sequence.

2 Synthesize the oligonucleotide probe and purify by gel electrophoresis (probes are usually synthesized by institutional or commercial synthesis facilities).

3 Set up the following 50-μl labeling reaction mixture:

25 pM probe
0.2 mM dig-11-dUTP or bio-16-dUTP (BMB)
1.0 mM **$CoCl_2$**
140 mM **potassium cacodylate**
30 mM Tris-HCl (pH 7.6)
0.1 mM **DTT** (dithiothreitol)
100 units terminal transferase (BMB)

CAUTION: $CoCl_2$; cacodylate; DTT (see Appendix 2 for Caution)

4 Incubate the reaction at 37°C for 60 minutes.

5 Purify the reaction products by gel filtration on 1 x 20-cm G50 Sephadex column.

Chemical Labeling of Oligonucleotide Probes with Fluorochromes

1 Synthesize, on a DNA synthesizer or otherwise, the oligonucleotide probe with amino-modified dT (Glen Research) or amino link II (Applied Biosystems).

2 Label the probe with fluorescein isothiocyanate (FITC), Texas Red (Molecular Probes), or Cy3 (GE Healthcare).

a Dissolve 0.01 μM of amino-modified oligonucleotide (calculated from number of free amines) in 200 μl of 0.2 M $NaHCO_3$, pH 9.0.

b Add 100 μl of freshly prepared fluorochrome (10 mg/ml in DMF [*N,N*-dimethylformamide]).

c Mix the reagents and leave at room temperature overnight in the dark.
Cy3 is water-soluble and does not require DMF.

3 Pass the reaction products through a 1 x 20-cm G50 Sephadex column.

4 Purify the labeled probe on a 10% polyacrylamide native gel.

5 Excise the band of labeled DNA and extract overnight.

Preparation and Fixation of Samples for Hybridization

Protocols are given for the preparation of cultured cells, frozen tissue, and paraffin-embedded tissue sections. For further details, see Chapter 4.

PREPARATION OF CULTURED CELLS FOR HYBRIDIZATION

1 Autoclave acid-washed (boiled in 0.1 M HCl) coverslips in 0.5% gelatin.

2 Plate the cells, ~10^6 cells per 100-mm dish containing six coverslips.

3 Culture the cells for appropriate times depending on the cell type being used; cells should not be confluent or too sparse.
For further details on preparation of coverslips, see Chapter 4.

PREPARATION OF FROZEN TISSUE SECTIONS

1 Coat the coverslip with Histostik (Accurate Chemical) or any other reagent to enhance adherence of tissue to the slide (see Chapter 4).
Histostik stock: Add 1 ml of Histostik per liter of H_2O, dip the coverslip in the solution and store at room temperature.

2 Cut frozen sections, 4–8 µm at –20°C (see Chapter 4).

3 Pick up sections with coverslip and leave at room temperature for 5 minutes.
For further details, see Chapter 4.

PREPARATION OF PARAFFIN-EMBEDDED TISSUE SECTIONS

1 Cut standard sections 5 µm in thickness (see Chapter 4).

2 Deparaffinize the sections with **xylene** (see Chapter 4) and proceed as for immunohistochemistry.
For further details, see Chapters 4, 16, and 18.

CAUTION: Xylene (see Appendix 2 for Caution)

In Situ Hybridization and Detection Protocols

The following protocols describe the hybridization of fixed cells to nick-translated cDNA probes or to DNA oligonucleotide probes for the localization of individual RNAs (i.e., actin, fibronectin, rRNA, XIST) or populations of RNAs (poly[A]$^+$ RNA). Techniques are provided for both fluorescence and enzymatic detection.

CONTROLS

As with any experiment, controls are important to validate the experimental result.

- Perform a mock hybridization without labeled probe.
- Perform a hybridization with a nonspecific or irrelevant labeled probe, i.e., vector DNA or sense oligonucleotide.
- Pretreat one sample with RNase A prior to hybridization.
- Hybridize with excess unlabeled probe followed by labeled probe and reporter.
- Hybridize with a probe known to work, as a positive control for your experimental protocol.
- Cells can be pretreated with α-amanitin, for 5 hours (50 μg/ml to inhibit RNA polymerase II or 300 μg/ml to inhibit RNA polymerases II and III) prior to fixing the sample. No signal should be detected unless the target RNA is stable.

mRNA IN SITU HYBRIDIZATION WITH NICK-TRANSLATED PROBES

1 Fix cells with freshly made 4% **formaldehyde** in PBS, pH 7.4, at room temperature for 15 minutes.

All solutions should be made in molecular biology grade ultrapure H_2O (no RNase). Wear gloves at all times and use sterile disposable pipettes and tips.

4% formaldehyde is made up fresh prior to use by dissolving the appropriate amount of EM-grade ***paraformaldehyde*** *(Prill form, Electron Microscopy Sciences) in PBS (phosphate-buffered saline) in a Pyrex bottle with a stir bar. The aldehyde goes into solution by heating on a hot plate in the hood at 60°C. Keep the bottle cap loosened so that pressure does not build up. Cool to 20°C and pH to 7.4.*

CAUTION: Formaldehyde; paraformaldehyde (see Appendix 2 for Caution)

2 After rinsing in PBS (three times for 10 min each), permeabilize cells with 0.5% Triton X-100 at 4°C for 5 minutes.

3 Rinse cells in PBS (three times for 10 min each) and then in 2x SSC (one time for 5 min).

4 Dry down ~50–100 ng of nick-translated probe and 20 μg of competitor *Escherichia coli* tRNA per coverslip in a Speed Vac (Savant).

For hybridization using oligonucleotide probes see the following protocol on p. 285.

5 Add 10 μl of deionized **formamide** to the dried DNA.

CAUTION: Formamide (see Appendix 2 for Caution)

6 Denature the probe and tRNA by heating at 90°C for 10 minutes. Chill the probe on ice immediately.

7 Add hybridization buffer containing 4 μl of 50% dextran sulfate, 4 μl of 5% BSA (bovine serum albumin) (molecular biology grade, RNase-free) and 2 μl of 20x SSC (standard saline citrate) to the denatured probe so that the final concentrations in the hybridization mixture are 5 ng/μl of probe, 50% formamide, 1 μg/μl of *E. coli* tRNA, 2x SSC, 1% BSA, and 10% dextran sulfate.

8 Place 20 μl of hybridization mixture/probe on each coverslip.

Remove the coverslip from the petri dish containing buffer, because removing it from a bufferless dish may result in breaking the coverslip because of surface tension effects. Wick excess buffer from the coverslip by removing it from the petri dish and carefully touching the edges of the coverslip with a piece of Whatman #1 filter paper and gingerly dry the top of the coverslip (i.e., the side with no cells). Place the hybridization cocktail on the coverslip and invert it onto a glass slide.

9 Invert coverslips onto a slide, seal with rubber cement, and incubate in a humid chamber at 37°C for 4–16 hours.

A mock hybridization should be performed as a control with each experiment.

A humidified chamber can be easily made using a 140-mm petri dish. Place a moist paper towel at the bottom of the dish, put two wooden dowels on top of the towel, and place the slides across the dowels so that they are not in direct contact with the towel. A 140-mm dish will hold four slides.

10 Rinse cells in 2x SSC/50% formamide at 37°C, 2x SSC, and 1x SSC at room temperature for 30 minutes each.

To remove the coverslip from the slide, flood the slide with buffer and, using your forceps, gently pull the coverslip straight toward you without drastically breaking the surface tension, as this would remove cells from the coverslip.

11 For detection of the hybridization signal there are two protocols, depending on the labeling reagent used in the hybridization probe.

For Biotin-labeled Probes

a Wash cells in 4x SSC, 0.1% Triton at room temperature for 5 minutes.

b Incubate the cells in 4x SSC/1% BSA/2 μg/ml FITC-avidin DCS (Vector Laboratories) for 30–60 minutes in a humid chamber at room temperature in the dark.

c Wash cells in 4x SSC two times for 10 minutes each, and then in 2x SSC for 10 minutes.

For Digoxigenin-labeled Probes

a Wash cells in PBS, 0.1% Triton at room temperature for 10 minutes.

b Incubate cells in 1x PBS and 1% BSA with anti-digoxigenin conjugated to a fluorochrome (0.5 μg/ml) (BMB) at 37°C for 30–60 minutes.

c Wash cells in three changes of PBS at room temperature with agitation.

BSA should be molecular biology grade (RNase-free).

12 Stain cells with **DAPI** (4′,6-diamidino-2-phenylindole)at a concentration of 0.1 μg/ml to label total DNA (see Chapter 5) or incubate with a primary antibody followed by a secondary antibody to localize a protein in the same cells (see Chapter 15).

CAUTION: DAPI (see Appendix 2 for Caution)

13 Mount coverslips in fluorescence mounting medium (see Chapter 4).

IN SITU HYBRIDIZATION TO POLY(A)$^+$ RNA USING AN OLIGONUCLEOTIDE PROBE

1 Fix cells with freshly made 2–4% paraformaldehyde in PBS, pH 7.4, at room temperature for 15 minutes. Wear gloves at all times and use sterile disposable pipettes and tips.

*2% **formaldehyde** is made up fresh prior to use by dissolving the appropriate amount of EM-grade **paraformaldehyde** (Prill form, Electron Microscopy Sciences) in PBS in a Pyrex bottle with a stir bar. The aldehyde goes into solution by heating on a hot plate in the hood at 60°C. Keep the bottle cap loosened so that pressure does not build up. Cool to 20°C and pH to 7.4.*

CAUTION: Formaldehyde; paraformaldehyde (see Appendix 2 for Caution)

2 Wash cells in PBS (three times for 10 min each).

3 Permeabilize cells with 0.5% Triton X-100 for 5 minutes on ice.

4 Wash cells in PBS (three times for 10 minutes each).

5 Rinse cells in 2x SSC (two times for 10 min each).

6 Hybridization buffer containing 2 µl of yeast tRNA (10 mg/ml), 2 µl of 20x SSC, 4 µl of 50% dextran sulfate is added to 1 µl of oligo (~100 ng of oligo dT_{50}-biotin or digoxigenin), 5 µl of deionized **formamide**, 2 µl of 10% BSA, and 4 µl of dH_2O. Final concentrations in the hybridization mixture are 5 ng/µl of probe, 25% formamide, 1 µg/µl yeast tRNA, 2x SSC, 10% dextran sulfate. The entire mixture is placed onto one coverslip.

CAUTION: Formamide (see Appendix 2 for Caution)

BSA should be molecular biology grade (RNase-free).

7 Invert coverslips onto a slide, seal with rubber cement, and incubate in a humid chamber at 42°C overnight (~16 hr).

A mock hybridization should be performed as a control with each experiment.

A humidified chamber can be easily made using a 140-mm petri dish. Place a moist paper towel at the bottom of the dish, put two wooden dowels on top of the towel, and place the slides across the dowels so that they are not in direct contact with the towel. A 140-mm dish will hold four slides.

8 Float the coverslips off the slide with 2x SSC and wash in 2x SSC (two times for 30 min each).

To remove the coverslip from the slide, flood the slide with buffer and, using your forceps, gently pull the coverslip straight toward you without drastically breaking the surface tension, as this would remove cells from the coverslip.

9 Wash cells in 1x SSC and 0.5x SSC (if background is a problem) for 15 minutes each.

10 For detection of the hybridization signal, there are two protocols depending on the labeling reagent used in the hybridization probe.

For Biotin-labeled Probes

a Wash cells in 4x SSC, 0.1% Triton X-100 at room temperature for 5 minutes.

b Incubate the cells in 4x SSC/1% BSA/2 µg/ml FITC-avidin DCS (Vector Laboratories), for 30–60 minutes in a humid chamber at room temperature in the dark.

c Wash cells in 4x SSC two times for 10 minutes each and then in 2x SSC for 10 minutes.

For Digoxigenin-labeled Probes

a Wash cells in PBS, 0.1% Triton at room temperature for 10 minutes.

b Incubate cells in 1x PBS and 1% BSA with anti-digoxigenin conjugated to fluorochromes (0.5 μg/ml) (BMB) for 30–60 minutes at 37°C.

c Wash cells in three changes of PBS at room temperature with agitation.
BSA should be molecular biology grade (RNase-free).

11 Mount coverslips in fluorescence mounting medium (see Chapter 4).

DETECTION OF mRNA USING AN ALKALINE PHOSPHATASE-CONJUGATED PROBE

1 Fix cells with freshly made 4% formaldehyde in PBS, pH 7.4, at room temperature for 15 minutes.
All solutions should be made in molecular-biology-grade ultrapure H_2O (no RNase). Wear gloves at all times and use sterile disposable pipettes and tips.
*2% **formaldehyde** is made up fresh prior to use by dissolving the appropriate amount of EM-grade **paraformaldehyde** (Prill form, Electron Microscopy Sciences) in PBS in a Pyrex bottle with a stir bar. The aldehyde goes into solution by heating on a hot plate in the hood at 60°C. Keep the bottle cap loosened so that pressure does not build up. Cool to 20°C and pH to 7.4.*
Caution: Formaldehyde; paraformaldehyde (see Appendix 2 for Caution)

2 Wash in PBS three times for 10 minutes each.

3 Permeabilize cells with 0.5% Triton-X in PBS at room temperature for 5 minutes.

4 Wash in PBS three times for 10 minutes each.

5 Prior to hybridization, incubate the cells in 50% formamide, 2x SSC, and 5 mM **NaH_2PO_4**, pH 7.0, at room temperature for 10 minutes.
CAUTION: NaH_2PO_4 (see Appendix 2 for Caution)

6 Dry down 50–100 ng of nick-translated probe and 20 μg of competitor *E. coli* tRNA per coverslip in a Speed Vac (Savant).

7 Add 10 μl of deionized **formamide** to the dried DNA.
CAUTION: Formamide (see Appendix 2 for Caution)

8 Denature the probe and tRNA by heating for 10 minutes at 90°C. Chill the probe on ice immediately.

9 Add 10 μl of 2x hybridization buffer to the probe/formamide solution, mix carefully, and apply to the coverslip. Seal coverslips with rubber cement and incubate at 37°C overnight in a humidified chamber.

2x Hybridization Buffer	
20 x SSC	20 μl
50% dextran sulfate	40 μl
dH_2O	40 μl

Start out by using ~50–100 ng of probe. Titrate probe to achieve maximum signal to noise.

A mock hybridization should be performed as a control with each experiment.

A humidified chamber can be easily made using a 140-mm petri dish. Place a moist paper towel at the bottom of the dish, put two wooden dowels on top of the towel, and place the slides across the dowels so that they are not in direct contact with the towel. A 140-mm dish will hold four slides.

10 After hybridization, wash coverslips at 37°C for 30 minutes in petri dishes containing 50% formamide and 2x SSC.

To remove the coverslip from the slide, flood the slide with buffer and, using your forceps, gently pull the coverslip straight toward you without drastically breaking the surface tensions, as this would remove cells from the coverslip.

11 Wash coverslips in 1x SSC, three times for 10 minutes each at room temperature with gentle agitation.

12 Equilibrate the cells in buffer 1 for 1 minute.

Buffer 1

150 mM NaCl
100 mM Tris-HCl
Final pH 7.5

13 Incubate the cells in 30 µl of buffer 1 containing 1% BSA and a 1:250 dilution of alkaline phosphatase conjugate anti-digoxigenin (Boehringer Mannheim) at 37°C for 30 minutes.

BSA should be molecular biology grade (RNase-free).

14 Wash cells two times for 15 minutes each time at room temperature in buffer 1 followed by equilibration in buffer 3 for 1 minute at room temperature.

Buffer 3

100 mM Tris-HCl, pH 9.5
100 mM NaCl
50 mM **$MgCl_2$**

CAUTION: $MgCl_2$ (see Appendix 2 for Caution)

15 Transfer each coverslip to a petri dish containing substrate solution (2 ml of nitroblue tetrazolium/bromochloroindolyl phosphate solution [3 mg **NBT** (nitroblue tetrazolium) and 2 mg **BCIP** (bromochloroindolyl phosphate) in 10 ml of buffer 3] [Boehringer Mannheim]).

CAUTION: NBT; BCIP (see Appendix 2 for Caution)

If endogenous alkaline phosphatase activity is a problem, levamisole (SP-5000, Vector Laboratories) can be added to the substrate solution.

16 Monitor color development on an inverted microscope and stop the reaction by adding buffer 3 when sufficient reaction product is observed.

17 Wash coverslips in buffer 3 for 1 minute and then mount the coverslips for viewing (see Chapter 4).

TROUBLESHOOTING NOTES

The following suggestions are provided should problems arise in the localization of a particular RNA species.

High Background

High background usually decreases the signal-to-noise ratio and is often caused by staining derived from nonspecific interactions between labeled probes and cellular structures. To improve the signal to noise ratio, the following points should be considered.

Probes

- *Probe concentration.* The concentration of probe should be titrated to reach the maximal signal to noise ratio. Too much probe will increase nonspecific background labeling.
- *The purity of probes.* The probe should be purified away from nonincorporated fluorochrome-conjugated free nucleotides, which may contribute to background staining. In addition, the sequence of interest, without vector sequences, should be used as the template in making the probe, because the vector sequences may result in nonspecific interactions with cellular components.
- Large probe size can increase the trapping of the probe on cellular structures nonspecifically. The probe size should be limited to a median of 200 bp.
- Direct labeling using a fluorochrome-conjugated probe can be used to eliminate possible background from the secondary detection reagents.
- Electrostatic attraction between the probe and basic proteins in the cell can be reduced by pretreatment of the cells or tissue section with 0.25% acetic anhydride in 0.1 M triethanolamine, pH 7.5, for 5 minutes. This treatment blocks basic groups by acetylation.

Prehybridization

Prehybridization in hybridization buffer without the probe is beneficial for decreasing nonspecific background. A saturation binding of the blocking reagent to the potential sites of nonspecific interaction prior to the addition of the probe could limit the nonspecific binding of the probe to these sites.

Hybridization and Wash Conditions

- Hybridization stringency is controlled by the percentage of formamide, the concentration of SSC, and the temperature during hybridization and washes. When background is a problem, a higher stringency that limits nonspecific interactions can be achieved by increasing the percentage of formamide or the temperature and/or lowering the concentration of SSC (for DNA–RNA interactions).
- The concentration of blocking reagents such as BSA, tRNA, or sheared salmon sperm DNA can be increased to block nonspecific interactions.
- Hybridization time can be reduced to 4–6 hours to reduce potential nonspecific stickiness of probe.

Secondary Detection

- The conjugation of the secondary reporter should be titrated to achieve the maximal signal-to-noise ratio. An oversaturated reporter can result in high background.
- The concentration of blocking reagents can be increased to block nonspecific stickiness of secondary reporters. In some cases, 0.1% casein and 0.1% Tween 20 have been used as blocking agents prior to incubation with antibody detection reagents. Casein forms negatively charged micelles, which saturate positively charged groups in the cells or tissue.

No Signal or Weak Signal

Probes

- When signals are not detected, one can increase the probe concentration as high as 400 ng/20 µl/coverslip. If detectable signals cannot be achieved at such a concentration, other alternatives should be sought (i.e., different probe type).
- The labeling efficiency of the probe may not be optimal. For example, nick translation labels less efficiently than polymerase chain reaction or random priming. When making oligonucleotide probes, one can increase the number of labeled nucleotides/oligo to increase the sensitivity of detection. However, in the case of oligos, too many labeled nucleotides may affect the hybridization efficiency.

RNase Activity

RNase inhibitors such as RNAsin (28 units/ml) (Boerhinger Mannhein) or vanadyl sulfate ribonucleotide complex (VRC) (2 µl of 200 mM stock) can be added to the hybridization buffer.

Secondary Detection

- Hybridization signals can be amplified by including several layers of reporting molecules. For example, when biotinylated nucleotides are used to label a probe, the hybridization signal can first be coupled to streptavidin, then incubated with biotinylated mouse IgG, followed by a fluorochrome-conjugated anti-mouse IgG antibody. Each step in the chain can amplify the signal significantly. However, nonspecific background may also be amplified.
- The intensity of the signal detected by enzyme-conjugated probes can be controlled by the length of the enzymatic reaction. A very weak signal can be intensified by a prolonged reaction time. Commonly used enzymatic reagents are alkaline phosphatase and peroxidase.
- More sensitive detectors such as cooled CCD cameras, with the ability to integrate the signal intensity over time, permit the detection of weak signals that may not be visualized by the eye or conventional film cameras.

Probe Accessibility to the Target Sequence

Because of extensive binding to proteins that modify and/or transport RNAs, and elaborate secondary structures, some RNA sequences may not be accessible for hybridization. When the precise cellular localization of the RNA is not important for the experiment, the following treatments can be considered to increase the accessibility of the probe to the target sequences.

- *Preextraction before fixation.* Preextraction in CSK buffer with 0.5% Triton X-100 for 3 minutes on ice before fixation in paraformaldehyde (Carmo-Fonseca et al. 1992) can increase the detection of the target sequences.
- *Types of fixation.* Methanol fixation at –20°C, or a 3.7% formaldehyde fixation with the addition of 5% acetic acid (Dirks et al. 1995) may increase accessibility to the target sequences.
- *Partial protease digestion.* A limited protease digestion (0.05–0.1% pepsin in 0.01 M HCl for 1–3 min) after the permeabilization step can strip away some of the RNA-binding proteins to expose the target sequences. However, the concentration of the protease and the duration of treatment should be carefully titrated so that cells are not completely destroyed.

REFERENCES

Carmo-Fonseca M., Pepperkok R., Carvalho M.T., and Lamond A.I. 1992. Transcription-dependent colocalization of the U1, U2, U4/U6 and U5 snRNPs in coiled bodies. *J. Cell Biol.* **117:** 1–14.

Carmo-Fonseca M., Tollervey D., Barabino S.M.L., Merdes A., Brunner C., Zamore P.D., Green M.R., Hurt E., and Lamond A.I. 1991. Mammalian nuclei contain foci which are highly enriched in components of the pre-mRNA splicing machinery. *EMBO J.* **10:** 195–206.

Carter K.C., Taneja K.L., and Lawrence J.B. 1991. Discrete nuclear domains of poly(A) RNA and their relationship to the functional organization of the nucleus. *J. Cell Biol.* **115:** 1191–1202.

Dirks R.W., Daniël K.C., and Raap A.K. 1995. RNAs radiate from gene to cytoplasm as revealed by fluorescence in situ hybridization. *J. Cell Sci.* **108:** 2565–2572.

Haramis A.G., Brown J.M., and Zeller R. 1995. The limb deformity mutation disrupts the SHH/FGF-4 feedback loop and regulation of 5′ HoxD genes during limb pattern formation. *Development* **121:** 4237–4245.

Huang S. and Spector D.L. 1996. Intron-dependent recruitment of pre-mRNA splicing factors to sites of transcription. *J. Cell Biol.* **133:** 719–732.

Huang S., Deerinck T.J., Ellisman M.H., and Spector D.L. 1994. In vivo analysis of the stability and transport of nuclear poly(A)$^+$ RNA. *J. Cell Biol.* **126:** 877–899.

Jiménez-García L.F. and Spector D.L. 1993. In vivo evidence that transcription and splicing are coordinated by a recruiting mechanism. *Cell* **73:** 47–59.

Latham V.M., Kislawskis E.H., Singer R.H., and Ross A.F. 1994. β-actin mRNA localization is regulated by signal transduction mechanisms. *J. Cell Biol.* **126:** 1211–1219.

Lawrence J.B. and Singer R.H. 1991. Spatial organization of nucleic acid sequences within cells. *Semin. Cell Biol.* **2:** 83–101.

Singer R.H., Lawrence J.B., and Villnave C. 1986. Optimization of in situ hybridization using isotopic and non-isotopic detection methods. *BioTechniques* **4:** 230–243.

Sundell C. and Singer R.H. 1991. Requirement of microfilaments in sorting of actin messenger RNA. *Science* **253:** 1275–1277.

Tautz D. and Pfeifle C. 1989. A nonradioactive in situ hybridization method for the localization of specific FRNAs in *Drosophila* embryos reveals translational control of the segmentation gene *hunchback*. *Chromosoma* **98:** 81–85.

Xing Y., Johnson C.V., Dobner P.R., and Lawrence J.B. 1993. Higher level organization of individual gene transcription and RNA splicing: Integration of nuclear structure and function. *Science* **259:** 1326–1330.

CHAPTER 18

Whole-Mount In Situ Detection of RNAs in Vertebrate Embryos and Isolated Organs

Anna Haramis[1] and Rolf Zeller[2]

[1]*European Molecular Biology Laboratory, Heidelberg, Germany*
[2]*University of Basel, Basel, Switzerland*

INTRODUCTION

Nonisotopic in situ hybridization using intact embryos or organs is a widely used method that enables fast determination of the spatial distribution of RNAs. Large numbers of samples can be analyzed simultaneously, which permits direct comparison of different developmental stages and/or genotypes. Furthermore, the sensitivity and reproducibility of whole-mount in situ hybridization facilitates the analysis of genes expressed in restricted patterns, that is, in small structures easily overlooked when using serial sections. In addition, the opportunity for inspecting histological details is not lost because the embryos can be sectioned after staining. Therefore, it is the method of choice to determine the expression patterns of gene products (Rosen and Beddington 1993).

Protocols enabling simultaneous detection of two different transcripts by whole-mount in situ hybridization have been developed. Probes labeled with digoxigenin and with fluorescein can be detected differentially by antibodies coupled to different enzymatic detection systems (Hauptmann and Gerster 1994; Jowett and Lettice 1994; see also Chapter 17 for a discussion of probes). The methods described here can also be used for detection of multiple gene products using digoxigenin-labeled riboprobes, provided that the target transcripts are localized in distinct structures/cell types of the embryo (see Fig. 18.3).

Before attempting to study the distribution of RNAs of interest in a particular tissue, it is important to establish their abundance and temporal distribution by Northern blot and/or RNase protection analysis. RNase protection assays should be used to select possible probes for subsequent whole-mount in situ hybridization. Several parameters affect the strength of the hybridization signal and signal-to-noise ratio. Both G-C- and A-T-rich sequences should be avoided, and the best probes are often derived from coding regions. However, when analyzing expression of members of highly conserved gene families, probes must be derived from least conserved regions to assure specific detection of RNAs. Sensitivity of detection and signal-to-noise ratios depend on probe length and concentration. However, probe length seems less critical when using digoxigenin (or other haptenes) than when using ^{35}S-labeled riboprobes. Riboprobes up to 2.5 kb in length have been successfully used without previous alkali hydrolysis. It has been reported that alkali hydrolysis of digoxigenin riboprobes may reduce sensitivity instead of aiding hybridization (Wilkinson 1993; Rosen and Beddington 1993). The protocol described uses rather stringent hybridization conditions (50% formamide at 70°C) for specific detection of different types of mRNAs. Therefore, it may be necessary to reduce hybridization temperatures to 63°C (or 55°C) when using A-T-rich probes (Harland 1991; Wilkinson 1993).

Before starting whole-mount in situ hybridizations, it is important that large amounts of tissue are collected for the preparation of acetone powder, and riboprobes should be tested by RNA analysis. In general, three overnight steps are required during the procedure (fixation, hybridization, antibody incubation). Linearization of the template DNA can be done overnight during the fixation step. The preparation of the riboprobes requires 4–5 hours. Pretreatment of samples before the hybridization requires 4 hours of work and an additional 3 hours for prehybridization. Posthybridization washes and antibody preblocking require a total of 8 hours. Postantibody washes require about 6 hours. Signal detection requires a minimum of 2 hours and could extend to overnight. If different samples are to be analyzed at the same time, they are stored at –20°C after fixation or prehybridization, to allow sample collection over several days to weeks.

The methods described have been used to localize different types of transcripts in mouse (Figs. 18.1 and 18.2), chicken (Fig. 18.3), and *Xenopus laevis* embryos (see, e.g., Haramis et al. 1995). However, the parameters discussed below might need optimization depending on particular tissues and target mRNAs. Furthermore, controls are suggested that should aid the adaptation of the protocols to "difficult" tissues or probes. In general, it is important to ensure that the target RNAs remain intact during fixation and subsequent treatments. Therefore, embryos or organs should be isolated at 4°C and fixed as fast as possible. It is important to use RNase-free solutions to avoid possible degradation of target RNAs. During initial experiments, it is advisable to perform the procedure exactly as described and to complete it without intermediate storage. Furthermore, a probe with a known distribution in the tissue(s) of interest should be included as a positive control. Omission of the riboprobe and/or anti-digoxigenin Fab fragments serves as a negative control and should reveal nonspecific labeling and/or activity of endogenous alkaline phosphatases. Alternatively, a sense-strand digoxigenin riboprobe can be synthesized and used as a negative control.

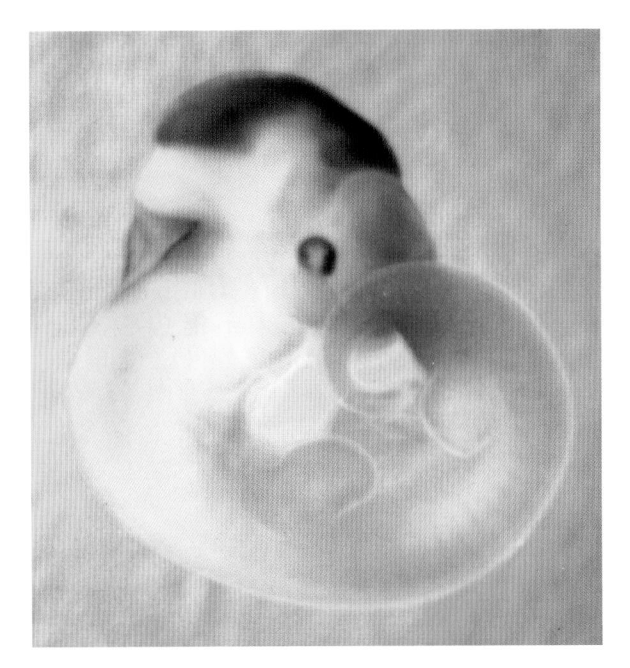

FIGURE 18.1

Whole-mount in situ hybridization using a digoxigenin-labeled Otx-2 antisense riboprobe (Simeone et al. 1993) on mouse embryo (embryonic day 11). Intense blue staining indicating Otx-2 expression is detected in the telencephalic, diencephalic, and mesencephalic regions of the brain and developing eye at this stage. (Photo provided by R. Dono, EMBL.)

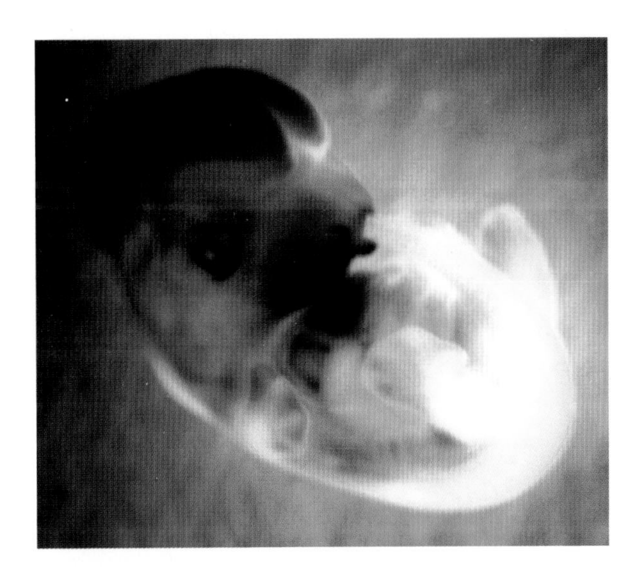

FIGURE 18.2

Whole-mount in situ hybridization using a digoxigenin-labeled Otx-1 antisense riboprobe (Simeone et al. 1993) on mouse embryo (embryonic day 13). Intense blue staining indicating Otx-1 expression is detected in the developing brain, ocular region, and nasal cavities. (Photo provided by R. Dono, EMBL.)

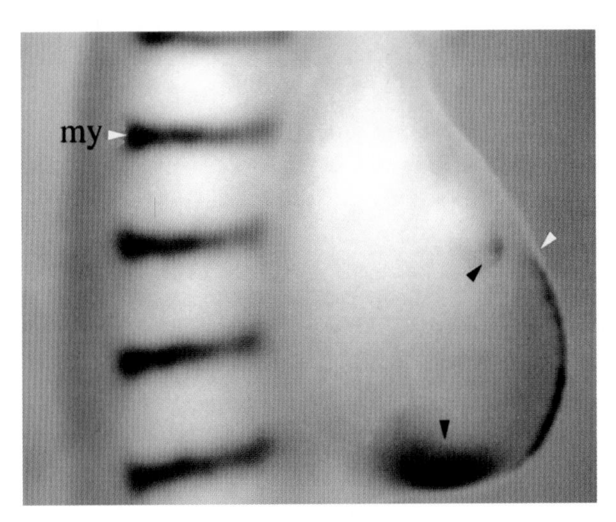

FIGURE 18.3

Whole-mount in situ hybridization using digoxigenin-labeled murine *Shh* (Echelard et al. 1993) and chicken *Fgf-4* (Niswander et al. 1994) riboprobes on a chicken wing bud containing an anterior graft of mouse polarizing region. Chicken *Fgf-4* expression is detected in the apical ectodermal ridge and myotomes (my; *white arrowheads*). *Shh* expression is detected in the grafted anterior mouse tissue (*black arrowhead*). The murine *Shh* riboprobe also cross-hybridizes to endogenous chicken transcripts located in the posterior wing bud mesenchyme (*black arrowhead*). Anterior is at the *top* and posterior at the *bottom* of the figure. (Photo provided by A.G. Haramis and R. Zeller, EMBL.)

Hybridization and Detection of RNA Transcripts

The protocols described below are based on protocols by Wilkinson (1993) and Harland (1991). These have been used successfully for the detection of mRNAs in mouse and chicken embryos. The method is also applicable to experimentally manipulated embryos, such as for detection of transcripts on mouse tissue grafted to chicken embryos (Haramis et al. 1995). The preparation of digoxigenin-labeled RNA probes is followed by protocols for fixation of embryos, hybridization, and detection of RNA hybrids.

Use deionized, distilled H_2O in all recipes and protocol steps. The solutions should be autoclaved where possible, or **DEPC** (diethyl pyrocarbonate)-treated to inhibit RNases.

CAUTION: DEPC (see Appendix 2 for Caution)

PREPARATION OF DIGOXIGENIN-LABELED RNA PROBES

Antisense riboprobes are synthesized as run-off transcripts from linearized templates, using bacteriophage polymerases. Therefore, the template DNA consists of the DNA fragment of interest cloned in a vector containing the promoter of the respective RNA polymerase (T3, T7, SP6). RNA synthesis is carried out incorporating a digoxigenin-substituted ribonucleotide. Nonradioactive probes have several advantages: They are easily synthesized in large quantities, are stable for several months, and can be reused up to three times. An additional advantage of RNA versus DNA probes is that they result in cleaner signals, because nonspecifically bound probe is removed during the ribonuclease treatment. The protocol described below is carried out essentially as described in Wilkinson (1993). All buffers and solutions should be made up with DEPC-treated H_2O.

1 Mix the reagents in the following order at room temperature.

sterile distilled H_2O	13 μl
10x transcription buffer	2 μl
0.2 M **DTT** (dithiothreitol)	1 μl
nucleotide mix	2 μl
linearized plasmid encoding target (1 μg/μl)	1 μl
RNasin (placental ribonuclease inhibitor [100 units/ml] [RNasin, Boehringer Mannheim])	1 μl
bacteriophage RNA polymerase (SP6, T3, or T7 bacteriophage RNA polymerase [10 units/μl])	1 μl

10x Transcription Buffer

400 mM Tris-HCl, pH 8.25
60 mM **$MgCl_2$**
20 mM spermidine (Boehringer Mannheim)

Nucleotide Mix, pH 8

10 mM GTP
10 mM ATP
10 mM CTP
6.5 mM UTP
3.5 mM **digoxigenin**-UTP (Boehringer Mannheim)

CAUTION: DTT; $MgCl_2$; digoxigenin (see Appendix 2 for Caution)

Although sensitive reagents should be kept on ice, the reaction must be mixed at room temperature, otherwise the spermidine in the transcription buffer will precipitate the DNA.

2 Incubate at 37°C for 2 hours.

3 Remove 1 µl of the reaction and analyze on a 1% agarose/TBE (Tris-borate/EDTA) gel to estimate amount of RNA synthesized.
The RNA band should be about ten times more abundant than the plasmid band, indicating that about 10 µg of riboprobe has been synthesized.

4 Add 2 µl of DNase I (RNase-free) to the transcription reaction and incubate at 37°C for 15 minutes.

5 Add 100 µl of TE (Tris-EDTA buffer), 10 µl of 4 M LiCl, 300 µl of 100% ethanol. Mix and incubate at –20°C for 30 minutes.
Tris-EDTA solution, pH 8
Li Cl precipitates RNA.

6 Spin in microfuge at 4°C for 10 minutes. Wash pellet with 70% ethanol/H_2O.

7 Air-dry pellet (do not use a speed vac for drying the pellet) and redissolve in DEPC-TE at 0.1 µg/µl.
Store at –20°C. Use 10 µl for each milliliter of hybridization solution.
Digoxigenin-labeled riboprobes are stable at –20°C for at least 2 years.

FIXATION, PRETREATMENT, AND HYBRIDIZATION OF SAMPLES

Embryos or organs are isolated and fixed to preserve their morphology. They are pretreated to make the target transcripts accessible to the riboprobe. The pretreatments and the prehybridization block nonspecific binding sites, thus reducing nonspecific hybridization of riboprobes and background staining during enzymatic detection of RNA hybrids by antibodies. Subsequently, the samples are hybridized to digoxigenin-labeled riboprobes.

Materials

Dissection tools (scissors, forceps)
Dissection microscope
20-ml snap-cap glass vials
Small spoon or spatula to transfer samples
1x PBS (phosphate-buffered saline)
1x PBS + 0.1% Tween-20 (PBT)
4% **formaldehyde** in PBS made fresh from **paraformaldehyde**, cooled to 4°C
25%, 50%, and 75% **methanol** in PBT (methanol/PBT)

100% methanol
6% hydrogen peroxide, in PBT (Aldrich)
2 mg/ml **glycine** (Merck) in PBT (freshly prepared)
4% formaldehyde containing 0.2% **glutaraldehyde** (Sigma, EM grade) in PBT

CAUTION: Formaldehyde; paraformaldehyde; methanol; glycine; glutaraldehyde (see Appendix 2 for Caution)

It is very important that the samples are always kept in liquid. Drying of samples during the procedure will cause high levels of background staining. Therefore, always leave the sample covered with a small quantity of liquid when exchanging solutions. This also prevents physical damage or loss of samples. All washes are carried out at room temperature for 5 minutes with gentle agitation, unless stated otherwise.

1 Dissect embryos in ice-cold PBS using appropriate dissection tools and a microscope, if necessary. Remove completely extraembryonic membranes and the amnion from embryos.

Trapping of riboprobes and antibodies in embryonic cavities is one of the major causes of nonspecific signals. It is therefore important to open (i.e., pierce) cavities such as heart chambers or brain ventricles before or after fixation (necessary for mouse embryos older than embryonic day 10 or chicken embryos older than embryonic stage 20). For advanced embryos (such as mouse embryos from embryonic day 14 onward) or postnatal stages, it is necessary to isolate the organs or tissues of interest to achieve the best result. Extraembryonic membranes can be stored at –80°C and used for preparation of genomic DNA to genotype embryos by polymerase chain reaction (PCR) or Southern analysis.

2 Transfer the embryos into 20-ml snap-cap glass vials containing about 10 ml of 4% formaldehyde fixative (4°C). Fix at 4°C for 4 hours to overnight.

The 4% formaldehyde fixative must be prepared from paraformaldehyde and stored frozen in aliquots at –20°C. An aliquot is defrosted and used once. Fixation of embryos and organs for whole-mount in situ hybridization is less critical than when using ^{35}S-labeled riboprobes on histological samples. Fixation times can vary from 4 hours to overnight without significant effects on signal-to-noise ratios. If desired, it is possible to fix several samples in one glass vial. For all steps carried out using glass vials (steps 2–12), use ~5–10 ml of solution.

3 Wash embryos twice in PBT, at 4°C.

It is advisable to perform the dehydration and rehydration steps (steps 4 and 5), even if the embryos are used immediately and not stored. It was found that these steps improve permeabilization of the tissues and increase the sensitivity. Alternatively, the embryos can be processed from step 8 onward.

4 Dehydrate embryos by successive washes in 25%, 50%, and 75% methanol/PBT at room temperature. Finally, wash twice in 100% methanol.

The samples can be stored in 100% methanol for up to 6 months at –20°C. Alternatively, the samples can be processed to step 14 and then stored at –20°C.

5 Rehydrate embryos using the methanol/PBT series in the reverse order and wash twice in PBT, at room temperature.

6 (*Optional*) Bleach embryos with 6% hydrogen peroxide in PBT at room temperature for 15 minutes.

This step seems to significantly reduce background staining. However, extensive bleaching may reduce the signal (Wilkinson 1993). The length of bleaching can be varied according to the tissue of interest. For example, treatment can be extended up to 1 hour if tissues are prone to high levels of background staining.

7 Wash the embryos three times in PBT.

8 Digest embryos with 10 µg/ml proteinase K in PBT at room temperature for 15 minutes. *The samples become very fragile during this step.*

10 μg/ml proteinase K in PBT, not predigested

Proteinase K digestion removes proteins, unmasking RNAs, and helps to permeabilize the tissues, allowing the riboprobe to access its target mRNAs during subsequent hybridization (step 15). The length of treatment and/or concentration of proteinase K needs to be optimized. For example, if the distribution of RNAs in embryonic ectodermal structures is studied, proteinase K digestion must be reduced to 5 μg/ml for 4 minutes or 10 μg/ml for 2 minutes to keep the ectoderm intact (Crossley and Martin 1995). Alternatively, proteinase K digestion can be substituted by detergent permeabilization (Rosen and Beddington 1993).

9 Stop digestion by washing in freshly prepared 2 mg/ml **glycine** in PBT.

CAUTION: Glycine (see Appendix 2 for Caution)

10 Wash twice in PBT.

11 Postfix embryos at room temperature for 20 minutes with freshly prepared 0.2% **glutaraldehyde**/4% **formaldehyde** in PBT.

CAUTION: Glutaraldehyde; formaldehyde (see Appendix 2 for Caution)

It is important that the 4% formaldehyde used for this step is freshly prepared. Inclusion of 0.2% glutaraldehyde in the fixative is optional but seems to enhance the signal-to-noise ratio.

Postfixation of samples is essential to retain unmasked RNAs and preserve morphology during subsequent hybridization.

12 Wash embryos three times in PBT.

13 Add 1 ml of prehybridization solution to glass vials and transfer the sample to 2-ml Eppendorf tubes.

Prehybridization Solution

50% **formamide** (deionized)
5x SSC, pH 4.5
2% blocking reagent (Boehringer Mannheim)
0.1%Tween-20
0.5% CHAPS
50 μg/ml yeast RNA (10 mg/ml stock solution, in distilled sterile H_2O)
5 mM EDTA
50 μg/ml heparin (10 mg/ml stock solution, in distilled sterile H_2O)

CAUTION: Formamide (see Appendix 2 for Caution)

2% Blocking Reagent

Dissolve the blocking powder (Boehringer Mannheim) in PBT at 65°C before use.

Embryos are best transferred using a small spoon or spatula. Alternatively, the sample with the prehybridization solution can be carefully poured into Eppendorf tubes. Depending on sample size, more than one sample can be hybridized in one tube. This facilitates direct comparative analysis of different samples. About 1.5–2 ml of solution should be used for all subsequent steps.

14 Remove the solution and add 2 ml of fresh prehybridization solution. Prehybridize at 65°C for 3 hours.

Eppendorf tubes are placed into a preheated heating block and secured. The heating block is placed sideways on a rocking platform. This setup enables optimal and gentle mixing of the prehybridization solution and is used for hybridization and all subsequent washes not carried out at room temperature. Alternatively, a shaking water bath may be used. The samples can be stored in prehybridization solution at –20°C for up to 6 months, either before or after completing step 14.

15 Replace prehybridization solution with 1 ml of hybridization solution and hybridize at 70°C overnight.

Hybridization Solution

Prehybridization solution containing 1 μg/ml digoxigenin-labeled probe (see above).

Prehybridization and hybridization solutions can be reused up to three times (stored at –20°C). See Troubleshooting at end of protocol for suggestions on recommended hybridization temperatures.

POSTHYBRIDIZATION WASHES AND IMMUNOLOGICAL DETECTION OF RNA HYBRIDS

Nonspecifically bound riboprobes are removed by extensive washes and RNase A digestion. Subsequently, the digoxigenin-labeled RNA hybrids are detected immunologically using anti-digoxigenin antibodies (Fab fragments conjugated to alkaline phosphatase). The immunocomplexes are detected by an alkaline phosphatase color reaction. All solutions should be prewarmed to the appropriate temperature before use. Remove solutions using pasteur pipettes and ensure that the samples always remain completely covered by a small quantity of liquid. All incubations and washes are carried out with gentle rocking (as described before; step 14) at the temperatures indicated.

1 Remove hybridization solution and add 800 μl of prehybridization solution. Wash at 70°C for 5 minutes.

2 Without removing the prehybridization solution, add 400 μl of 2x SSC (standard saline citrate), pH 4.5, to each tube, and wash at 70°C for 5 minutes. Repeat the addition of 2x SSC and washing twice.

The successive addition of 2x SSC to the prehybridization solution leads to a stepwise dilution of the prehybridization solution to 25% prehybridization solution/75% 2x SSC.

20x SSC Stock Solution

A. Dissolve 175.3 g of NaCl and 88.2 g of sodium citrate in 800 ml DH_2O.
B. Adjust pH to 7.0 with 10 N NaOH.
C. Bring up to 1 liter with DH_2O and autoclave.

3 Remove the mix and wash at 70°C twice for 30 minutes each time in 0.1% CHAPS (3-([3-cholamidopropyl]-dimethyammonio)-1-propanesulfonate)/2x SSC.

4 Digest samples with 20 μg/ml RNase A in 0.1% CHAPS/2x SSC, at 37°C for 1 hour.

The conditions of the RNase A digestion might have to be adjusted. This step eliminates nonspecific riboprobes and is essential to obtain the best signal-to-noise ratios. The specifically hybridized RNA hybrids are stable, but RNase concentrations might have to be lowered or digestion omitted when detecting heterologous or rare mRNAs. Pilot experiments using varying RNase concentrations and/or digestion times might be necessary to establish the best conditions.

5 Wash at room temperature twice, 10 minutes each time, in maleic acid buffer. Wash at 70°C twice, 30 minutes each time, in maleic acid buffer.

Maleic Acid Buffer

100 mM **maleic acid** (Sigma)
150 mM NaCl
Prepare from stock solutions, using sterile distilled H_2O. Adjust pH to 7.5 with **HCl**.
Autoclave. The solution is stable at room temperature for months.

CAUTION: Maleic acid; HCl; acids and bases that are concentrated (see Appendix 2 for Caution)

This step significantly decreases nonspecific signals, probably by blocking nonspecific antibody-binding sites.

6 Wash twice at room temperature, 10 minutes each time, in PBS.

7 Wash at room temperature 5 minutes in PBT.

8 Incubate samples 2–3 hours in blocking solution, at room temperature.

Blocking Solution

10% sheep serum (heat-inactivated)
1% BSA (bovine serum albumin) in PBT

Sheep Serum

Thaw the sheep serum in a 37°C water bath.
Once liquid, heat-inactivate at 70°C, for 30 minutes.
Spin at 4°C for 15 minutes at 4000 rpm.
Take the supernatant and make aliquots of 2 ml that can be stored at –20°C.
Do not refreeze aliquots.

This step blocks nonspecific antibody-binding sites.

9 During steps 5–8, preabsorb the Fab fragments as described in the protocol below.

10 Replace the blocking solution with 2 ml of preabsorbed Fab fragment solution and incubate at 4°C overnight, with gentle agitation.

The purpose of the subsequent washes is to remove nonspecifically bound antibody, preventing nonspecific reaction with chromogenic substrates.

11 Remove the Fab fragment solution and replace with 0.1% BSA in PBT. Transfer the samples to 20-ml snap-cap glass vials.

About 10 ml of solution is used for all subsequent steps, and all washes are carried out with gentle rocking, at room temperature.

12 Wash five times at room temperature, 45 minutes each time, in 0.1% BSA in PBT.

13 Wash twice, 30 minutes each time, in PBT.

14 Wash three times, 10 minutes each time, in NTMT buffer.

Alkaline Phosphatase Buffer, pH 9.5 (NTMT)

100 mM NaCl
100 mM Tris-HCl, pH 9.5
50 mM **$MgCl_2$**
0.1% Tween-20

CAUTION: $MgCl_2$ (see Appendix 2 for Caution)

Make NTMT from concentrated stocks on the day of use, because the pH will decrease on storage, because of absorption of carbon dioxide (Wilkinson 1993). For the *Xenopus* protocol, it is necessary to filter the solution before use.

Levamisole (Sigma, 2 mM final) may be included in the NTMT buffer to inhibit endogenous alkaline phosphatases.

15 Incubate in 3 ml of BM purple solution (undiluted), or 3 ml of **NBT/BCIP** (nitroblue tetrazolium/bromochloroindolyl phosphate) substrate solution. Make sure that samples are completely submerged in staining solution. Cover glass vials with aluminium foil and gently rock them for the first 20 minutes of the developing reaction.

CAUTION: NBT; BCIP (see Appendix 2 for Caution)

> BM Purple AP substrate (Boehringer Mannheim)
>
> ***NBT/BCIP Substrate Solution (prepare fresh each time; optional)***
>
> Add 220 µl of NBT solution (75 mg/ml in **DMF** (*N,N*-dimethylformamide); 330 µg/ml final concentration) to 50 ml of alkaline phosphatase buffer, pH 9.5, and mix gently (do not vortex).
>
> Add 170 µl of BCIP solution (50 mg/ml in DMF; 170 µg/ml final) and mix gently.
>
> **CAUTION:** DMF (see Appendix 2 for Caution)

The BM Purple and the NBT/BCIP staining solution are light sensitive, therefore it is important to keep the samples in the dark. Monitor progression of staining with minimal exposure to light.

16 Monitor the staining by eye or using a microscope. After the signal has developed to the desired extent, stop the staining reaction by washing the samples at least six times with PBT.

The samples can be stored for a 2–3 months in PBT (containing 100 mM EDTA) at 4°C, without significant effects on signal strength.

Signal strength depends on abundance of the target RNA, hybridization conditions, and additional factors (see Troubleshooting at end of protocol). Developing times of up to 48 hours are required for some RNAs, whereas other probes give strong signals after only 20 minutes. However, for most probes, incubation times of about 2 hours are required to obtain convincing results. The staining solution should be exchanged for overnight incubation. It is also possible to restain samples by repeating the procedure from step 14 onward.

17 (*Optional*) Fix the samples overnight in 4% formaldehyde in PBT at 4°C. Subsequently, wash samples several times and store in PBT at 4°C.

This step fixes the signal, but might affect overall morphology. Furthermore, samples cannot be restained after fixation.

18 Document your results using a photomicroscope as soon as possible.

Samples must be photographed submerged in PBT (in a petri dish) to avoid reflections and drying artifacts. It is best to illuminate the sample from top or side using glass-fiber optics. Use tungsten slide film or black and white negative film to obtain best results.

PREPARATION OF EMBRYONIC POWDER AND PREABSORPTION OF FAB FRAGMENTS

Preabsorbing the detecting antibody using acetone powder prepared from the tissues under study reduces nonspecific binding. The following protocols describe the preparation of acetone powder (adapted from Wilkinson 1993) and preabsorption of the Fab fragments.

Preparation of (Mouse) Acetone Powder

For 1 ml of embryo powder, at least 50 mouse embryos of embryonic day 12 are needed. For chicken embryonic powder, chickens of stage 24–26 can be used and the yield of powder is high.

1 Collect mouse embryos of embryonic day 10–12, remove the extraembryonic membranes, freeze the embryos in liquid nitrogen, and store at –70°C until needed. To make the powder, homogenize the embryos in **liquid nitrogen** using a pestle and mortar.

CAUTION: Liquid nitrogen (see Appendix 2 for Caution)

2 After liquid nitrogen has evaporated, add 4 volumes of ice-cold **acetone**, mix well, and incubate on ice for 30 minutes.

CAUTION: Acetone (see Appendix 2 for Caution)

3 Spin at 10,000*g* for 10 minutes at 4°C and remove supernatant.

4 Wash pellet with cold acetone and spin again.

5 Spread the pellet on a sheet of filter paper and grind it to a fine powder and allow to air-dry.

6 Store in an airtight Eppendorf tube at –20°C.
The powder is stable at –20°C for several months. Just before use, the embryo powder should be heat-inactivated at 70°C for 30 minutes.

Preabsorption of Fab Fragments

The Fab fragments can bind nonspecifically to tissues. Therefore, it is crucial that the Fab fragments are properly preabsorbed using acetone powder prepared from tissues of interest. The Fab fragment solution can be reused up to three times within less than 6 weeks. Store the solution at 4°C and check for possible contamination before use.

1 For a final volume of 2 ml per tube, weigh out 3 mg of mouse embryo powder (prepared as described in the above protocol) and heat-inactivate in 1 ml of PBT at 70°C for 30 minutes.

2 Cool on ice and then add BSA (1% final concentration), heat-inactivated sheep serum (10% final concentration), and 1 μl of antibody.

3 Rock the solution at 4°C for 2–3 hours.

4 Centrifuge the solution at 4°C for 10 minutes.

5 Dilute the supernatant to 2 ml with PBT containing 10% sheep serum, 1% BSA, producing a final dilution of antibody of 1:2000.

▪ *Notes: Troubleshooting*

- *Low or no hybridization signal.* It is likely that the target RNA was not detected due to its low abundance, degradation, or problems associated with the riboprobe (see critical parameters). However, several steps of the protocol should be carefully checked.

a Overfixation of tissue is an unlikely possibility but cannot be completely excluded. If overfixation is suspected, the optimal fixation time should be determined in pilot experiments.

b Alternatively, overdigestion with proteinase K will cause loss of sensitive structures, especially in young embryos. Proteinase K concentration and/or incubation times should be reduced.

c It is quite possible that hybridization temperatures are too stringent or the probe concentration is too low. Increase probe concentration and/or lower the hybridization temperature (e.g., to 55°C).

d Excessive RNase digestion will reduce signal strength. Either RNase concentration or the length of digestion can be reduced. The digestion step can also be performed at room temperature, or completely omitted.

- *High levels of nonspecific signal.* There are several parameters that may affect the signal-to-noise ratio.

a It is likely that the hybridization conditions are not sufficiently stringent or that the probe concentration is too high. Reduce probe concentrations or increase the stringency of hybridization and washes (temperature, ionic strength, and incubation times). In the case of *Xenopus* embryos, increasing salmon sperm DNA concentration was found to significantly improve specificity.

b Alternatively, nonspecifically bound probe could be incompletely digested by the RNase. RNase A concentration can be increased up to 100 μg/ml.

c Another common problem is the residual activity of endogenous alkaline phosphatases, which would react with the chromogenic substrates during development of the signals. In such cases, levamisole (2 mM final for mouse/chicken and 5 mM for *Xenopus* embryos) should be freshly added to the NTMT buffer. Levamisole inhibits endogenous phosphatases without major effects on Fab-conjugated alkaline phosphatases.

d The chromogenic substrates used for developing can age and cause increased background. BM Purple AP substrate solution (Boehringer Mannheim) is a ready-to-use stabilized developing solution that often results in superior signal-to-noise ratios.

- *Loss of morphology.* Whole-mount in situ hybridization involves multiple manipulations, which might lead to damage or even complete loss of samples. Handle samples very carefully and check for possible damage following each handling or transfer step. Insufficient fixation or postfixation (after proteinase K digestion) will invariably cause loss of morphology. Furthermore, extensive hydrogen peroxide treatment should be avoided when very young embryos are studied (E6–E9).

REFERENCES

Crossley P.H. and Martin G.R. 1995. The mouse *Fgf8* gene encodes a family of polypeptides and is expressed in regions that direct outgrowth and patterning in the developing embryo. *Development* **121:** 439–451.

Echelard Y., Epstein D.J., St-Jacques B., Shep L., Mohler J., McMahon J.A., and McMahon A.P. 1993. Sonic hedgehog, a member of a family of putative signaling molecules, is implicated in the regulation of CNS polarity. *Cell* **75:** 1417–1430.

Haramis A.G., Brown J.M., and Zeller R. 1995. The *limb deformity* mutation disrupts the SHH/FGF-4 feedback loop and regulation of 5′ *HoxD* genes during limb pattern formation. *Development* **121:** 4237–4245.

Harland R.M. 1991. In situ hybridization: An improved whole-mount method for *Xenopus* embryos. *Methods Cell Biol.* **36:** 685–695.

Hauptmann G. and Gerster T. 1994. Two-color whole-mount in situ hybridization to vertebrate and *Drosophila* embryos. *Trends Genet.* **10:** 266.

Jowett T. and Lettice L. 1994. Whole mount *in situ* hybridizations on zebrafish embryos using a mixture of digoxigenin- and fluorescein-labelled probes. *Trends Genet.* **10:** 73–74.

Lamb T.M., Knecht A.K., Smith W.C., Stachel S.E., Economides A.N., Stahl N., Yancopolous G.D., and Harland R.M. 1993. Neural induction by the secreted polypeptide noggin. *Science* **262:** 713–718.

Niswander L., Jeffrey S., Martin G.R., and Tickle C. 1994. A positive feedback loop coordinates growth and patterning in the vertebrate limb [see comments]. *Nature* **371:** 609–612.

Rosen B. and Beddington S.P. 1993. Whole mount *in situ* hybridization in the mouse embryo: Gene expression in three dimensions. *Trends Genet.* **9:** 162–167.

Schulte-Merker S. 1993. In situ hybridization. In *The zebrafish book* (ed. M. Westerfield), pp. 9.16–9.21. University of Oregon Press, Eugene.

Simeone A., Acampora D., Mallamaci A., Stornaiuolo A., D'Apice M.R., Nigro V., and Boncinelli E. 1993. A vertebrate gene related to orthodenticle contains a homeodomain of the bicoid class and demarcates anterior neuroectoderm in the gastrulating mouse embryo. *EMBO J.* **12:** 2735–2747.

Wilkinson D.G. 1993. In situ hybridisation. In *Essential developmental biology: A practical approach* (ed. C.D. Stern and P.W.H. Holland), pp. 257-276. IRL Press, Oxford.

CHAPTER 19

Image Production Using Transmission Electron Microscopy

Mark H. Ellisman

CRBS, University of California, San Diego, School of Medicine, La Jolla, California

INTRODUCTION

In the late 19th century, it was realized that because of diffraction, there was a limit to the resolution achievable with an optical microscope that was dependent upon the wavelength of the illuminating medium. The mathematician Ernst Abbé expressed the diffraction limit on resolution in the following equation:

$$d = \frac{0.612\lambda}{n \sin \alpha}$$

where d is the distance between two points that can just be resolved, or the effective resolution; 0.612 is Abbé's constant; λ is the wavelength of the image-forming radiation; n is the index of refraction of the medium between the specimen and lens; and α is the half-angle of the cone of light from the specimen plane accepted by the front surface of the lens.

From this equation, it becomes obvious that to get the best possible resolution, one should have the smallest possible λ and the largest possible $n \sin \alpha$ (also referred to as the numerical aperture [NA] of a lens). Using light with a wavelength of 500 nm, immersion oil so that $n = 1.55$, and a 63x objective with a half-acceptance angle of 65° (NA = 1.4), the best resolution we can hope to obtain with the light microscope is about 200 nm. Because the maximum half-acceptance angle cannot be greater than 90° and media with a refractive index greater than 1.6 are unknown, it is nearly impossible to have an NA much greater than 1.4. From Abbé's equation, one can see that not much can be done to improve resolution except to use a shorter wavelength. However, light with a wavelength shorter than 400 nm approaches the UV range, which cannot be seen with the eye.

Microscope design was such that light microscopes achieved the resolution limit around the turn of the century. Abbé himself remarked that, "It is poor comfort to hope that human ingenuity will find ways and means to overcome this limit." (Note: New solutions to get around this limit of light microscopy resolution include the ongoing development of approaches such as 4Pi microscopy.)

Fortunately for microscopists, based on the work of Einstein and Planck, the French mathematician Louis de Broglie in 1924 advanced the theory that a beam of electrons had a wavelength and offered this classic equation to calculate it:

$$\lambda = \frac{h}{mv}$$

where λ is the wavelength, h is Planck's constant, m is the mass of an electron, and v is the velocity of the electron.

Electron microscopes use a beam of electrons rather than light to form an image of a specimen. Substituting known values into the above equation and taking into account relativistic effects, a 100,000-V potential electron beam would have a wavelength of ~0.0038 nm, almost five orders of magnitude shorter than visible light. According to Abbé's equation, and where $n = 1$ (because the inside of the electron microscope is under vacuum) and sin α is about equal to α (because this angle is very small in the electron microscope, $\sim 10^{-2}$ radians), we see a predicted resolution of about 0.24 nm or an improvement in resolution of about three orders of magnitude over the light microscope. This is sufficient to achieve atomic resolution and is a tremendous improvement over the light microscope. However, because of limitations in specimen preparation, this resolution is not approached with most biological specimens but is usually on the order of 2 nm.

PARTS OF THE ELECTRON MICROSCOPE

Transmission electron microscopes consist of the following major components: a high-voltage generator, an electron gun, a series of electromagnetic lenses and deflection coils that form the column, a vacuum system, and the associated electronics to control these components. (See Fig. 19.1.)

The high-voltage generator provides the electron potential between the cathode and the anode that is used to accelerate the electrons. Conventional transmission electron microscopes typically operate with an accelerating potential in the range of 60–100 keV. The electrons are supplied by an electron gun consisting of a filament, which upon heating acts as an electron emitter. The filament is typically composed of a thin tungsten wire, but more exotic materials such as lanthanum hexaboride (LaB_6) are also used. The filament is surrounded by a cathode cap (also called the Wehnelt cylinder) that is biased at a slightly different voltage potential from the filament and acts both to focus and to control the number of electrons in the beam. Together they are collectively called the cathode because the high voltage is applied to both. Below the cathode is a metal disk with a hole in the center that is held at ground relative to the cathode and acts as the anode. The high-voltage potential between the cathode and the anode determines the energy of the electron beam and will directly influence the penetrating power of the electrons and the contrast of the resultant image.

Below the anode, a series of electromagnetic lenses are used to control the beam. Electromagnetic lenses consist of a circular casing containing a wire coil through which a current is passed to generate a magnetic field in the core. In the center of each lens is a pole piece that acts to greatly concentrate the magnetic field. Controlling the function of each lens is accomplished by changing the current flowing through the lens coils and thus changing the strength of its magnetic field. In reality, each lens may consist of a series of smaller lenses that act in concert.

The arrangement of lenses in an electron microscope is shown in Figure 19.1. From the top of the column to the bottom, they are the condenser lens, the objective lens, the intermediate lens, and the projector lenses. The condenser lens regulates the convergence of the beam and thus controls its brightness. The objective lens is used to form and focus the image of the specimen. The specimen is actually positioned inside this lens. Below this is the intermediate lens, which is the primary magnifying lens. Last, the projector lens effectively projects the image onto a screen. Because electrons cannot be seen with the eye, they are projected onto a fluorescent screen and the resulting photons are viewed. To record an image, the electrons are allowed to strike a piece of film or a charge-coupled device (CCD) camera.

To propagate the electron beam down the column, a high vacuum, better than 10^{-4} torr, is required to prevent air molecules from interacting with the beam. To reach this high vacuum, two different types of pumps are employed. Most electron microscopes use a mechanical rotary pump to go from atmosphere down to about 10^{-2} torr, then use an oil diffusion pump to go from 10^{-2} torr to below 10^{-4} torr.

IMAGE FORMATION WITH ELECTRONS

There are two principal phenomena involved in image formation (i.e., contrast generation) that relate to how electrons in the beam interact with the atoms in the specimen. These two phenomena are referred to as elastic and inelastic scattering. It is the combination of both elastic and inelastic electron–specimen interactions that forms the overall final image.

Elastic Scattering

Elastic scattering occurs when beam electrons encounter the nuclei of atoms in the specimen and are deflected through relatively large angles without the loss of energy. The degree of this elastic

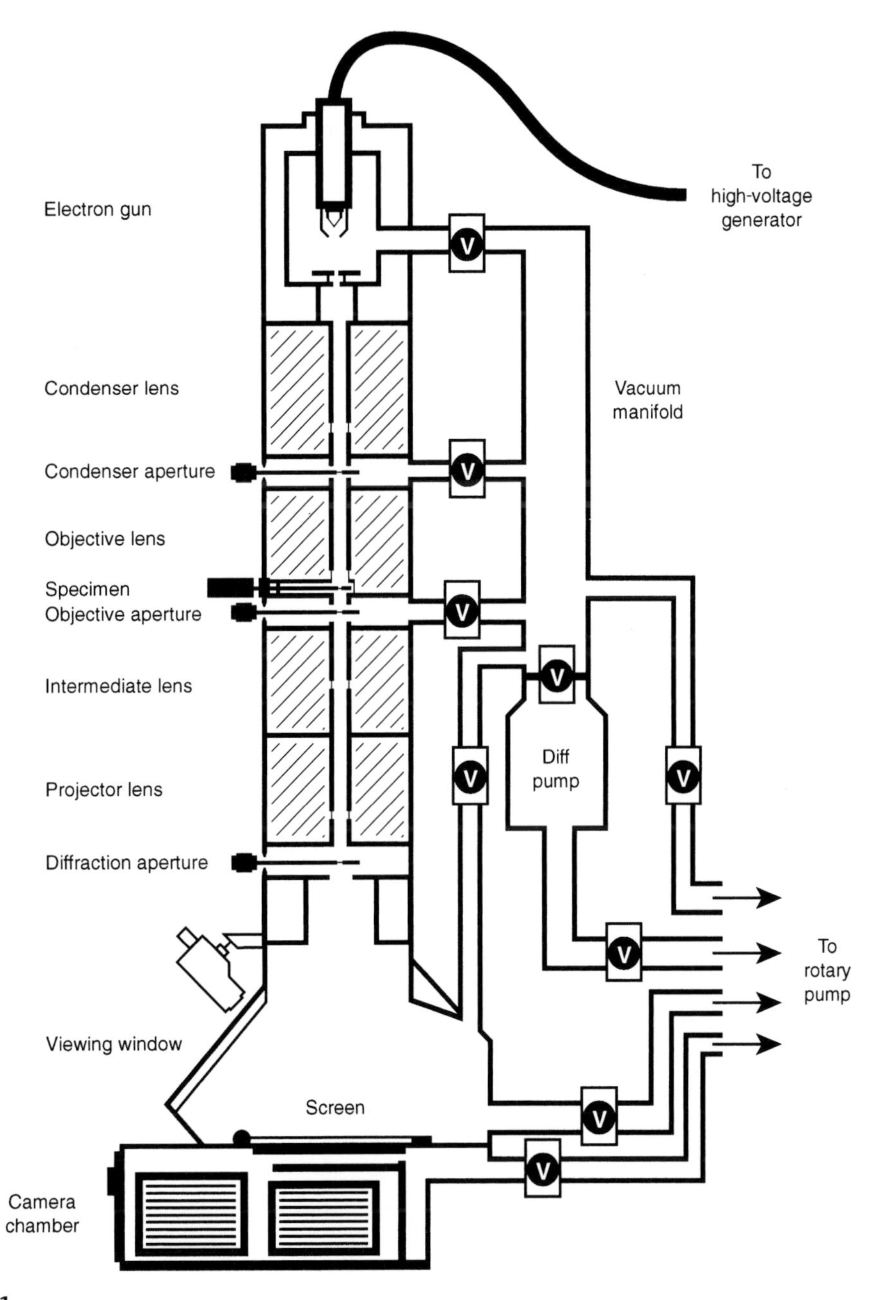

FIGURE 19.1

Components of a transmission electron microscope.

scattering is directly related to the size of the atom's nucleus or its *Z* number. The larger the nuclei, the greater the likelihood that the electrons are going to be elastically scattered at a wider angle.

To improve image contrast, biological specimens are stained with heavy metals to increase elastic scattering in certain regions of the specimen. For this purpose, heavy-metal stains that contain lead, uranium, and osmium are routinely used. In electron micrographs, dark regions are simply areas that have bound more of these heavy-metal atoms. With most stained biological specimens, elastic scattering (referred to as amplitude contrast) is the main source of contrast.

The positioning of an aperture immediately below the specimen (the objective aperture, see Fig. 19.1) to block electrons that have been elastically scattered in wide angles results in reduced numbers of electrons in the image in regions with higher *Z* number and more electrons in areas with lower *Z* number. Thus, the objective aperture can be used to increase specimen contrast.

Higher-energy electrons will be scattered through less of an angle than lower-energy electrons. From this, one can understand that the energy of the electron beam is an important determinant in the overall contrast of the image. In general, when imaging a sample of a given thickness, higher voltages provide less contrast and lower voltages more contrast.

Inelastic Scattering

Inelastic scattering occurs when a beam electron interacts with an orbital electron in the specimen. Because both beam and specimen electrons have similar mass, the beam electrons are not deflected when the two interact so much as they suffer an energy loss. This produces an effect called phase contrast.

Phase contrast arises from the wavelike properties of the electron beam. When the incident electrons lose energy through inelastic scattering, these electrons have experienced a change in wavelength. When these waves produced by unscattered and scattered electron waves recombine, if the resultant wave has a different amplitude from the unscattered wave, phase contrast occurs. This effect is mostly noticeable in thin, low-contrast specimens at higher magnifications.

INTERMEDIATE- AND HIGH-VOLTAGE ELECTRON MICROSCOPY

One of the major limitations of conventional electron microscopes is that the specimens used must be extremely thin—on the order of 50–250-nm thick. This is due to the inability of electrons in the 60–100-keV range to penetrate specimens thicker than this without being either absorbed or suffering substantial energy loss and thus severely degrading resolution by causing chromatic aberration. These very thin specimens result in a substantial loss of 3D information for most specimens. To overcome this limitation, higher accelerating voltages can be used. Intermediate-voltage electron microscopes (IVEMs) operating in the 200–500-keV range and high-voltage electron microscopes (HVEMs) operating in the 500 keV–3 MeV range can be used to image much thicker specimens. IVEMs are commonly used to image specimens in the 0.5–3-µm-thick range and HVEMs have been used to image specimens greater than 5 µm thick. These instruments can be used to create 3D images via stereopairs and to create 3D computer representations by electron tomography. Detailed introductions to the principles and practice of EM may be found in Meek (1976), Hayat (1989), and Dykstra (1992).

REFERENCES

Dykstra M.J. 1992. *Biological electron microscopy*. Plenum Press, New York.

Hayat M.A. 1989. *Principles and techniques of electron microscopy: Biological applications*, 3rd edition. CRC Press, Boca Raton, Florida.

Meek G.A. 1976. *Practical electron microscopy for biologists*, 2nd edition. John Wiley, New York.

CHAPTER 20

Preparative Methods for Transmission Electron Microscopy

Thomas J. Deerinck,[1] Maryann Martone,[1] and Mark H. Ellisman[2]

[1]*NCMIR, University of California, San Diego, La Jolla, California*
[2]*CRBS, University of California, San Diego, School of Medicine, La Jolla, California*

INTRODUCTION

Although the principles of the electron microscope are similar to those of the light microscope, specimen preparation requirements are quite different. First, because of the superior resolution of the electron microscope, specimens must be optimally fixed. Many of the fixatives commonly used for light microscopy (LM), e.g., ethanol and formalin, are not suitable for electron microscopy (EM) because they result in structural damage that is detectable at the ultrastructural level. Second, because the specimens are imaged under a high vacuum, they must be dehydrated. An exception is high-resolution cryomicroscopy, where the specimen is kept frozen during imaging. Third, because specimens must be thin enough to pass electrons (usually no more than 250 nm for conventional transmission electron microscopes [TEMs]), they must be embedded in a plastic resin in order to withstand the rigors of sectioning and bombardment by the electron beam. Fourth, because most biological structures possess little inherent contrast within the electron microscope (i.e., they allow most electrons to pass through), they must be stained with compounds that impart contrast, usually heavy metals. This chapter presents a detailed overview of specimen preparation for TEM including:

- Fixatives and fixation protocols
- Buffer vehicles
- Dehydration protocols
- Embedding tissue blocks
- Flat embedding and mounting of tissue sections
- Embedding cultured cells
- Ultramicrotomy
- Poststaining
- High voltage electron microscopy (HVEM)
- Specimen preparation for intermediate voltage electron microscopy (IVEM)/HVEM
- Three-dimensional analysis of HVEM/IVEM specimens

Fixation for Electron Microscopy

As with LM, the purpose of tissue fixation is to retain the tissue in a state as near to life as possible by rapidly terminating all enzymatic activity to minimize postmortem changes. In addition, fixation for EM seeks to immobilize cellular constituents with minimum changes in morphology and to maintain the maximal degree of morphological preservation during subsequent dehydration, embedding, and irradiation by the electron beam. Choice of fixative depends on the tissue used and the objectives of the study. Fixation techniques presented in this section are designed for maximum structural preservation. Some types of tissue may be more difficult to preserve, e.g., lung and marine invertebrates, and so may require special procedures (see Bullock 1983; Hayat 1989a). In some cases, such as EM immunocytochemistry and histochemistry, suboptimal fixation must often be employed in order to retain enzyme activity or antigenicity. Fixation for EM immunocytochemistry is presented in Chapter 21. A thorough discussion of the types of fixatives available and their possible reaction mechanisms are provided in Glauert (1975), Bullock (1983), and Hayat (1989a). Specimen preparation protocols for cells and tissues are presented in Table 20.1.

TABLE 20.1 TEM specimen preparation schedule for tissues and cells

Step	Tissues	Cells
1. Fixation in 2% glutaraldehyde/ 2% formaldehyde in 0.15 M sodium cacodylate buffer	a. perfusion or b. immersion for 1–4 hrs at 4°C	15 min
2. Postfixation in same fixative	for perfusion-/fixed tissues only: 1 hr at 4°C. Cut tissue into 1-mm cubes	not applicable
3. Wash in 0.15 M sodium cacodylate buffer at 4°C	3×, 10 min	3×, 5 min
4. Fix in 1% OsO_4 in cacodylate buffer at 4°C	1 hr	30 min
5. Rinse in dH_2O at 4°C	3×, 5 min	3×, 2 min
6. Stain with 1% aqueous UA at 4°C	30 min–overnight	30 min
7. Dehydrate tissue	20% acetone 10 min at 4°C 50% acetone 10 min at 4°C 70% acetone 10 min at 4°C 90% acetone 10 min at 4°C 100% acetone 15 min at RT 100% acetone 15 min at RT	20% ethanol 2 min at 4°C 30% ethanol 2 min at 4°C 50% ethanol 2 min at 4°C 70% ethanol 2 min at 4°C 90% ethanol 2 min at 4°C 100% ethanol 2 min at RT 100% ethanol 2 min at 4°C
8. Infiltrate with Epon/Araldite at RT with agitation	a. 1:1 acetone:resin overnight b. 100% complete resin: 2 changes over 3 hr	1:1 ethanol:resin 60 min 100% resin 60 min 100% resin 60 min
9. Embed tissue and polymerize	24–48 hr at 60°C	24–48 hr at 60°C
10. Section		
11. Poststain with UA and lead if necessary		
12. View and record micrographs		

UA, uranyl acetate; RT, reverse transcriptase.

The most widely used fixatives for TEM are chemical cross-linkers such as aldehydes, e.g., formaldehyde, glutaraldehyde, and acrolein. These compounds react with primary amine groups and stabilize proteins by cross-linking them into a gel. Coagulant fixatives such as ethanol and acetone are not suitable for EM because they cause considerable changes in protein structure. For biological specimens, glutaraldehyde alone or in combination with formaldehyde is the most efficient and reliable fixative.

Not all chemical fixatives react equally well with all cellular components, and thus a combination of fixatives is typically employed. In addition to proteins, aldehydes alone or in combination with other compounds also appear to retain some carbohydrates, especially glycogen. Nucleic acids are most effectively stabilized with formaldehyde. Although aldehydes may interact with phospholipids containing free amino groups, most lipids are not retained by aldehyde fixation. Therefore, secondary fixation with osmium tetroxide is usually employed. Osmium tetroxide is a heavy-metal compound that reacts with unsaturated fatty acids and is the best fixative for preservation of membrane structure. Osmium is now routinely used as a secondary fixative following primary fixation with aldehydes. However, osmium is capable of destroying enzymatic activity and antigenicity more so than aldehydes. Therefore, in tissue that will be used for postembedding immunolabeling or histochemistry, it may be necessary to omit osmium fixation prior to embedding (see Chapter 21).

Although chemical fixation is extremely useful and is the most widely employed type of fixation, it has its limitations. The rate of fixation is not fast enough for investigation of many dynamic subcellular processes or to prevent some anoxic changes or redistribution of ions from occurring within tissue. In addition, since the interaction of the fixative with the tissue can produce physiological changes, e.g., release of neurotransmitters and muscle contraction, chemical fixation will "fix the response of the cell to the fixative and not the living state" (Hayat 1989a). Other types of fixation have been employed to speed up the rate of immobilization of cellular constituents either alone or in combination with chemical fixation. These include microwave fixation and cryofixation followed by freeze substitution. Detailed reviews of these procedures may be found in Hayat (1989a), Reid and Beesley (1991), and Kok and Boon (1992).

PREPARATION AND HANDLING OF CHEMICAL FIXATIVES

Glutaraldehyde

Glutaraldehyde is a dialdehyde and is the fixative of choice for ultrastructural studies. Its main disadvantage is its slow rate of penetration into tissues: 1.5 mm/24 hours for soft tissues at room temperature (Hayat 1989a). Glutaraldehyde also has a more deleterious effect on antigenicity in immunocytochemical studies than formaldehyde. Monomeric distilled glutaraldehyde should be purchased in sealed glass vials at a concentration of no more than 50%. Purified glutaraldehyde is stable at 4°C for several months in sealed vials. Over time, solutions of glutaraldehyde will spontaneously polymerize and increase in impurities, as evidenced by a drop in pH to below 3 and the development of a yellow color. Such solutions should be discarded. Exposure to room temperature or to oxygen will increase the rate of deterioration. The use of freshly opened glutaraldehyde for each experiment is recommended to minimize variability between experiments.

Formaldehyde

Formaldehyde is a monoaldehyde and is a less effective cross-linker than glutaraldehyde. Because it is a milder fixative, it is not used alone for structural studies but is extensively used in histochemical and immunocytochemical studies. Many of the cross-links induced by formaldehyde fixation are reversible and, therefore, tissue fixed in formaldehyde alone will often unfix when exposed to isotonic buffers (Tokuyasu 1984). When in solution, formaldehyde forms polymers of varying lengths that may affect the extent of cross-linking and rate of penetration. To minimize variability from experiment to experiment, formaldehyde should be prepared freshly from paraformaldehyde powder. Commercial solutions of formaldehyde or formalin should not be used because they contain considerable amounts (up to 15%) of methanol to inhibit polymerization.

Paraformaldehyde/Glutaraldehyde Fixative

A good standard fixative, introduced by M.J. Karnovsky, is a mixture of glutaraldehyde (1–2%) and formaldehyde (2–4%), made fresh from paraformaldehyde, in sodium cacodylate or sodium phosphate buffer. This combination exploits the advantages and minimizes the disadvantages of each fixative. Formaldehyde fixes less well but penetrates into tissue quickly to stabilize cellular constituents, whereas glutaraldehyde fixes more thoroughly but penetrates more slowly.

Preparation of 2% Paraformaldehyde/2% Glutaraldehyde in Sodium Cacodylate Buffer

1 Prepare 2x concentration of buffer (0.3 M **sodium cacodylate buffer**, pH to 7.2 with **HCl**).

2 Prepare 25 ml of 8% **paraformaldehyde** just before use:

   a Heat 20 ml of ddH_2O to 60°C.

   b Weigh out 2 g of paraformaldehyde powder in a fume hood and add to heated H_2O while stirring (solution will be cloudy).

   c Add 2–3 drops of 1 N **NaOH** and stir until solution is no longer cloudy.

   d After paraformaldehyde is dissolved, cool, and filter through Whatman #1 filter paper.

   e Bring volume to 25 ml with dd H_2O.

3 Add in the following sequence:

2x buffer	50 ml
8% paraformaldehyde	25 ml
25% **glutaraldehyde**	8 ml
dH_2O	17 ml

CAUTION: Cacodylate; paraformaldehyde; NaOH; acids and bases that are concentrated; glutaraldehyde (see Appendix 2 for Caution)

Paraformaldehyde may also be purchased as small granules (Polysciences), which are easier to handle than the powder.

Acrolein

Acrolein is an aldehyde that is highly reactive, penetrates the tissue readily (1 mm/hr in rat liver) and fixes well even at low temperatures (–80°C) (Hayat 1989a). Because of its rapid penetration, it is useful for fixing large pieces of tissue, plant cells, and microorganisms with cell walls (Glauert 1975). It is usually used in combination with either glutaraldehyde or formaldehyde. Fixation in 3.75% acrolein + 2% formaldehyde has been used for immunocytochemical investigations in the nervous system (King et al. 1983).

CAUTION: Acrolein (see Appendix 2 for Caution)

Solutions of acrolein may be neutralized in a 10% bisulfite solution (Hayat 1989a).

Osmium Tetroxide

OsO_4 is a heavy-metal compound that serves both as a fixative and a stain. It cross-links some proteins, but its main use is for stabilization of lipids. OsO_4 oxidizes double bonds in unsaturated fatty acids and is itself reduced. It is also reduced by ethanol during tissue dehydration. The reduction of OsO_4 results in the formation of osmium blacks ($OsO_4 \cdot nH_2O$) at the site of reduction. These impart a black or dark brown color to the tissue and confer electron density to membranes, thus contributing to the formation of specimen contrast in electron micrographs. Osmium also acts as a mordant for subsequent heavy-metal staining.

OsO_4 is most commonly used as a secondary fixative after primary fixation with aldehydes. The combination of glutaraldehyde primary fixation and OsO_4 secondary fixation is unsurpassed for ultrastructural preservation. OsO_4 is typically used in concentrations of 1–2% diluted in either sodium phosphate or sodium cacodylate buffer at pH 7.2–7.4. The rate of penetration of OsO_4 into tissue depends on the density of the tissue, but is usually around 0.5 mm/hour for most specimens (Hayat 1989a). Prolonged fixation in OsO_4 is usually not desirable because it results in extraction of tissue components, particularly proteins. Small tissue blocks (<1 mm thick) should be used to keep the fixation time to a minimum (usually 0.5–2 hr for tissue blocks; 20 min for single cells and particulate specimens).

OsO_4 is expensive and can be purchased as a 4% or 8% solution in sealed glass ampoules, although it may also be purchased as a crystal. OsO_4 crystals have a low solubility in H_2O, so stock solutions made from crystals are usually prepared in advance and stored in the refrigerator.

CAUTION: OsO_4 (see Appendix 2 for Caution)

▪ *Notes*

- Store osmium stocks in a double jar in a refrigerator designated for toxic chemicals. Any osmium vapors that escape the container will blacken the inside of the refrigerator and may contaminate other contents. Osmium can penetrate plastics, so a glass jar must be used. The use of glass-stoppered bottles is not recommended as osmium vapors will escape.
- Stock solutions should be a pale yellow color. Osmium solutions are easily contaminated by the presence of organic matter and exposure to light. Any contamination will cause reduction of osmium and turn solutions gray-purple and finally black. Such solutions should be discarded. Avoid all contact with metals.

Additives to Fixatives

The addition of various substances to either the primary or secondary fixative solution has proven useful for stabilization of certain cellular components, particularly membranes, and also for imparting additional contrast to membranes. See Glauert (1975), Bullock (1983), and Hayat (1989a) for additional information.

Nonelectrolytes

The addition of nonelectrolytes such as sucrose, dextran, or **PVP** to the primary fixative may help to minimize extraction of certain cellular components. See Glauert (1975), Bullock (1983), and Hayat (1989a), for discussion.

CAUTION: PVP (see Appendix 2 for Caution)

Electrolytes

The addition of various ions, particularly **$CaCl_2$** and **$MgCl_2$**, has been reported to reduce the extraction of certain constituents and to help stabilize membranes and cytoskeletal elements. Addition of $CaCl_2$ has been reported to reduce or eliminate myelin-like figures, believed to be due to a reorganization of phospholipids within the tissue during fixation (Glauert 1975; Bullock 1983). The final concentration should be 1–3 mM. The use of divalent cations with phosphate buffers is not recommended.

CAUTION: $CaCl_2$; $MgCl_2$ (see Appendix 2 for Caution)

Tannic Acid

Tannic acid (1–2%) may be used in the primary glutaraldehyde fixative (or glutaraldehyde-formaldehyde mixtures) both as a supplemental fixative and to improve membrane contrast. Tannic acid acts as a mordant for subsequent staining with OsO_4, lead salts, and UA, and may help to reduce extraction of cell components during dehydration and embedding. Low-molecular-weight EM-grade tannic acid should be used to facilitate penetration. Solutions of tannic acid and glutaraldehyde should be made up immediately prior to use to avoid the formation of precipitates. Check the pH of the fixative solution after addition of tannic acid.

CAUTION: Tannic acid (see Appendix 2 for Caution)

Potassium Ferrocyanide

Addition of **potassium ferrocyanide** ($Fe(CN)_6^{4-}$) or **ferricyanide** ($Fe(CN)_6^{3-}$) to the secondary osmium fixation solution results in enhanced staining of membranes and glycogen. Typically, 0.8% either ferrocyanide or ferricyanide is added to 1% OsO4 in cacodylate buffer, and tissues are incubated from 1 to 2 hours. The ferrocyanide-OsO_4 solution will turn dark brown, whereas the ferricyanide-OsO_4 solution is a pale yellow. $CaCl_2$ (5 mM) is typically added to the aldehyde fixative and washes to improve membrane contrast. In some cases, this procedure results in electron-dense deposits within the lumen of the endoplasmic reticulum (Forbes et al. 1977).

CAUTION: Potassium ferrocyanide; potassium ferricyanide (see Appendix 2 for Caution)

Buffer Vehicles for Electron Microscopy

The two most common buffers in use for standard TEM specimen preparation are sodium cacodylate and sodium phosphate buffer. **Cacodylate** has superior buffering properties and results in fewer precipitates than does phosphate buffer, but it is more toxic and expensive than phosphate. The osmolarity of the buffer should be isotonic or slightly hypertonic to physiological fluids to prevent swelling. Cells are still osmotically active after fixation with aldehydes, so isotonicity must be maintained through fixation with OsO_4. It has been debated whether glutaraldehyde contributes to the final osmolarity of the fixative solution (for discussion, see Bullock 1983; Hayat 1989a). It is generally thought that the osmolarity of the vehicle (buffer plus any additives) represents the osmolarity of the fixative solution, provided that low concentrations of glutaraldehyde (≤2%) are used. See Hayat (1989a) for recommended osmolarities for various tissues.

Sodium cacodylate buffer at a concentration of 0.15 M (~305 mOsM) and a pH of 7.2 is usually used to buffer the primary fixative. Sorenson's phosphate buffer is used at a concentration of 0.15 M (300 mOsM) at a pH of 7.2–7.4. Buffers may be prepared as double-strength stocks and stored in the cold.

CAUTION: Cacodylate (see Appendix 2 for Caution)

MODES OF PRIMARY FIXATION FOR TISSUES

The three main routes of chemical fixation for intact tissue are (1) vascular perfusion, (2) dripping or injection into an organ, and (3) immersion. Since the fine structure of many tissues is dependent on a continuous blood supply, fixation of such tissues should be carried out in vivo if possible. In general, vascular perfusion is the preferred route because it allows rapid and uniform penetration of the fixative into all parts of the tissue prior to any mechanical manipulation. It is indispensable for good fixation of highly vascular tissue such as central nervous system, kidney, and liver.

Whole-Body Vascular Perfusion of Small Animals

Vascular perfusion delivers fixative to tissue through the animal's own vasculature. Generally, the animal is first perfused with oxygenated Ringer's solution (see below) to remove blood from the vasculature, followed by perfusion with the fixative solution. The fixative vehicle should be isotonic with the animal's fluids and should be maintained at body temperature. For ultrastructural studies, the use of oxygenated Ringer's solution is recommended to minimize anoxic changes in the tissue prior to introduction of the fixative. Although perfusion is routinely performed using gravity-based systems, mechanical pumps (e.g., Masterflex, Cole-Parmer, Inc.) are easier to set up and control. A system should be set up whereby two input lines are available, one for the flush and one for the fix. It should be possible to switch back and forth between the two lines with a stopcock to avoid the introduction of any bubbles into the tubing when changing solutions. To prepare for perfusion, first fill the input line with fixative, then switch the stopcock and fill the entire line (input and output) with Ringer's solution. Allow the Ringer's solution to flow for a few seconds to remove residual fixative from the line. Make sure that all air bubbles are removed from the line before placing a needle (25 gauge is suitable for most small rodents) or some other cannula on the end of the line.

Detailed protocols for perfusion fixation of individual tissues are given in Hayat (1989a).

1. After the animal is deeply anesthetized, lay it on its back on a surgical board.
2. Make a midline incision down the whole length of the thorax and much of the abdomen, and reflect the skin.
3. Expose the heart by cutting the ribs and intercostal muscles along each side and raising the flaps so formed.
   At this point, the diaphragm has been cut and the animal can no longer breathe. Work quickly to avoid anoxia.
4. Cannulate the left ventricle of the heart and clip the right atrium.
5. Start the flow of Ringer's solution (see below) at a predetermined flow rate until the atrial outflow is clear (usually 1–2 minutes) and the liver is blanched.
6. Switch to the fixative. Body tremor and rigor should begin soon (<1 min) after the fixative enters the animal.
7. Run perfusate through for 10 minutes or about 100–200 ml, whichever is first.
8. Remove the target tissue and place into a dish of cold fixative for further dissection.
9. Cut the tissue into slabs 1 mm thick and transfer the dissected tissues to a vial of cold fixative for one additional hour.

Preparation of Rat Ringer's Solution

1 Prepare the following stock solutions.

	m.w.	Stock solutions	Final ringers
a. NaCl	58.44	79.8 g/l	99 ml/l (140 mM)
b. **KCl**	74.55	37.5 g/l	10 ml/l (5 mM)
c. **$MgCl_2 \cdot 6H_2O$**	203.30	20.0 g/	10 ml/l (1 mM)
d **Na_2HPO_4**	141.96	18.0 g/l	10 ml/l (1.3 mM)
e. $NaHCO_3$	84.01	50.0 g/l	25 ml/l (10 mM)
f. **$CaCl_2 \cdot 2H_2O$**	146.90	30.0 g/l	10 ml/l (2 mM)

CAUTION: KCl; $MgCl_2$; Na_2HPO_4; $CaCl_2$ (see Appendix 2 for Caution)

2 Add 700 ml of ddH_2O to a 1-liter graduated cylinder.

3 Add the amounts given in column 4 for stock solutions a–e. Do not add the $CaCl_2 \cdot 2H_2O$ as it will precipitate.

4 Bubble the solution for 10 minutes with a mixture of 95% air and 5% CO_2. This will adjust the pH and aerate the solution.

5 While the solution is bubbling, add 2 g of glucose.

6 After 10 minutes, add the $CaCl_2 \cdot 2H_2O$ and make up to 1 liter with ddH_2O. Warm to 37°C in a water bath before use.

■ *Notes*

- Stock solutions should be stored at 4°C, with the exception of the Na_2HPO_4, which should be stored at room temperature.
- Xylocaine (0.2%) may be added to relax blood vessels and heparin (250 units/ml) to prevent blood clotting. These may be added along with the glucose during bubbling.

Dripping or Injection of Fixative In Vivo

The following procedure is recommended by Hayat (1989a).

1 Anesthetize the animal and expose the tissue, with care taken not to damage the target organ or its blood supply.

2 Immediately flood the tissue with a generous supply of chilled fixative.
Once exposed, the tissue must not be allowed to dry out.

3 Add fresh fixative every few seconds for 2–3 minutes.

4 Place a thin cotton pad over the tissue for an additional 20 minutes while fixation and hardening continue.

5 Remove the organ and place in fresh fixative.

6 Wash away fresh blood and other fluids with excess fixative to avoid dilution of the fixative. Because only the surface of the organ is fixed, only the outer layers of tissue will be suitable for

EM examination. The internal structure of organs can be preserved by injecting fixative directly into the organ with a micropipette or hypodermic needle. Injection should be slow and steady and should proceed for 2–20 minutes.

Immersion Fixation

Immersion fixation is not the preferred method for ultrastructural examination due to the uneven and slow penetration of fixative into the tissue block. In addition, the unfixed tissue may sustain mechanical damage and undergo anoxic and postmortem changes while being dissected from the animal.

1 To minimize these effects, dissect small samples of tissue with as little trauma as possible.

2 Immediately place the samples into cold fixative (4°C).

3 Leave the tissues in fixative for 1–4 hours, depending on the tissue type.

Only the outer layers of tissue will show optimum fixation, since deeper layers will be reached by fixative only after morphological changes have begun to occur.

FIXATION OF CULTURED CELLS

In general, fixation of cell monolayers is identical to that used for tissue blocks, except that the times are reduced by ~50%. See Table 20.1 for a typical processing schedule. Prior to fixation, it may be necessary to wash the cells in buffer very briefly to remove culture medium that contains components that may be cross-linked by chemical fixation. The fixative should be the same temperature as the medium. Fixation of cell suspensions may be accomplished by adding an equal volume of double-strength fixative to the suspension, or they may be fixed after pelleting (see embedding protocols below).

EN BLOC STAINING WITH UA

UA is a heavy metal that is used both as a stain to impart additional contrast to the tissue and as a fixative for stabilization of phospholipids prior to dehydration. En bloc staining with UA is usually performed after osmium treatment and before dehydration but may also be performed during dehydration. Both phosphate and cacodylate buffers react with UA to form precipitates. After OsO_4, the tissue is less affected by osmotic pressure, so it should be rinsed several times in ddH_2O (three changes in 15 mins) prior to staining with UA to avoid precipitates. If a buffered UA solution is desired, veronal acetate or maleate can be used as a buffer. Following the H_2O rinses, the specimen is immersed in 1% UA for 0.5–2 hours. En bloc staining with UA is not recommended when glycogen is to be visualized, because it interferes with staining of this compound. Tissue blocks may be left in UA overnight if necessary.

To Prepare a 1% Aqueous Solution of UA

1 Heat 50 ml of ddH_2O to 60°C.

2 Under a fume hood, add 0.5 g of **UA** to the warmed H_2O and mix until dissolved.

CAUTION: Uranyl acetate; radioactive substances (see Appendix 2 for Caution)

The UA should dissolve quickly.

3 Allow the solution to cool.

Notes

- UA solutions are fairly stable when protected from light but, over time, may form precipitates that will be deposited on the tissue during staining. For best results, prepare fresh solutions of UA and filter through a 0.22-μm Millipore filter just prior to use. When removing stain from stored solutions, do not agitate the contents of the bottle, and remove stain from just below the surface to avoid particulate material near the bottom.
- UA is radioactive and toxic and should be handled with care. Powders should be weighed out and solutions handled in a fume hood. Individual batches of UA differ in their amount of radioactivity. As a precaution, bottles of UA should be sealed in metal containers and should not be stored near work areas.

Embedding

Before sectioning, tissue is usually embedded in plastic to provide support during sectioning and stability under the electron beam. With few exceptions, most of the useful resins for EM embedding are not miscible with H_2O. Therefore, before infiltration with plastic resin, tissues must be dehydrated in an organic solvent.

DEHYDRATION

Tissues are typically dehydrated through a graded series of ethanol, methanol, or acetone prior to infiltration with resin. If ethanol or methanol is used, either 100% acetone or propylene oxide should be used as a transition solvent before embedment, because many resins are not as readily miscible in alcohol as in these other solvents. Acetone and ethanol are routinely used for dehydration of most samples: Propylene oxide and methanol are toxic and are usually not required. Acetone is reported to cause less shrinkage than ethanol and thus is the solvent of choice for most applications. Two exceptions are (1) when using the acrylic resin LR White, which will not polymerize in the presence of acetone, or (2) when using plastic culture plates or coverslips that would dissolve in acetone. In these cases, ethanol may be used satisfactorily.

Dehydration results in extraction of lipids, primarily at higher solvent concentrations, and also of proteins, primarily at lower solvent concentrations. Fixation in OsO_4 and UA help to minimize this extraction, but even so dehydration times should be kept as short as possible and be performed at 4°C. If dehydration in the cold is employed, the specimen is warmed to room temperature when in 90% or 95% solvent, and the final dehydration steps in 100% acetone are carried out at room temperature to avoid condensation of H_2O in the cold solvent. Dehydration is easily carried out in glass scintillation screw-top vials. Vials should be filled at least two thirds full and care should be taken when removing solvent not to let the tissue blocks dry out. A standard dehydration schedule is given in Table 20.1. Acetone should be reagent grade and kept dry. Shorter times can be used if necessary, but in this case, smaller blocks of tissue should be used and dehydration should take place at room temperature under agitation.

RESINS

The two major classes of resins in use today for specimen embedding are the epoxides (Epon 812, Araldite, Spurr's) and the acrylics (LR White, LR Gold, and the Lowicryls). Other resins are available, and these are discussed in Hayat (1989a), Harris (1991), and Dykstra (1992). The choice of resin depends on the type of tissue and the intended procedure.

Epoxides

Epoxides show less shrinkage and tend to be more stable under the electron beam than the acrylic resins and are thus the resins of choice for structural studies. These resins are generally not miscible with H_2O; therefore the tissue must be thoroughly dehydrated before infiltration. Incomplete dehydration usually results in a polymerized block with poor sectioning qualities. These resins are fairly impermeable to aqueous solutions and thus are more difficult to use for postembedding immunolabeling or histochemical studies. The most popular of the epoxide resins are Epon 812,

Araldite, and Spurr's. Epon 812 was one of the first epoxy resins to be used and is no longer manufactured, but the name Epon 812 is still used to describe several currently available resins that are similar (although not identical) to the original, e.g., Poly/Bed 812 (Polysciences), EMbed812 (Electron Microscopy Sciences), Eponate 12, and Medcast (Pelco, Redding, California). Epon 812, Araldite, and Spurr's each differ in their viscosity (lowest, Spurr's; highest, Araldite), which directly affects the ease of infiltration and amount of shrinkage during polymerization. It has become popular following Mollenhauer (1964) to use mixtures of these resins, e.g., Epon-Araldite, to gain the advantages of both.

Acrylic Resins

The London resins (LR White and LR Gold) and the low-temperature Lowicryl resins are the most common acrylic resins in use today. These resins are less stable under the electron beam than epoxide resins, but they are less viscous and more hydrophilic. These are the resins of choice for postembedding immunolabeling, as they are readily permeable to the immunolabeling solutions. They also permit only partial dehydration of the tissue prior to infiltration, thereby reducing the denaturation and extraction of antigenic sites that may occur during dehydration in higher solvent concentrations.

LR Gold and Lowicryl are both low-temperature acrylic resins and can be infiltrated into specimens at temperatures as low as –80°C. LR Gold was designed to be used on unfixed specimens. Detailed protocols for its use can be found in Hayat (1989a). Embedding at low temperatures appears to preserve the three-dimensional structure of proteins better than does embedding at temperatures above 4°C, and also appears to reduce the amount of damage caused by dehydration and embedding. Thus, less extensive cross-linking of the tissue is required during primary fixation, which may be beneficial in certain immunolabeling and histochemical protocols. The Lowicryls (K4M, HM20, K11M, and HM23) do not become very viscous at low temperatures (<35°C) and can be used to embed specimens at very low temperatures. Each resin possesses different degrees of polarity and hydrophilia (K4M > K11M > HM20 > HM23) and a different freezing point. Detailed discussions of the Lowicryls can be found in Hayat (1989a) and Newman and Hobot (1993). Embedding at low temperatures is facilitated by the use of a special freezing apparatus (e.g., Balzers Union LTE 020 Low Temperature Embedding Apparatus or the Leica Reichert CS Auto) that permits incremental lowering/raising of temperature and UV light polymerization.

Preparation and Handling of Epoxide Resins

Various additives are included in epoxy formulations. These include anhydride hardeners such as dodecenyl succinic anhydride (DDSA) or nadic methyl anhydride (NMA) plasticizers or flexibilizers such as dibutyl phthalate (DBP) to improve the sectioning characteristics and prevent the block from becoming too brittle (primarily for Araldite), and amine accelerators such as DMP-30 (2,4,6-tridimethylamino methyl phenol) or BDMA (benzyl dimethylamine) to speed up polymerization. Selection of an appropriate block hardness is dependent on tissue type and is usually the result of experimentation with several different hardness grades. The hardness of the block is generally controlled by the ratio of NMA to DDSA: DDSA alone produces the softest block, and addition of NMA increases the hardness. Individual components of the resin mixture have a relatively long shelf life if stored in tightly sealed bottles in a cool place away from light. Stock mixtures of resins should not be mixed with anhydrides in advance, because these mixtures slowly polymerize at room temperature. Epoxy stocks may be prepared in advance without anhydrides or accelerators and stored in the freezer to retard polymerization. Frozen stocks may be stored in large polypropylene syringes or tubes. If frozen, stocks must be allowed to warm up to room temperature before use to prevent condensation of H_2O.

Because of their high viscosity, it is preferable to measure components by weight rather than by volume. After measuring out any component, carefully wipe off the mouth of the bottle before capping to ensure a tight seal and to prevent the cap from being glued to the bottle. Stocks are easily prepared in disposable polyethylene tripour beakers. Glassware may be cleaned with acetone, but it is messy.

Because of their relatively high viscosity, epoxy resins require lengthy infiltration times. Using a graded series of resin:solvent mixtures generally helps the embedding process. The number of such steps required depends on the type of resin used and the tissue sample. Most tissues can be successfully infiltrated using the schedule in Table 20.1. Specimen vials should be uncovered or loosely capped during 100% resin steps to permit evaporation of any traces of solvent. Infiltration is improved if performed with gentle agitation. For more difficult-to-embed specimens, a more gradual infiltration can be performed with acetone:resin steps of 2:1, 1:1, 1:3, for several hours each, followed by 100% resin for up to 16 hours (overnight). Carrying out infiltration under vacuum may also help. Tissue blocks are then placed in a suitable mold (see pp. 319–321) and polymerized in a 60°C oven for 24 hours. Resin may also be polymerized at 50°C for 48 hours, which may be useful for postembedding immunolabeling in cases where the antigen is heat sensitive. Polymerization in a vacuum oven is sometimes recommended in order to remove any bubbles that may form around the tissue during polymerization. However, vacuum polymerization is not recommended for Spurr's because of the volatility of some of the components.

Recipes for Epon/Araldite

A good general-purpose resin for most tissues is a combination of Epon 812 and Araldite 502. A stock solution containing Epon, Araldite, and DBP can be made up in advance. Stock solutions may be stored for several days at room temperature provided they are tightly sealed. During the final steps of dehydration (i.e., the first 100% solvent step), the stock is mixed with the hardeners and accelerators. Recipes for a series of hardness follow.

Preparation of stock solution of Epon/Araldite

Mix thoroughly in a polypropylene tripour beaker

Epon 812 equivalent	30.76 g
Araldite 502	22.19 g
DBP	1.0 g

Recipes for complete resin

	Increasing hardness →				
Component	A	B	C	D	E
Stock (Epon/Araldite)	9.44 g	9.44 g	9.44 g	9.44 g	9.44 g
DDSA	20.00 g	15.00 g	10.00 g	5.00 g	0.00 g
NMA	0.00 g	3.33 g	6.67 g	10.00 g	13.33 g
DMP-30	0.2 ml	0.2 ml	0.2 ml	0.2 ml	0.2 ml

▪ *Notes*

- Resin formulation A is best for soft animal tissues.

- DMP-30 has been largely replaced by BDMA because it is less toxic and has a longer shelf life. However, according to Hayat (1989a), DMP-30 produces better cutting quality than does BDMA. If BDMA is used, add 0.4 ml in place of the DMP-30.
- When preparing stocks and final solutions, great care must be taken to mix the components thoroughly. Incomplete mixing is probably the major cause of poor embedding. Mixing can be done with a Teflon-coated stirring rod or a glass rod. Stir at least 100 times in each direction. If using a glass stirrer, make sure that it will not leave glass chips (pasteur pipettes are not recommended) behind because these will interfere with sectioning and can severely damage glass and diamond knives.
- The Araldite resin Durcopan ACM (Fluka, available from Polysciences) may be used in place of Epon/Araldite. It comes in kit form.

Embedding in Spurr's Resin

Because of its low viscosity, Spurr's resin is useful for difficult-to-embed specimens, e.g., plant and insect tissues. Spurr's resin should be used only if necessary because of its high toxicity. A kit containing the required components may be purchased from Pelco.

Recipe for complete Spurr's resin

Component	Amount
Vinylcyclohexene dioxide (**VCD**)	10 g
Diglycidyl ether of polypropylene glycol (DER 736)	6 g
Nonenyl succinic anhydride (NSA)	26 g
Dimethylaminoethanol (DMAE)	0.4 g

CAUTION: VCD (see Appendix 2 for Caution)

■ *Notes*

- The infiltration times are typically shorter for Spurr's than for Epon/Araldite, e.g., 1:1 resin:solvent, 30 minutes, 100% resin, two times for 30 minutes, followed by overnight polymerization at 60°C.
- Rapid polymerization can be performed in 3 hours at 70°C if the amount of catalyst (DMAE) is increased to 1 g.
- The hardness of the block may be adjusted by varying the amount of DER 736. Increase the amount of DER 736 to 7 g to produce softer blocks, and decrease it to 4 g to produce harder blocks. Dykstra (1993) recommends using 6.3 g of DER routinely to avoid rapid dulling of glass knives.
- Hardeners and accelerators should never be mixed together by themselves because of the danger of explosion. Epoxy components, particularly DMP-30, are toxic and should not be allowed to contact skin. Gloves should be worn, but do not necessarily afford complete protection against skin contact because many reagents will penetrate gloves over time. Particular care should be taken when handling Spurr's resin, since the low viscosity of its components allows them to penetrate the skin more rapidly than Epon/Araldite. In the event of contact with skin, wash well with soap and water. Do not use solvents to clean epoxy from skin. All epoxy solutions should be prepared in the fume hood, as some components are volatile and possibly carcinogenic. Waste solu-

tions should be polymerized after evaporation of any solvent and then disposed of in normal trash. After hardening, epoxy resins are generally inert, although the inhalation or ingestion of small chips and dust produced during sectioning should be avoided. Unpolymerized waste should be collected and handled by qualified waste disposal personnel. Do not pour down the sink!

Preparation and Handling of LR White

LR White is used at room temperature and comes in three grades: soft, medium, and hard. The hard grade is usually recommended for biological specimens. Upon receipt, LR White can be aliquotted into small containers (e.g., glass scintillation vials) and stored tightly capped in the refrigerator. LR White may be polymerized by heat, UV light, or chemical reaction. The presence of oxygen will interfere with polymerization, so tissue is embedded in tightly sealed gelatin capsules to protect the resin from air. LR White resin comes complete with catalyst and requires no additives for heat polymerization. An additional catalyst (benzoyl methyl ether) can be added at a concentration of 0.5% for UV light polymerization. Polymerization is generally complete within 24 hours. Rapid polymerization of LR White can be achieved within 20–30 minutes using a chemical accelerator provided with the LR White kit. This method is often recommended for postembedding immunocytochemical studies. Resin is prepared by adding 15 μl of accelerator/10 ml of resin and stirring vigorously prior to use. The reaction is exothermic and is typically performed on ice. If the chemical accelerator is used, tissues should not be osmicated prior to dehydration, as the dark color causes focal heat accumulation, and tissue temperature can rise to over 60°C. Acetone will inhibit the polymerization of LR White, so ethanolic dehydration must be employed. Since the presence of H_2O does not interfere with polymerization, tissue blocks may be passed directly from 70% ethanol to resin. This feature is advantageous for immunocytochemical investigations and cases where higher concentrations of alcohol would cause excessive extraction of lipids. LR White should not be polymerized in polyethylene BEEM capsules as the plastic may be dissolved by the LR White.

Although LR White is thought to be less toxic than epoxide resins, contact with skin should still be avoided. Waste can be polymerized and disposed of in the regular trash.

Typical schedule for LR White embedding

30% ethanol	10 minutes
50% ethanol	10 minutes
70% ethanol	10 minutes
95% ethanol	10 minutes
100% LR White	overnight
100% LR White	three times for 1 hour

Embed in gelatin capsules, polymerize overnight at 60°C.

If tissues are going to be infiltrated from 70% ethanol, they should be removed from 70% ethanol and put into a 1:1 mixture of LR White:70% ethanol for 1 hour followed by infiltration with 100% LR White as above.

EMBEDDING TISSUE BLOCKS

There are numerous molds available for embedding tissue blocks. Some of the more popular molds are illustrated in Figure 20.1.

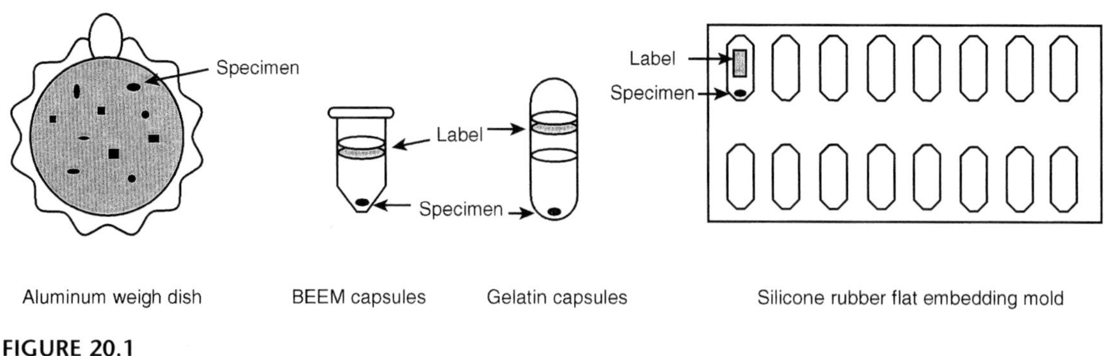

FIGURE 20.1

Types of embedding molds.

Embedding in Aluminum Foil Dishes and Mounting of Specimen Blocks

1. Label aluminum dishes by pressing into the bottom with a pen or sharp stick. The label will be embossed in the plastic upon removal from the dish.
2. Place the tissue blocks well separated into the bottom of the dish.
3. Cover the blocks with fresh 100% complete resin.
4. Place dishes into a 60°C oven for 24–48 hours.
5. Remove the foil from around the polymerized resin.
6. Clamp the polymerized resin wafer firmly into a vise and cut carefully around the block with a fine jeweler's saw.
7. Place the block onto a blank chuck and place a drop of cyanoacrylate glue next to the block. The glue is drawn under the block by capillary action. Blanks may be made from a 7.5-mm-diameter Lucite rod that is cut into 1.5-cm lengths.
8. If desired, the tissue block can be further stabilized by placing epoxy glue around the outside of the block (e.g., 5-Minute Epoxy).

Embedding in Preformed Molds

Tissue blocks may also be placed in preformed molds such as polyethylene BEEM capsules, gelatin capsules, or shallow silicon rubber molds (Fig. 20.1). Gelatin capsules are useful for resins, such as LR White, that are sensitive to oxygen. Tissue blocks embedded in BEEM or gelatin capsules may be labeled prior to addition of resin by writing in pencil on a thin strip of white paper and inserting the paper into the capsule, using the blunt end of a pasteur pipette.

1. Place capsules upright in a commercially available holder or in a handmade cardboard holder.
2. Place the tissue at the bottom of the capsule as illustrated in Figure 20.1.
3. Fill with fresh 100% complete resin and polymerize for 24–48 hours at 60°C.

4 Once it is polymerized, remove the resin block from the mold.
Blocks are now ready for sectioning.

FLAT EMBEDDING AND MOUNTING OF TISSUE SECTIONS

Several methods are available for embedding tissue sections so that they remain flat. This is especially useful for stained tissue where the stained area is confined to the tissue surface. If thin sections (<50 μm) are employed, it is usually best to cut them into small pieces (1–2 mm) prior to dehydration in order to prevent curling. After osmication and dehydration, tissue sections are very brittle and will break when flattened.

Embedding between Plastic Coverslips

1 Place tissue section in a fresh drop of resin onto a small square plastic coverslip.

2 Cut a second coverslip into four smaller squares and place one of the squares on top of the tissue section.

3 Place the coverslips onto aluminum foil and polymerize in a 60°C oven overnight.

4 After polymerization, cut out the tissue block with a pair of scissors and glue the block, still between the coverslips, to a specimen chuck. The plastic coverslip can be sectioned through with a dry glass knife, or it may be carefully removed with a razor blade prior to sectioning.

Embedding between Glass Slides

This method, introduced by Olschowka (1988), produces extremely flat sections.

1 Prepare coated glass slides. Dip clean glass slides up to the frosted end into liquid release agent (Mould Release Compound, Electron Microscopy Sciences). Place slides upright for 2 hours at room temperature in a dust-free environment to allow excess liquid to drain; then dry in a 60°C oven overnight.

2 Place tissue sections along with a small drop of fresh resin evenly spaced onto a coated glass slide (frosted side up) (see Fig. 20.2). Do not place sections too close to the edge.

3 Place a second coated slide over the sections, forming a sandwich (see Fig. 20.2). To ensure flatness, small binder clips may be placed around the edges, although this may result in some tissue compression.

4 Place slides overnight in a 60°C oven.

5 Separate slides by inserting a razor blade gently between the two slides. Too much pressure will break the slides. Safety goggles should be worn during this step because small glass chips sometimes break off. The sections will remain in a thin layer of polymerized resin on one of the slides. Sections can be viewed and photographed under the light microscope to select the desired area or to evaluate the degree of staining.

6 To mount specimens, selected areas may be excised with a scalpel, removed from the slide with

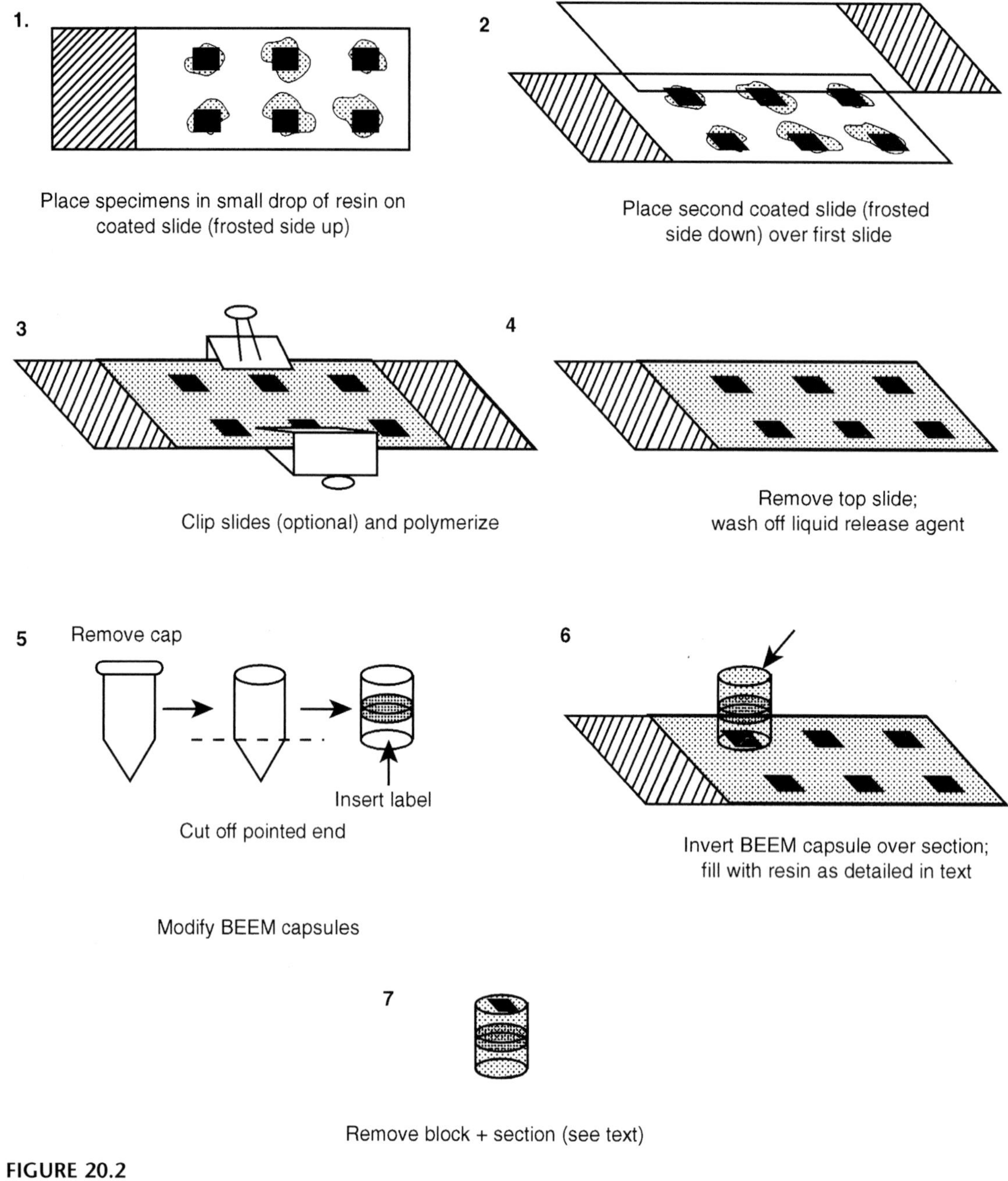

FIGURE 20.2

Flat embedding tissue sections.

a fine forceps, and glued onto blank specimen blocks. Alternatively, the following procedure may be used, as illustrated in Figure 20.2.

a Remove the caps from BEEM capsules and cut off the pointed end with a razor blade to form a tube. Labels may be inserted into BEEM capsules as described on pp. 322–323.

b Position the flat end of the capsule over the section and place a drop of complete resin into the cut end. Return the slides to the oven, taking care that the capsule remains over the tissue section, and allow the resin to polymerize for at least 4–5 hours in order to form a seal around the bottom of the capsule.

c Fill the capsule to the top with complete resin and allow to polymerize overnight.

d To remove the block from the slide, first slit the side of the BEEM capsule and remove it from around the resin block. Apply gentle pressure to the resin block. The block along with the tissue should snap off the slide. If it does not, place the slides on a hot plate set at medium high heat. After the slides have heated for a few minutes, remove the slide and gently slide a new razor blade under the section. Make sure to use downward pressure on the blade. The razor blade should slide easily under the tissue section; if not, return the slide to the hot plate and repeat.

EMBEDDING CULTURED CELLS

Cell Suspensions

For cell suspensions or particulate specimens (e.g., bacteria, organelles, subcellular fractions), samples are usually embedded in a matrix such as agar, agarose, gelatin, or BSA (bovine serum albumin) prior to dehydration and embedding. This step is not necessary if the sample forms a pellet of sufficient cohesiveness following centrifugation to withstand subsequent processing intact. When working with cell suspensions, care must be taken not to introduce glass particles into the suspension, as these can interfere with sectioning and may destroy diamond knives. Plastic pipettes or fire-polished glass pipettes should be used for all steps.

The following protocol describes embedding cell suspensions in agar (Dykstra 1993).

1 Make up 10 ml of 3–4% agar in distilled H_2O in a scintillation vial by placing in a larger beaker containing dH_2O on a 45°C hot plate until the agar melts.

2 Pellet the sample with a centrifuge. Carefully remove the fixative from the pelleted sample and replace with 1 ml of an appropriate buffer.

3 Resuspend the pellet by vortexing gently, transfer to a fresh microcentrifuge tube, and let stand for 15 minutes. Repellet and repeat the wash.

4 Pellet sample and remove the buffer.

5 Quickly heat a pasteur pipette by drawing up and then expelling some heated H_2O from the agar bath. Immediately draw up approximately 1 ml of molten agar and place into the microfuge tube containing the sample. The pipette tip should be placed at the bottom of the pelleted sample.

6 Working quickly, pellet the sample for 30 seconds. Centrifugation times longer than 60 seconds may result in shearing artifacts.

7 Remove the tube and allow the agar to solidify. The tube may be placed on ice to speed the process.

8 Use a single-edged razor blade to remove the microfuge tube carefully from around the agar block. Cut the agar block into 1-mm slices. These slices can be handled like tissue blocks for osmication, dehydration, and embedding.

Monolayers

Cells grown in monolayers may be processed for EM by removing them from the substrate with a rubber policeman, suspending them in a suitable buffer, and treating them as described above. To take intact monolayers through the embedding process, cells should be plated onto plastic coverslips that are resistant to organic solvents, e.g., Thermanox (Nunc) or Aclar (Ted Pella, Inc.). These coverslips may be placed in the bottom of glass petri dishes and dehydrated and infiltrated with resin as described in Table 20.1. Cells grown on glass coverslips are more difficult to embed because the resin tends to adhere to the glass. An exception is cells grown on special culture dishes available from MatTek (glass bottom uncoated microwells, MatTek Corp., Ashland, Massachusetts) that have a cover glass incorporated into the bottom. Cells may be grown on these coverslips, dehydrated, and infiltrated with resin. After embedding in Epon/Araldite, this coverslip is easily removed and the cells remain embedded in the resin. If these dishes are employed, dehydration should be carried out in ethanol only, as acetone will dissolve the plastic culture plate.

Because cell monolayers are very thin, the incubation times can be greatly reduced. Finer dehydration steps are recommended for cultured cells. The dehydration solution should be introduced into the culture dish with one pipette, while the previous solution is withdrawn simultaneously with a second pipette so that the cell layer does not dry out. A typical embedding schedule is given in Table 20.1.

Embedding Cells Grown on Coverslips in Epoxy Resin

1. Label an aluminum weigh dish and place two sticks parallel to each other at a distance slightly less than the size of the coverslip (Fig. 20.3).
2. Fill the dish with fresh 100% resin until it just covers the sticks.
3. Place the coverslip over the sticks, cell side down.

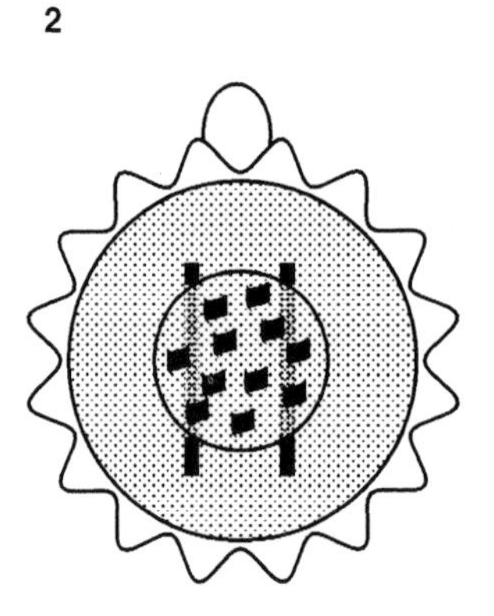

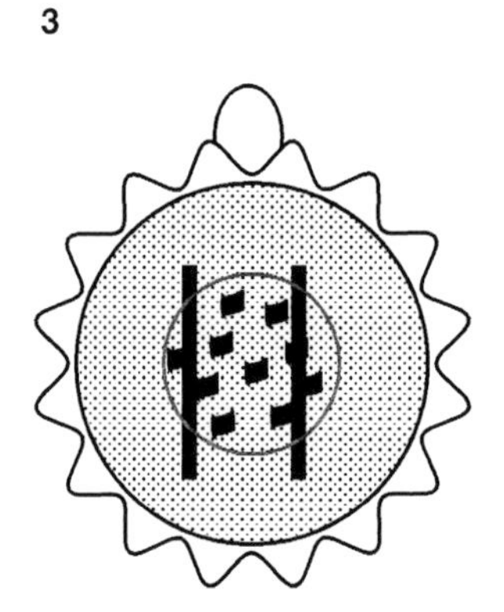

FIGURE 20.3

Embedding coverslips.

4 After polymerization, remove the coverslip by gently peeling it off. The cells will remain in the resin block.

5 Cells may be mounted on specimen blocks as described above.

Embedding Cells Grown on Coverslips in LR White

Embedding monolayers in LR White may be somewhat tricky because of the need to protect the resin from oxygen.

1 For embedding in LR White, place the coverslip cell-side-up onto the bottom of the aluminum weigh dish between two parallel wooden sticks.

2 Generously fill the dish with resin.

3 Cover the top of the dish with a sheet of Aclar or other film to protect it from oxygen.

4 After polymerization, strip off the Aclar and gently peel the coverslip away from the resin.
Usually, there is a thin film of LR White over the coverslip that must be removed first.

5 Cut out the blocks and mount as previously described.

▪ *Notes: Problems with Embedding*

- Poor embedding can result in difficulty in sectioning and instability under the beam (for further discussion, see Mollenhauer 1988). One cause is incomplete polymerization, in which case blocks can be returned to the oven for further polymerization. Another cause of poor sectioning is incomplete infiltration of the specimen. Indications of poor infiltration include the tissue appearing softer than the surrounding plastic, and disintegration of section shortly after cutting. Infiltration of difficult-to-embed specimens may be improved by (1) increasing infiltration times (up to several days, if necessary); (2) using a lower-viscosity resin such as Spurr's; (3) using a more gradual infiltration by increasing the number of solvent:resin steps; or (4) infiltrating at 50°C. Most problems with embedding result from improper mixing of reagents, so it is important that components be thoroughly mixed prior to infiltration and embedding.
- Poorly embedded samples are generally discarded. Sometimes poorly infiltrated tissue blocks may be rescued by returning the blocks to 100% resin for several days at room temperature and then polymerizing as usual. Additional methods for re-embedding are presented in Hayat (1989a).

Ultramicrotomy

Once the specimen is embedded, it must be sectioned thin enough (usually on the order of 50–250 nm) to be viewed under the electron microscope. Cutting of ultrathin sections requires the following specialized equipment and materials: an ultramicrotome, a knife maker, and glass strips for making glass knives and EM grids. Access to a vacuum evaporator is useful for making thin carbon films. Of all aspects of EM specimen preparation, cutting and collecting ultrathin sections requires the most skill. Detailed tutorials on cutting and mounting ultrathin sections are provided in Hayat (1989a) and Dykstra (1992, 1993).

Grids

Ultrathin sections are collected on metal grids for viewing under the electron microscope. Grids come in many sizes, shapes, and materials. Four common types are illustrated in Figure 20.4. The type of grid used depends on the type of specimen to be examined and the type of procedure to be performed. Copper grids are the most common because they are inexpensive, are good electrical and thermal conductors, and are nonmagnetic. However, copper grids may react with many of the solutions needed for enzyme histochemistry or immunocytochemistry. Gold or nickel grids are usually preferred for these types of procedures.

The amount of specimen that will be visible under the electron microscope is dependent on the mesh size. Grids are assigned a number between 50 and 1000, which refers to the mesh size (holes/inch). The larger the mesh size, the smaller the holes and the smaller amount of open space. Higher-mesh grids provide more specimen support and can handle smaller specimens than lower-mesh grids, but result in less of the specimen being visible. The most common sizes for routine work are 200–400: 200 mesh grids have approximately 60–70% open area. Hexagonal mesh and thin-bar high-transmission grids are also available to increase the amount of open space while still providing good specimen support (Fig. 20.4). Grids without bars, termed slot grids (Fig. 20.4), are used when the maximum amount of open area is desired; e.g., for serial section work and for correlated LM and EM examination of the same feature. Slot grids must be coated with a suitable electron-transparent substrate to act as a specimen support (see p. 329).

Grids have both a shiny and a dull side. According to Hayat (1989a), specimens should be collected on the dull side, as this side is rougher and thus holds the section more securely than does the shiny side. However, there is no general agreement as to which side is superior. It is good practice to be consistent so that the side with the sections can be readily identified.

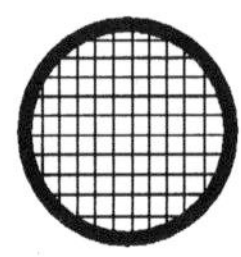

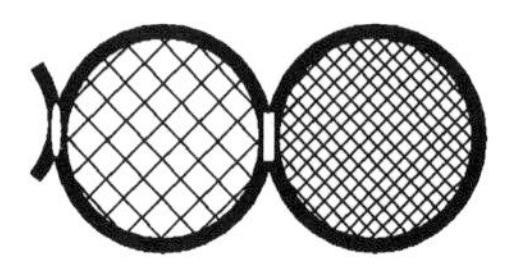

FIGURE 20.4

Types of grids.

CLEANING OF GRIDS

Empty grids into a small beaker and sonicate in technical grade acetone (1 min), followed by reagent grade acetone (1 min) and 95% ethanol (1 min). Empty out the ethanol and invert the beaker over a glass petri dish with filter paper on the bottom. As the grids dry, they will fall from the beaker onto the paper. This cleaning procedure may not be necessary for new grids but should be employed for previously used grids. If the used grids had been coated with a substrate such as Formvar (see below), they should be sonicated in a suitable solvent, e.g., ethylene dichloride, for two times for 1 minute before beginning the cleaning sequence outlined above.

COATING OF SLOT GRIDS

Coating of grids with various electron-transparent support films, usually some type of plastic, adds stability to delicate specimens and provides support for sections collected on slot grids. Support films are also employed for negative staining of particulate samples (e.g., bacteria, microsomes). If grids of sufficiently small mesh size are used, such supports are usually not necessary for epoxide-embedded sections, but may be required for some acrylic resins that are less stable under the beam. Since substrates will decrease resolution and specimen contrast, they should not be used if not needed. One of the most popular supports is Formvar (polyvinyl formaldehyde) dissolved in ethylene dichloride. Other substrates include collodion, Butvar, and carbon. Protocols for the preparation of support films are described below.

Cast on Glass Method

Films of even texture of either Formvar or Butvar can be made using the "cast on glass" method, illustrated in Figure 20.5. Formvar may be purchased as a solution or as a powder that needs to be dissolved in ethylene dichloride before use. Butvar may be purchased prediluted in chloroform. Commercially prepared solutions generally result in a film with fewer imperfections than do those prepared from powders. Formvar solutions are typically employed at a concentration of 0.25–1%. The more concentrated the Formvar, the thicker the film. To prepare 0.25% Formvar from powder, add 0.125 g of Formvar powder to 50 ml of ethylene dichloride in a glass-stoppered bottle. It is recommended that the ground-glass stopper be wrapped with Teflon tape to prevent the Formvar solution from gluing the stopper to the neck of the stock bottle (Dykstra 1993). After swirling the solution to mix the contents, let the mixture sit at room temperature for ~8 hours.

To Prepare Films

1. Assemble the following materials: Fisher precleaned uncoated microscope slides, a Coplin jar, a flotation dish (75 x 150-mm culture dish or finger bowl), a glass rod, and dH_2O. All surfaces should be very clean.

2. Clean microscope slides by gently breathing on their surface and wiping dry with a Kimwipe. Dip slides into a Coplin jar filled to 1.5 inches with Formvar solution, lift slides out, and let dry in a dust-free environment. The drying time will increase with an increase in the relative humidity.

3. Fill the flotation dish to overflowing with dH_2O. Sweep the surface with a glass rod to remove any dust particles.

4. With a clean sharp razor blade, score the film 1–2 mm from the edge of the slide on all sides (Fig. 20.5). Breathe on the film so that it becomes fogged.

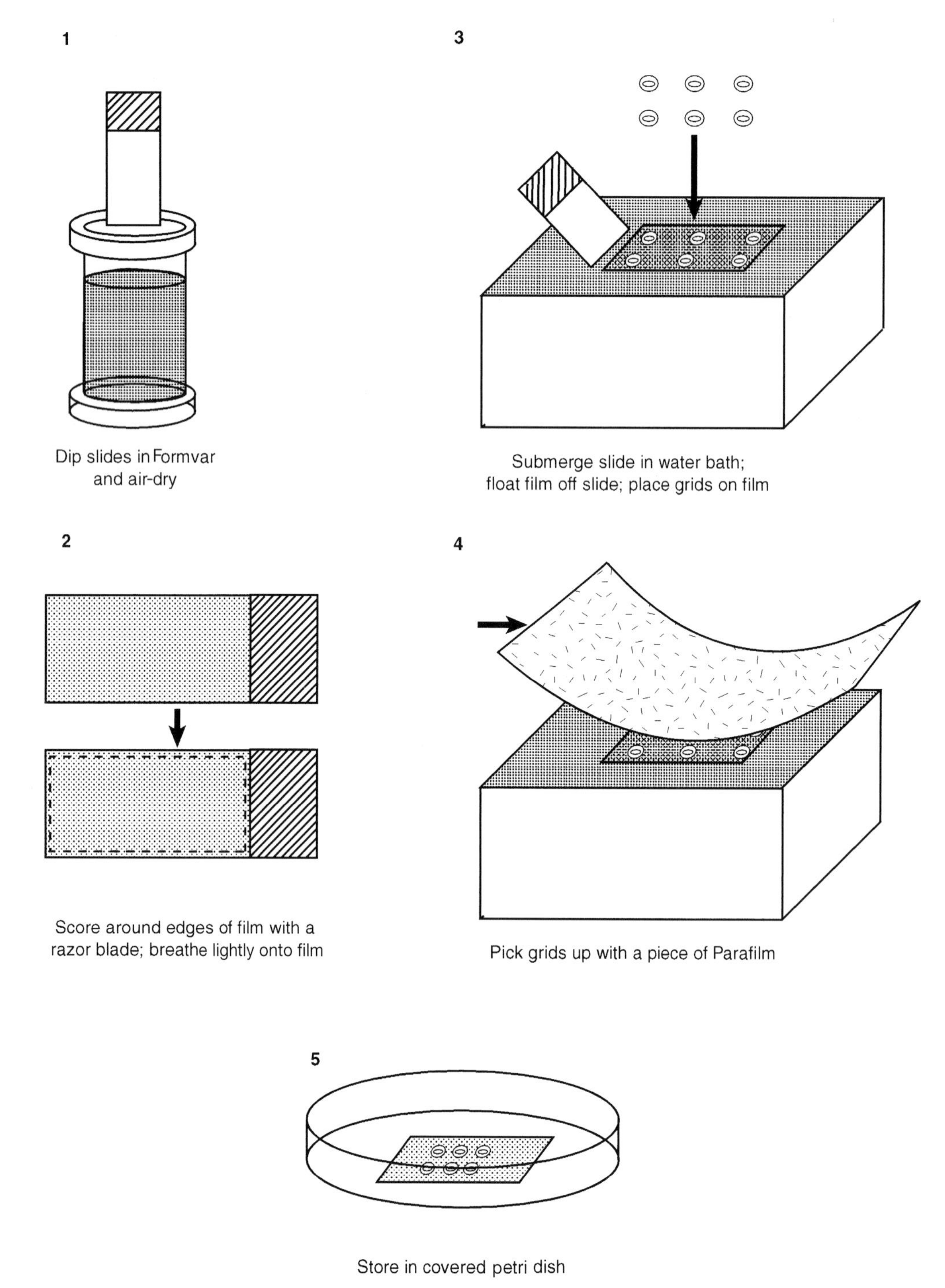

FIGURE 20.5

Formvar coating slot grids.

5 Introduce the slide slowly and steadily into the H_2O at an angle of 30°–45° as illustrated. The surface tension of the H_2O will float the film as the slide is submerged. When the film is free, allow the slide to drop to the bottom of the dish.

6 Examine the film for variations in thickness. It should be a uniform gray-silver in color (50 nm). If bright interference colors, dust, or wrinkles are present, discard the film.

7 Using EM (jeweler's) forceps, carefully transfer clean grids, dull side down, onto the plastic film "raft" floating on the H_2O. Grids should not be too crowded.

8 After all of the grids have been placed, take a piece of Parafilm and hold it as illustrated in Figure 20.5. Gently contact one end of the floating grid film and roll the strip over the remaining grid film. Lift the Parafilm strip with its attached grid film out of the flotation dish and place grid-side-up on a piece of filter paper in a petri dish.

9 To remove grids from the grid film, gently perforate the plastic around the grid with the sharp tips of a forceps. Carefully lift the coated grid and examine it under a dissecting scope. The film should be flat, smooth, and free of gross contamination or tears.

Notes

- Butvar has been reported to be less hydrophobic and stronger than Formvar.
- Coated grids can be used immediately or after stabilization with a thin coat of carbon from a vacuum evaporator.
- Substrated grids may be stored for several weeks, but over time the quality of the support film degrades and it tends to detach from the grid more easily during section processing.
- Substrates tend to be quite hydrophobic, which can present difficulties in picking up sections onto the grid. The hydrophobicity increases the longer the grids are stored. This problem can be reduced by glow-discharging the grids prior to use and also by using freshly prepared grids.
- Unwanted holes may occur in the film if dipping is performed on a humid day.
- If films show a tendency to fall off of the grid during processing, grids can be cleaned in ethylene dichloride, dipped in Formvar, and dried before being placed on the film.
- Ethylene dichloride is flammable. Avoid contact of Formvar solutions with skin. Prepare Formvar films in a well-ventilated area.

Cast on Water

Plastic films may be produced by dropping the liquid substrate directly on the H_2O and allowing it to spread across the surface. This method is most useful for collodion films (nitrocellulose dissolved in amyl acetate), as Formvar drops tend to sink before spreading. In this method, 2 or 3 drops of a 2% collodion solution from a pasteur pipette are placed onto a clean H_2O surface. After a few minutes, the amyl acetate will evaporate, leaving a silver to gold film of collodion covering the surface of the dish. Inspect the film for areas free of wrinkles and of the desired thickness (usually areas with a silver-gray color) and place grids, dull side down, on the film. Remove the film from the H_2O with a piece of Parafilm as above.

Preparation of Holey Films

Special Formvar films with many small holes ("holey" films) are employed for the diagnosis of astigmatism in the microscope and may also be used as support films for particulate specimens. In this latter usage, the holes act as small grid squares for viewing specimens. This type of film is often produced inadvertently when coating grids on a humid day. It may be produced deliberately through the introduction of a small amount of H_2O into the liquid Formvar by breathing on the surface of the film while it is still wet. Holey films should be stabilized with carbon prior to use.

Preparation of Carbon Films

Carbon films are prepared using a vacuum evaporator. Thin carbon films are highly transparent to electrons because of the low molecular weight of carbon atoms. They are also very stable and possess a very fine structure and good electrical conductivity. This is the most common type of film used for high-resolution work. Their main disadvantages are their difficult preparation and their hydrophobicity. Carbon films are sometimes deposited directly on specimens or plastic films to impart stability under the electron beam. Carbon support films may be prepared by coating a sheet of mica approximately 12 x 15 mm in size. Mica sheets are available from most EM suppliers.

1. Cleave the mica sheet to expose a new clean surface and place in the vacuum evaporator.
2. Make a greasy fingerprint on a piece of white filter paper and place the paper in close proximity to the mica. The fingerprinted area will not darken as quickly as the surrounding area as the carbon is deposited and can be used to gauge the relative thickness of the film.
3. Make sure the vacuum in the bell jar is better than 10^{-4} torr before applying current to the carbon electrodes.
4. Deposit carbon until the fingerprint is just clearly visible for thin films, and longer for thicker films.
5. Remove the mica from the vacuum evaporator. The carbon film is separated from the mica sheet by gradually immersing it in H_2O, similar to the method used to float Formvar films (see Fig. 20.5).
6. To coat a grid, take a single grid with a fine forceps and submerge it under the floating film. Lift the film with the grid and remove the drop of H_2O clinging to the bottom of the grid with a piece of filter paper.

Poststaining

STAINING SURVEY SECTIONS FOR LIGHT MICROSCOPY

During block trimming for ultracryotomy, it is often useful to examine a semithin section (0.25–2 μm) under the light microscope to assess the quality of fixation, to orient the tissue properly, and to ensure that the feature of interest is correctly positioned within the final block face. Semithin sections may be viewed under a phase-contrast microscope, or they can be stained with a cytochemical stain to reveal cellular detail. The most popular stain for epoxy-embedded sections is toluidine blue, which will stain epoxy-embedded tissue but not the plastic. To prepare a stock solution of toluidine blue, dissolve 1% toluidine blue in 1% sodium borate.

To Stain Semithin Sections

1. Transfer semithin sections using a platinum loop onto a drop of H_2O placed on a clean microscope slide.
2. Dry the sections on a 60°C hot plate. Leave thicker sections (2 μm) on the hot plate for at least 30 minutes.
3. Cover the sections while they are still on the hot plate with a few drops of toluidine blue filtered through a 0.45 μm Millipore filter. Stain sections until the edge of the drop of stain begins to dry and take on a metallic green color. Thicker sections require less staining than do thin sections.
4. Rinse excess stain off the slide with distilled H_2O, followed by 95% ethanol to destain and differentiate. Air-dry and check under microscope.
5. Sections may be coverslipped with mineral oil and sealed with nail polish. Permount may also be used, although the stain will fade over time. Fading is reported not to occur in Polymount (Polysciences) (Dykstra 1992).

■ ***Note***

- If the sections show excessive numbers of wrinkles, make sure that the drop of H_2O on which the sections are placed is large enough to allow the section to stretch during heating. Drying the sections longer and raising the temperature of the hot plate prior to staining may also help. If there is a lot of debris, the stain probably overdried.

STAINING OF ULTRATHIN SECTIONS FOR EM

Biological specimens are made up mostly of low-atomic-weight elements such as carbon, oxygen, hydrogen, and nitrogen. These elements do not effectively stop electrons, and thus biological specimens possess very low contrast under the EM in their native states. Embedding media are also typically composed of the same elements, further reducing the specimen contrast. Thus, prior to viewing under the EM, the specimen is usually stained with electron-dense substances to impart contrast to cellular structures. These may be applied either prior to embedding (en bloc) or after sectioning (en grid). In addition to providing contrast, such substances serve to protect the sections from beam damage and to impart clarity to the fine structure of cells. To be an effective stain for EM, the substance must either be electron dense or act as a mordant for a subsequent electron-dense stain. Most of the stains employed are metals with high molecular weights, because these are most effective in stopping electrons. The most common stains are UA and lead citrate. Reduced OsO_4 is also electron dense and, furthermore, acts as a mordant for subsequent heavy-metal staining. Some structures will only stain after fixation with OsO_4 whereas others, e.g., compact myelin, will stain in the absence of osmium. It should be remembered, however, that no contrasting agent currently in use stains all structures equally well. In addition, although the combination of osmium fixation with UA and lead staining provides staining of most cellular structures, there may be some structures within the specimen that are not stained at all.

UA Staining

UA may be used either as an en bloc stain (see above) or as a poststain. UA is particularly good for DNA-containing structures and membranes. The preparation and handling of aqueous UA is described above. In some instances, alcoholic solutions of UA are employed, particularly to aid in penetration for staining of thick sections or sections embedded in Spurr's resin. Alcoholic solutions of UA (1–2%) may be prepared in 95% ethanol or 50–100% methanol.

Procedure for Grid Staining with UA

1. Prepare a smooth, clean, staining surface by placing a piece of Parafilm or dental wax in the bottom of a glass petri dish.
2. Place one drop of **UA** stain for each grid onto the Parafilm. Float grid on top of stain with side containing sections face down using a fine forceps.

   CAUTION: Uranyl acetate; radioactive substances (see Appendix 2 for Caution)
3. Cover the dish with the lid and stain for 3–10 minutes.
4. Set up a series of three 50-ml glass beakers containing dH_2O. Pick the grid out of the stain droplet with fine forceps, moving very quickly to prevent the stain from drying on the grid. Agitate the grid several times up and down through each beaker.
5. Drain off excess H_2O with filter paper. Place grids section-side-up on filter paper to dry fully before additional staining with lead.

   If there is any fluid on the tip of the forceps, the grid will be drawn up into the forceps. To avoid this, use a filter paper wedge to remove any H_2O between the tips prior to releasing the grid.

Lead Staining

Lead stains are particularly good for contrasting ribosomes, glycogen, and cytoplasmic ground substance. Lead staining is typically employed after UA staining, since UA acts as a mordant for lead. Lead also has a strong affinity for structures containing reduced osmium. Most lead stains are employed en grid, but some en bloc stains have been developed. There are several formulations of lead stains in existence. Recipes for lead citrate and a triple lead stain are given below (for additional recipes, see Hayat 1989a). Lead reacts with CO_2 in air to form a lead carbonate precipitate. Care should be taken to minimize contact of lead solutions with air. All H_2O should be boiled to remove excess CO_2 before making lead solutions. According to Hayat (1989a), lead solutions should be stored in plastic rather than glass bottles for long-term storage, as lead can leach silicon from glass.

Preparation of Reynold's Lead Citrate

This is the most popular lead stain because it is easy to prepare and is less prone to lead carbonate formation than other formulations.

1 Place in a 50-ml volumetric flask

lead nitrate, $Pb(NO_3)_2$	1.33 g
Sodium citrate, $Na_3\ (C_6H_5O_7) \cdot 2H_2O$	1.76 g
Freshly boiled, cooled ddH_2O	30 ml

CAUTION: Lead nitrate (see Appendix 2 for Caution)

2 Shake intermittently and vigorously for 30 minutes. A heavy, white precipitate will form.

3 Add 8.0 ml of 1 N **NaOH** and dilute to 50 ml with boiled and cooled dH_2O.

CAUTION: NaOH (see Appendix 2 for Caution)

4 Stopper the flask and mix by repeated inversion until the precipitate dissolves. The solution should clear with no evidence of precipitate. The final pH is 12.

Preparation of Simple Lead Citrate (Veneble and Coggeshall 1965)

This lead solution can be made up simply from lead citrate. Add 0.01–0.04 g of lead citrate to 10 ml of freshly boiled and cooled H_2O in a screw-top vial. Add 0.1 ml of 10 N NaOH, close vial, and shake vigorously until lead dissolves.

Preparation of Triple Lead Stain (Sato 1968)

This is more robust lead stain than lead citrate.

1 Weigh out into a clean 100-ml volumetric flask

lead nitrate	1 g
lead citrate	1 g
lead acetate	1 g
sodium citrate	2 g

CAUTION: Lead nitrate; lead acetate (see Appendix 2 for Caution)

2 Add 82 ml of very clean, double-distilled, freshly boiled and cooled H_2O and sonicate for 30–60 minutes or until the stain solution takes on a milky appearance.

3 Add 18 ml of freshly made 1 N NaOH (5 g NaOH/125 ml ddH_2O). The solution should clear instantly. Sonicate again if a few grains of salt remain. The final solution should be clear.

4 Cover the flask with aluminum foil. Let the stain stand overnight and come to room temperature before use. The solution should be stable at room temperature for ~6 months.

Notes

- The stain will keep for several months if kept stoppered. When removing stain, remove it from just below the surface and do not mix the contents during removal. Discard the stain when a precipitate appears or when excessive contamination is found on the grids.
- Lead is extremely toxic and should be handled with the utmost caution. Do not let either solutions or powders come into contact with skin. Do not inhale lead-containing powders. Place all waste solutions in properly labeled bottles that can be collected by qualified waste disposal personnel.

Grid Staining with Lead

Grids may be stained with lead immediately after staining with UA (make sure the grid is dried thoroughly before staining with lead).

1. Construct a CO_2-free staining chamber by pressing a sheet of Parafilm or dental wax into the bottom of a petri dish to make a smooth surface. Place **NaOH** pellets around the edges of the Parafilm to remove CO_2 from the chamber, and keep the dish covered whenever possible. Avoid breathing into the chamber before and during staining.

   CAUTION: NaOH (see Appendix 2 for Caution)

2. Prepare three wash beakers as described above for UA staining.

3. Filter lead solutions through a 0.22-μm Millipore filter prior to use. This is easily done using a 10-ml syringe fitted with a filter. Discard the first drop of stain coming through the filter. Place one drop of stain for each grid to be stained on the Parafilm. It is best not to stain more than five grids at a time using this method.

4. Float the grids, section-side-down, for 1–3 minutes. With the Sato lead, 30 seconds to 1 minute is usually sufficient. Keep the dish covered whenever possible.

5. Immediately wash the grids as described above for UA, taking care not to let the stain dry on the grids. Grids may then be dried on clean, dry, filter paper.

Notes: Problems in Poststaining with UA and Lead

- The biggest problems associated with grid staining are the presence of excessive amounts of dirt and the formation of electron-dense precipitates that obscure structural detail. To determine whether these precipitates are the result of grid staining or specimen preparation, look for an area under the electron microscope that contains only plastic and not tissue, e.g., the lumen of a blood vessel or the edge of the tissue section. If precipitate is found over tissue-free areas, it was likely introduced as a result of grid staining. To minimize dirt formation, filter all solutions, including H_2O used for rinses before use, and make sure that the sections are not allowed to dry out on the grid during staining. To minimize precipitate formation, use fresh staining solutions that are filtered just before use. According to Hayat (1989a), UA precipitates can be removed from sections by rinsing the sections with warm distilled H_2O. Lead precipitates can be removed by exposing the sections to 2% aqueous UA for 2–8 minutes or 10% acetic acid for 1–5 minutes.
- When specimens are collected onto coated grids, grid staining may sometimes produce excessive wrinkling of the tissue. This may be a problem when trying to track the same structure in series

of sections. If poststaining is required, it is possible to float the sections directly on drops of staining solutions prior to collecting them onto grids. Alternatively, for material that is to be serially sectioned, UA and lead staining may be performed en bloc.

En Bloc Lead Staining

Although lead solutions are most commonly applied after sectioning, in some cases tissue may require staining en bloc. This may be necessary to achieve adequate depth of staining for high-voltage microscopy or when poststaining presents a problem; for example, for long series of sections collected on slot grids. However, en bloc staining with lead is reported to decrease the quality of ultrastructural preservation and is not recommended unless necessary (Hayat 1989a). The use of lead aspartate for en bloc staining was introduced by Walton (1979). Lead aspartate staining may be used alone or after en bloc staining with UA. Commercially prepared solutions of lead aspartate are available from Polysciences. Lead aspartate may also be prepared as follows.

1. Prepare a stock solution containing 0.998 g of aspartic acid in 250 ml of ddH_2O. The aspartic acid goes into solution more rapidly if the pH is raised to 3.8. The stock solution is stable for 1-2 months in the refrigerator.

2. Prepare a staining solution of 0.02 M **lead nitrate** and 0.03 M aspartic acid by adding 0.06 g of lead nitrate to 10 ml of the stock solution. Adjust the pH of the staining solution to 5.5 with 1 M **KOH**.

   CAUTION: Lead nitrate; KOH (see Appendix 2 for Caution)

3. Heat the staining solution to 60°C prior to use to ensure stability. Discard the solution if any precipitate forms.

4. Place tissue blocks that have been fixed in aldehydes and OsO_4 in a stoppered vial and cover with staining solution for 30–60 minutes at 60°C.

Other Stains

Although lead and UA are the most common general stains used in EM, numerous other compounds have been employed either as general stains or to stain specific compounds; e.g., bismuth, phosphotungstic acid (PTA), ruthenium red, lanthanum, and silver. A thorough discussion of types of stains and staining mechanisms is provided in Hayat (1989a, 1993).

High-Voltage Electron Microscopy

The use of higher accelerating voltages (up to 1,000,000 eV) provides several advantages for electron microscopists. First, higher accelerating voltages result in increased resolution, mainly because of the reduction in chromatic aberration. Second, the higher the accelerating voltage, the thicker the specimen that can be viewed under the electron microscope. High-voltage scopes may image specimens up to 6–8-µm thick without a severe loss in resolution. This permits three-dimensional analysis of biological specimens with subcellular resolution. Excessive specimen heating with increasing thickness limits the maximum specimen thickness to ~10 µm.

High-voltage electron microscopes are classified as either high voltage (HVEM), operating from 600,000–3,000,000 eV, or intermediate high voltage (IVEM), operating in the range of 200,000–400,000 eV. In contrast to earlier generations of HVEMs, the new generation of IVEMs is much easier to use and provides adequate accelerating voltage for most biological applications. They operate much like a conventional TEM and often are equipped with computer-driven stage and specimen controls and with CCD technology to aid in focusing and image analysis. Because of the expense of these instruments, specialized centers for HVEM and IVEM have been set up in the United States for use by biological and material scientists. A good introduction to principles of high-voltage electron microscopy is provided by Glauert (1974).

SPECIMEN PREPARATION FOR IVEM/HVEM

The overall protocol for specimen preparation for HVEM/IVEM is similar to that for conventional TEM. Specimens up to 0.5 µm in thickness may be embedded and stained as for conventional TEM, although poststaining times must usually be increased. However, for specimens thicker than 1 µm, selective staining protocols must often be employed because of the large depth of field of the electron microscope. Because all structures throughout the depth of the specimen will be in focus at the same time, the image produced by a TEM is a projection of the 3D specimen onto 2D film. With thin specimens, the degree of overlap between cellular structures is minimal, and image interpretation is usually not difficult. However, in thicker specimens, the amount of superimposition of structures within the micrograph hampers meaningful interpretation of many specimens. Since most biological structures possess very little inherent contrast, selectively contrasting only a few structures within the section limits the problem of structure overlap and reveals the 3D organization of stained structures at high resolution.

Embedding

Specimens may be prepared and embedded as for conventional TEM. For very thick specimens, it may be advantageous to use a resin with low electron-scattering properties. The araldite resin Durcupan ACM (Fluka, available from Polysciences; not to be confused with the water-miscible Durcupan) appears to be superior to Epon/Araldite for IVEM. Recipes are given in the manufacturer's handout for hard, medium, and soft formulations. The medium grade is good for most specimens. The embedding protocol provided in Table 20.1 works well for soft tissues. More difficult-to-embed specimens may require a more lengthy infiltration, as specified in the manufacturer's instructions.

Sectioning

The maximum thickness of specimen viewable is dependent on the accelerating voltage, the type of specimen, and the type of stain used. Denser tissues, e.g., muscle, may require thinner sections than less dense tissues. Cutting thick sections (2–5 μm) with a glass knife may sometimes produce pitting on the tissue block surface, especially if several sections are cut in succession. This results in unwanted holes being formed in the sections. To avoid this damage, several ultrathin sections are cut between each thick section until the block face is again mirror smooth (Favard and Carasso 1973). Diamond knives are usually not recommended for cutting thick sections as the edge may be damaged. However, less expensive diamond knives produced for light microscopy (e.g., Diatome histoknife) may be used for sections 0.5–2 μm thick with excellent results, because knife marks are generally less noticeable at high accelerating voltages because of the decrease in specimen contrast.

Thicker sections (>3 μm) are typically cut with a dry glass knife and then flattened in warm H_2O. Sections may be mounted as usual with the aid of a dissecting scope. Thick sections are easily dislodged from grids. To check for the stability of the sections, use an eyelash manipulator to try to remove a test section from a grid. If it is easily dislodged, it may be better to collect sections on a clamshell grid (Fig. 20.4) or to use either substrated grids or grid glue.

To Prepare Grid Glue

1 Soak 2 inches of Scotch tape in 20 ml of **xylene** for several minutes.

CAUTION: Xylene (see Appendix 2 for Caution)

2 Remove the tape backing from the solution: The adhesive will stay in the xylene.

3 Allow the adhesive to dissolve completely.

4 Dip grids into solution and air-dry.

They are now sticky.

Staining

Viewing of semithin sections (0.25–1 μm) usually does not require selective staining. However, the higher the voltage, the fewer electrons are scattered by the specimen and, thus, the lower the image contrast. Therefore, contrasting with UA, lead, or PTA is usually necessary. Penetration of stains throughout the entire thickness of sections >1-μm thick can be a problem. This may be overcome by (1) staining en bloc with UA, PTA, or lead prior to embedding (N.B.: if using PTA prior to embedding, propylene oxide should not be used as a transition solvent as it removes PTA); (2) using alcoholic rather than aqueous solutions of UA and PTA for poststaining; (3) increasing the staining times and/or temperature for poststaining, e.g., staining with alcoholic UA at 60°C for up to 24 hours allows penetration up to 15 μm (Favard and Carasso 1973); (4) staining the section from both sides (for nonsubstrated grids) by immersing the grid in the staining solution; (5) addition of a detergent (0.3% Tween 80) or DMSO (dimethyl sulfoxide) to the staining solution.

▪ *Note*

- Lengthy staining times and increased temperatures result in a greater chance of UA and lead precipitates depositing on the section surface. These may be removed by treatment with 0.05% citric acid (Hayat 1989a) or 0.05% nitric acid (Favard and Carasso 1973) for 1 minute.

Selective Staining

Examples of stains selective for various organelles or cellular structures are given in Favard and Carasso (1973). Several of the enzyme techniques, given in Chapter 27, produce selective staining of one or more structures and may be used for HVEM. Immunocytochemical localization with DAB (diaminobenzidine tetrahydrochloride) may also be used, provided that sufficient penetration of the labeling reagents is achieved.

3D ANALYSIS OF HVEM/IVEM SPECIMENS

The use of 3D analytical techniques may help to resolve some of the structural overlap present within thick sections and may also be employed to extract quantitative information about specimen structure from a thick section.

Stereopairs

To take advantage of the information present within thick sections, the specimen is usually photographed at two different tilt angles and viewed in 3D as a stereopair with a stereoviewer. Some authors have used such stereopairs to obtain estimates of surface area and volume of biological structures, e.g., T-tubules in muscle (Arii and Hama 1987). The optimum tilt angles for viewing in 3D depends on the specimen thickness and the overall magnification (Hudson and Makin 1970). The lower the magnification and thinner the specimen, the larger the tilt increment required for optimal 3D. A good practice is to take tilt images in 3s (e.g., +5°, 0°, –5°) so that a range of disparities is available.

Tomography

A second method that has been employed more often in recent years is electron tomography. In this method, a 3D volume reconstruction is obtained by incrementally tilting the specimen through a range of tilts, usually about 60°. A volume reconstruction is then obtained by back-projecting each of the tilt images. Although the resolution of the final volume is inferior to that of the individual micrographs, this method is useful for examining the 3D structure of objects without resorting to collecting and aligning large numbers of serial sections. It is also useful for 3D analysis, since the resulting volumes can be sectioned along any arbitrary plane using computer algorithms. Tomography has recently been combined with serial sectioning techniques to produce volume reconstruction of large structures, e.g., dendrites, at EM resolution (Soto et al. 1994).

Tomography requires a microscope fitted with a eucentric stage and access to specialized digitizing and computational software. Details of tomographic procedures and theory are presented in Frank (1992).

SUGGESTED READINGS

Bendayan M. and Puvion E. 1984. Ultrastructural localization of nucleic acids through several cytochemical techniques on osmium-fixed tissues: Comparative evaluation of the different labelings. *J. Histochem. Cytochem.* **32:** 1185–1191.

Bendayan M. and Stephens H. 1984. Double labelling cytochemistry applying the protein-A gold technique. In *Immunolabeling for electron microscopy* (ed. J.M. Polak and I.M. Varndell), pp. 143–154. Elsevier, New York.

Benhamou N. 1989. Preparation and application of lectin-gold complexes. In *Colloidal gold: Principles, methods and applications* (ed. M.A. Hayat), pp. 96–143. Academic Press, New York.

Eldred W.D., Zucker C., Karten H.J., and Yazulla S. 1983.

Comparison of fixation and penetration enhancement techniques for use in ultrastructural immunocytochemistry. *J. Histochem. Cytochem.* **31:** 285–292.
Glauert A.M., ed. 1972–1992. *Practical methods in electron microscopy*, volumes 1–14. Elsevier, New York.
Hayat M.A. 1989. *Principles and techniques of electron microscopy: Biological applications*, 3rd edition. CRC Press, Boca Raton, Florida.
Hayat M.A., ed. 1970–1978. *Principles and techniques of electron microscopy*, Volumes 1–9. Van Reinhold, New York.
———. 1993. *Stains and cytochemical methods.* Plenum Press, New York.
King J.C., Lechan R.M., Kugel G., and Anthony E.L.P. 1983. Acrolein: A fixative for immunocytochemical localization of peptrides in the central nervous system. *J. Histochem. Cytochem.* **31:** 62–68.
Krang R.F.E. and Klomparens K.L. 1988. *Artifacts in biological electron microscopy.* Plenum Press, New York.
Polak J.M. and Varndell I.M. 1984. *Immunolabeling for electron microscopy.* Elsevier, New York.

REFERENCES

Arii T. and Hama K.. 1987. Method of extracting three-dimensional information from HTVEM stereoimages of biological materials. *J. Electron Microsc.* **36:** 177–195.
Bullock G.R. 1983. The current status of fixation for electron microscopy: A review. *J. Microsc.* **133:** 1–15.
Dykstra M.J. 1992. *Biological electron microscopy.* Plenum Press, New York.
———. 1993. *A manual of applied techniques for biological electron microscopy.* Plenum Press, New York.
Favard P. and Carasso N. 1973. The preparation and observation of thick biological sections in the high voltage electron microscope. *J. Microsc.* **97:** 59–81.
Forbes M.S., Plantholt B.A., and Sperelakis N. 1977. Cytochemical staining procedures selective for sarcotubular systems of muscle: Modifications and applications. *J. Ultrastruct. Res.* **60:** 306–327.
Frank, J. 1992. *Electron tomography.* Plenum, New York.
Glauert A.M. 1974. The high voltage electron microscope in biology. *J. Cell Biol.* **63:** 717–748.
———. 1975. *Fixation, dehydration and embedding of biological specimens.* Elsevier, New York.
Harris J.R. 1991. *Electron microscopy: A practical approach.* Oxford University Press, New York.
Hayat M.A. 1989a. *Principles and techniques of electron microscopy: Biological applications*, 3rd edition. CRC Press, Boca Raton, Florida.
———. 1989b. *Colloidal gold: Principles, methods and applications.* Academic Press, New York.
Hayat M.A. and Miller S.E. 1990. *Negative staining.* McGraw-Hill, New York.
Hudson B. and Makin M.J. 1970. The optimum tilt angle for electron stereo-microscopy. *J. Physics E* **3:** 311.
Kok L.B. and Boon M.E. 1992. *Microwave cookbook for microscripts: Art and science of visualization.* Coloumb Press Leyden, Leiden, The Netherlands.
Mollenhauer H.H. 1964. Plastic embedding mixtures for use in electron microscopy. *Stain Tech.* **39:** 111–114.
———. 1988. Artifacts caused by dehydration and epoxy embedding in transmission electron microscopy. In *Artifacts in biological electron microscopy* (ed. R.F.E Krang and K.L. Klomparens), pp. 43–64. Plenum Press, New York.
Newman G.R. and Hobot J.A. 1993. *Resin microscopy and on-section immunocytochemistry.* Springer-Verlag, New York.
Olschowka J.A. 1988. A simple embedding method for optimizing the thin sectioning of preembedded immunocytochemically stained tissue. *J. Electron Microsc. Tech.* **10:** 373–374.
Reid N. and J.E. Beesley. 1991. *Sectioning and cryosectioning for electron microscopy. Practical methods in electron microscopy,* volume 13. Elsevier, New York.
Sato T. 1968. A modified method for lead staining of thin sections. *J. Electron Microsc.* **17:** 158–159.
Soto G.E, Young S.J., Martone M.E., Deerink T.J., Lamont S., Carragher B.O., Hama K., and Ellisman M. 1994. Serial section electron tomography: A method for three-dimensional reconstruction of large structures. *Neuroimage* **1:** 230–243.
Tokuyasu K.T. 1984. Immuno-cryoultramicrotomy. In *Immunolabeling for electron microscopy* (ed. J.M. Polak and I.M. Varndell), pp. 71–82. Elsevier, New York.
Venable J.H. and Coggeshall R. 1965. A simplified lead citrate stain for use in electron microscopy. *J. Cell Biol.* **25:** 407–408.
Walton J. 1979. Lead aspartate, an en bloc contrast stain particularly useful for ultrastructural enzymology. *J. Histochem. Cytochem.* **27:** 1337–1342.

CHAPTER 21

Immunoelectron Microscopy

Robert Ochs

Pittsburgh, Pennsylvania

INTRODUCTION

Immunoelectron microscopy (IEM) can be easily subdivided into the two general techniques of preembedding and postembedding IEM. With each technique, preparation methods and detection reagents may vary somewhat; these issues are dealt with separately where appropriate. As is usually the case, the choice of method depends on the particular needs and demands of the experiment as well as the tools available. Only the most commonly used methodologies are described here, but this does not imply that other techniques might not also work just as well. In fact, a number of excellent review chapters and books have been devoted to this topic (Bullock and Petrusz 1982; Varndell and Polak 1987; Raska et al. 1990). For more in-depth information, the reader is encouraged to consult these or other sources.

Each section begins with a short theoretical discussion and a list of advantages and disadvantages of the technique, followed by several different detailed protocols. Where appropriate, vendors and reagents are identified by name. This does not imply an endorsement, only that these reagents have been used successfully in IEM experiments. In general, it is advisable to first conduct labeling experiments on cells using immunofluorescence (see Chapters 6–12) before attempting IEM. This is a more feasible approach to work out fixation conditions and antibody labeling procedures for each cell and antigen-antibody system in preparation for IEM. Furthermore, performing immunofluorescence first will also enable the investigator to anticipate the type of labeling pattern or distribution that should be encountered using the electron microscope.

Preembedding IEM

Preembedding IEM is just that: All fixation, permeabilization, and antibody labeling steps are performed prior to embedding cells or tissue. Before colloidal gold secondary probes and hydrophilic resins became widely available, this was the most common technique in use for IEM, and it is still the preferred method for many applications. Some of the advantages and disadvantages of this procedure are as follows.

Advantages

- It provides direct comparison to immunofluorescence procedures.
- It allows high labeling intensity (density).
- After antibody labeling, material can be additionally fixed with glutaraldehyde or osmium tetroxide, affording quality preservation.
- It is technically easy to perform.

Disadvantages

- Cells must be permeabilized, resulting in a loss of membrane integrity and in more soluble elements.
- Specimens must be prepared anew for each additional experiment or antibody labeling procedure.
- Penetration of primary antibodies and secondary detecting reagents can be a problem in well-fixed cells.
- Images can be different from conventional EM, sometimes making interpretations difficult.
- Relatively large amounts of antibody are required.

Postembedding IEM

Postembedding IEM involves antibody labeling of thin sections of embedded cells and tissues on EM grids. In recent years, this has become the most popular method for IEM studies.

Advantages

- Because cells are not permeabilized, preservation is better than for preembedding IEM, especially for membrane-bounded compartments.
- It can be utilized for tissues as well as cells.
- Embedded samples can be resectioned and used for different antibody labeling studies at some later time.
- Small amounts of antibody are required for labeling.
- Images can be comparable to those obtained by conventional EM.

Disadvantages

- Antibodies can have difficulty penetrating sections—only cut surfaces are labeled.
- Labeling intensity may be low.
- Resins used for postembedding can be difficult to polymerize and section.
- Special equipment and unusual techniques are required for resin polymerization.
- Some resins are toxic or allergenic, making them hazardous and making disposal a problem.

PREPARATION OF CELLS OR TISSUES FOR ANTIBODY LABELING

Processing Cells or Tissues

For preembedding IEM, cells can be processed attached as a monolayer or in suspension. Preembedding IEM is not recommended for tissues. Monolayer cells should be grown on a type of plastic that will stand up to the organic solvents generally used for dehydration. Most plastics will tolerate ethanol but not acetone or propylene oxide. Special Lux Permanox dishes (Electron Microscopy Sciences) are available that are resistant to all of these solvents and have the added advantage that polymerized epoxy resin can easily be separated from the plastic surface. After fixation or labeling, it is also possible to scrape cells with a plastic cell scraper or rubber policeman and then form a cell pellet by centrifugation.

For postembedding IEM, cultured cells can be scraped or trypsinized to produce a cell pellet, fixed, embedded in warm 1–2% agar, cooled to solidify, and then cut into small pieces and processed like tissue. Pieces, whether tissue or agar-containing cells, should be no larger than 1–2 mm^3 to allow for maximum penetration of resin. Alternatively, cells can be grown on Thermanox coverslips that are cut to fit into JB-4 molds (Polysciences) for polymerization.

Fixation

The goal of fixation in any immunocytochemistry procedure is to preserve structure as close to the "living state" as possible and yet allow specific antibody labeling without denaturation of the antigen in question. The simultaneous achievement of these goals is, of course, difficult and therefore any fixation method is really a trade-off between preserving structure and retaining antigenicity for antibody binding. Basically, one should start with better fixed material and proceed to less well fixed samples—to the point where labeling is specific without significant background.

The preferred fixative for IEM is phosphate-buffered 2–4% **formaldehyde**, which is made fresh from dry paraformaldehyde (a polymer of formaldehyde) by heating in distilled water in a fume hood, adding NaOH to clear, and then adjusting the volume with a concentrated (10x) PBS (phosphate-buffered saline) stock solution (to make 1x PBS) and adjusting the pH to 7.2–7.4. Fixation may be performed at 4°C or room temperature and in general for 15 minutes to 1 hour. Formaldehyde is a monoaldehyde, and aldehydes fix proteins by cross-linking amine groups by introducting inter- and intramolecular methylene bridges. The dialdehyde **glutaraldehyde** can be used with formaldehyde in low concentrations of 0.01–0.5%. Higher concentrations of glutaraldehyde cross-link so efficiently that proteins are fully denatured.

CAUTION: Formaldehyde; glutaraldehyde (see Appendix 2 for Caution)

Alcohols (methanol or ethanol) or acetone are generally unusable as primary fixatives for IEM except where antigens are extremely fixation-sensitive or where their distribution is artifactually changed by aldehyde fixation. These organic solvents fix proteins in place by precipitation. Their actions are usually enhanced by lower temperatures of 4°C or –20°C. Denaturation is mild, antigenicity is high, and some shrinkage occurs.

Permeabilization

For preembedding IEM labeling of intracellular antigens, it is necessary to permeabilize the plasma membrane of fixed cells with nonionic detergents or cold organic solvents. Detergent permeabilization of unfixed cells is not recommended because of the danger of artifactually changing the distribution of intracellular antigens before fixation. Commonly used detergents are Triton X-100 at 0.01–0.5% and the milder detergents, saponin or digitonin.

Reducing Agents

After fixation, it is sometimes necessary to reduce nonreactive aldehyde groups, which otherwise will bind antibodies nonspecifically, resulting in high background staining. Reducing agents include 50 mM NH_4Cl, 35 mM glycine, and 0.1% $NaBH_4$.

Blocking Solutions

To prevent antibodies from binding nonspecifically to cellular or tissue components, different blocking solutions have been employed. In general, both primary and secondary antibodies are also diluted in these blocking solutions. Typical blocking agents include normal serum of the same species as was used to generate the secondary antibody (i.e., normal goat serum if the secondary antibody was produced in a goat) at a concentration of 1–10%, BSA (bovine serum albumin) (Fraction V) at 1–10%, 1–10% cold water fish skin gelatin (available from Electron Microscopy Sciences), or 1–5% nonfat dried milk (also called "Blotto").

ANTIBODIES

Primary antibodies should be well characterized. It is essential whenever possible to prove antibody specificity by immunoblotting and subsequent affinity purification (Olmsted 1981). Monoclonal antibodies have the advantage (or sometimes disadvantage) of recognizing a single epitope, whereas polyclonal antibodies are often of higher titer and avidity and may recognize multiple epitopes, some of which may be more resistant to denaturation by fixation. When dealing with antigens for which antibody production is difficult or impossible (because the immunogen is unavailable or the antigen is not very immunogenic), autoantibodies are sometimes the only antibodies available. This is especially true for many nuclear proteins (see Raska et al. 1990). Care must be taken because many autoantibodies may not be monospecific. A cautionary note should also be given in regard to antipeptide antibodies, which may cross-react with several proteins because of similar primary amino acid sequences. Whichever antibody is chosen, specificity is of the utmost importance. Antibodies of a defined specificity should be used in the range of 5–50 μg/ml.

Controls

Positive and negative antibody controls are extremely important, especially when localizing unknown antigens. Positive controls, which test for proper working of the method and reagents, should consist of previously characterized antibodies whose localization is known. Negative controls ideally should consist of no primary antibody and preabsorption of specific antibody with an excess of purified antigen.

ANTIBODY LABELING PROCEDURES AND DETECTION REAGENTS (PROBES)

Besides the choice of pre- or postembedding, antibody labeling methods are usually of only three types: direct, indirect, or amplified. For each of these, antibody detection can be based on either a soluble probe (usually enzyme-based) or a particulate probe (most commonly colloidal gold). Each of these methods is discussed in the context of the most commonly employed procedures.

As with any good experiment, the importance of proper controls cannot be stressed enough! At the minimum, negative controls should include preabsorbed primary antibody or labeling with secondary antibody (or probe) alone. Especially when dealing with previously uncharacterized antibodies or antigens, positive controls should be included to assure proper working of the reagents, etc. Whenever possible, affinity-purified primary and secondary antibody reagents should be used, and these should be previously characterized by immunoblotting.

Indirect Peroxidase

For preembedding IEM, indirect labeling with secondary antibodies coupled to horseradish peroxidase is one method of choice because the antibodies penetrate fixed and permeabilized cells and tissues with ease, making this a very sensitive technique. There are, however, a number of disadvantages to this approach. For example, because there is no fixed end point to the peroxidase reaction, the intensity of label is arbitrary, endogenous peroxidases are a concern, the peroxidase reaction product obscures the object being labeled (Fig. 21.1), quantitation is not possible, and secondary stains such as lead citrate and uranyl acetate cannot be used because they may mask the electron-dense peroxidase reaction product.

ABC (Avidin-Biotin Complex)

Another commonly used procedure for preembedding IEM is the so-called ABC peroxidase method (Hsu et al. 1981). (This technique has also gained popularity for use in the light microscopic immunocytochemical staining of paraffin-embedded tissues.) In this three-step amplified procedure, primary antibody is followed by biotinylated secondary antibody, which is then followed by the peroxidase-coupled preformed avidin-biotin complex (the ABC reagent). This method has been standardized and is available in a kit from Vector Labs. Because it is an amplified technique, labeling intensity is very heavy and, consequently, primary antibodies can be highly dilute, saving reagents and diluting out any nonspecific reactivities that may also be present. Aside from the drawbacks mentioned above for the indirect peroxidase procedure, other disadvantages include high background (perhaps due to endogenous biotin found in culture media, etc.) and the necessity for a different biotinylated secondary antibody for each different species of primary antibody (but the ABC reagent is universal). In addition, the detecting moiety in this sandwich technique is somewhat distant spatially from the site of primary antibody interaction.

Colloidal Gold

Although not considered particularly useful for preembedding applications because of poor penetration, colloidal gold probes have been utilized for both pre- and postembedding IEM. A conven-

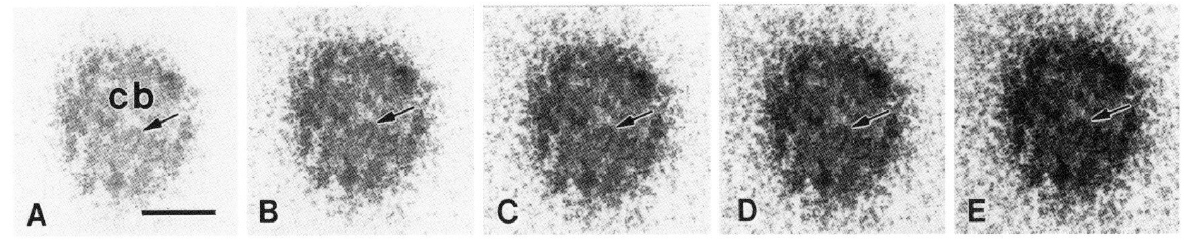

FIGURE 21.1

Preembedding indirect immunoperoxidase labeling of a nuclear coiled body (cb) using an antibody to the coiled body-specific protein p80-coilin. This photographic series at different exposure times illustrates how the peroxidase reaction product, although resulting in heavy labeling, obscures the structure(s) being labeled—in this case, fibrils of the coiled body (*arrows*). (*A*) 10-sec exposure; (*B*) 15-sec exposure; (*C*) 20-sec exposure; (*D*) 25-sec exposure; (*E*) 30-sec exposure. Bar, 0.5 µm. (Reprinted, with permission, from Ochs et al. 1994.)

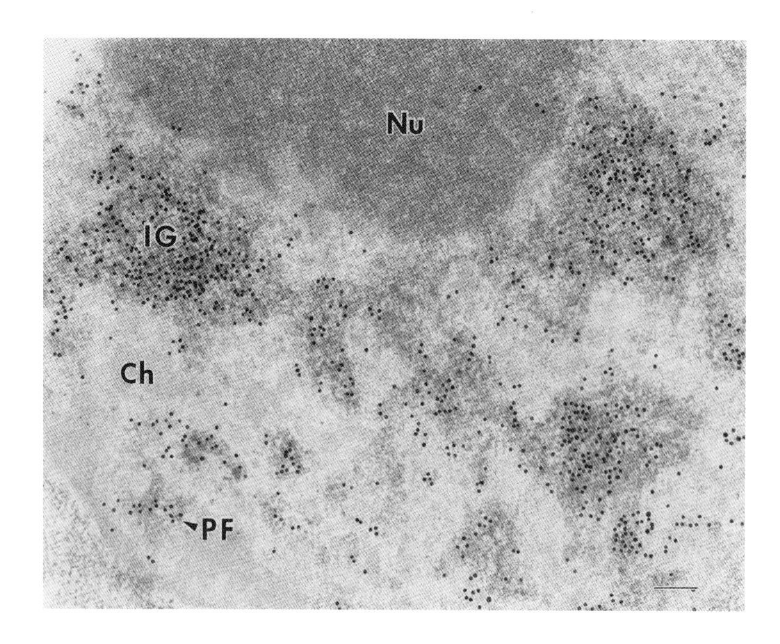

FIGURE 21.2

Cell section embedded in LR White and immunogold labeled by the postembedding method using a 15-nm colloidal gold-conjugated secondary antibody. The pre-mRNA splicing factor SC35 is localized to clusters of interchromatin granules (IG) and perichromatin fibrils (PF). The nucleolus (Nu) as well as the chromatin (Ch) shows little to no immunolabeling. Bar, 200 nm. (Photo provided by D. L. Spector, Cold Spring Harbor Laboratory.)

ient range of sizes is commercially available, down to 1 nm in diameter. Being particulate, very electron-dense, and uniform in size, colloidal gold probes are quantifiable and do not obscure the structure being labeled (Fig. 21.2). Furthermore, they can be easily visualized after secondary staining with lead and uranyl ions. Multiple labeling is also possible using gold of different sizes. Despite these advantages, there are also significant disadvantages. In many cases, labeling intensity is low, especially in postembedding techniques, primarily because only the cut surface of a thin section is available for antibody binding. Another concern stems from the fact that antibodies are electrostatically adhered to the surface of the colloidal gold particles (Geoghegan and Ackerman 1977) and, as a result, there is the potential for bound antibodies to be released from the surface of the particles, destabilizing the colloid and rendering the probe ineffective for specific antibody interactions. Additionally, advances in technology have made 1-nm colloidal gold probes routinely available (Slot and Geuze 1985). These ultrasmall particles result in increased label density and better penetration, and they have found limited use in some preembedding applications. These smaller probes can also be silver-enhanced in situ to make a larger and more visible gold particle.

Nanogold

Nanogold (available from Nanoprobes) is a 1.4-nm gold particle to which antibodies are covalently bound. Since it is not a colloid, Nanogold is more hydrophilic than colloidal gold preparations, with correspondingly better penetration properties. There is also the added advantage that bound antibodies or proteins will not detach from the probe itself. Because of its good penetrating ability and small size, Nanogold has even been used successfully in preembedding IEM localization of nuclear proteins (Fig. 21.3) (Ochs et al. 1994). Disadvantages include the fact that Nanogold must be silver-enhanced for visibility and, as a result, multiple labeling is not possible.

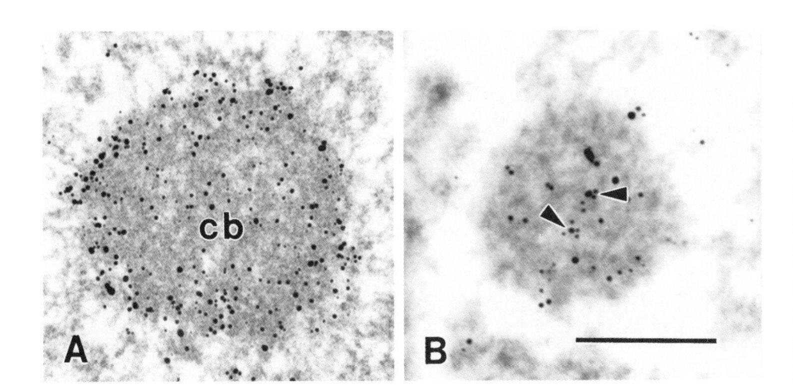

FIGURE 21.3

Preembedding (*A*) and postembedding (*B*) indirect immunogold labeling of coiled bodies (cb) with 1.4-nm Nanogold enhanced with silver. Note that the preembedding technique (*A*) results in much heavier and more uniform labeling compared to the postembedding technique (*B*), which labels only exposed surface antigens in a nonuniform distribution (*arrowheads* in *B*). Bar, 1 µm. (Photo provided by R. Ochs.)

Amplification of Gold Probes: Biotin-Avidin or Biotin-Antibiotin

Various methods are available for signal amplification. This is especially important in postembedding IEM where the number of gold particles is inherently low. The most commonly used procedures involve biotinylated secondary antibodies followed by avidin-gold or anti-biotin-gold. Antibiotin can bind all over the biotin molecule, allowing the most amplification, and, as alluded to previously, the resultant gold particle is localized some distance from the actual antigenic site. As with all amplification techniques, correspondingly higher background labeling is also possible.

Immunolabeling for EM Using Fluorescent Labels

Although fluorescent molecules cannot be visualized directly in the electron microscope, there exist several methods for correlated fluorescent and electron microscopy. In the first method, photooxidation, a fluorophore is excited in the presence of DAB (diaminobenzidine tetrahydrochloride) and molecular oxygen, and the fluorescence is gradually replaced by oxidized DAB. The resulting reaction product is a very fine precipitate that provides excellent resolution under the electron microscope. Deerinck et al. (1994) have determined that the most commonly used fluorophores for fluorescent immunolabeling, fluorescein and rhodamine, oxidize DAB very poorly. However, they demonstrated that eosin, a brominated fluorescein derivative, is an effective oxidizer of DAB and also exhibits reasonable fluorescence at the light microscopic level. Eosin-IgG conjugates have an absorbance maxima at 525 nm and an emission maxima at 545 nm. This method provides superior resolution at the electron microscopic level compared to peroxidase methods because the DAB does not appear to diffuse very far from the site of deposition. In addition, the small size of the eosin molecule compared to ABC, PAP (peroxidase-antiperoxidase), or colloidal gold results in greater penetration of immunolabeling reagents into the tissue. For detailed methods for conjugation of eosin to IgG and for photooxidation of immunolabeled samples, see Deerinck et al. (1994).

Although fluorescein is not useful for photooxidation of DAB, fluorescein-labeled specimens may be viewed under the electron microscope by using antifluorescein antibodies conjugated to either peroxidase or colloidal gold (van Dam et al. 1991). In addition, fluorescein and goat-anti-mouse Fab′ fragments have been covalently coupled directly to Nanogold (Powell et al. 1997). This reagent allows direct visualization of the localization of an antigen, first by fluorescence microscopy and subsequently by electron microscopy (Powell et al. 1997).

RESINS

Resins employed in postembedding IEM are mostly hydrophilic in nature, usually acrylic polymers. Although hydrophilic resins are good for retaining antigenicity and penetration of aqueous antibody solutions, polymerization and sectioning can be problematic.

Epon

Epoxy resins are hydrophobic in nature, and the temperature necessary for their polymerization (60°C) denatures most antigens, making this resin generally unacceptable for postembedding IEM. Etching of the resin can be accomplished with 10% H_2O_2, saturated sodium metaperiodate, or saturated ethoxide, but additional morphological damage and protein denaturation can occur because of the etching compounds themselves.

LR White

LR White is a hydrophilic acrylic polymer when polymerized by heating at 60°C. The polymerization temperature may denature some antigens. Optimum polymerization can be tricky, with blocks being either "sticky" (underpolymerized) or "brittle" (overpolymerized). For optimum results, oxygen should be excluded because it inhibits polymerization, and embedding in gelatin capsules is therefore recommended. Thin sections of LR White-embedded material demonstrate good stability in the electron beam. LR White has the added advantage of low toxicity and irritability, especially when compared to the Lowicryl resins.

Lowicryl

Lowicryl K4M is a low-temperature hydrophilic embedding medium that is polymerized at about –20°C with UV light. Since polymerization is achieved at low temperature, this embedding medium avoids the heat denaturation that can occur with other resins. For best polymerization, it is recommended to buy a special UV polymerization chamber (available from Ted Pella) that achieves low temperatures using dry ice. Like LR White (Fig. 21.2), Lowicryl sections are stable in the beam and images are very comparable to stained Epon sections. As with LR White, the optimum degree of polymerization may be somewhat difficult to achieve, requiring adjustments in temperature of polymerization, length of polymerization, and amount of cross-linker.

CAUTION: Lowicryl (see Appendix 2 for Caution)

Preembedding IEM: Indirect Peroxidase and ABC Peroxidase

This procedure is modified from Spector et al. (1991) and Ochs and Press (1992). (See Fig. 21.1.)

1. Place a single coverslip in a 35-mm diameter petri dish with 3 ml of medium.
   Cells are usually grown on 22-mm square No. 1-1/2 glass coverslips.

2. After 2–3 days in culture, rinse monolayer cells with PBS and fix at room temperature for 30 minutes with 3% **formaldehyde** (made fresh from paraformaldehyde) and varying amounts of **glutaraldehyde** (0.01–0.5%) buffered with PBS (pH 7.4).

   CAUTION: Formaldehyde; glutaraldehyde (see Appendix 2 for Caution)

   The acceptable concentration of glutaraldehyde to allow detection of each antigen will vary, but the higher the concentration of glutaraldehyde, the better the fixation. One sample should be run without any glutaraldehyde as a positive control, since most antigens are accessible with this fixation. The fixative should be used at the pH of the medium in which the cells are grown.

3. Rinse the cells with PBS three times for 10 minutes each and quench with 35 mM glycine or 1 mg/ml $NaBH_4$ made in PBS three times for 5 minutes each.

4. Permeabilize the cells with 0.2% Triton X-100 in PBS at room temperature for 5 minutes.

5. Rinse the cells in PBS three times for 5 minutes each.

6. Block the cells with 1% normal goat serum in PBS (NGS [normal goat serum]/PBS) at room temperature for 30 minutes.
   If endogenous peroxidase activity is a problem, aldehyde-fixed cells or tissue can be treated with 0.01 M periodic acid for 10 minutes followed by 0.1 mg/ml $NaBH_4$ in H_2O for 10 minutes. Wash sample in PBS and proceed to incubate in primary antibody.

7. Incubate cells with primary antibody diluted in 1% NGS (normal goat serum)/PBS at room temperature for 1 hour or at 4°C overnight.

8. Rinse the cells with PBS three times for 5 minutes each.

9. Block the cells with 1% NGS/PBS for 30 minutes.
   For the ABC procedure, an additional secondary antibody incubation step is included here consisting of affinity-purified biotinylated antibody reactive to the species of the primary antibody (i.e., biotinylated anti-rabbit IgG if the primary antibody was made in a rabbit). As with the other antibody steps, the antibody is diluted in 1% NGS/PBS, incubated at room temperature for 1 hour, rinsed in PBS three times for 5 minutes each, and then blocked with 1% NGS/PBS for 30 minutes.

10. Incubate the cells with affinity-purified secondary antibody coupled to peroxidase diluted 1:50 in 1% NGS/PBS at room temperature for 1 hour.
    For the ABC procedure, the ABC reagent (Vector Labs) is substituted for the peroxidase-conjugated secondary antibody.

11. Rinse the cells in PBS three times for 5 minutes each.

12. Fix the cells with 1% glutaraldehyde buffered with PBS for 30 minutes at room temperature.

13 Rinse the cells in PBS three times for 5 minutes each.

14 Rinse the cells in 50 mM Tris-HCl (pH 7.6) (Tris buffer) three times for 5 minutes each.

15 Incubate the cells in 1 mg/ml **DAB** (available in preweighed vials from Polysciences)/0.03% H_2O_2 (made from 30% stock) in Tris buffer. The incubation time will vary depending on the antigen. Place the petri dish on the stage of an inverted microscope and observe the cells using bright-field illumination (not phase or Nomarski, as they will impart contrast that may be confused with the reaction product) every 5 minutes up to a total of 30 minutes. Keep the cells covered with foil between observations. The reaction should be stopped when a specific brown reaction product is observed. The longer the incubation time in DAB, the greater the possibility of increased background signal.

CAUTION: DAB (see Appendix 2 for Caution)

16 Rinse the cells with Tris buffer three times for 5 minutes each.

17 Postfix the cells with 2% OsO_4 in 0.1 M cacodylate buffer, pH 7.3, for 30 minutes.

Osmium tetroxide has a high vapor pressure and should always be handled in an efficient fume cabinet. OsO_4 is one of the most dangerous fixatives as it is capable of fixing in the vapor form as well as in the liquid form. Be particularly careful of your eyes and do not breathe the vapor. Solutions of OsO_4 should be made up only in glass containers because OsO_4 can penetrate plastic. Proper disposal of osmium tetroxide requires changing the oxidative state of the osmium so that it no longer is in the tetroxide configuration. This is achieved by a two-step process. First, NaOH is added to the OsO_4 until a pink color appears and then methanol is added to a blue end point. Methanol acts as an electron pair donor to the osmium, changing the reactive state of the metal. Check with your safety department for any special handling procedures that they may require.

Cacodylate buffer contains arsenic and can be absorbed through the skin and/or by droplet inhalation. Reaction with reducing agents can produce volatile arsenical compounds.

18 Rinse the cells in distilled H_2O three times for 5 minutes each.

If necessary, cells can be kept at this step overnight at 4°C.

19 Dehydrate the cells in a graded series of ethanol and embed in Polybed 812 (Polysciences) (see Chapter 20).

20 Examine cells: *Embedded cells* can be "popped out" of the dish, examined in the light microscope, have specific areas marked for sectioning, have marked areas cut or punched out, be remounted on a blank Epon block, and be thin-sectioned parallel to the substrate as monolayers. *Sections* should be examined unstained in the electron microscope.

Preembedding IEM: Nanogold

This procedure is modified from Ochs et al. (1994). (See Fig. 21.3.)

1 Process monolayer cells as in steps 1–9 of the preceding procedure for preimbedding IEM to fix, block, and incubate overnight in primary antibody.

2 Rinse the cells in PBS three times for 5 minutes each.

3 Block cells in 1% NGS/PBS for 30 minutes.

4 Incubate the cells in 1.4-nm Nanogold (Nanoprobes) covalently linked to anti-human, anti-mouse, or anti-rabbit Fab′ diluted 1/100 in 1% NGS/PBS for 1 hour at room temperature with shaking.

5 Rinse the cells in PBS three times for 5 minutes each.

6 Fix the cells in 1% **glutaraldehyde**/PBS for 30 minutes.

CAUTION: Glutaraldehyde (see Appendix 2 for Caution)

7 Rinse the cells in PBS three times for 5 minutes each.

8 Rinse the cells well in distilled H_2O three times for 5 minutes each.

9 Silver-enhance the samples with HQ Silver (according to manufacturer's directions [Nanoprobes]) for 3–4 minutes in the dark at room temperature.

Longer times will result in larger silver-coated gold particles.

10 Rinse the samples with H_2O three times for 5 minutes each to stop the silver reaction.

11 Osmicate the samples with 1% OsO_4 in 0.1 M cacodylate buffer, pH 7.3, for 30 minutes.

Osmium tetroxide has a high vapor pressure and should always be handled in an efficient fume cabinet. OsO_4 is one of the most dangerous fixatives as it is capable of fixing in the vapor form as well as in the liquid form. Be particularly careful of your eyes and do not breathe the vapor. Solutions of OsO_4 should be made up only in glass containers because OsO_4 can penetrate plastic. Proper disposal of osmium tetroxide requires changing the oxidative state of the osmium so that it no longer is in the tetroxide configuration. This is achieved by a two-step process. First, NaOH is added to the OsO_4 until a pink color appears and then methanol is added to a blue end point. Methanol acts as an electron pair donor to the osmium, changing the reactive state of the metal. Check with your safety department for any special handling procedures that they may require.

Cacodylate buffer contains arsenic and can be absorbed through the skin and/or by droplet inhalation. Reaction with reducing agents can produce volatile arsenical compounds.

12 Rinse the samples with H_2O three times for 5 minutes each.

13 Dehydrate the samples with ethanol and embed in Polybed 812 (see Chapter 20).

14 Examine thin sections stained with lead citrate and uranyl acetate (see Chapter 20) or unstained.

Gold particle labeling may be quantitated by planametric analysis of EM negatives and expressed as number of gold particles/μm^2.

Postembedding IEM: LR White or Lowicryl K4M

LR White and Lowicryl K4M are both hydrophilic resins used for postembedding IEM. Sample fixation, dehydration, resin infiltration, sectioning, and antibody labeling are identical, but polymerization conditions differ substantially.

POSTEMBEDDING IEM OF MONOLAYER CELLS GROWN IN FLASKS

Process Cells for Postembedding IEM with Lowicryl K4M

1 Rinse the cells rapidly with PBS and fix in 2–4% **formaldehyde**/PBS containing varying amounts of **glutaraldehyde** (0.01–0.5%) at room temperature for 30 minutes to 1 hour.

CAUTION: Formaldehyde; glutaraldehyde (see Appendix 2 for Caution)

2 Rinse the cells with PBS three times for 5 minutes each and quench with 35 mM glycine/PBS three times for 5 minutes each.

3 Rinse the cells with PBS three times for 5 minutes each, scrape the cells from the bottom of the flask with a cell scraper, and centrifuge into a pellet.

4 Embed cell pellet in warm 1% agar/PBS, recentrifuge, and cool on ice.

5 Cut cell pellet into small 1–2 mm^3 pieces.

6 Dehydrate the pieces in ethanol according to the following schedule:

30% at 4°C for 30 minutes
50% at –20°C for 60 minutes
70% at –20°C for 60 minutes
95% at –20°C for 60 minutes
100% at –20°C for 60 minutes two times

7 Infiltrate the pieces with ethanol/resin mixture according to the following schedule:

1:1 at –20°C for 60 minutes
1:2 at –20°C for 60 minutes
100% resin at –20°C for 60 minutes
100% resin at –20°C overnight

8 Fill gelatin capsules to the top with pure resin, transfer pieces of tissue or cells to the capsules, and then place the top on to exclude as much air as possible.

9 Polymerize.

Place gelatin capsules in the specially built **UV** cryochamber (Pelco, Ted Pella) precooled to –30°C with four blocks of dry ice. Polymerize with UV at –20°C for 2 days and at room temperature for 1 day.

CAUTION: UV radiation (see Appendix 2 for Caution)

Soft blocks can be polymerized for longer periods of time.

10 Collect thin sections on the dull side of gold Veco grids (Electron Microscopy Sciences).

Immunolabel

11 Float individual grids (with sections down) for 30 minutes on the surface of small drops of blocking solution consisting of 1–10% BSA or NGS in PBS.

For best results, immunolabeling is performed on a large sheet of parafilm on a flat surface.

12 Transfer grids (without blotting) to drops containing primary antibody made in blocking buffer at room temperature for 2 hours or at 4°C overnight.

13 Rinse grids with PBS five times for 5 minutes each.

14 Block in 1% NGS/PBS for 30 minutes.

15 Incubate grids in secondary antibody linked to colloidal gold or Nanogold at room temperature for 1 hour.

If colloidal gold is used, a particle size of 5–30 nm can be visualized at most magnifications. If 1-nm colloidal gold or Nanogold is used, silver enhancement is recommended as described previously in the Nanogold preembedding IEM procedure (see p. 353).

As noted in the chapter introduction, a number of different modifications may be made in the procedure. These include amplification of signal with multistep biotin-avidin or biotin-antibiotin procedures or multiple labeling with gold particles of different sizes.

16 Rinse grids in PBS five times for 5 minutes each.

17 Fix in 1% glutaraldehyde/PBS for 30 minutes at room temperature.

18 Rinse grids in PBS three times for 5 minutes each.

19 Rinse grids in distilled H_2O three times for 5 minutes each.

20 Counterstain with **uranyl acetate** and lead citrate (see Chapter 20), rinse in distilled H_2O, and blot dry.

CAUTION: Uranyl acetate (see Appendix 2 for Caution)

POSTEMBEDDING IEM OF MONOLAYER CELLS GROWN ON COVERSLIPS AND EMBEDDED IN LR WHITE

This procedure is modified from Spector et al. (1991). (See Fig. 21.2.)

Process Cells for Postembedding IEM

1 Grow cells on Thermanox coverslips cut to fit into JB-4 molds (Polysciences) (12 mm x 16 mm x 5 mm).

Coverslips are cleaned with 100% ethanol (dip and wipe on filter paper three times and then dip and air-dry).

2 Fix the cells in 2% **formaldehyde** and varying concentrations of **glutaraldehyde** (0.01–0.5%) at room temperature for 20 minutes.

Caution: Formaldehyde; glutaraldehyde (see Appendix 2 for Caution)

3 Wash the cells in PBS containing 0.3 M glycine three times for 10 minutes each.

4 Wash the cells in PBS one time for 10 minutes.

5 Dehydrate in the following series of graded ethanol:

70% one time at room temperature for 7 minutes
80% one time at room temperature for 7 minutes
90% one time at room temperature for 7 minutes
95% two times at room temperature for 7 minutes
100% two times at room temperature for 7 minutes

6 Transfer coverslips to glass petri dish or Thermanox petri dish.

7 Infiltrate samples on shaker according to the following routine:

LR White (hard) 50:50 with 100% ethanol for several hours
change to 100% LR White for overnight
change LR White several times the next day

8 Place coverslip in JB-4 mold cells up, add fresh LR White, and position chuck in mold.
Before using chuck, back off screw so it is flush with the inside of the chuck wall. Cut off outside portion of screw that is not needed and seal the remaining portion of screw in place with Krazy glue. This will provide a sealed chuck so that no air enters the LR White mold.

9 Polymerize samples at 60°C for at least 2 days.

10 Remove block and take coverslip off block with a razor blade. Clean chucks with 100% **methanol**.

Caution: Methanol (see Appendix 2 for Caution)

11 Remount portions of the block containing cells on blank Epon blocks and face and then thin-section regions containing cells of interest with a diamond knife.

Antibody labeling

12 Collect LR White sections on the dull side of gold Veco grids (Electron Microscopy Sciences).
If blocking is necessary for nonspecific labeling, buffer is supplemented with BSA and NGS. Start with a 2% block and increase if necessary up to 10%. Grids are floated face down on drops of blocking solution in Tris-buffered saline with 1.0% Tween-20 and incubated at room temperature for 60 minutes.

13 Transfer grids to drops of appropriately diluted primary antibody in Tris-buffered saline (pH 7.6) with 1.0% Tween-20. Incubate grids in a humidified chamber overnight in the refrigerator, approximately 4°C.

Tris-buffered Saline (pH 7.6)

0.02 M Tris
0.15 M sodium chloride
0.02 M **sodium azide**
1.0% Tween-20
Adjust pH to 7.6.

CAUTION: Sodium azide (see Appendix 2 for Caution)

14 Allow the grids and solutions to come to room temperature after overnight incubation.

15 Wash grids ten times for 1 minute each (or longer and more frequently, if needed) in Tris-buffered saline.

16 Transfer grids to appropriate colloidal gold-labeled secondary antibody diluted 1:20 in Tris-buffered saline with 1.0% Tween-20.

Microfuge diluted gold-labeled secondary antibody for 30 seconds to 1 minute at full speed to get rid of gold clumps.

17 Incubate grids for 60 minutes at room temperature.

18 Wash grids ten times for 1 minute each in Tris-buffered saline.

19 Wash grids ten times for 1 minute each in ddH_2O, and then counterstain (if desired) for electron microscopy with **uranyl acetate** and lead citrate (see Chapter 20).

Caution: Uranyl acetate (see Appendix 2 for Caution)

DOUBLE LABELING WITH COLLOIDAL GOLD

Sections may be double labeled with primary antibodies raised in two different species by using secondary antibodies conjugated to different sizes of colloidal gold, e.g., 5 nm versus 10 nm. The same procedure is followed as for single labeling, but a cocktail containing the two primary antibodies is used in step 13 and a cocktail of the two secondary antibodies is used in step 16 of the above protocol. Species-specific secondary antibodies should be used to avoid cross-reactivity. Proper controls to ensure that no cross-reactivity occurs, e.g., incubating each primary with the incorrect secondary, should also be performed. If two primary antibodies from the same species need to be used, sections may be double labeled by labeling one side of the section with the first antibody and one size colloidal gold and the other side with the second antibody and a different size colloidal gold (Bendayan and Stephens 1984). The first label is applied by floating the grid on a drop of antibody, taking care that the grid does not sink. After labeling, the grid is blotted dry and labeled on the second side.

REFERENCES

Bendayan M. and Stephens H. 1984. Double labelling cytochemistry applying the protein-A gold technique. In *Immunolabeling for electron microscopy* (ed. J.M. Polak and I.M. Varndell), pp. 143–154. Elsevier, New York.

Bullock G.R. and Petrusz P. 1982. *Techniques in immunocytochemistry*, vol. 1. Academic Press, London.

Deerinck T.J., Martone M.E., Lev-Ram V., Green D.P.L., Tsien R.Y., Spector D.L., Huang S., and Ellisman M. 1994. Fluorescence photooxidation with eosin: A method for high resolution immunolocalization and in situ hybridization detection for light and electron microscopy. *J. Cell. Biol.* **126:** 901–910.

Geoghegan W.D. and Ackerman G.A. 1977. Adsorption of horseradish peroxidase, ovomucoid and anti-immunoglobulin to colloidal gold for the indirect detection of concanavalin A, wheat germ agglutinin and goat anti-human immunoglobulin G on cell surfaces at the electron microscopic level: A new method, theory and application. *J. Histochem. Cytochem.* **25:** 1187–1200.

Hsu S.M., Raine L., and Fanger H. 1981. Use of avidin-biotin-peroxidase complex (ABC) in immunoperoxidase techniques: A comparison between ABC and unlabelled antibody (PAP) procedures. *J. Histochem. Cytochem.* **29:** 577–580.

Ochs R.L. and Press R.I. 1992. Centromere auto-antigens are associated with the nucleolus. *Exp. Cell Res.* **200:** 339–350.

Ochs R.L., Stein, Jr., T.W., and Tan E.M. 1994. Coiled bodies in the nucleolus of breast cancer cells. *J. Cell Sci.* **107:** 385–399.

Olmsted J.B. 1981. Affinity purification of antibodies

from diazotized paper blots of heterogeneous protein samples. *J. Biol. Chem.* **256:** 11955–11957.

Powell R.D., Halsey C.M.R., Spector D.L., Kaurin S.L., McCann J., and Hainfeld J.F. 1997. A covalent fluorescent-gold immunoprobe: Simultaneous detection of a pre-mRNA splicing factor by light and electron microscopy. *J. Histochem. Cytochem.* **45:** 947–956.

Raska I., Ochs R.L., and Salamin-Michel L. 1990. Immunocytochemistry of the cell nucleus. *Electron Microsc. Rev.* **3:** 301–353.

Slot J.W. and Geuze H.J. 1985. A new method of preparing gold probes for multiple-labeling cytochemistry. *Eur. J. Cell Biol.* **38:** 87–93.

Spector D.L., Fu X.-D., and Maniatis T. 1991. Associations between distinct pre-mRNA splicing components and the cell nucleus. *EMBO J.* **10:** 3467–3481.

van Dam G.J., Bogitsh B.J., Fransen J.A.M., Kornelis D., van Zuyl R.J.M., and Deelder A.M. 1991. Application of the FITC-anti-FITC-gold system to ultrastructural localization of antigens. *J. Histochem. Cytochem.* **39:** 1725–1728.

Varndell I.M. and Polak J.M. 1987. EM immunolabelling. In *Electron microscopy in molecular biology* (ed. J. Sommerville and U. Scheer), pp. 179–200. IRL Press, Oxford.

APPENDIX 1

Microscopy: Lenses, Filters, and Emission/Excitation Spectra

TABLE A1.1 Excitation and emission maxima for typical fluorochromes

Fluorochrome	Exc (nm)	Em (nm)	Fluorochrome	Exc (nm)	Em (nm)	Fluorochrome	Exc (nm)	Em (nm)
Acridingelb	470	550	BAO (Ruch)	280	460	Cyanine Cy3	575	605
Acridinorange + DNA	502	526	BCECF	430/480	520	Cyanine Cy5	640	705
Acridinorange + RNA	460	650	Berberinsulfat	430	550	Dansylchlorid	380	475
Acriflavin-Feulgen	480	550–600	BFP (Blue Fluoresc. Protein)	380	440	DAPI + DNA	359	461
Alexa 350	346	445	BOBO-1	462	481	DiBAC$_4$ Dibutylbarbitursäure	439	516
Alexa 405	401	421	BOBO-3	570	602	Trimethinoxonol		
Alexa 430	428	545	BODIPY	503	512	DIDS Thiocyanatostilbene	342	418
Alexa 488	494	519	BODIPY 581/591 phalloidin	584	592	DilC$_{3/4/12/22}$ Indocarbocyanine	540–560	556–575
Alexa 532	525	555	BODIPY 630/650	625	650	DiOC$_{7/16/2}$ Carbocyanine	550	580
Alexa 546	554	570	Calcein	495	500–550	DiSC$_{1/2/3/6}$ Thiacarbocyanine	559	585
Alexa 555	550	565	Calcein blue	375	420–450	Diaminonaphtylsuäfonsäure	340	525
Alexa 568	578	603	Calcium Crimson	588	611	DsRed	558	583
Alexa 594	590	615	Calcium Green	554	575	DTAF	495	528
Alexa 633	632	650	Calcium Orange	506	531	Ethidiumbromid + DNA	510	595
Alexa 647	650	668	Calcofluor White	440	500–520	Euchrysin	430	540
Alexa 660	663	690	Cascade blue	376,399	423	Evans Blue	550	611
Alexa 680	679	702	Catecholamine	410	470	Feulgen	480	560
Alizarinkomplexon	595		CFP	434	477	Fluo-3	480	520
Allphycocyanin	630	660	Chromomycin/Mitramycin	436–460	470	Fluoresceindiacetat FDA	499	
AMCA	345	425	Coriphosphin	460	575	Fluoresceinisothiocyanat FITC	490	525
7-Amino-Actinomycin D	555	655	CPM	385	471	Fluoro-Gold	350–395	530–600
Amino-Methylcumarin	354	441	CTC 5-cyano-2,3-ditolyl-	602		Blue FluoSpheres	360	415
6-Amino-Quinolin	360	443	Tetrazolium-Chlorid			Crimson FluoSpheres	625	645
Auramin	460	550	Cyanine Cy2	489	505	Red FluoSpheres	580	605

Dark Red FluoSpheres	650	690	NBD - Amine	460–485	534–542	Rhodol green	500	525
Yellow-Green FluoSpheres	490	515	NBD - Chlorid	480	510–545	Säurefuchsin	540	630
Fura Red (Ca)	436	657	Nil Rot	485	525	SBFI	340/380	420
FURA-2	340/380	500/530	Olivomycin	350–480	470–630	SNARF	480	600/650
GFP green	471	503	Oregon Green	493	520	Stilben SITS, SITA	365	460
GFP Mutant W	458	480	Pararosanilin-Feulgen	560	625	Sulfaflavin	380–470	470–580
GFP Wild Type	396	508	PBFI	340/380	420	SYTO 18	468	533
HcRed	590	614	Phosphin	465	565	SYTO 60	652	678
Hoechst 33258 (Bisbenzimid)	365	480	C-Phycocyanin	605	645	Tetracyclin	390	560
Hoechst 33342 (Bisbenzimid)	355	465	B-Phycoerythrin	546/565	575	Tetrametylrhodamin	540	566
Indo-1	360	410/480	R-Phycoerythrin	480–550	578	Texas Red	595	620
Life/Dead Viability/Cytotoxicity Kit			POPO-1	434	456	Thioflavin S	430	550
Lissamin-Rodamin B	535	580	POPO-3	534	570	Thiazinrot	510	580
Lucifer Yellow	428	540	Primulin	410	550	TOTO-1 + DNA	509	533
LysoTracker Green (DNA)	504	511	Propidiumjodid	536	617	TOTO-3 + DNA	642	661
LysoSensor Blue	374	424	Pyren	343	380–400	TRITC	540	580
Magdalrot	540	570	Pyronin	490–580	530–610	XRITC	560	620
Marina Blue	362	459	QUIN 2	340/365	490	YFP	514	527
Merocyanin	555	578	Quinacrine mustard	440	510	YOYO-1	491	509
4-Metylumbelliferon	360	450	Resorufin	571	585	YOYO-3	642	660
Mithramycin	420	575	Rhodamin	540–560	580	Xylenolorange	377	610
Mito Tracker Green	490	516	Rhodamine 123	540–560	580			
Mito Tracker Red	578	599	Rhodamine phalloidin	550	575			

TABLE A1.2 Filter specifications

Fluorochrome[a]	Filter manufacturer	Filter set
Single dye sets		
Cy5	Chroma	41008[b]
DAPI	Zeiss	01 or 02
	Nikon	UV-1A, UV-2A, or UV-2B
	Olympus	U-Excitation
	Leitz	A or A2
	Chroma	31000
FITC	Zeiss	10
	Nikon	B-2H, B-1H
	Olympus	B + G520 or IB + G520
	Leitz	L3 or L3.1
	Chroma	31001
	Chroma	41001
GFP	Chroma	41017
HcRed	Chroma	41043
Propidium iodide	Zeiss	15 or 14
Rhodamine	Nikon	G-1B or G-2A
Texas red	Olympus	G-Excitation
	Nikon	M2, N2, or N2.1
	Chroma	31002, 31004, 31005
SpectrumGreen	Zeiss	17 or 10
	Nikon	B-2E, B-1E, B-2H, or B-1H
	Olympus	B/G520 or IB/G520
	Leitz	L3 or L3.1
SpectrumOrange	Zeiss	15 or 14
	Nikon	G-1B or G-2A
	Olympus	G-Excitation
	Leitz	M2, N2, or N2.1
TRITC	Chroma	41002
YFP	Chroma	41028
Dual dye sets		
CFP + YFP	Chroma	51017
Cy3 + Cy5	Chroma	51007
DAPI + FITC	Chroma	51000
	Omega	XP50
DAPI + propidium iodide	Chroma	51002
DAPI + Texas red	Chroma	51003
FITC + Texas red	Chroma	51006
	Omega	XP53
FITC + Cy5	Chroma	51008
FITC + propidium iodide	Chroma	51005
SpectrumGreen + SpectrumOrange	Zeiss	23
SpectrumGreen + propidium iodide	Zeiss	23, 19, 16, 11, or 09
	Nikon	B-3A, B-2A, or B-1A
	Olympus	B or IB
	Leitz	H3, I2/3, or K3
SpectrumOrange + DAPI	Imagenetics	DAPI/IO4, IO2/IO4, or DAPI/IO4c/O10c
Triple dye sets		
CFP + YFP + Cy5	Chroma	61009
DAPI + Cy3 + Cy5	Chroma	61004
DAPI + FITC + Cy5	Chroma	61003
DAPI + FITC + TRITC	Chroma	61000
DAPI + FITC + propidium iodide	Chroma	61001
DAPI + FITC + Texas red	Chroma	61002
	Omega	XP56
FITC + Cy3 + Cy5	Chroma	61005
SpectrumGreen SpectrumOrange + DAPI	Imagenetics	DAPI/IO4c/IO10c

Reprinted, with permission, from Bieber (1994; ©John Wiley & Sons).
[a]SpectrumGreen and SpectrumOrange are available from GIBCO/BRL. A variety of other filters are available from the respective manufacturers.
[b]Most people cannot see Cy5 emission by eye.

TABLE A1.3 Properties of objective lenses

Objective lens	Magnification	NA	WD (mm)	Comments	Manufacturer
Achromat	4x	0.1	16		N
	10x	0.25	6.1		O
	20x	0.40	3.0		O
	40x	0.65	0.45		O
	60x	0.80	0.15		O
	100x	1.25	0.13	oil immersion	O
Alpha Plan Fluar	100x	1.45	–	TIRF, oil	Z
Plan Achromat	0.5x	0.02	7.0		N
	1x	0.04	3.2		N
	2x	0.05–0.06	5–7.5		N,O
	4x	0.10	22–30		N,O
	10x	0.25	10.5		N,O
	20x	0.40	1.30		N,O
	40x	0.65	0.56		N,O
	50x	0.90–0.50	0.20–0.40	oil immersion	N,O
	100x	1.25	0.15–0.17	oil immersion	N,O
	100x	1.25	0.23	oil	N
Plan Fluor (Neo Fluor)	1.25x	0.04	3.5		Z
	2.5x	0.075	9.3		Z
	4x	0.16	13.0		N,O
	5x	0.15	13.6		Z
	10x	0.30	5.6–10.0		N,O,Z
	20x	0.50	1.3–1.6		N,O,Z
	20x	0.75	oil, 0.35; glycerin, 0.34; water, 0.33	multi-immersion, oil, glycerin, water	N
	40x	0.75	0.33–0.51		N,O,Z
	40x	1.3	0.2		Z
	60x	1.25	0.10		N,O
	63x	1.25	0.10	oil immersion	Z
	100x	1.30	0.05–0.10	oil immersion	N,O,Z
Epiplan-Neofluor	1.25x	0.035	3.0		Z
	5x	0.15	13.7	dark-field WD 4 5.5	Z
	10x	0.30	5.7	dark-field WD 4 3.5	Z
	20x	0.50	1.4	dark-field WD 4 1.2	Z
	20x	0.65	0.29	oil, polarizing	Z
	50x	0.80	0.58		Z
	50x	1.0	0.29	oil, polarizing	Z
	100x	0.90	0.24	dark-field WD 4 0.27	Z
	100x	1.3	0.13	oil, polarizing	Z
Plan Apochromat	1.25x	0.04	5.10		O
	2x	0.08	6.20		N,O
	4x	0.16	12.2		N,Z
	10x	0.32	1.9		N,Z
	10x	0.45	2.8	optimum resolution	N,Z
	20x	0.60	0.45		N,Z
	20x	0.75	0.61	optimum resolution	N,Z
	40x	0.95	0.13–0.16	correction	N,O,Z
	40x	1.00	0.31	oil immersion	N,Z
	60x	1.40	0.10	oil immersion	N,O
	60x	1.45	0.13	TIRF, sperical aberration correction, oil	N
	60x	1.45	0.15	oil, TIRF	O
	63x	1.40	0.09	oil immersion	Z
	100x	1.40	0.09–0.10	oil immersion	N
	100x	1.45	0.13	TIRF, oil	N,O,Z
	100x	1.65	0.10	TIRF, oil, need coverglass with $n = 1.78$, special oil (diiodomethane)	O
Plan Neo Fluar	63x	1.3	0.17	Live cell imaging, water or glycerin, correction collar for immersion medium, and setting for temperature (24 or 37°C)	Z
	63x	1.25	0.1	oil	Z

This information was collected from Zeiss (Z), Olympus (O), and Nikon (N). NA indicates the numerical aperture of the lens; WD indicates the working distance of the lens. For specialty objectives, contact the respective manufacturers.

FILTERS USED IN FLUORESCENCE MICROSCOPY

The primary filtering element in the epifluorescence microscope is the set of three filters housed in the fluorescence *filter cube*, also called the *filter block*: the *excitation* filter, the *emission* filter, and the *dichroic beam splitter*. A typical filter cube is illustrated schematically in Figure A1.1.

- The **excitation** filter (also called the *exciter*) is a color filter that transmits only those wavelengths of the illumination light that efficiently excite a specific dye. Common filter blocks are named after the type of excitation filter: UV or U (ultraviolet) for exciting DAPI, Indo-1, etc.; B (blue) for exciting FITC; and G (green) for exciting TRITC, Texas Red®, etc. Although short-pass filter designs were used in the past, band-pass filter designs are now used.
- The **emission** filter (also called the *barrier filter* or *emitter*) is a color filter that attenuates all of the light transmitted by the excitation filter and very efficiently transmits any fluorescence emitted by the specimen. This light is always of longer wavelength (more to the red) than the excitation color. These can be either band-pass filters or long-pass filters. Common barrier filter colors are blue or pale-yellow in the U-block; green or deep-yellow in the B-block; and orange or red in the G-block.
- The **dichroic beam splitter** (also called the *dichroic mirror* or *dichromatic beam splitter*[1]) is a thin piece of specially coated glass (the *substrate*) set at an angle 45° to the optical path of the microscope. This coating has the unique ability to reflect one color (the excitation light) but transmit another color (the emitted fluorescence). Current dichroic beam splitters achieve this with great

[1] The term "dichroic" is also used to describe a type of crystal or other material that selectively absorbs light depending on the polarization state of the light. (Polaroid® plastic film polarizer is the most common example.) To avoid confusion, the term "dichromatic" is sometimes used.

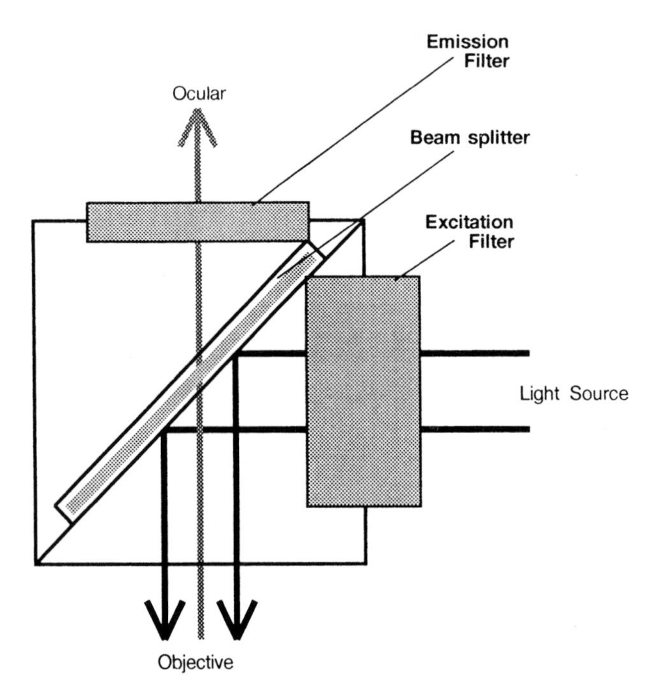

FIGURE A1.1 Schematic of a fluorescence filter cube.

efficiency, i.e., with greater than 90% reflectivity of the excitation along with ~90% transmission of the emission. This is a great improvement over the traditional gray half-silvered mirror, which reflects only 50% and transmits only 50%, giving only about 25% efficiency.

Most microscopes have a slider or turret that can hold from two to four individual filter cubes. It must be noted that the filters in each cube are a matched set, and mixing filters and beam splitters should be avoided unless the complete spectral characteristics of each filter component are known.

Other optical filters can also be found in fluorescence microscopes:

- A **heat filter** (also called a *hot mirror*) is incorporated into the illuminator collector optics of most but not all microscopes. It attenuates infrared light (typically wavelengths longer than 800 nm) but transmits most of the visible light.
- **Neutral-density filters**, usually housed in a filter slider or filter wheel between the collector and the aperture diaphragm, are used to control the intensity of illumination.
- Filters used for techniques other than fluorescence, such as color filters for transmitted-light microscopy and linear polarizing filters for polarized-light microscopy, are sometimes installed. (Information courtesy of Chroma Technology Corp.)

Care and Cleaning of Optical Equipment

PURPOSE OF CLEANLINESS

Microscopes provide optimal images only if lens surfaces are kept clean. Cleanliness requirements are most stringent in polarizing and interference microscopes, but a fingerprint on any lens of a microscope can spoil the image. Similarly, it is important to use clean, unscratched, high-quality slides and coverslips and to keep fingerprints and all other dirt from their surfaces.

HOW TO CLEAN A LENS (from S. Inoué 1986)

1 Never clean any optic dry.

2 Use only high-quality lens paper.

3 **Do not touch anything** (even lens paper) **to a lens surface** except as a last resort. Never rub a lens with lens paper or other items such as Q-Tips, commercial facial tissue, or Kimwipes. Small dust particles are always present in the air and, if rubbed into a microscope lens, these dust particles can cause microscopic scratches in the lenses, and the image will be degraded. Commercial facial tissue or bathroom tissue contains diatom frustules (glass) as a filler. These must be avoided for lens cleaning. Lens coatings must be treated with respect.

4 Only use oil on lenses designed to be used with oil. It can be time-consuming to clean a low-power lens that has accidentally been immersed in oil. Dry lenses may become soiled. Soak an unused Q-Tip with a solvent and very gently roll over the surface of the lens once. Pass a second Q-Tip with solvent over the lens gently—barely touch the surface. Various solvents are used. The glass cleaner Sparkle seems to work well and is inexpensive. If the dirt is water soluble, place a drop of distilled water on the lens surface and then gently blot it up with lens paper (good quality) or other tissue; however, the lens should not be contacted directly.

5 Inspect the lens using an inverted ocular from the microscope as a magnifier. Observe the objective lens in the reflected light from a lamp (generally in the ceiling). To observe microscopic imperfections and dirt, place the lens to be inspected and cleaned under a dissecting microscope and focus on the lens. The lens will need to be illuminated from above.

6 If Sparkle or distilled water does not remove the dirt, then try ether. Xylene and benzene have been used, but these solvents are very carcinogenic and might also harm the lens cements. Use a 1:1:1 mixture of chloroform, water, and alcohol (well shaken). Ether also works quite well, but it is ***very flammable*** and should be used in a well-ventilated area with no flames nearby.

KEEPING LENSES CLEAN

1 Always clean oil off lens after each session.

2 Keep objectives that are not in use in the screw-top containers in which they came. Always inspect lens surfaces with an inverted ocular before putting them away.

3 Some microscopists find it useful not to clean the lens between slides if they are continually using an oil immersion lens with a series of slides. Instead, wipe off excess oil (especially important in inverted microscopes) with lens paper and place new slide with a bit of new oil onto the old oil, being careful to avoid air bubbles.

4 If using seawater, be especially careful to remove it from all metallic and lens surfaces as it will corrode metals and etch away some lens antireflex coatings.

5 Keep condensers, compensators, etc., in boxes, plastic "baggies," or petri dishes until used.

6 Prohibit all smoking in microscopy laboratories because tobacco tar accumulates on lens surfaces and reduces their efficiency.

PRECAUTIONS THAT HELP AVOID DAMAGE

1 Keep seawater and other corrosive fluids away from microscopes.

2 Avoid placing microscopes in vulnerable locations, e.g., near the edge of a table. Place high-quality microscopes on air-cushioned optical tables to avoid image deterioration due to microscope movement.

3 When removing objectives, etc., use two hands, and hold the objective over the table so that if an accident occurs, the distance the lens has to fall is minimized.

PERSONAL SAFETY

1 Never open an ether bottle if there is a flame in the room.

2 Observe the no smoking regulation.

3 Avoid damage to the retina of your eyes by protecting them from unfiltered arc sources (mercury, xenon, etc.), strobe lamps, and especially lasers.

4 If a high-pressure mercury arc lamp should explode or even stop working, ***evacuate the room without delay.*** If possible, open a window as you exit and then notify your institution's safety office.

REFERENCES

Bieber F. 1994. Microscope and image analysis. In *Current protocols in human genetics* (ed. N.C. Dracopoli et al.), pp. 4.4.5. John Wiley, New York

Inoué S. 1986. *Video microscopy.* Plenum Press, New York.

APPENDIX 2

Cautions

GENERAL CAUTIONS

Please note that the Cautions Appendix in this manual is not exhaustive. Readers should always consult their local safety office and individual manufacturers for current and specific product information.

The following general cautions should always be observed.

- **Become completely familiar with the properties of substances used before** beginning the procedure.
- **The absence of a warning** does not necessarily mean that the material is safe, because information may not always be complete or available.
- **If exposed** to toxic substances, contact your local safety office immediately for instructions.
- **Use proper disposal procedures** for all chemical, biological, and radioactive waste.
- **For specific guidelines on appropriate gloves**, consult your local safety office.
- **Handle concentrated acids and bases** with great care. Wear goggles and appropriate gloves. A face shield should be worn when handling large quantities.

  Do not mix strong acids with organic solvents as they may react. Sulfuric acid and nitric acid especially may react highly exothermically and cause fires and explosions.

  Do not mix strong bases with halogenated solvent as they may form reactive carbenes which can lead to explosions.
- **Handle and store pressurized gas containers** with caution as they may contain flammable, toxic, or corrosive gases; asphyxiants; or oxidizers. For proper procedures, consult the Material Safety Data Sheet that must be provided by your vendor.
- **Never pipette** solutions using mouth suction. This method is not sterile and can be dangerous. Always use a pipette aid or bulb.
- **Keep halogenated and nonhalogenated** solvents separately (e.g., mixing chloroform and acetone can cause unexpected reactions in the presence of bases). Halogenated solvents are organic solvents such as chloroform, dichloromethane, trichlorotrifluoroethane, and dichloroethane. Some nonhalogenated solvents are pentane, heptane, ethanol, methanol, benzene, toluene, *N,N*-dimethylformamide (DMF), dimethyl sulfoxide (DMSO), and acetonitrile.
- **Laser radiation**, visible or invisible, can cause severe damage to the eyes and skin. Take proper precautions to prevent exposure to direct and reflected beams. Always follow manufacturer's safety guidelines and consult your local safety office. See caution below for more detailed information.
- **Flash lamps**, because of their light intensity, can be harmful to the eyes. They also may explode on occasion. Wear appropriate eye protection and follow the manufacturer's guidelines.
- **Photographic fixatives and developers** also contain chemicals that can be harmful. Handle them with care and follow manufacturer's directions.
- **Power supplies and electrophoresis equipment** pose serious fire hazard and electrical shock hazards if not used properly.

- **Microwave ovens and autoclaves** in the lab require certain precautions. Accidents have occurred involving their use (e.g., to melt agar or bacto-agar stored in bottles or to sterilize). If the screw top is not completely removed and there is not enough space for the steam to vent, the bottles can explode and cause severe injury when the containers are removed from the microwave or autoclave. Always completely remove bottle caps before microwaving or autoclaving. An alternative method for routine agarose gels that do not require sterile agar is to weigh out the agar and place the solution in a flask.
- **Ultrasonicators** use high-frequency sound waves (16–100 kHz) for cell disruption and other purposes. This "ultrasound," conducted through air, does not pose a direct hazard to humans, but the associated high volumes of audible sound can cause a variety of effects, including headache, nausea, and tinnitus. Direct contact of the body with high-intensity ultrasound (not medical imaging equipment) should be avoided. Use appropriate ear protection and display signs on the door(s) of laboratories where the units are used.
- **Use extreme caution when handling cutting devices** such as microtome blades, scalpels, razor blades, or needles. Microtome blades are extremely sharp! Use care when sectioning. If unfamiliar with their use, have someone demonstrate proper procedures. For proper disposal, use the "sharps" disposal container in your lab. Discard used needles *unshielded*, with the syringe still attached. This prevents injuries (and possible infections; see Biological Safety) while manipulating used needles because many accidents occur while trying to replace the needle shield. Injuries may also be caused by broken pasteur pipettes, coverslips, or slides.

GENERAL PROPERTIES OF COMMON CHEMICALS

The hazardous materials list can be summarized in the following categories.

- Inorganic acids, such as hydrochloric, sulfuric, nitric, or phosphoric, are colorless liquids with stinging vapors. Avoid spills on skin or clothing. Spills should be diluted with large amounts of water. The concentrated forms of these acids can destroy paper, textiles, and skin as well as cause serious injury to the eyes.
- Inorganic bases such as sodium hydroxide are white solids that dissolve in water and under heat development. Concentrated solutions will slowly dissolve skin and even fingernails.
- Salts of heavy metals are usually colored powdered solids which dissolve in water. Many of them are potent enzyme inhibitors and therefore toxic to humans and to the environment (e.g., fish and algae).
- Most organic solvents are flammable volatile liquids. Avoid breathing the vapors, which can cause nausea or dizziness. Also avoid skin contact.
- Other organic compounds, including organosulphur compounds such as mercaptoethanol or organic amines, can have very unpleasant odors. Others are highly reactive and should be handled with appropriate care.
- If improperly handled, dyes and their solutions can stain not only your sample, but also your skin and clothing. Some of them are also mutagenic (e.g., ethidium bromide), carcinogenic, and toxic.
- All names ending with "ase" (e.g., catalase, β-glucuronidase, or zymolase) refer to enzymes. There are also other enzymes with nonsystematic names like pepsin. Many of them are provided by manufacturers in preparations containing buffering substances, etc. Be aware of the individual properties of materials contained in these substances.

- Toxic compounds are often used to manipulate cells. They can be dangerous and should be handled appropriately.
- Be aware that several of the compounds listed have not been thoroughly studied with respect to their toxicological properties. Handle each chemical with the appropriate respect. Although the toxic effects of a compound can be quantified (e.g., LD_{50} values), this is not possible for carcinogens or mutagens where one single exposure can have an effect. Also realize that dangers related to a given compound may also depend on its physical state (fine powder vs. large crystals/diethylether vs. glycerol/dry ice vs. carbon dioxide under pressure in a gas bomb). Anticipate under which circumstances during an experiment exposure is most likely to occur and how best to protect yourself and your environment.

HAZARDOUS MATERIALS

Note: In general, proprietary materials are not listed here. Kits and other commercial items as well as most anesthetics, dyes, fixatives, and stains are also not included. Anesthetics also require special care. Follow the manufacturer's safety guidelines that accompany these products.

Acetic acid (glacial) is highly corrosive and must be handled with great care. Liquid and mist cause severe burns to all body tissues. It may be harmful by inhalation, ingestion, or skin absorption. Wear appropriate gloves and goggles and use in a chemical fume hood. Keep away from heat, sparks, and open flame.

Acetone causes eye and skin irritation and is irritating to mucous membranes and upper respiratory tract. Do not breathe the vapors. It is also extremely flammable. Wear appropriate gloves and safety glasses.

Acrolein is extremely toxic and volatile. It may be harmful by inhalation, ingestion, or skin absorption. Wear appropriate gloves and use in a chemical fume hood.

Ammonium chloride, NH_4Cl, may be harmful by inhalation, ingestion, or skin absorption. Wear appropriate gloves and safety glasses and use in a chemical fume hood.

BCIP, *see* **5-Bromo-4-chloro-3-indolyl-phosphate**

Biotin may be harmful by inhalation, ingestion, or skin absorption. Wear appropriate gloves and safety glasses and use in a chemical fume hood.

BrdU, *see* **5-Bromo-2′-deoxyuridine**

5-Bromo-4-chloro-3-indolyl-phosphate (BCIP) is toxic and may be harmful by inhalation, ingestion, or skin absorption. Wear appropriate gloves and safety glasses. Do not breathe the dust.

5-Bromo-2′-deoxyuridine (BrdU) is a mutagen. It may be harmful by inhalation, ingestion, or skin absorption. It may cause irritiation. Avoid breathing the dust. Wear appropriate gloves and safety glasses and always use in a chemical fume hood.

Bromophenol blue may be harmful by inhalation, ingestion, or skin absorption. Wear appropriate gloves and safety glasses and use in a chemical fume hood.

$CaCl_2$, *see* **Calcium chloride**

Cacodylate contains arsenic, is highly toxic, and may be fatal if inhaled, ingested, or absorbed through the skin. Wear appropriate gloves and safety glasses and use in a chemical fume hood. *See also* **Potassium cacodylate; Sodium cacodylate.**

Calcium chloride, $CaCl_2$, is hygroscopic and may cause cardiac disturbances. It may be harmful by inhalation, ingestion, or skin absorption. Do not breathe the dust. Wear appropriate gloves and safety goggles.

$C_7H_7FO_2S$, *see* **Phenylmethylsulfonyl fluoride**

Cobalt chloride, $CoCl_2$, may be harmful by inhalation, ingestion, or skin absorption. Wear appropriate gloves and safety glasses.

$CoCl_2$, *see* **Cobalt chloride**

DAB, see **3,3′-Diaminobenzidine tetrahydrochloride**

DABCO, *see* **1,4-Diazabicyclo-[2,2,2]-octane**

DAPI, *see* **4′,6-Diamidine-2′phenylindole dihydrochloride**

DEPC, *see* **Diethyl pyrocarbonate**

4′,6-Diamidine-2′phenylindole dihydrochloride (DAPI) is a possible carcinogen. It may be harmful by inhalation, ingestion, or skin absorption. It may also cause irritation. Avoid breathing the dust and vapors. Wear appropriate gloves and safety glasses and use in a chemical fume hood.

3,3′-Diaminobenzidine tetrahydrochloride (DAB) is a carcinogen. Handle with extreme care. Avoid breathing vapors. Wear appropriate gloves and safety glasses and use in a chemical fume hood.

1,4-Diazabicyclo-[2,2,2]-octane (DABCO) may be harmful by inhalation, ingestion, or skin absorption. Wear appropriate gloves and safety glasses and use in a chemical fume hood.

Diethyl pyrocarbonate (DEPC) is a potent protein denaturant and is a suspected carcinogen. Aim bottle away from you when opening it; internal pressure can lead to splattering. Wear appropriate gloves, safety goggles, and lab coat and use in a chemical fume hood.

Digoxigenin may be fatal if inhaled, ingested, or absorbed through the skin. Wear appropriate gloves and safety glasses and use in a chemical fume hood.

***N,N*-Dimethylformamide (DMF), $HCON(CH_3)_2$,** is a possible carcinogen and is irritating to the eyes, skin, and mucous membranes. It can exert its toxic effects through inhalation, ingestion, or skin absorption. Chronic inhalation can cause liver and kidney damage. Wear appropriate gloves and safety glasses and use in a chemical fume hood.

Dimethyl sulfoxide (DMSO) may be harmful by inhalation or skin absorption. Wear appropriate gloves and safety glasses and use in a chemical fume hood. DMSO is also combustible. Store in a tightly closed container. Keep away from heat, sparks, and open flame.

Dithiothreitol (DTT) is a strong reducing agent that emits a foul odor. It may be harmful by inhalation, ingestion, or skin absorption. When working with the solid form or highly concentrated stocks, wear appropriate gloves and safety glasses and use in a chemical fume hood.

DMF, *see* ***N,N*-Dimethylformamide**

DMSO, *see* **Dimethyl sulfoxide**

Dinitrophenol (DNP) may be fatal by inhalation, ingestion, or skin absorption. Wear appropriate gloves and safety glasses and use only in a chemical fume hood.

DNP, *see* **Dinitrophenol**

DTT, *see* **Dithiothreitol**

Ethidium bromide is a powerful mutagen and is toxic. Consult the local institutional safety officer for specific handling and disposal procedures. Avoid breathing the dust. Wear appropriate gloves when working with solutions that contain this dye.

Ferricyanide, *see* **Potassium ferricyanide**

Ferrocyanide, *see* **Potassium ferrocyanide**

Formaldehyde, HCHO, is highly toxic and volatile. It is also a possible carcinogen. It is readily absorbed through the skin and is irritating or destructive to the skin, eyes, mucous membranes, and upper respiratory tract. Avoid breathing the vapors. Wear appropriate gloves and safety glasses and always use in a chemical fume hood. Keep away from heat, sparks, and open flame.

Formamide is teratogenic. The vapor is irritating to the eyes, skin, mucous membranes, and upper respiratory tract. It may be harmful by inhalation, ingestion, or skin absorption. Wear appropriate gloves and safety glasses and always use a chemical fume hood when working with concentrated solutions of formamide. Keep working solutions covered as much as possible

Formic acid, HCOOH, is highly toxic and extremely destructive to tissue of the mucous membranes, upper respiratory tract, eyes, and skin. It may be harmful by inhalation, ingestion, or skin absorption. Wear appropriate gloves and safety glasses (or face shield) and use in a chemical fume hood.

Glacial acetic acid, *see* **Acetic acid (glacial)**

Glutaraldehyde is toxic. It is readily absorbed through the skin and is irritating or destructive to the skin, eyes, mucous membranes, and upper respiratory tract. Wear appropriate gloves and safety glasses and always use in a chemical fume hood.

Glycine may be harmful by inhalation, ingestion, or skin absorption. Wear gloves and safety glasses. Avoid breathing the dust.

HCl, *see* **Hydrochloric acid**

HCHO, *see* **Formaldehyde**

H_3COH, *see* **Methanol**

Heptane may be harmful by inhalation, ingestion, or skin absorption. Wear appropriate gloves and safety glasses. It is extremely flammable. Keep away from heat, sparks, and open flame.

HNO_3, *see* **Nitric acid**

$HOCH_2CH_2SH$, *see* **β-Mercaptoethanol**

Hydrochloric acid, HCl, is volatile and may be fatal if inhaled, ingested, or absorbed through the skin. It is extremely destructive to mucous membranes, upper respiratory tract, eyes, and skin. Wear appropriate gloves and safety glasses and use with great care in a chemical fume hood. Wear goggles when handling large quantities.

KCl, *see* **Potassium chloride**

KOH, *see* **Potassium hydroxide**

Lead, Pb, is a toxic metal. It presents a long-term danger, because lead accumulates in the liver and interferes with its function. Avoid contact with skin and wear appropriate gloves when handling.

Lead nitrate, $Pb(NO_3)_2$, and all other lead salts may be harmful by inhalation, ingestion, or skin absorption. Wear appropriate gloves and safety glasses and always use in a chemical fume hood.

Liquid nitrogen (LN_2) can cause severe damage because of extreme temperature. Handle frozen samples with extreme caution. Do not breathe the vapors. Seepage of liquid nitrogen into frozen vials can result in an exploding tube upon removal from liquid nitrogen. Use vials with O-rings when possible. Wear cryo-mitts and a face mask. No not allow the liquid nitrogen to spill onto your clothes. Do not breathe the vapors.

Lowicryl is an irritant and may be harmful by inhalation, ingestion, or skin absorption. Wear appropriate gloves and safety glasses.

Magnesium chloride, $MgCl_2$, may be harmful by inhalation, ingestion, or skin absorption. Wear appropriate gloves and safety glasses and use in a chemical fume hood.

Maleic acid is toxic and harmful by inhalation, ingestion, or skin absorption. Reaction with water or moist air can release toxic, corrosive, or flammable gases. Do not breathe the vapors or dust. Wear appropriate gloves and safety glasses.

MeOH or **H_3COH,** *see* **Methanol**

β-Mercaptoethanol (2-Mercaptoethanol), $HOCH_2CH_2SH$, may be fatal if inhaled or absorbed through the skin and is harmful if ingested. High concentrations are extremely destructive to the mucous membranes, upper respiratory tract, skin, and eyes. β-Mercaptoethanol has a very foul odor. Wear appropriate gloves and safety glasses and always use in a chemical fume hood.

Methanol, MeOH or **H_3COH,** is poisonous and can cause blindness. It may be harmful by inhalation, ingestion, or skin absorption. Adequate ventilation is necessary to limit exposure to vapors. Avoid inhaling these vapors. Wear appropriate gloves and goggles and use only in a chemical fume hood.

$MgCl_2$, *see* **Magnesium chloride**

$NaBH_4$, *see* **Sodium borohydride**

Na_2HPO_4, *see* **Sodium hydrogen phosphate**

NaN_3, *see* **Sodium azide**

NaOH, *see* **Sodium hydroxide**

NBT, *see* **4-Nitro blue tetrazolium chloride**

NH_4Cl, *see* **Ammonium chloride**

Nitric acid, HNO_3, is volatile and must be handled with great care. It is toxic by inhalation, ingestion, and skin absorption. Wear appropriate gloves and safety goggles and use in a chemical fume hood. Do not breathe the vapors. Keep away from heat, sparks, and open flame.

4-Nitro blue tetrazolium chloride (NBT) may be harmful by inhalation, ingestion, or skin absorption. Wear appropriate gloves and safety glasses.

OsO_4, *see* **Osmium tetroxide**

Osmium tetroxide (osmic acid), OsO_4, is highly toxic if inhaled, ingested, or absorbed through the skin. Vapors can react with corneal tissues and cause blindness. There is a possible risk of irreversible effects. Wear appropriate gloves and safety goggles and always use in a chemical fume hood. Do not breathe the vapors.

Paraformaldehyde is highly toxic. It is readily absorbed through the skin and is extremely destructive to the skin, eyes, mucous membranes, and upper respiratory tract. Avoid breathing the dust. Wear appropriate gloves and safety glasses and use in a chemical fume hood. Paraformaldehyde is the undissolved form of formaldehyde.

Pb, *see* **Lead**

$Pb(NO_3)_2$, *see* **Lead nitrate**

Pepsin may be harmful by inhalation, ingestion, or skin absorption. Wear appropriate gloves and safety glasses.

Phenol is extremely toxic, highly corrosive, and can cause severe burns. It may be harmful by inhalation, ingestion, or skin absorption. Wear appropriate gloves, goggles, protective clothing, and always use in a chemical fume hood. Rinse any areas of skin that come in contact with phenol with a large volume of water and wash with soap and water; do not use ethanol!

Phenylenediamine may be harmful by inhalation, ingestion, or skin absorption. Wear appropriate gloves and safety glasses and use in a chemical fume hood.

Phenylmethylsulfonyl fluoride (PMSF), $C_7H_7FO_2S$, is a highly toxic cholinesterase inhibitor. It is extremely destructive to the mucous membranes of the respiratory tract, eyes, and skin. It may be fatal by inhalation, ingestion, or skin absorption. Wear appropriate gloves and safety glasses and always use in a chemical fume hood. In case of contact, immediately flush eyes or skin with copious amounts of water and discard contaminated clothing.

PI, *see* **Propidium iodide**

PMSF, *see* **Phenylmethylsulfonyl fluoride**

Polyvinylpyrrolidone (PVP) may be harmful by inhalation, ingestion, or skin absorption. Wear appropriate gloves and safety glasses and use in a chemical fume hood.

Potassium cacodylate, *see* **Cacodylate**

Potassium chloride, KCl, may be harmful by inhalation, ingestion, or skin absorption. Wear appropriate gloves and safety glasses.

Potassium ferricyanide, $K_4Fe(CN)_6$, may be fatal by inhalation, ingestion, or skin absorption. Wear appropriate gloves and safety glasses and always use with extreme care in a chemical fume hood. Keep away from strong acids.

Potassium ferrocyanide, $K_4Fe(CN)_6 \cdot 3H_2O$, may be fatal by inhalation, ingestion, or skin absorption. Wear appropriate gloves and safety glasses and always use with extreme care in a chemical fume hood. Keep away from strong acids.

Potassium hydroxide, KOH and **KOH/methanol,** is highly toxic and may be fatal if swallowed. It may be harmful by inhalation, ingestion, or skin absorption. Solutions are corrosive and can cause severe burns. It should be handled with great care. Wear appropriate gloves and safety goggles.

Propidium iodide (PI) may be harmful by inhalation, ingestion, or skin absorption. It is irritating to the eyes, skin, mucous membranes, and upper respiratory tract. It is mutagenic and possibly carcinogenic. Wear appropriate gloves, safety glasses, protective clothing, and always use with extreme care in a chemical fume hood.

PVP, *see* **Polyvinylpyrrolidone**

Quinacrine may be fatal by inhalation, ingestion, or skin absorption. Wear appropriate gloves and safety glasses and use in a chemical fume hood.

Radioactive substances: When planning an experiment that involves the use of radioactivity, include the physicochemical properties of the isotope (half-life, emission type, and energy), the chemical form of the radioactivity, its radioactive concentration (specific activity), its total amount, and its chemical concentration. Order and use only as much as really needed. Always wear appropriate gloves, lab coat, and safety goggles when handling radioactive material. **X-rays** and **gamma rays** are electromagnetic waves of very short wavelengths either generated by technical devices or emitted by radioactive materials. They may be emitted isotropically from the source or may be focused into a beam. Their potential dangers depend on the time period of exposure, the intensity experienced, and the wavelengths used. Be aware that appropriate shielding is usually of lead or other similar material. The thickness of the shielding is determined by the energy(s) of the X-rays or gamma rays. Consult the local safety office for further guidance in the appropriate use and disposal of radioactive materials. Always monitor thoroughly after using radioisotopes. A convenient calculator to perform routine radioactivity calculations can be found at

http://www.graphpad.com/calculators/radcalc.cfm

SDS, *see* **Sodium dodecyl sulfate**

Silane is extremely flammable and corrosive. It may be harmful by inhalation, ingestion, or skin absorption. Keep away from heat, sparks, and open flame. The vapor is irritating to the eyes, skin, mucous membranes, and upper respiratory tract. Wear appropriate gloves and safety goggles and always use in a chemical fume hood.

Sodium azide, NaN_3, is highly poisonous. It blocks the cytochrome electron transport system. Solutions containing sodium azide should be clearly marked. It may be harmful by inhalation, ingestion, or skin absorption. Wear appropriate gloves and safety goggles and handle it with great care. Sodium azide is an oxidizing agent and should not be stored near flammable chemicals.

Sodium borohydride, $NaBH_4$, is corrosive and causes burns. It may be harmful by inhalation, ingestion, or skin absorption. Wear appropriate gloves and safety goggles and use in a chemical fume hood.

Sodium cacodylate may be carcinogenic and contains arsenic. It is highly toxic and may be fatal by inhalation, ingestion, or skin absorption. It also may cause harm to the unborn child. Effects of contact or inhalation may be delayed. Do not breathe the dust. Wear appropriate gloves and safety goggles and use only in a chemical fume hood. *See also* **Cacodylate**.

Sodium dodecyl sulfate (SDS) is toxic, an irritant, and poses a risk of severe damage to the eyes. It may be harmful by inhalation, ingestion, or skin absorption. Wear appropriate gloves and safety goggles. Do not breathe the dust.

Sodium hydrogen phosphate, Na_2HPO_4, (sodium phosphate, dibasic) may be harmful by inhalation, ingestion, or skin absorption. Wear appropriate gloves and safety glasses and use in a chemical fume hood.

Sodium hydroxide, NaOH, and **solutions containing NaOH** are highly toxic and caustic and should be handled with great care. Wear appropriate gloves and a face mask. All other concentrated bases should be handled in a similar manner.

Tannic acid is an irritant. Large amounts may cause liver and kidney damage. It may be harmful by inhalation, ingestion, or skin absorption. Wear appropriate gloves and safety glasses. Do not breathe the dust.

UA, *see* **Uranyl acetate**

Uranyl acetate (UA) is toxic if inhaled, ingested, or absorbed through the skin. Wear appropriate gloves and safety glasses and use in a chemical fume hood.

UV light and/or **UV radiation** is dangerous and can damage the retina. Never look at an unshielded UV light source with naked eyes. Examples of UV light sources that are common in the laboratory include handheld lamps and transilluminators. View only through a filter or safety glasses that absorb harmful wavelengths. UV radiation is also mutagenic and carcinogenic. To minimize exposure, make sure that the UV light source is adequately shielded. Wear protective appropriate gloves when holding materials under the UV light source.

VCD, *see* **Vinylcyclohexene dioxide**

Vinylcyclohexene dioxide (**VCD**) is irritating to the eyes, mucous membranes, and upper respiratory tract and is a carcinogen. Wear appropriate gloves and safety glasses and use in a chemical fume hood.

Xylene is flammable and may be narcotic at high concentrations. It may be harmful by inhalation, ingestion, or skin absorption. Wear appropriate gloves and safety glasses and use only in a chemical fume hood. Keep away from heat, sparks, and open flame.

Index